THE PRIMATE VISUAL SYSTEM

METHODS & NEW FRONTIERS IN NEUROSCIENCE

Series Editors
Sidney A. Simon, Ph.D.
Miguel A.L. Nicolelis, M.D., Ph.D.

Published Titles

Apoptosis in Neurobiology
Yusuf A. Hannun, M.D., Professor of Biomedical Research and Chairman/Department of Biochemistry and Molecular Biology, Medical University of South Carolina
Rose-Mary Boustany, M.D., tenured Associate Professor of Pediatrics and Neurobiology, Duke University Medical Center

Methods for Neural Ensemble Recordings
Miguel A.L. Nicolelis, M.D., Ph.D., Professor of Neurobiology and Biomedical Engineering, Duke University Medical Center

Methods of Behavioral Analysis in Neuroscience
Jerry J. Buccafusco, Ph.D., Alzheimer's Research Center, Professor of Pharmacology and Toxicology, Professor of Psychiatry and Health Behavior, Medical College of Georgia

Neural Prostheses for Restoration of Sensory and Motor Function
John K. Chapin, Ph.D., Professor of Physiology and Pharmacology, State University of New York Health Science Center
Karen A. Moxon, Ph.D., Assistant Professor/School of Biomedical Engineering, Science, and Health Systems, Drexel University

Computational Neuroscience: Realistic Modeling for Experimentalists
Eric DeSchutter, M.D., Ph.D., Professor/Department of Medicine, University of Antwerp

Methods in Pain Research
Lawrence Kruger, Ph.D., Professor of Neurobiology (Emeritus), UCLA School of Medicine and Brain Research Institute

Motor Neurobiology of the Spinal Cord
Timothy C. Cope, Ph.D., Professor of Physiology, Emory University School of Medicine

Nicotinic Receptors in the Nervous System
Edward D. Levin, Ph.D., Associate Professor/Department of Psychiatry and Pharmacology and Molecular Cancer Biology and Department of Psychiatry and Behavioral Sciences, Duke University School of Medicine

Methods in Genomic Neuroscience
Helmin R. Chin, Ph.D., Genetics Research Branch, NIMH, NIH
Steven O. Moldin, Ph.D, Genetics Research Branch, NIMH, NIH

Methods in Chemosensory Research
Sidney A. Simon, Ph.D., Professor of Neurobiology, Biomedical Engineering, and Anesthesiology, Duke University
Miguel A.L. Nicolelis, M.D., Ph.D., Professor of Neurobiology and Biomedical Engineering, Duke University

The Somatosensory System: Deciphering the Brain's Own Body Image
Randall J. Nelson, Ph.D., Professor of Anatomy and Neurobiology, University of Tennessee Health Sciences Center

New Concepts in Cerebral Ischemia
Rick C. S. Lin, Ph.D., Professor of Anatomy, University of Mississippi Medical Center

DNA Arrays: Technologies and Experimental Strategies
Elena Grigorenko, Ph.D., Technology Development Group, Millennium Pharmaceuticals

Methods for Alcohol-Related Neuroscience Research
Yuan Liu, Ph.D., National Institute of Neurological Disorders and Stroke, National Institutes of Health
David M. Lovinger, Ph.D., Laboratory of Integrative Neuroscience, NIAAA

In Vivo Optical Imaging of Brain Function
Ron Frostig, Ph.D., Associate Professor/Department of Psychobiology, University of California, Irvine

Primate Audition: Behavior and Neurobiology
Asif A. Ghazanfar, Ph.D., Primate Cognitive Neuroscience Lab, Harvard University

Methods in Drug Abuse Research: Cellular and Circuit Level Analyses
Dr. Barry D. Waterhouse, Ph.D., MCP-Hahnemann University

Functional and Neural Mechanisms of Interval Timing
Warren H. Meck, Ph.D., Professor of Psychology, Duke University

Biomedical Imaging in Experimental Neuroscience
Nick Van Bruggen, Ph.D., Department of Neuroscience Genentech, Inc., South San Francisco
Timothy P.L. Roberts, Ph.D., Associate Professor, University of Toronto

THE PRIMATE VISUAL SYSTEM

Edited by

Jon H. Kaas, Ph.D.

Distinguished Professor, Department of Psychology, Vanderbilt University, Nashville, Tennessee

Christine E. Collins, Ph.D.

Research Associate, Department of Psychology, Vanderbilt University, Nashville, Tennessee

CRC PRESS

Boca Raton London New York Washington, D.C.

Cover: Center: The phylogenetic relationships of extant primate groups. The size of each circle is proportional to the number of species in the group. (Drawing by T.M. Preuss; see Chapter 10 for details.) Dorsolateral views of the brain of a prosimian primate, *Galago garnetti*; New World owl monkey, *Aotus trivirgatus*; and Old World macaque monkey, *Macaca mulatta,* showing the locations of proposed visual areas and nonvisual cortical areas. (Drawings by B. Martin; see Kaas, Chapter 6, for details.) Lower right: A human brain with colored regions denoting approximate locations of several motion-responsive areas. (See Blake et al., Chapter 13, for details.) (Composite illustration by Barbara Martin of Graphics Services at The John F. Kennedy Center, Vanderbilt University.)

Library of Congress Cataloging-in-Publication Data

The Primate visual system / edited by Jon H. Kaas and Christine E. Collins.
p. cm. — (Methods & new frontiers in neuroscience)
Includes bibliographical references and index.
ISBN 0-8493-1243-4 (alk. paper)
1. Visual pathways. 2. Visual cortex. 3. Primates—Physiology. I. Kaas, Jon H. II. Collins, Christine E. (Christine Elaine) III. Methods & new frontiers in neuroscience series.

QP383.15.P755 2003
573.8′8198—dc21 2003046139

Direct all inquiries to CRC Press LLC, 2000 N.W. Corporate Blvd., Boca Raton, Florida 33431.

International Standard Book Number 0-8493-1243-4
Library of Congress Card Number 2003046139
Printed in the United States of America 1 2 3 4 5 6 7 8 9 0
Printed on acid-free paper

Series Preface

Our goal in creating the **Methods & New Frontiers in Neuroscience** series is to present the insights of experts on emerging experimental techniques and theoretical concepts that are, or will be, at the vanguard of neuroscience. Books in the series cover topics ranging from methods to investigate apoptosis, to modern techniques for neural ensemble recordings in behaving animals. The series also covers new and exciting multidisciplinary areas of brain research, such as computational neuroscience and neuroengineering, and describes breakthroughs in classical fields like behavioral neuroscience. We want these books to be the books every neuroscientist will use in order to get acquainted with new methodologies in brain research. These books can be given to graduate students and postdoctoral fellows when they are looking for guidance to start a new line of research.

Each book is edited by an expert and consists of chapters written by the leaders in a particular field. Books are richly illustrated and contain comprehensive bibliographies. Chapters provide substantial background material relevant to the particular subject. Hence, they are not only "methods books," but they also contain detailed "tricks of the trade" and information as to where these methods can be safely applied. In addition, they include information about where to buy equipment and about web sites helpful in solving both practical and theoretical problems

We hope that as the volumes become available, the effort put in by us, by the publisher, by the book editors, and by individual authors will contribute to the further development of brain research. The extent that we achieve this goal will be determined by the utility of these books.

Sidney A. Simon, Ph.D.
Miguel A.L. Nicolelis, M.D., Ph.D.
Series Editors

Preface

The chapters in this book focus on the primate visual system, but not the entire primate visual system. The emphasis is on parts of the system that seem specialized in primates and thus notably different from those in other taxa of mammals. This specialization starts in the retina, where major modifications in the proportion of ganglion cell classes, their distribution across the retina, and their central targets are essential for the emphasis in primates on detailed frontal vision (Silveira, Chapter 2; Casagrande and Royal, Chapter 1). In particular, vast numbers of ganglion cells are concentrated in the central retina, and most of these cells (roughly 80%) are of the parvocellular (P cell) class, which mediates form vision and object recognition. Basically, all of these P cells, together with the magnocellular (M cell) and koniocellular (K cell) classes, project to the dorsal lateral geniculate nucleus (LGN) of the visual thalamus. The LGN has a characteristic pattern of lamination, based on the segregation of inputs according to ganglion cell class and eye of origin, so that an experienced investigator can easily identify the LGN as from a primate by appearance alone. In addition, modifications of the overall pattern of lamination allow one to identify the structure as belonging to one of the three major branches of the primate radiation (prosimian primates, tarsiers, and anthropoid primates). In diurnal anthropoid primates (monkeys, apes, and humans), a great expansion of the P cell layers, reflecting the importance of detailed central vision and color vision, is obvious.

Just medial and caudal to the LGN, the pulvinar complex of the visual thalamus is also greatly expanded compared with the pulvinar (often termed the lateral posterior nucleus or complex) of other mammals. Parts of the pulvinar complex receive visual inputs from the superior colliculus, and some visual inputs come directly from the retina, but the uniqueness of the pulvinar complex in primates arises from its relationship with the visual cortex. Because most of the inputs of the visual pulvinar are from subdivisions of visual cortex and because all of the outputs are to subdivisions of visual cortex, the complexity and size of the visual pulvinar are closely tied to the great expansion and complexity of the visual cortex in primates, especially in the anthropoid primates. Recently, there has been much progress in determining how the pulvinar complex is divided into nuclei, and how these nuclei are interconnected with areas of visual cortex. These recent findings are nicely reviewed by Stepniewska (Chapter 3).

In primates, nearly all projections of the LGN are to primary visual cortex (V1), and this area is responsible for directly or indirectly activating most of the rest of visual cortex. Although much could be said about the many studies of the internal organization of V1, the response properties of neurons in V1, and the roles of local circuits in producing these response properties, such extensive coverage of even a major visual area is not possible in this book. Instead, we concentrate on the very recent progress in our understanding of how the neurons in V1 of monkeys emerge in development (Chino et al., Chapter 4). Surprisingly, neurons have many adultlike properties soon after birth, suggesting that the development of these properties does not depend on postnatal visual experience. In addition, inputs, outputs, internal organization, and other features of V1 are covered in other chapters (Casagrande and Royal, Chapter 1; Stepniewska, Chapter 3; Roe, Chapter 5; Kaas, Chapter 6; Bullier, Chapter 8; Preuss, Chapter 10; Rosa and Tweedale, Chapter 11).

In proportional size, the second visual area, V2, is the next largest (after V1) visual area, and great progress has been made in understanding the nature of the modular organization of V2, how this organization relates to the processing of types of visual information, and how these modules receive different distributions of inputs from V1 and project to other visual areas, especially DL (V4) and MT (V5). Anna Roe's studies (Chapter 5) have been in the forefront of those producing this progress.

Primates clearly have a large number of cortical areas that are involved in processing visual information. Some investigators estimate that macaque monkeys have as many as 35 areas that are predominantly visual in function. This number is likely to vary across primate taxa, with prosimian primates having fewer visual areas and humans having more, but most visual areas have not been well defined (see Kaas, Chapter 6), and the exact number of visual areas is not known for any primate. However, it is generally recognized that the visual areas can be assigned to levels in a processing hierarchy, with V1 in the early distribution center, V2 as the major cortical target of V1 at the second level, areas with direct inputs from V1 and V2 at a third level, and so on. The problem is that the large number of visual areas and the extensive connection framework they produce are so complex that different hierarchies can be constructed from the data, depending on the assumptions of the investigator and the data considered. As such hierarchies provide a useful framework for considering how visual cortex mediates functions, a chapter is specifically devoted to discussing visual hierarchies, including parallel components (Bullier, Chapter 8; also see Casagrande and Royal, Chapter 1). Areas early in the hierarchy are considered in one chapter (Kaas, Chapter 6), these and other areas are described further in a chapter especially devoted to the mapping features of visual cortex (Rosa and Tweedale, Chapter 11), and higher level areas of inferotemporal cortex of the ventral stream of processing for object vision are described in a third chapter (Tanifuji, Chapter 14). The classical visuomotor stream of processing involves subdivisions of posterior parietal cortex and visuomotor areas of the frontal lobe (Schall et al., Chapter 9). Thus, the major organizational features of visual cortex in primates are covered by a collection of chapters by leading investigators.

Other chapters are devoted to specific features of the visual cortex. In general, investigators have been concerned with the driving, feedforward connections in the visual cortex, but visual areas also project back to the areas that provide their inputs. Overall, there has been growing interest in determining what these feedback connections do. Fortunately, we have an excellent chapter (Rockland, Chapter 16) by a pioneer who first described the major anatomical features of cortical feedback connections. A major cell type in cortex is the pyramidal cell, and it came as a great surprise to many that the dendritic arbors and other aspects of pyramidal cell morphology vary across visual areas and across primate species. As such variations in cell morphology reflect specialization for different functional roles, studies of pyramidal cells in visual cortex provide uniquely new insights into cortical processing (Elston, Chapter 15). The related chapter by Preuss (Chapter 10) addresses the issue of specialization in the retina, LGN, and areas of visual cortex. More specifically, Preuss demonstrates that there are marked differences in the internal organizations of these structures in humans and monkeys, clearly suggesting that the visual systems of these primates do not function in exactly the same ways.

Another issue that has drawn the attention of researchers recently is the plasticity of the mature visual system. Lesions of the retina and cortex are followed by compensations in the visual system that we are just now beginning to describe and understand (Collins and Kaas, Chapter 7). Such recoveries may relate to the perceptual process of "filling-in" of blind spots in the visual field, and the preservation and recovery of visual abilities, possibly including some of those described as "blindsight."

Finally, great progress has been made in understanding the organization and functions of the human visual system. Preuss (Chapter 10) describes some of the anatomical specializations of the human visual system, and Blake et al. (Chapter 13) outline how visual motion is processed in human visual cortex. Humans are especially good at identifying faces, and a region of the human temporal lobe called the "fusiform face area" has been identified as an area involved in this ability. Gauthier addresses the intriguing issue of what learning about face processing in the visual cortex tells us about how the visual system mediates object recognition (Chapter 12).

Overall, these chapters by leading investigators provide a current and extensive review of the neuronal mechanisms of visual perception and action in the visual system of primates. Interesting species differences and specializations are considered, and the functional and anatomical adaptiveness of the mature system is reviewed, as are new findings on the development of visual cortex. Of course, such an effort means that some topics have not been covered, or have not been covered adequately. In particular, we have omitted parts of the subcortical visual system that are not directly involved in influencing visual cortex and, thus, are not involved in visual perception. The visual hypothalamus, the pregeniculate nucleus, the visual sector of the reticular nucleus and the claustrum, the visual components of the basal ganglia, the accessory optic system, the pretectum, the superior colliculus, the visuomotor nucleus of the pons, and the visual cerebellum all deserve discussion, but this must occur elsewhere. Fortunately, a complete volume in this series is planned on the superior colliculus. Enjoy this introduction to the primate visual system.

Jon H. Kaas and Christine E. Collins

Editors

Jon H. Kaas, Ph.D., received a B.A. in psychology from Northland College in Ashland, Wisconsin in 1959. He did his graduate training with I. T. Diamond in psychology at Duke University, and received a Ph.D. in 1965. From 1965 through 1968, he was a postdoctoral fellow in the Laboratory of Neurophysiology at the University of Wisconsin under the supervision of C. N. Woolsey. After serving as Assistant Professor in the laboratory from 1968 through 1972, he became Associate Professor of Psychology at Vanderbilt University. At Vanderbilt, he was promoted to Professor in 1978, became Centennial Professor in 1987, and Distinguished Professor in 2001.

Dr. Kaas has received the Earl Sutherland Prize for Achievement in Research, the Javits Neuroscience Investigator Award, the Krieg Cortical Discoverer Award, and the American Psychological Association Distinguished Scientific Contribution Award. He has been Visiting Scientist in the Department of Physiology and Pharmacology, University of Queensland; the Neurosciences Institute, New York; the Department of Physiology at the University of California at San Francisco; and the Center for Advanced Study in the Behavioral Sciences at Stanford University. He is an elected member of the National Academy of Sciences, the American Academy of Arts and Sciences, the Society of Experimental Psychologists, and the American Association for the Advancement of Science. His research interests have focused on how sensory perceptual systems are organized in mammalian brains, especially primates. This research has resulted in over 225 research papers, 130 book chapters and reviews, and the editorship of 4 previous books.

Christine E. Collins, Ph.D., received a B.S. in biology in 1988 and an M.S. in biology in 1991 from the University of Illinois at Urbana-Champaign. While at the University of Illinois, she became involved in biopsychology research in the laboratory of Dr. W. T. Greenough, where she worked on experience-dependent plasticity in sensorimotor cortex of rats trained in a reaching task. Her master's thesis in the Greenough laboratory was focused on the time course of the astrocytic response to injury in the superior colliculus following unilateral enucleation in adult and aged rats. She continued her graduate work in psychology at Cornell University, Ithaca, New York, where she studied developmental plasticity in the song system of zebra finches with Dr. T. J. DeVoogd. She finished her Ph.D. in 1998 and moved to Vanderbilt University where she has worked as a postdoctoral research associate in the laboratory of J. H. Kaas. There she has studied the functional organization of vision in primates and reorganization in visual cortex following direct damage to cortex.

Contributors

Hua Bi
College of Optometry
University of Houston
Houston, Texas, U.S.A.

Narcisse P. Bichot
Laboratory of Neuropsychology
National Institute of Mental Health
National Institutes of Health
Bethesda, Maryland, U.S.A.

Randolph Blake
Vanderbilt Vision Research Center
Vanderbilt University
Nashville, Tennessee, U.S.A.

Jean Bullier
Centre de Recherche Cerveau et Cognition
CNRS-UPS UMR
Toulouse, France

Vivien A. Casagrande
Department of Cell and Developmental Biology
Vanderbilt University Medical School
Nashville, Tennessee, U.S.A.

Yuzo M. Chino
College of Optometry
University of Houston
Houston, Texas, U.S.A.

Christine E. Collins
Department of Psychology
Vanderbilt University
Nashville, Tennessee, U.S.A.

Guy N. Elston
Vision, Touch & Hearing Research Centre
Department of Physiology and Pharmacology
School of Biomedical Sciences
The University of Queensland
Brisbane, Australia

Isabel Gauthier
Center for Integrative and Cognitive Neuroscience
Vanderbilt Vision Research Center
Department of Psychology
Vanderbilt University
Nashville, Tennessee, U.S.A.

Emily Grossman
Department of Psychology
Harvard University
Cambridge, Massachusetts, U.S.A.

Jon H. Kaas
Department of Psychology
Vanderbilt University
Nashville, Tennessee, U.S.A.

Aditya Murthy
National Brain Research Centre
New Delhi, India

Todd M. Preuss
Division of Neuroscience and Center for Behavioral Neuroscience
Yerkes National Primate Center
Emory University
Atlanta, Georgia, U.S.A.

Kathleen S. Rockland
Laboratory for Cortical Organization and Systematics
RIKEN Brain Science Institute
Saitama, Japan

Anna W. Roe
Department of Neurobiology
Yale University School of Medicine
New Haven, Connecticut, U.S.A.

Marcello G. P. Rosa
Department of Physiology
Monash Unversity
Melbourne, Australia

David W. Royal
Center for Molecular Neuroscience and Center for Integrative and Cognitive Neuroscience
Vanderbilt University
Nashville, Tennessee, U.S.A.

Takashi R. Sato
Center for Integrative and Cognitive Neuroscience
Vanderbilt Vision Research Center
Department of Psychology
Vanderbilt University
Nashville, Tennessee, U.S.A.

Jeffrey D. Schall
Center for Integrative and Cognitive Neuroscience
Vanderbilt Vision Research Center
Department of Psychology
Vanderbilt University
Nashville, Tennessee, U.S.A.

Robert Sekuler
Volen Center for Complex Systems
Brandeis University
Waltham, Massachusetts, U.S.A.

Luiz Carlos de Lima Silveira
Departamento de Fisiologia
Centro de Ciências Biológicas
Universidade Federal do Pará
Pará, Brazil

Iwona Stepniewska
Department of Psychology
Vanderbilt University
Nashville, Tennessee, U.S.A.

Manabu Tanifuji
Laboratory for Integrative Neural Systems
RIKEN Brain Science Institute
Saitama, Japan

Kirk G. Thompson
Laboratory of Sensorimotor Research
National Eye Institute
National Institutes of Health
Bethesda, Maryland, U.S.A.

Rowan Tweedale
Vision, Touch & Hearing Research Centre
The University of Queensland
Brisbane, Australia

Bin Zhang
College of Optometry
University of Houston
Houston, Texas, U.S.A.

Contents

1 Parallel Visual Pathways in a Dynamic System

Vivien A. Casagrande and David W. Royal

CONTENTS

1.1 INTRODUCTION

The idea that information from different sensory modalities is processed in parallel can be traced to the 1800s when Johannes Müller put forth the "law of specific nerve energies."[1] The law in essence states that perceptions are determined by which nerve fibers are activated, not by how the nerve fibers are activated. For example, mechanical pressure to the eye produces a sensation of light, and activating axons in the auditory nerve by an electric shock gives rise to a sensation of sound. Today, we recognize that there are specific receptor cells, tuned to be sensitive to different forms of energy in the environment and that these receptor cells connect to specific nerves. Two other ideas about parallel processing of sensory information are well established. First, it is well accepted that sensory qualities within a modality, such as light touch vs. pain and temperature within the somatosensory system, are carried by separate, parallel pathways. This form of *functional parallelism* extends to other

0-8493-1243-4/04/$0.00+$1.50

sensory systems including the visual system, the subject of this chapter. Second, within modalities, such as vision, audition, and somesthesis, information from different locations in the periphery is transmitted in parallel to the brain to maintain knowledge about spatial location. In other words, different locations on the skin, the cochlear membrane, and the retina send redundant signals about sensory qualities in parallel (*topographic parallelism*) to the brain to create maps of these sensory sheets. In the last case, the same sensory qualities are transmitted in parallel to allow for appreciation of these qualities at different spatial locations. Parallel processing of the type described implies that sensory experience is initially broken down into basic elements, which are transmitted in parallel, and that the reconstruction of the whole occurs at some central brain location. Historically, the idea was that there are sensory areas where separate senses are appreciated within the cortex and that these were later combined in an "association cortex." The problem with this idea is that it now appears that most of the cortex is occupied by separate sensory areas or modules within areas (at least 32 visual areas have been recognized in macaque monkeys) leaving little room in cortex for association areas.[2,3] According to the current view, each higher order sensory area performs a separate specialized function or set of functions. It can be argued that parallel processing and modular specialization have the advantage that localized damage does not cause the entire system to malfunction. Also modular systems are easier to improve from an evolutionary standpoint since changes are not required within the entire network. Nevertheless, sensory modules need to receive input from somewhere and need to communicate their computational achievements to other parts of the system so the independence of these units can only be relative. Also, specialization is expensive because it requires dedicated units and there are not enough resources to have every sensation, thought, and action produced by separate cells, pathways, or modules. Additionally, effective behavior of the system as a whole requires smooth cooperation of components over very short time periods. Thus, the system must either be more integrated than it appears or have some means of tightly coordinating relevant tasks.

This chapter explores the question of parallel processing and the problem of integration in the primate visual system. Section 1.2 presents a brief history of parallel processing in the visual system showing how earlier views have channeled our thinking. Next, we consider how messages are defined within parallel channels and the degree to which these parallel channels beginning at the periphery are truly functionally specialized. We argue that the way messages are coded by channels is still a matter of debate and that separate channels likely evolve only under conditions where messages either are incompatible if carried by a single channel or result in loss of important information and that each channel carries more than one message. Section 1.4 considers the question of whether the signatures that define the parallel input channels from the lateral geniculate nucleus (LGN) can be traced to cells in primary visual cortex (i.e., V1 or striate cortex) or beyond this level to extrastriate visual areas. We argue that such LGN pathway signatures are difficult to recognize beyond the LGN and that V1 output pathways are not segregated according to the rules governing LGN parallel pathway inputs. Section 1.5 explores data that demonstrate that parallel visual channels carry information not just about vision but also about the other senses, as well as about eye movements and cognitive state. Section 1.6 discusses the dynamics and functional implications of parallel visual pathways. In particular, we explore the issue of differences in the timing of messages sent by different pathways and the impact of feedback on the messages that are sent. Timing clearly plays an important role and is critical for the subsequent integration of visual signals. Feedback also can alter feedforward messages. In these sections we also explore the differences and similarities between the organization of the visual system and other sensory systems as a way to uncover functional roles. The final section provides a summary and a list of unanswered questions about parallel visual system organization.

1.2 HOW ARE PATHWAYS DEFINED?

How we think about parallel processing in the visual system is a product of several distinct approaches to the problem and ways of conceptualizing brain function. At one end of the spectrum we can conceive of cells, pathways, modules, and areas in the brain as dedicated to one specific function. For example, the law of specific nerve energies implies that cells, pathways, and areas connected to the optic nerve will provide sensory qualities related to vision. At the other end of the spectrum are brain models that define functions through the activity of networks where cells contribute to a number of functions depending on which network is active (see References 4 and 5). The truth likely lies between the two extremes, as it is clear that in a basically segmented body plan like ours, neurons belonging to different segments specialize to perform different functions; however, these segments are not isolated but connected intimately to a larger network that coordinates purposeful behavior.

Discussed below are two key lines of investigation using different approaches that have strongly affected our views of parallel processing in the visual system.[6] In the first approach, parallel visual processing is treated as an engineering problem. In the mid-1960s, Enroth-Cugell and Robson[7] proposed that at its lowest level the visual system could work as a series of spatial filters, namely, as spatial frequency analyzers. In this model, cells tuned to different ranges of spatial frequencies respond to the appropriate frequency within the visual image and transmit this information centrally in parallel. Enroth-Cugell and Robson[7] used this linear systems approach to subdivide cat retinal ganglion cells into two types: those that summed luminance changes linearly across their receptive fields (referred to as X cells) and those that did not (referred to as Y cells). X cells were considered the "interesting" cells because they followed the logic of the model. This general approach led to numerous physiological and psychophysical studies based on the idea that the visual system's response to any pattern could be predicted from its response to more basic temporal and spatial filtering components. In their original work, Enroth-Cugell and Robson,[7] however, did not argue that these cells limited their analysis to one spatial dimension or one attribute. They also described other properties that distinguished Y from X cells including the higher conduction velocities, sensitivity to higher speeds and lower contrasts, lower spatial frequency cutoffs, larger average receptive field center sizes, and more transient responses of Y vs. X cells.[7] These observations were important from the standpoint of parallel processing because they identified a collection of properties, not a single property (e.g., a single wavelength or single spatial frequency), that distinguished X from Y cells. Almost 20 years of studies on X and Y cells followed, showing that X and Y cells also could be distinguished based on morphology, retinal distribution, central targets, and receptive field properties (see Reference 8). From the constellation of traits defining each of these cell classes it was proposed that X cells were part of a channel to cortex subserving high-resolution pattern vision whereas Y cells were part of a channel subserving crude form and motion vision (see Reference 8). Also during this period, other cell types were discovered in the cat retina and LGN collectively referred to as W cells. The W cell category referred to those cells that investigators could not classify as either X or Y cells. Not surprisingly W cells were found to vary widely in properties, sharing in common only the attributes of low conduction velocity and relatively large receptive field sizes[8] (for review, see Reference 9). Because many W cells have heavy projections to the midbrain targets it was proposed that they subserve a more primitive (subcortical) type of vision referred to as "ambient vision." X and Y cells, by contrast, provided "focal vision" or more highly evolved vision that required cortex. *Ambient vision* was seen as preconscious vision used by the earliest vertebrates to aid in spatial orientation and navigation relying on peripheral cues whereas *focal vision* was seen as the conscious, mostly foveal, vision used to identify and classify objects (the dominant form of vision in primates). The analysis of X, Y, and W cells in cats also led later to a similar set of investigations on LGN parvocellular (P), magnocellular (M), and koniocellular (K) cells in primates, where both similarities and differences between cats and primates were uncovered[6] (for review, see Reference 10).

The ambient/focal vision or the "two visual systems" hypothesis was actually linked to a second very influential set of studies begun in the 1960s by Gerald Schneider. Schneider published a key article[11] in which he proposed that there was an anatomical separation between visual coding of the location (where) of a stimulus and its identification (what). Based on behavioral/lesion work in hamsters he argued that there were basically two visual systems supported by two separate pathways from the retina, the "where" pathway involving the superior colliculus and a "what" pathway involving the primary visual cortex (striate cortex or V1)[11] (Figure 1.1A). The "where" vs. "what" or ambient vs. focal pathways were subsequently modified and described as independent pathways to separate cortical targets, one involving a pathway from colliculus to pulvinar to extrastriate cortex and the other from the LGN to V1 (see References 12 and 13 and Figure 1.1B). The idea that these pathways were capable of independent parallel operation was demonstrated clearly in tree shrews where complete removal of V1 (and resulting complete degeneration of the LGN) or removal of the cortical target of the colliculopulvinar pathway does not impair simple pattern discrimination or acuity.[14,15] The retinocolliculopulvinar pathway to cortex has been offered as an explanation for the "blindsight" exhibited by humans in the absence of V1[16] (see, however, Reference 17). In 1982, a different version of the "where" vs. "what" two visual systems hypothesis was proposed. Ungerleider and Mishkin[18] argued based on a combination of prior clinical observations and their own lesion/behavior work in macaque monkeys that visual object identification (what) depended on the temporal cortex whereas object location

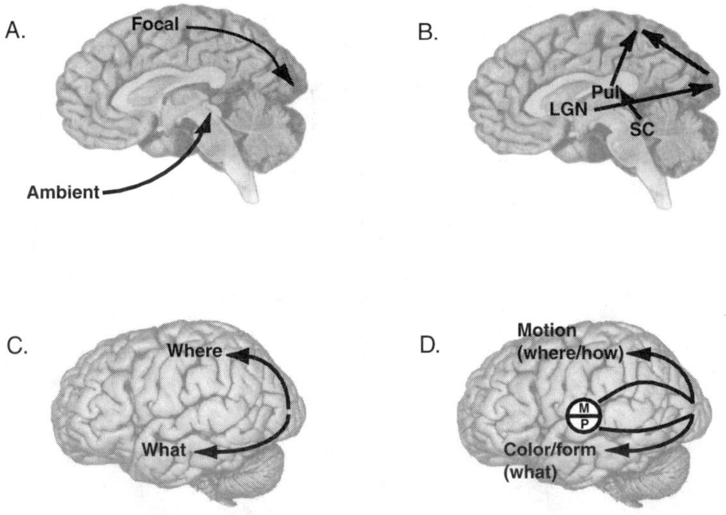

FIGURE 1.1 Evolution of the two visual systems hypothesis. (A) A diagram of the original two visual systems hypothesis proposed by Schneider.[11] Each system is supported by its own pathway, with the superior colliculus serving as the critical integration site for the "where" pathway (ambient vision) and visual cortex serving as the main target for the "what" pathway (focal vision). (B) Diamond and Hall[13] subsequently modified and described the pathways as both having important cortex targets, with one pathway involving a channel from colliculus to pulvinar to extrastriate cortex and the other from the LGN to V1. (C) Ungerleider and Mishkin[18] modified the where and what hypothesis substantially by suggesting that object identification (what) depended on temporal cortex whereas object location (where) depended on parietal cortex. They also suggested that *both* pathways require LGN and V1. (D) Two current models were suggested by Livingstone and Hubel[21] and Goodale and Milner.[19,20] Livingstone and Hubel proposed that the "where" channel linked M retinal and LGN cells to the "where" (dorsal stream) hierarchy of visual areas terminating within the parietal lobe. P retinal and LGN cells are linked in their model to the "what" (ventral stream) hierarchy of visual areas terminating within the temporal lobe. Goodale and Milner proposed a modification of this view in which the "where" stream becomes the "how" stream involved with unconscious vision for action and the "what" stream remains the conscious visual hierarchy of areas involved in object identification.

(where) required the parietal cortex (Figure 1.1C). They also suggested that both areas required primary visual cortex (V1). The cortical version of the "what" vs. "where" hypothesis suggested that if the two visual systems originated subcortically they must both pass through the LGN. More recently, Goodale and Milner[19] (see also Reference 20) outlined a cortical version of the two visual systems hypothesis that combines features of both the original ambient/focal two visual systems hypothesis with the Ungerleider and Mishkin proposal. In Goodale's[19] model, the parietal cortical hierarchy of visual cortical areas is specialized to handle an ambient-like preconscious "vision for action" used to interact with objects and move around in the environment whereas the temporal lobe hierarchy of visual areas is, like focal vision, conscious vision allowing for the recognition of objects.

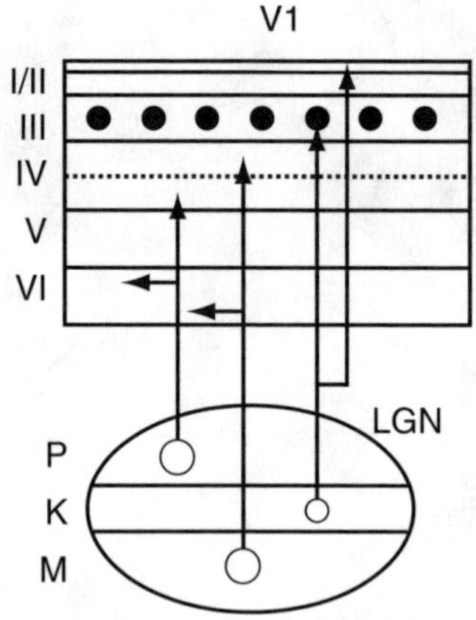

FIGURE 1.2 Information flow from LGN to striate cortex. P cells send input primarily to upper layer VI and lower layer IV (IVC of Brodmann). M cells send input primarily to lower layer VI and upper layer IV. K cells send input primarily to layer I and the CO-rich areas (blobs) of layer III.

Until Livingstone and Hubel[21] proposed a link between the parallel behavior of subcortical cells and pathways and the two cortical visual streams, no effort had been made to link parallel LGN pathways with parallel visual cortical pathways. Livingstone and Hubel[21] outlined their hypothesis that different attributes such as form, color and motion were segregated within the layers and cytochrome oxidase (CO) blob compartments of V1. According to this model the P retinogeniculocortical pathway (form and color) projects ultimately to the "what" hierarchy of visual areas ending in the temporal lobe and the M retinogeniculocortical pathway (motion) to the "where" hierarchy of visual areas ending in the parietal lobe (Figure 1.1D). Evidence to support the links between the P pathway and form/color and the M pathway and motion came primarily from physiology and connectional anatomy. Physiological studies had shown that P LGN cells exhibit chromatic opponency and have high spatial resolution and that M cells are not selective for wavelength but exhibit high temporal resolution[21] (reviewed in Reference 22). Livingstone and Hubel and others provided evidence that linked the P pathway to the CO-blob and interblob compartments in cortical layer III of V1 with appropriate output pathways to the "what" hierarchy of extrastriate visual areas, as well as evidence that the M pathway projected to the "where" hierarchy of visual areas via connections within V1 layer IVB (Figure 1.2). The K pathway was ignored, in part, because it did not fit well with this form of the two visual systems model (see References 23 and 24). The two visual systems model of Livingstone and Hubel[21] assumes that two peripheral pathways (M and P) are specialized to transmit a collection of properties that support one or the other of the two systems (motion vision or object vision). Because no "third" visual system had been suggested that meant that the K pathway was an orphan. When it was discovered years later that a small percentage of K cells transmit S cone signals (Blue-ON) in marmosets,[25] this fact was jumped on as an explanation for the existence of this entire pathway because it fit nicely with the two visual systems model. The problem is that the vast majority of K cells studied in all primates, so far, do not respond to S cones and share many achromatic spatial and temporal properties in common with both M and P LGN cells.[26–28] This fact, as well as the general heterogeneity of characteristics of cells within the K pathway, makes it harder to shoehorn them into one of the two visual systems boxes. These issues raise more profound questions concerning parallel visual pathways and these questions bear on hypotheses concerning the link between visual physiology and behavior. In other words, what messages are being transmitted in parallel; are these qualitatively distinct, and do they directly support specific types of visual behavior? These are issues we address in the next sections concerned with what exactly is being transmitted by parallel visual pathways.

1.3 DO LGN CHANNELS CARRY DISCRETE MESSAGES?

Each of the parallel LGN pathways in primates is believed to relay a distinct set of messages to cortex. This belief rests primarily on single unit recordings in which the main messages are defined based on a "rate code." In other words the message is defined as the relative magnitude of the response of the cell in terms of number of spikes averaged over trials lasting typically 500 ms. From studies using this criterion in a number of primates, it has been shown, as mentioned earlier, that P cells are selective to higher spatial frequencies, lower temporal frequencies, and lower contrasts than M cells with K cells (at least those that can be driven by traditional grating stimuli) falling in between.[25–27,29,30] Some P and K cells (in diurnal primates) are selective for wavelength, whereas M cells are not.[25] It is from these data, as well as anatomical links to the two cortical visual systems (see above), that hypotheses about pathways devoted to color, form, and motion were developed.

1.3.1 DEFINING THE CODE

The problem, of course, with a simple rate code is that LGN cells respond to a variety of stimuli. In other words, a P cell that is selective for color can also respond well to an achromatic grating stimulus of appropriate spatial frequency as well as to these same stimuli flashing or moving at a preferred temporal frequency. The difficulty based on such a rate code is distinguishing between different combinations of stimuli that could potentially produce the same rate of response per defined time block such as a grating presented at the preferred spatial frequency but a higher-than-preferred temporal frequency vs. a non-optimal spatial frequency presented at the preferred temporal frequency or even a grating presented at the preferred spatial and temporal frequency but at a much lower contrast. All of the latter conditions could potentially result in an identical rate code. Clearly, these situations do not cause perceptual confusion. The potential problem of producing the same rate of response to different combinations of stimuli is compounded for central target cells of these parallel pathways where other properties such as direction selectivity, orientation selectivity, etc. are added to the collection of earlier stimulus attributes.

1.3.2 A Temporal Code?

There have been a number of suggestions for how neurons, or groups of neurons, could disambiguate problems presented by a strict rate code. It is not the purpose of this chapter to present a detailed treatise on this issue[4,31,32] but simply to summarize a few key points. One hypothesis holds that information lies in the pattern of spikes over time, not simply in the number of spikes. We know, for example, that P cells tend to show sustained responses to standing contrasts, M cells show transient responses to the same stimulus, and K cells can behave either way. It also is clear that temporal information such as the frequency of drift of a grating stimulus is reflected in the responses of most LGN cells of all classes, which, for optimal temporal frequencies, fire with bursts of spikes in synchrony with the stimulus as the bars pass over the receptive field center. Similar response patterns can be seen in the simple cell cortical targets of LGN cells.[33] At the level of the LGN, Reinagel and Reid[34] have argued that temporal coding is very precise and reproducible with many individual spikes timed with better than 1-ms precision. This precision appears to be conserved within a cell class at least for cat X cells.[34] A key question, however, is whether the pattern of spikes also contains additional information about stimulus quality that would help to resolve ambiguity, as some have argued for both LGN and cortical neurons.[5,35,36] Several groups contend that the spike patterns of LGN and cortical cells do carry more information about spatial patterns than is contained in a simple rate code,[35,36] although these results are still controversial.[37] The strongest argument that the temporal pattern of spikes in single cells carries sensory information comes from work showing that when thalamic relay cells fire in bursts (i.e., burst mode) they are extremely effective in producing cortical spikes, increasing the chances of producing cortical spikes by more than 200% over the nonburst condition (i.e., tonic mode).[38] Although it has been argued that thalamic relay cells fire in rhythmic bursts mainly when animals are asleep,[39–41] Sherman and colleagues examining P and M LGN cells (K cells were not mentioned) in awake macaque monkeys and Swadlow and colleagues[38] examining ventrobasal somatosensory relay cells in awake rabbits have shown that thalamic cells also burst irregularly under awake conditions. Under awake conditions, bursts occur mainly when animals are inattentive or drowsy. In the LGN Sherman[42] has suggested that the very powerful activation of cortex by bursts of spikes may be used as a "wake-up call" alerting the animal to relevant stimuli that then are further analyzed when these cells are in tonic mode.

Concerning temporal codes, it recently has been shown by Oram et al.[32] that response latency following stimulus onset can carry information that is distinct from numbers of spikes or response magnitude.[32] The latter authors argue that response latency could provide independent information about stimulus contrast, for example. The key point is that they find that latency can code information that is distinct from spike rate. Clear differences in the onset latencies of K, M, and P LGN cells (M > P > K) to the presentation of an optimum target have been demonstrated (see, for example, Reference 29). These differences are magnified slightly by the differences in axon conduction latencies of these pathways to cortex. Although it is difficult to predict the cortical impact of these differences because stimulus attributes, stimulus intensity, and degree of convergence of inputs on the postsynaptic cortical cell all affect response latency,[43] it remains plausible that V1 takes advantage of these timing differences to increase the probability that cells reach threshold. For example, one could imagine that direct input from the much slower K axons might arrive in layer IIIB of cortex at the same time as indirect signals from the faster M and P pathways that must traverse several synapses in the layers below to reach the same cells in cortical layer III. P and M cell signals might arrive at the same time in layer III because the larger number of P cells could potentially bring target layer 4 cells to threshold more rapidly through convergence than the smaller number of converging M cells.[43] In this way, cells in layer III could combine several attributes about a stimulus carried by each of the parallel pathways. The timing differences may also be important in combining sensory signals from the feedforward parallel pathways with the multiple pathways that feed back to V1 from higher cortical areas.

1.3.3 Single Cells or Populations?

The coding problem also may be solved by population coding. Large numbers of LGN neurons project to single cortical cells so must work together to have an impact on the target. The question is whether visual cortical cells represent information as a pattern across elements or simply combine signals from LGN cells that act as independent conveyors of information. Experiments in which the P or M LGN layers were ablated and macaque monkeys tested for their ability to discriminate stimulus contrast showed a surprising result relative to the properties of these pathways. Prior work had shown that individual M cells in all primates studied have much better contrast sensitivity than individual P cells.[10] Following ablation of the M pathway, however, monkeys showed no discernible deficit in contrast sensitivity whereas following ablation of the P pathway monkeys showed a marked deficit in contrast sensitivity. This result suggests that perceptual appreciation of contrast does not relate to the relative sensitivity displayed by individual M and P cells but presumably reflects a pooling of signals from many P cells[44,45] (Figure 1.3). This result does not speak to the issue of whether this pooling involves coordination of the patterns of spikes within a population or whether it involves precisely coordinated synchrony across populations. These examples simply illustrate the difficulty in identifying how sensory messages are coded by the parallel pathways in a way that results in meaningful behavior.

1.3.4 Coding under Different Levels of Illumination

Even if one assumes that the main sensory messages carried by parallel pathways are coded by the relative rate of spike production in individual neurons, the messages carried by separate parallel pathways vary with visual conditions. For example, it has been shown that, although the P pathway is often equated with the high acuity and wavelength selectivity of the cone pathway, both P and M cells are active under scotopic conditions where only rods are active.[46] No detailed studies have been made of rod input to the K pathway although W cells apparently do carry both rod and cone signals.[47] This means that under scotopic conditions the spatial, temporal, and wavelength selectivities of both pathways are quite different, with the P pathway

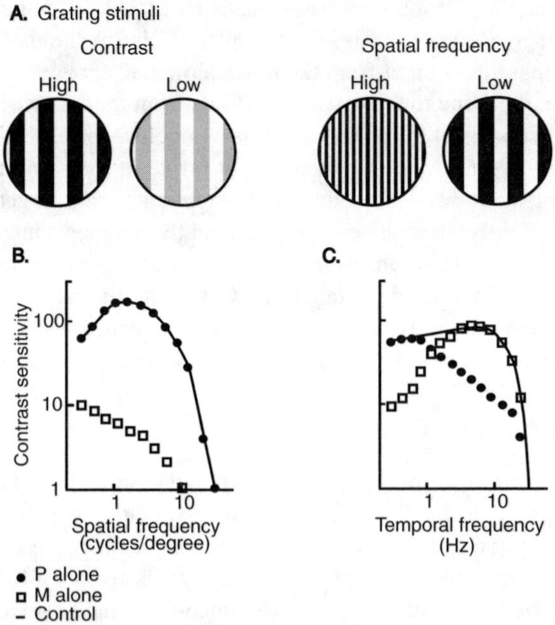

FIGURE 1.3 Visual losses after selective ablation of the magnocellular (M) and parvocellular (P) layers of the lateral geniculate nucleus in monkeys. (A) Luminance contrast is the difference between the brightest and darkest parts of the grating. Spatial frequency is the number of light and dark bars (cycles) in the grating per degree of visual angle. Temporal frequency (not shown) is how fast the stationary grating is turned on and off per second (Hz). (B) Contrast sensitivity is the inverse of the lowest stimulus contrast that can be detected. Contrast sensitivity for all spatial frequencies is reduced when only the M pathway remains after P ablation. The solid line in B and C shows sensitivity of the normal monkey; filled circles show the contribution of the P pathway (after M layer ablations) and open squares the contribution of the M pathway (after P layer ablations). (C) Contrast sensitivity to a grating with low spatial frequency is reduced at lower temporal frequencies when only M cells remain and at higher frequencies when only P cells remain. (From Kandel, E.R. et al., *Principles of Neural Science*, 4th ed., McGraw-Hill, New York, 2000, 531. With permission.)

being achromatic and shifting to a preference for lower spatial frequencies and the M no longer carrying high temporal frequency signals.

The above examples indicate that parallel visual pathways such as the K, M, and P pathways of primates, and individual cells within these pathways, are not dedicated or specialized in the sense of labeled lines from the periphery. Instead, each pathway or cell within a pathway is selective to a range of stimuli that shifts depending on environmental conditions. Why then is segregation of pathways maintained through the layers of the LGN to the first synapses within V1 or possibly beyond V1? Two reasons are likely. The first is that sensory attributes carried by parallel pathways may simply be incompatible. Retinal ganglion cells cannot have both extensive and confined dendrites at the same time or thin and thick axons at the same time. Cells with confined dendrites (P) can have high acuity but sacrifice sensitivity. Cells with thick axons (M) will send signals at a faster rate. Cells with sparse dendrites (K) may be able to sample from unique combinations of retinal inputs but be limited in the messages they can send. By creating separate pathways, incompatible stimulus attributes present at the same moment can be sent in parallel. The second reason, as argued by many, is that parallel pathways avoid fusion of important signals that would be useful to combine in unique ways at later stages. An example could be parallel pathways related to the left and right eye. Taken together, these examples emphasize that parallel visual pathways from retina through the LGN to cortex carry multiple messages about the distribution of spatial and temporal frequencies and wavelengths and, as discussed below, other nonvisual information. At present, it is evident that coding within parallel visual pathways must involve more than transmission of spike rate in individual neurons within a given pathway to avoid ambiguity. Evidence suggests that additional information exists in the form of a temporal code either within individual cells or, more likely, across populations of cells.

1.4 CAN LGN CHANNEL SIGNATURES BE TRACED BEYOND THE LGN?

This section explores the degree to which separate LGN pathways remain separate within V1. In other words do M, P, and K pathways bear signatures that can be traced through to the V1 output cells that will send messages into the proposed separate hierarchies of visual areas? In the original two visual systems model of Livingstone and Hubel,[21] signals from P and M LGN channels were pictured as being sent via separate networks in V1 to separate classes of output cells that, in turn, provided signals to the "what" (ventral stream) vs. "where" (dorsal stream) hierarchies of visual areas. There are several difficulties in determining whether LGN channel signals remain separate in V1. First, although K, M, and P pathways terminate to an extent within separate layers or sublayers of V1 (see Figure 1.2) there is tremendous opportunity for signal mixing following the first synapse. Second, V1 cell properties such as orientation selectivity, direction selectivity, binocularity, end stopping, etc. are distinct from those that specifically define the K, M, and P input pathways. This means that, with the possible exception of chromatic signals, there are no LGN pathway signatures that can easily be identified in the vast majority of V1 cells. Differences in threshold spatial and temporal frequency selectivity or contrast sensitivity could indicate that one pathway provides input but does not rule out the contribution of other pathways given that cortical cells respond to a range of spatial and temporal frequencies and contrasts. Latency to respond can only be used as a signature for cells receiving direct input and then only for the initial spikes; beyond this time point latency differences will reflect a mix of inputs.[43] Nevertheless, the idea that some output cells of V1 are dominated by or provide exclusive conduits for LGN input pathways persists (see Reference 48). In fact, the main cortical target area of the LGN, V1, is often pictured as a giant railroad switching station. In this model, trains that come from LGN may be switched to different tracks but remain recognizable as they leave the V1 station.

Is this really the case? To answer this question we need to describe briefly the V1 output pathways, their connections, and their targets. As in all cortical areas, input to V1 terminates primarily in the middle layers (IV and III), whereas output to extrastriate cortical areas exits from the upper layers (II/III) and output to subcortical structures exits from the lower layers (V and VI). V1 in primates has a number of cortical and subcortical targets. There are four major cortical targets of V1: these include V2, V3, the dorsal medial area (DM)/V3a, and the middle temporal (MT) visual area with smaller projections to the dorsal lateral or fourth visual area (DL/V4) and contralateral V1 (at least along the vertical meridian representation).[49] The bulk of the output from V1 goes to V2 and arises from cells in interblob columns in layer III[50,51] (see, however, Reference 52). This projection terminates primarily within the CO pale bands and CO thick bands of V2. A second projection arises from the CO-rich blobs of V1 and terminates in the CO-rich thin bands of V2.[50,53] In macaque monkeys V3 receives the second largest projection from V1, which also arises from layer III and possibly IVB.[54] Connections to area DM/V3a arise from CO blobs in layer III and cells below these compartments in IVB, which also project to MT.[55–57] All these output cortical targets of V1 and additional higher-order visual areas provide major feedback projections to V1, which can directly impact the output cells in the superficial layers and also influence activity within the deeper layers of V1.[58] The deeper layers of V1, layers V and VI, send axons back to the thalamus and to the midbrain and pons. Layer VI is unique in that cells in this layer send both direct and indirect (via the thalamic reticular nucleus) feedback to the LGN and provide major pathways for V1 to regulate its own input. Cells in layer VI also send axons to the visual sectors of the claustrum, which appears to modulate the responses of V1 neurons via feedback. Cells in layer V provide the major driving input to many cells in the pulvinar nucleus of the thalamus in monkeys; the pulvinar, in turn, provides input to a number of extrastriate areas, including V2, V3, DM/V3a, and MT, that also feed signals back to V1. In addition, cells in layer V send a major projection to the superficial layers of the superior colliculus and other midbrain areas such as the pretectum, as well as nuclei in the pons that are concerned with eye movements. Thus, V1 is in a position to inform these structures of its activities and be informed by them indirectly through connections with the LGN or through feedback from extrastriate areas (see Reference 59 for overview).

The anatomical links between the input and output pathways are complex with enormous opportunity for mixing of input signals before these reach output cells. Nevertheless, functional assays of these pathways suggest that some output cells are driven specifically by one input pathway.[48,60] For example, Yabuta et al.[60] found that when they measured the sources of excitatory input to cells in layer IVB using local uncaging of glutamate in slice preparation, this input depended on cell type with pyramidal IVB cells receiving strong input from both IVCβ (P) and IVCα (M) layers but spiny stellate cells receiving strong input only from layer IVCα (M). Because stellate cells in IVB mainly project to area MT, this finding argues for a specialized projection from the M pathway to area MT. Because areas V2, V3, and DM/V3a also receive input mainly from pyramidal cells in layer IVB, it seems likely that these other areas receive signals that reflect a mixture of P and M pathway input. The idea that the output pathway to MT may be uniquely dominated by the M pathway is also supported by earlier studies in which it was shown that following selective GABA blockade of M and P LGN layers (along with neighboring K pathways) most MT cells were dependent on input from the M pathway.[61] In spite of this finding, cells projecting to MT clearly do not resemble M cells, being binocular, direction and orientation selective.[62] Also, other studies have clearly demonstrated that cells in area MT respond to the movement of isoluminant red/green and blue/yellow stimuli showing that signals from P or K pathways must reach this area and are capable of driving cells.[63] In contrast to the studies of input to area MT most other studies either attempting to examine for signatures of M and P pathways or examining the impact of blocking these pathways on output pathways to other V1 targets have concluded the V1 cells receive a mixed input from LGN pathways, suggesting that potential M domination of one output pathway to MT is the exception not the rule.[61,64,65] Moreover, anatomically much of the output to the ventral stream leaves from layer IIIA, which receives no direct input from layer IVC, but receives signals only after they have been processed in other layers.[66] Thus, both the wiring and physiology suggest that considerable integration of signals takes place in V1 before the signals are transmitted to the majority of output cells in V1. This is not to say that output cells combine input signals in the same way; laminar differences in the physiological properties of V1 cells argue that input signals are combined in distinct ways to support the next steps in analysis. The contributions of input pathways with regard to individual cells is also likely to be dynamically regulated depending on the stimulus content.[64]

1.5 ARE VISUAL PATHWAYS PURELY VISUAL?

Once retinal signals from the parallel visual pathways arrive within the LGN they are in a position to be influenced by a wide variety of other inputs as well as the complex feedforward and feedback inhibitory circuits. These inputs include not only other visual inputs from visual cortical areas, superior colliculus, pretectum, parabigeminal nucleus, and the visual sector of the thalamic reticular nucleus, but also nonvisual inputs via some of the latter nuclei as well as brain stem cholinergic, noradrenergic, serotonergic pathways and hypothalamic histaminergic pathways[67] (Figure 1.4). The number of synapses provided by these nonretinal inputs greatly outnumber the synapses made by the parallel pathways coming from the retina and the signals these nonretinal inputs provide are controlled in complex ways by a variety of transmitter receptors. In addition, evidence suggests that many nonretinal inputs may be parallel pathway specific, suggesting that the separate visual pathways to V1 (e.g., K, M, and P pathways) may be modulated independently. For example, glutamatergic input from the superior colliculus and cholinergic input from the parabigeminal nucleus appear to terminate mainly on K cells in primates and on W cells in cats.[68,69] In contrast, GABAergic input from the pretectum has been shown to terminate mainly on P cells in primates and X cells in the A layers of cats.[70,71] Finally, there is some evidence in primates that cholinergic input from the brain stem targets mainly P layers in some primates and M layers in others.[72,73] Taken together, these facts suggest that the parallel visual pathways from the LGN to V1 are not strictly visual in content and may be modulated in different ways by their

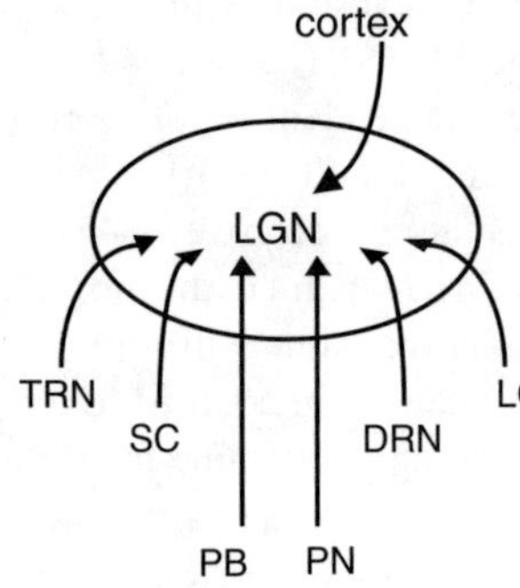

FIGURE 1.4 Schematic diagram illustrating some areas that provide input to the LGN: thalamic reticular nucleus (TRN), superior colliculus (SC), parabigeminal nucleus (PB), parabrachial nucleus (PN), dorsal raphe nucleus (DRN), the locus coeruleus (LC), and cortex.

nonretinal inputs. In spite of the overwhelming amount of data suggesting that visual pathways are not strictly visual, however, it has been difficult to identify what behaviorally relevant impact these extraretinal inputs might have. The difficulties relate, in part, to the fact that the vast majority of studies of LGN cell properties have been performed in anesthetized paralyzed animals where many of these modulatory inputs may be silent.

1.5.1 PGO Waves

Nevertheless, it has been recognized for decades that the temporal structure of LGN cell activity can change with the state of the animal, particularly during sleeping and waking. As mentioned earlier, LGN cells in sleeping animals often exhibit a firing pattern that is characterized by rhythmic bursting.[41,74] Since the late 1950s it was known that animals entering the rapid-eye-movement (REM) phase of sleep exhibit prominent waves of activity known as pontogeniculo-occipital (PGO) waves.[75] These bursts of activity not only were shown to be temporally related to the characteristic eye movements seen in REM sleep, but also were identified in awake cats in these early studies[76] and were shown to be induced by blocking serotonin input to the pons.[77] Whether PGO waves identified in animals in REM sleep relate just to changes in the state of the animal (e.g., changes from a nondreaming to a dreaming, or REM, state), or reflect signals related to eye movements or both still remains unclear. Most investigators have argued that PGO waves reflect activation of alerting mechanisms related to changes in state. The similarity in the burst structure found in PGO waves and bursts identified in LGN in awake animals, which have been proposed to be part of a system designed to aid in attention to novel stimuli (see above and Reference 42), suggests that these waves reflect more of a change in state.

1.5.2 Saccadic Eye Movements

Eye movement influences on the LGN have been examined independently under different conditions. Although there is considerable controversy over the effects of eye movements on LGN cell activity, the bulk of the evidence indicates that eye movements alter the signals sent via LGN cells. Saccadic eye movements are ballistic movements that are particularly well developed in primates, allowing primates to bring the fovea to regions of interest within the visual scene. These movements are also evident in other mammals with frontally placed eyes such as cats. Because such ballistic eye movements sweep stimuli across the entire retina at high speed, it is advantageous to suppress the potentially disrupting smear of visual signals that must occur during saccadic eye movements. In fact, saccadic eye movements are known to modify perception in several different ways including raising the threshold for some stimuli.[78] This is probably why we cannot see our own eyes move when looking in a mirror but are perfectly capable of seeing someone else's eyes make saccadic movements. Psychophysical studies suggest that saccades do not affect the visibility of all stimuli equally (see Reference 78 for review). Suppression, which occurs both prior to and during a saccade, is strongest for detecting displacements of visual stimuli and for detecting stimuli containing low spatial frequencies[78] (see, however, Reference 79). Enhancements of visual sensitivity to some stimuli (e.g., color) are also seen following the conclusion of a saccade.[80] The latter results combined with data that indicate that detecting colors of equiluminant stimuli is not affected by saccadic eye movements indicates that saccadic eye movements should show differential effects on signals carried by M vs. P, and possibly K, visual pathways.

Conflicting data have been reported on the degree to which LGN cell activity is modulated during saccadic eye movements. In awake-behaving cats, Lee and Malpeli[81] showed that both X and Y LGN cells (which have been equated with P and M channels in primates, respectively) are suppressed equally during saccades with enhancement of responses following saccades. In macaque monkeys, however, results have argued for a variety of effects including very limited effects on LGN cells,[82] effects limited to the M pathway,[83] effects in both M and P pathways,[84] or (as we have reported) effects on all three LGN cell classes.[85] Even within one pathway there is no general agreement on the effect of saccadic eye movements on LGN cells. Maunsell et al.[83] found saccadic suppression limited only to M LGN cells. In contrast, Ramcharan et al.[86] found that M cell activity is enhanced during saccades. In agreement with Maunsell et al.,[83] Ramcharan et al.[86] found no changes in P cell activity. Interestingly, however, these investigators found that both M and P cells show a significant suppression of burst firing during saccades, suggesting that the temporal structure of the message is changed during saccades, perhaps increasing the visual threshold in this manner. Finally, Reppas et al.[84] report that M cells show saccadic suppression followed by significant enhancement of response, whereas P cells show mainly enhancement of responses following saccades. The conflicts between some of these studies may relate to the different methods used to examine for saccadic suppression in LGN. Comparisons across the different designs used to measure saccadic suppression in the above studies suggest that different tasks and analyses should be compared to disentangle potential confounds such as direct stimulation of the LGN receptive field, potential floor effects, and the potential that averaging over long time blocks has for washing out significant changes in activity. Both our own work in awake-behaving macaque monkeys and the study of Lee and Malpeli[81] provide evidence that LGN cell activity can be suppressed by saccades made in total darkness, suggesting that the suppression mechanism does not require retinal input to operate.[85]

If saccadic suppression does occur in LGN cells, what is the circuit responsible for the suppression? Intracellular recordings in rabbits[87] helped identify and outline a potential circuit responsible for saccadic suppression in the LGN. This circuit involves a projection from the deep layers of the superior colliculus (SC) to thalamic reticular nucleus (TRN) via the central lateral (CL) nucleus of the thalamus. Furthermore, the overall time course of activation in the SC and suppression in the LGN fits with the time course of saccadic suppression. There are many other pathways that could provide for such suppression. The superior colliculus provides direct input to the LGN but only from the superficial layers that project, in primates, primarily to the K layers[69] of the LGN. Thus far, however, too few K cells have been examined to determine if their responses to eye movements differ from the responses of P or M cells (Reference 85 and unpublished results). Other possible input sources include the midbrain reticular formation, pontine cholinergic cells, and the pretectum (see Reference 67). Regarding the pretectogeniculate pathway Schmidt[88] has shown that pretectogeniculate cells are excited during saccades. Schmidt and colleagues[88,89] have argued that pretectogeniculate cells inhibit LGN interneurons, thus causing excitation of LGN relay cells. The latter circuit would be entirely appropriate to explain a postsaccadic enhancement of activity given the timing reported. However, others[90] have provided evidence in cats that pretectal activity suppresses LGN activity and thus contributes directly to saccadic suppression.

From the perspective of this chapter, however, the main point is that, although controversy exists, the bulk of the evidence supports the idea that the signals LGN cells send to cortex can be modulated significantly by both behavioral state and by eye movements. In addition, evidence indicates that signals carried about eye movements may differ between the parallel visual pathways.

In addition to gross changes in behavioral state and information about eye movements, there is growing evidence that information from other senses as well as cognitive activities affect the messages that the LGN parallel pathways send to cortex. Even in anesthetized primates it has been demonstrated that both auditory and somatosensory information can directly modulate LGN cells. In fact, at least one study suggests that almost 100% of K LGN cells and close to 50% of M and P LGN cells show significant increases or decreases in spontaneous activity simply to auditory and tactile stimuli under the same visual stimulation conditions.[29]

1.5.3 Cognitive Aspects of Vision

Early stages of the visual system such as the LGN and V1 have until recently been regarded as being unaffected by cognitively related activities and attention. This view stems largely from studies that did not find any changes in V1 activity related to task performance or attention.[91] With the advent of different investigative techniques, however, evidence is accumulating that early stages in the visual pathway are more sophisticated and "active" than thought originally. Furthermore, newer findings suggest that, in addition to sensory signals and information related to eye movements, V1 and LGN cells might signal cognitively related aspects of vision reflecting task relevance and attentional shifts.[92–97] This novel view is easier to appreciate when one considers that the visual system is dynamic and that areas at all levels (with the exception of the retina) are receiving both feedforward and feedback information continuously. Although the LGN is at almost the lowest level in the feedforward hierarchy, it is potentially at the *highest* level in the feedback hierarchy (Figure 1.5). Although there is only a small amount of data (Reference 95 and unpublished results) directly demonstrating the existence of LGN modulation in relation to task relevance (see below), fMRI studies have shown that the LGN exhibits significant activity changes in a number of attentional tasks.[99] In fact, O'Connor et al.[99] found that attentional effects were *more* pronounced in the LGN than in V1. Also using fMRI it was found that the LGN showed significant activation in relation to imagined visual stimuli without any real stimuli being present[100] (however, see also Reference 97). There is also growing evidence that cells in V1 are modulated by changes in arousal, attention, and perceptual context.[101] In addition, manipulations designed to increase the size of the evoked responses of neurons in the middle temporal (MT) visual area pharmacologically have been shown to cause significant changes in activity in M, P, and K LGN cells in anesthetized macaque monkeys.[102] Because the same effects are seen in V1 and because only a tiny projection from MT to LGN cells has ever been identified (see Reference 59), it was assumed in the latter study that the MT influence on LGN cell activity was carried via V1 cells. Nevertheless, the latency for these changes was fast enough that it would be possible under these circumstances for P or K cells to be influenced by M pathway motion–driven effects from MT *before* the P or K cells send signals forward to V1. Thus, sensory signals and top-down attentional signals could come together as early as the LGN and then together with V1 evoke percepts that are then used to guide behavioral responses. Heeger and colleagues[103] have shown how trial-to-trial variability in the subjects' attentional state is reflected in the baseline fMRI signal in V1. This, in turn, causes variability in detection performance, the outcome of which is reflected in the signal as well. Super and colleagues[96] have shown how complex mental operations, associated with shifts of visual attention, are reflected by the activity of topographically appropriate neurons in area V1 of macaque monkeys. In their experiments attentional modulations occur after much longer latency than is required for the visual response itself (~200 ms). This latency is similar to the latency to modulate LGN activity in relationship to a behaviorally relevant cue that we found in awake-behaving monkeys (Reference 95 and unpublished results). It seems likely that the LGN and V1 form part of a dynamic loop that allows the system to enhance relevant stimulus components selectively based on behavioral relevance or to prepare for stimulus events that have behavioral significance. In keeping with this view, Paradiso and colleagues have shown that the appreciation of the attributes of a stimulus are better correlated with delayed components of V1 neural responses than with the initial transient response (reviewed in Reference 94). Surprisingly, neurons sometimes show similar delayed response components when the stimulus is *not presented*, but merely expected by the animal. Finally, Lamme and Roelfsema[104] have evidence that this late component of neural activity in V1 correlates with the perception of visual stimuli. They argue that perception does not take place without recurrent feedback to V1. We would argue

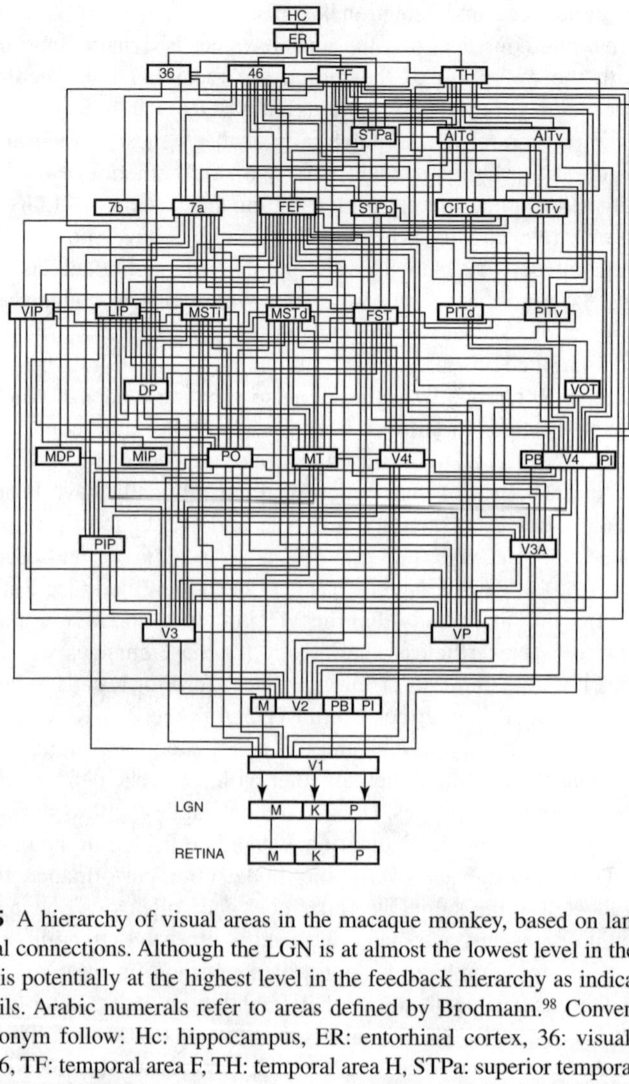

FIGURE 1.5 A hierarchy of visual areas in the macaque monkey, based on laminar patterns of anatomical connections. Although the LGN is at almost the lowest level in the feedforward hierarchy, it is potentially at the highest level in the feedback hierarchy as indicated here. See text for details. Arabic numerals refer to areas defined by Brodmann.[98] Conventional names for each acronym follow: Hc: hippocampus, ER: entorhinal cortex, 36: visual area 36, 46: visual area 46, TF: temporal area F, TH: temporal area H, STPa: superior temporal polysensory area (anterior), AITd: anterior inferior temporal area (dorsal), AITv: anterior inferior temporal area (ventral), 7a: visual area 7a, 7b: visual area 7b, STPp: superior temporal polysensory area (posterior), CITd: central inferior temporal area (dorsal), CITv: central inferior temporal area (ventral), VIP: ventral intraparietal area, LIP: lateral intraparietal area, MSTi: medial superior temporal area (inferior), MSTd: medial superior temporal area (dorsal), FST: fundus of the superior temporal area, PITd: posterior inferior temporal area (dorsal), PITv: posterior inferior temporal area (ventral), DP: dorsal prelunate area, VOT: ventral occipitotemporal area, MDP: medial dorsal parietal area, MIP: medial intraparietal area, PO: parietal-occipital area, MT: middle temporal area, V4t: visual areal 4 (temporal), PB: P blob, PI: P interblob, V4: visual area 4, PIP: posterior intraparietal area; V3A: visual areal 3A, V3: visual area 3, VP: ventral posterior area, V2: visual area 2, M: magnocellular, P: parvocellular, K: koniocellular. (From Felleman, D.J. and Van Essen, D.C., *Cereb. Cortex,* 1, 1, 1991. With permission.)

that this feedback must also reach the LGN, which works together with V1 to establish what is perceived. These data indicate that in addition to global changes in state, eye movements, nonvisual sensory information, and basic visual sensory signals, the LGN also carries information about the behavioral relevance of tasks being performed.

1.6 PARALLEL PATHWAYS IN A DYNAMIC SYSTEM: FUNCTIONAL IMPLICATIONS

Time never stops, and neither does visual information processing. As emphasized earlier in this chapter, the flow of sensory information is traditionally pictured as feedforward especially at the early stages of processing. In most diagrams, the flow of visual information processing is from retina to LGN to V1 to higher cortical areas and finally to motor areas that give rise to behavioral action. Yet, as shown in Figure 1.5, based on just number of synapses, the flow of information should actually be shown in reverse direction with the bulk of signals feeding down from higher to lower areas. Since time does not stop, this means that the past will always influence the future at least to the level of the thalamus. In fact, in the auditory system, this analogy goes all the way to the periphery because feedback from cortex reaches all the way back to the first input synapse in the brain stem.[105]

Another important issue that we have considered only in passing is that parallel sensory processing involves many more pathways than those that relay information through the main sensory relay nuclei of the thalamus. In the visual system, the retina sends signals directly to the hypothalamus (e.g., suprachiasmatic nucleus) for control of circadian rhythms, to several areas of the ventral thalamus and midbrain (e.g., the pregeniculate nuclei, the pretectal nuclei, the accessory optic nuclei, and the superior colliculus) that are involved with pupillary and accommodative responses, vestibular ocular reflexes, and head and eye movements, among other behaviors.[106] These areas also receive direct input from cortex and many feed their signals, in turn, back to the thalamus. In fact, it is often the case that the same retinal ganglion cells send signals to several of these locations at the same time via collaterals (see Reference 107 for review). This degree of complexity in parallel processing raises additional questions about the dynamics and timing of information processing within the system because the whole system must remain coordinated for appropriate behavioral outcomes to occur. Given the speed at which perceptual decisions can be made (<200 ms), this also means that some decisions must be made without involving cortex. Clearly, we are unaware of many visual behaviors such as the constant adjustments taking place in the size of our pupils and the shape of our lens as we view the environment. Simple movements involve constant unconscious coordination between the visual system and the vestibular system as becomes clear when observing the crippling effects of loss of vestibular function on normal everyday movements such as walking even when the visual system is completely intact (described in Reference 108). Such individuals must learn to use conscious visual monitoring to learn simply to stand and walk (only awkwardly) again. Even the saccadic eye movements one makes constantly while reading this page appear not to require conscious guidance.

In fact, it would be very difficult for readers to guess the number of eye movements they made just in reading this paragraph. Nevertheless, readers can consciously take their eyes off the printed page any time they wish, presumably relying on top-down signals from higher cortical areas (supplementary motor or anterior cingulate areas) that aid in the decision-making process.[109] These top-down signals must override subcortical visual/motor pathways that guide motor programs. The decision to take one's eyes off the page could, in turn, be driven by incoming visual or auditory signals such as the motion of a person entering the room or the sound of a kettle whistling. Cognitive processing of these signals has the potential, as described earlier, of influencing LGN to V1 signals that have yet to arrive.

The fact that we can perform cognitively guided perceptual judgments such as reading that require complex visually guided oculomotor (saccadic eye movements) and vestibular/ocular reflexes (to keep ourselves centered and upright) suggests that, in addition to parallel input pathways that can carry multiple messages forward to cortex, there must be a parallel network of pathways operating subcortically. The speed at which judgments, such as comprehending written words, take place indicates that there must be a degree of independence of processing on the part of some of these parallel pathways on both the sensory and motor side. One could also argue based on the fact that many nonmammalian vertebrates (e.g., frog; see Reference 110) are able to perform complex motor programs including negotiating barriers, avoiding predators, and capturing prey, and lose these abilities following subcortical lesions that many motor programs are guided fairly independently by parallel pathways that project subcortically. As noted recently by Guillery and Sherman,[107] these subcortical visual pathways can represent collaterals of parallel pathways that are projecting via the LGN to cortex. Such a system would remain efficient because the same signals sent to subcortical targets to guide basic motor programs would be sent via the LGN to V1 and then on to extrastriate areas for more complex perceptual/cognitive functions. Just as different subcortical areas perform different motor programs, there is strong evidence to suggest that different higher-order visual areas take messages from V1 and utilize these in different ways within the proposed dorsal and ventral visual hierarchies. Many of these visual cortical areas have been shown to have independent access to subcortical visual areas including nuclei in the thalamus, the superior colliculus, pretectum, and pontine eye movement nuclei (see Reference 59). At present, it is unclear how these separate cortical pathways cooperate, especially because they do not necessarily terminate on the same cells in each of these subcortical targets (e.g., different cortical areas terminate within different layers of the colliculus[111]). Again, the timing of the messages that are sent is likely critical given the speed at which normal judgments are performed and given the likelihood that cortical messages are integrated within the complex subcortical target networks controlling motor programs.

1.7 CONCLUSIONS AND UNANSWERED QUESTIONS

We conclude by making four major points. First, parallel retinogeniculocortical and retinosubcortical pathways are likely to operate in parallel to control complex visual behaviors. In the beginning of the chapter we considered different views of parallel visual processing starting with early views in which Ingle and Schneider[110] argued that subcortical vision makes an important contribution to visual behavior. This idea has basically been ignored as the focus has shifted more and more to parallel processing at the cortical level with the implicit idea that in "higher" mammals visual functions that were performed subcortically in lower vertebrates migrate in evolution to be performed by cortical areas. As argued in the last section this idea seems unlikely. Given the requirement that visuomotor behaviors must be performed very rapidly and efficiently, it is more likely that visual signals from retinal ganglion cells send basic sensory information in parallel to both cortical and subcortical destinations. Complex, but routine, visuomotor programs could then be carried out relatively independently with input from the cortex only at decision junctions, much as walking is carried out by motor programs in the spinal cord without the necessity for constant supervision by the cortex.[112]

A second major point is that parallel pathways from the retina (e.g., K, M, and P pathways in primates) carry more than a single message. Specialization is expensive and so likely arises only where messages to be carried are incompatible or must remain separate to be combined in different ways either cortically or subcortically. These retinal pathways cannot be linked to specific complex visual or visuomotor functions either cortically or subcortically but can carry nonsensory signals including cognitive and motor information even at the level of the LGN so are always multipurpose. Sensory messages carried by parallel retinal pathways also are utilized in a variety of ways. In other words, some K cells may carry signals relevant to identification of forms at later cortical stages or to accommodation at subcortical levels.

A third point is that coding within parallel visual pathways must involve more than transmission of spike rates in individual neurons to avoid ambiguity. Evidence suggests that additional information exists in the form of a temporal code either within individual cells or, more likely, across populations of cells. What these codes are and whether the codes are the same at different levels of the system remains an open question.

Finally, it is important to reemphasize the fact that vision is not a unidirectional process that goes from sensation through perception to action. Instead, the system is dynamic such that what happened in the past will literally "color" or change the sensory landscape in the future.

Several questions remain:

- What advantage is derived from maintaining a latency difference between the K, M, and P pathways?
- What do achromatic K cells contribute that is unique?
- How many output pathways to each extrastriate area leave from V1 and how distinct are these pathways?
- How are messages encoded by cells within the parallel pathways?
- Do coding strategies differ between thalamus and cortex?
- How do task demands regulate what sensory messages are sent via thalamocortical pathways?
- How independent are subcortical visuomotor programs from motor control?

ACKNOWLEDGMENTS

We thank Yuri Shostak, Julia Mavity-Hudson, and Yuliya Sokolova for helpful suggestions and support throughout the development of the manuscript; Julia Mavity-Hudson for help with the figures; and Shirin Pulous and James L. Enlow for clerical assistance. The work for this chapter was supported by Grants F31 NS44691-01 (DWR), EY01778 (VAC), and Core Grants EY08126 and HD15052.

REFERENCES

1. Ackerman, D.A., *Natural History of the Senses*, Random House, New York, 1990.
2. Felleman, D.J. and Van Essen, D.C., Distributed hierarchical processing in the primate cerebral cortex, *Cereb. Cortex,* 1, 1, 1991.
3. Kaas, J.H., The organization of neocortex in mammals: implications for theories of brain function, *Annu. Rev. Psychol.*, 38, 129, 1987.
4. Singer, W. and Gray, C.M., Visual feature integration and the temporal correlation hypothesis, *Annu. Rev. Neurosci.,* 18, 555, 1995.
5. Samonds, J., Allison, J., Brown, H., and Bonds, A., Cooperation between area 17 neuron pairs enhances fine discrimination of orientation, *J. Neurosci.*, in press.
6. Casagrande, V.A. and Xu, X., Parallel visual pathways: a comparative perspective, in *The Visual Neuroscience*, Chalupa, L. and Werner, J.S., Eds., MIT Press, Cambridge, MA, in press.
7. Enroth-Cugell, C. and Robson, J.G., The contrast sensitivity of retinal ganglion cells of the cat, *J. Physiol.* (London), 187, 517, 1966.
8. Stone, J., *Parallel Processing in the Visual System: The Classification of Retinal Ganglion Cells and Its Impact on the Neurobiology of Vision*, Plenum Press, New York, 1983.
9. Kaplan, E., The receptive field structure of retinal ganglion cells in cat and monkey, in *The Neural Basis of Visual Function,* Leventhal, A.G., Ed., Vol. 4 of *Vision and Visual Dysfunction*, Cronley-Dillon, J.R., Ed., Macmillan Press, London, 1991, 41.
10. Casagrande, V.A. and Norton, T.T., The lateral geniculate nucleus: a review of its physiology and function, in *The Neural Basis of Visual Function,* Leventhal, A.G., Ed., Vol. 4 of *Vision and Visual Dysfunction*, Cronley-Dillon, J.R., Ed., Macmillan Press, London, 1991, 41–84.
11. Schneider, G.E., Two visual systems: brain mechanisms for localization and discrimination are dissociated by tectal and cortical lesions, *Science*, 163, 895, 1969.
12. Casagrande, V.A., Harting, J.K., Hall, W.C., Diamond, I.T., and Martin, G.F., Superior colliculus of the tree shrew: a structural and functional subdivision into superficial and deep layers, *Science,* 177, 444, 1972.
13. Diamond, I.T. and Hall, W.C., Evolution of neocortex, *Science,* 164, 251, 1969.
14. Ware, C.B., Casagrande, V.A., and Diamond, I.T., Does the acuity of the tree shrew suffer from removal of striate cortex? A commentary on the paper by Ward and Masterton, *Brain Behav. Evol.*, 5, 18, 1972.
15. Ware, C.B., Diamond, I.T., and Casagrande, V.A., Effects of ablating the striate cortex on a successive pattern discrimination: further study of the visual system in the tree shrew (*Tupaia glis*), *Brain Behav. Evol.*, 9, 264, 1974.
16. Rafal, R.D., Smith, J., Krantz, J., Cohen, J., and Brennan, C., Extrageniculate vision on hemianopic humans: saccade inhibition by signals in the blind field, *Science*, 250, 118, 1990.

17. Collins, C.E., Lyon, D.C., and Kaas, J.H., Response of neurons in MT after long-standing lesions of V1 in adult New World monkeys. *J. Neurosci.,* 23, 2251, 2003.
18. Ungerleider, L.G. and Mishkin, M., Two cortical visual systems, in *Analysis of Visual Behavior,* Ingle, D.J., Goodale, M.A., and Mansfield, R.J., Eds., MIT Press, Cambridge, MA, 1982, 549.
19. Goodale, M.A. and Milner, A.D., Separate visual pathways for perception and action, *TINS,* 15, 20, 1992.
20. Milner, D.A. and Goodale, M.A., *The Visual Brain in Action,* in Oxford Psychology Series, Vol. 27, Oxford University Press, New York, 1995.
21. Livingstone, M.S. and Hubel, D.H., Do the relative mapping densities of the magno- and parvocellular systems vary with eccentricity? *J. Neurosci.,* 8, 4334, 1988.
22. DeYoe, E.A. and Van Essen, D.C., Concurrent processing streams in monkey visual cortex, *TINS,* 11, 219, 1988.
23. Casagrande, V.A., A third parallel visual pathway to primate area V1, *TINS,* 17, 305, 1994.
24. Hendry, S.H. and Reid, R.C., The koniocellular pathway in primate vision, *Annu. Rev. Neurosci.,* 23, 127, 2000.
25. Martin, P.R., White, A.J.R., Goodchild, A.K., Wilder, H.D., and Sefton, A.E., Evidence that blue-on cells are part of the third geniculocortical pathway in primates, *Eur. J. Neurosci.,* 9, 1536, 1997.
26. Xu, X., Ichida, J.M., Allison, J.D., Boyd, J.D., Bonds, A.B., and Casagrande V.A., A comparison of koniocellular, magnocellular and parvocellular receptive field properties in the lateral geniculate nucleus of the owl monkey (*Aotus trivirgatus*), *J. Physiol.,* 531, 203, 2001.
27. White, A.J., Solomon, S.G., and Martin P.R., Spatial properties of koniocellular cells in the lateral geniculate nucleus of the marmoset *Callithrix jacchus, J. Physiol.,* 533, 519, 2001.
28. Solomon, S.G., White, A.J., and Martin, P.R., Temporal contrast sensitivity in the lateral geniculate nucleus of a New World monkey, the marmoset *Callithrix jacchus, J. Physiol.,* 517, 901, 1999.
29. Irvin, G.E., Norton, T.T., Sesma, M.A., and Casagrande, V.A., W-like response properties of interlaminar zone cells in the lateral geniculate nucleus of a primate (*Galago crassicaudatus*), *Brain Res.,* 362, 254, 1986.
30. Norton, T.T., Casagrande, V.A., Irvin, G.E., Sesma, M.A., and Petry, H.M., Contrast-sensitivity functions of W-, X-, and Y-like relay cells in the lateral geniculate nucleus of bush baby (*Galago crassicaudatus*), *J. Neurophysiol.,* 59, 1639, 1988.
31. von der Malsburg, C., Binding in models of perception and brain function, *Curr. Opin. Neurobiol.,* 5, 520, 1995.
32. Oram, M.W., Xiao, D., Dritschel, B., and Payne, K.R., The temporal resolution of neural codes: does response latency have a unique role? *Philos. Trans. R. Soc. Lond. B,* 987, 2002.
33. Skottun, B.C., De Valois, R.L., Grosof, D.H., Movshon, J.A., Albrecht, D.G., and Bonds, A.B., Classifying simple and complex cells on the basis of response modulation, *Vision Res.,* 31, 1079, 1991.
34. Reinagel, P. and Reid, R.C., Precise firing events are conserved across neurons, *J. Neurosci.,* 22, 6837, 2002.
35. McClurkin, J.M., Gawne, T.J., Optican, L.M., and Richmond, B.J., Lateral geniculate neurons in behaving primates. II: Encoding of visual information in the temporal shape of the response, *J. Neurophysiol.,* 66, 794, 1991.

36. McClurkin, J.W., Optican, L.M., and Richmond, B.J., Cortical feedback increases visual information transmitted by monkey parvocellular lateral geniculate nucleus neurons, *Vis. Neurosci.*, 11, 601, 1994.
37. Tovée, M.J., Rolls, E.T., Treves, A., and Bellis, R.P., Information encoding and the responses of single neurons in the primate temporal cortex, *J. Neurophysiol.*, 70, 640, 1993.
38. Swadlow, H.A. and Gusev, A.G., The impact of "bursting" thalamic impulses at a neocortical synapse, *Nat. Neurosci.*, 4, 402, 2001.
39. Steriade, M. and Llinas, R.R., The functional states of the thalamus and the associated neuronal interplay, *Phys. Rev.*, 68, 649, 1988.
40. Steriade, M., Jones, E.G., and Llinas, R.R., *Thalamic Oscillations and Signaling*, Wiley, New York, 1990.
41. Steriade, M., McCormick, D.A., and Sejnowski, T.J., Thalamocortical oscillations in the sleeping and aroused brain, *Science*, 262, 679, 1993.
42. Sherman, S.M., Tonic and burst firing: dual modes of thalamocortical relay, *TINS*, 24, 122, 2001.
43. Maunsell, J.H.R., Ghose, M., Assad, J.A., McAdams, C.J., Boudre, C.E., and Noerager, B.D., Visual response latencies of magnocellular and parvocellular LGN neurons in macaque monkeys, *Vis. Neurosci.*, 16, 1, 1999.
44. Schiller, P.H., Logothetis, N.K., and Charles, E.R., Role of the color-opponent and broad-band channels in vision, *Vis. Neurosci.*, 5, 321, 1990.
45. Merigan, W.H. and Maunsell, J.H.R., Macaque vision after magnocellular lateral geniculate lesions, *Vis. Neurosci.*, 5, 347, 1990.
46. Wiesel, T.N. and Hubel, D.H., Spatial and chromatic interactions in the lateral geniculate body of the rhesus monkey, *J. Neurophysiol.*, 29, 1115, 1966.
47. Cleland, B.G., and Levick, W.R., Properties of rarely encountered types of ganglion cells in the cat's retina and an overall classification, *J. Physiol.*, 240, 457, 1974.
48. Vidyasagar, T.R., Kulikowski, J.J., Lipnicki, D.M., and Dreher, B., Convergence of parvocellular and magnocellular information channels in the primary visual cortex of the macaque, *Eur. J. Neurosci.*, 16(5), 945, 2002.
49. Kandel, E.R., Schwartz, J.H., and Jessell, T.M., *Principles of Neural Science*, 4th ed., McGraw-Hill, New York, 2000, 531.
50. Sincich, L.C. and Horton, J.C., Divided by cytochrome oxidase: a map of the projections from V1 to V2 in macaques, *Science*, 295, 1734, 2002.
51. Sincich, L.C. and Horton, J.C., Pale cytochrome oxidase stripes in V2 receive the richest projection from macaque striate cortex, *J. Comp. Neurol.*, 447, 18, 2002.
52. Livingstone, M.S. and Hubel, D.H., Connections between layer 4B of area 17 and the thick cytochrome oxidase stripes of area 18 in the squirrel monkey, *J. Neurosci.*, 7, 3371, 1987.
53. Livingstone, M.S. and Hubel, D.H., Anatomy and physiology of a color system in the primate visual cortex, *J. Neurosci.*, 4, 309, 1984.
54. Lyon, D.C. and Kaas, J.H., Evidence for a modified V3 with dorsal and ventral halves in macaque monkeys, *Neuron*, 33, 453, 2002.
55. Boyd, J.D., Guerguerian, R., and Casagrande, V.A., Extrastriate efferents distinguish a tripartite modular organization of V2 in the prosimian primate *Galago*, *Soc. Neurosci. Abstr.*, 23, 845, 1997.
56. Boyd, J.D. and Casagrande, V.A., Relationships between cytochrome oxidase (CO) blobs in primate primary visual cortex (V1) and the distribution of neurons projecting to the middle temporal area (MT), *J. Comp. Neurol.*, 409, 573, 1999.

57. Weisenhorn, D.M.V., Illing, R.B., and Spatz, W.B., Morphology and connections of neurons in area-17 projecting to the extrastriate areas MT and 19dm and to the superior colliculus in the monkey *Callithrix jacchus*, *J. Comp. Neurol.*, 362, 233, 1995.
58. Shao, Z. and Burkhalter, A., Different balance of excitation and inhibition in forward and feedback circuits of rat visual cortex, *J. Neurosci.*, 16, 7353, 1996.
59. Casagrande, V.A. and Kaas, J.H., The afferent, intrinsic, and efferent connections of primary visual cortex in primates, in *Cerebral Cortex*, in Vol. 10 of *Primary Visual Cortex of Primates,* Peters, A. and Rockland, K.S., Eds., Plenum Press, New York, 1994, 201–259.
60. Yabuta, N.H., Sawatari, A., and Callaway, E.M., Two functional channels from primary visual cortex to dorsal visual cortical areas, *Science*, 292, 297, 2001.
61. Nealey, T.A. and Maunsell, J.H., Magnocellular and parvocellular contributions to the responses of neurons in macaque striate cortex, *J. Neurosci.*, 14, 2069, 1994.
62. Movshon, J.A. and Newsome, W.T., Visual response properties of striate cortical neurons projecting to area MT in macaque monkeys, *J. Neurosci.*, 16, 7733, 1996.
63. Thiele, A., Henning, P., Kubischik, M., and Hoffmann, K.P., Neural mechanisms of saccadic suppression, *Science*, 295, 2460, 2002.
64. Allison, J.D., Melzer, P., Ding, Y., Bonds, A.B., and Casagrande, V.A., Differential contributions of magnocellular and parvocellular pathways to the contrast response of neurons in bush baby primary visual cortex (V1), *Vis. Neurosci.,* 17, 71, 2000
65. Sawatari, A. and Callaway, E.M., Convergence of magno- and parvocellular pathways in layer 4B of macaque primary visual cortex, *Nature*, 380, 442, 1996.
66. Lachica, E.A., Beck, P.D., and Casagrande, V.A., Intrinsic connections of layer III of striate cortex in squirrel monkey and bush baby: correlations with patterns of cytochrome oxidase, *J. Comp. Neurol.*, 329, 163, 1993.
67. Sherman, S. and Guillery, R., *Exploring the Thalamus,* Academic Press, San Diego, CA, 2001.
68. Harting J.K., van Lieshout, D.P., Hashikawa, T., and Weber, J.T.L., The parabigeminogeniculate projection: connectional studies in 8 mammals, *J. Comp. Neurol.*, 305, 559, 1991.
69. Feig, S. and Harting, J.K., Ultrastructural studies of the primate lateral geniculate nucleus: morphology and spatial relationships of axon terminals arising from the retina, visual cortex (area 17), superior colliculus, parabigeminal nucleus, and pretectum of *Galago crassicaudatus*, *J. Comp. Neurol.,* 343, 17, 1994.
70. Bickford, M., Ramcharan, E., Godwin, D., Erisir, A., Gnadt, J., and Sherman, S., Neurotransmitters contained in the subcortical extraretinal inputs to the monkey lateral geniculate nucleus, *J. Comp. Neurol.,* 424, 701, 2000.
71. Harting, J.K., Hashikawa, T., and van Lieshout, D., Laminar distribution of tectal, parabigeminal and pretectal inputs to the primate dorsal lateral geniculate nucleus: connectional studies in *Galago crassicaudatus*, *Brain Res.*, 366, 358, 1986.
72. Fitzpatrick, D. and Diamond, I.T., Distribution of acetylcholinesterase in the geniculo striate system of *Galago senegalensis* and *Aotus trivirgatus*: evidence for the origin of the reaction product in the lateral geniculate body, *J. Comp. Neurol.,* 194, 703, 1980.
73. Wilson, J., Manning, K., Forestner, D., Counts, S., and Uhlrich, D., Comparison of cholinergic and histaminergic axons in the lateral geniculate complex of the macaque monkey, *Anat. Rec.*, 255, 295, 1999.
74. Steriade, M. and McCarley, R.W., *Brainstem Control of Wakefulness and Sleep*, Plenum Press, New York, 1990.

75. Jouvet, M., Mechanisms of the states of sleep: a neuropharmacological approach, in *Sleep and the Altered States of Consciousness*, Kety, S.S., Evarts, E.V., and Williams, H.L., Eds., Lippincott/Williams & Wilkins, Baltimore, 1967, 86.
76. Brooks, D.C., Waves associated with eye movement in the awake and sleeping cat, *Electr. Clin. Neurophysiol.*, 24, 532, 1968.
77. Brooks, D.C., Gershon, M.D., and Simon, R.P., Brain stem serotonin depletion and ponto-geniculo-occipital wave activity in the cat treated with reserpine, *Neuropharmacology*, 11, 511, 1972.
78. Ross, J., Morrone, M.C., Goldberg, M.E., and Burr, D.C., Changes in visual perception at the time of saccades, *TINS*, 24(2), 113, 2001.
79. Castet, E. and Masson, G.S., Motion perception during saccadic eye movements, *Nat. Neurosci.*, 3, 177, 2000.
80. Burr, D., Morrone, M., and Ross, J., Selective suppression of the magnocellular visual pathway during saccadic eye movements, *Nature*, 371, 511, 1994.
81. Lee, D. and Malpeli, J., Effects of saccades on the activity of neurons in the cat lateral geniculate nucleus, *J. Neurophysiol.*, 79, 922, 1998.
82. Buttner, U. and Fuchs, A., Influence of saccadic eye movements on unit activity in simian lateral geniculate and perigeniculate nuclei, *J. Neurophysiol.*, 36, 127, 1973.
83. Maunsell, J.H., Nealey, T.A., and DePriest, D.D., Magnocellular and parvocellular contributions to responses in the middle temporal visual area (MT) of the macaque monkey, *J. Neurosci.*, 10, 3323, 1990.
84. Reppas, J.B., Usrey, W.M., and Reid, R.C., Saccadic eye movements modulate visual responses in the lateral geniculate nucleus, *Neuron*, 35, 961, 2002.
85. Ichida, J., Royal, D., Sáry, Gy., Xu, X., Shostak, Y., Schall, J., and Casagrande, V., Evidence for suppression of activity in both parvocellular (P) and magnocellular (M) lateral geniculate nucleus (LGN) cells during saccadic eye movements, *Soc. Neurosci. Abstr.*, 2001.
86. Ramcharan, E.J., Gnadt, J.W., and Sherman, S.M., The effects of saccadic eye movements on the activity of geniculate relay neurons in the monkey, *Vis. Neurosci.*, 18, 253, 2001.
87. Zhu, J. and Lo, F.S., Time course of inhibition induced by a putative saccadic suppression circuit in the dorsal lateral geniculate nucleus of the rabbit, *Brain Res. Bull.*, 41, 281, 1996.
88. Schmidt, M., Neurons in the cat pretectum that project to the dorsal lateral geniculate nucleus are activated during saccades, *J. Physiol.*, 76, 2907, 1996.
89. Schmidt, M. and Hoffman, K.-P., Physiological characterization of pretectal neurons projecting to the lateral nucleus in the cat, *Eur. J. Neurosci.*, 4, 318, 1996.
90. Funk, K. and Eysel, U., Pharmacological inactivation of pretectal nuclei reveals different modulatory effects on retino-geniculate transmission by X and Y cells in cat, *Vis. Neurosci.*, 12, 21, 1995.
91. Wurtz, R.H. and Mohler, C.W., Enhancement of visual responses in monkey striate cortex and frontal eye fields, *J. Neurophysiol.*, 39, 766, 1976.
92. Lee, T.S., Yang, C.F., Romero, D.R., and Mumford, D., Neural activity in early visual cortex reflects behavioral experience and higher-order perceptual saliency, *Nat. Neurosci.*, 5, 589, 2002.
93. Martinez, A., Allo-Vento, L., Sereno, M.I., Frank, L.R., Buxtion, R.B., Dubovitz, D.J., Wong, E.C., Hinrichs, H., Heinze, H.J., and Hillyard, S.S., Involvement of striate and extrastriate visual cortical areas in spatial attention, *Nat. Neurosci.*, 2, 364, 2002.
94. Paradiso, M.A., Perceptual and neuronal correspondence in primary visual cortex, *Curr. Opin. Neurobiol.*, 12, 155, 2002.

95. Sáry, Gy.M., Royal, D.W., Schall, J.D., and Casagrande, V.A., Task related modulation of lateral geniculate nucleus cell activity in the macaque monkey, submitted.
96. Super, H., Spekreijse, H., and Lamme, V.A., A neural correlate of working memory in the monkey primary visual cortex, *Science*, 293, 120, 2001.
97. Vanduffel, W., Tootell, R., and Orban, G., Attention-dependent suppression of metabolic activity in the early stages of the macaque visual system, *Cereb. Cortex*, 10, 109, 2000.
98. Brodmann, K., Localization in the cerebral cortex, in *Vergleichen Lokalisationslehre der Grosshirnrinde in ihren Prinzipien dargestellt auf Grund des Zellenbaues* (Leipzig, 1909, Johann Ambrosius Barth), translated and edited by L.J. Garey, Smith-Gordon, London, 1994.
99. O'Connor, D.H., Fukui, M., and Kastner, S., Attentional response modulation in the human LGN and visual cortex, *Soc. Neurosci. Abstr.,* 418.1, 2002.
100. Chen, W., Kato, T., Zhu, X., Ogawa, S., Tank, D., and Ugurbil, K., Human primary visual cortex and lateral geniculate nucleus activation during visual imagery, *Neuro Report,* 9, 3669, 1998.
101. Ito, M. and Gilbert, C.D., Attention modulates contextual influences in the primary visual cortex of alert monkeys, *Neuron*, 22, 593, 1999.
102. Jones, H.E., Wang, W., Andolina, I.M., Salt, T.E., and Sillito, A.M., MT feedback effects in the primate LGN, *Soc. Neurosci. Abstr.,* 658.17, 2002.
103. Heeger, D., Boynton, G., Demb, J., Seidemann, E., and Newsome, W., Motion opponency in visual cortex, *J. Neurosci.*, 19, 7162, 1999.
104. Lamme, V.A. and Roelfsema, P.R., The distinct modes of vision offered by feedforward and recurrent processing, *TINS*, 23, 571, 2000.
105. Weedman, D.L. and Ryugo, D.K., Pyramidal cells in primary auditory cortex project to cochlear nucleus in rat, *Brain Res.*, 706, 97, 1996.
106. Giolli, R.A. and Tigges, J., *The Primary Optic Pathways and Nuclei of Primates in the Primate Brain*, Noback, C.R. and Montagna, W., Eds., Appleton Century Crofts, New York, 1970, 29.
107. Guillery, R.W. and Sherman, S.M., The thalamus as a monitor of motor outputs, *Philos. Trans. R. Soc.*, 357, 1647, 2002.
108. Rodieck, R.W., *The First Steps in Seeing,* Sinauer Associates, Sunderland, MA, 1998, 306–307.
109. Schall, J.D., Neural basis of deciding, choosing and acting, *Nat. Neurosci. Rev.*, 2, 33, 2001.
110. Ingle, D.J. and Schneider, G.F., Brain mechanisms and vision: subcortical systems, *Science*, 168, 1493, 1970.
111. Harting, J.K., Updyke, B.V., and Van Lieshout, D.P., Corticotectal projections in the cat: anterograde transport studies of twenty-five cortical areas, *J. Comp. Neurol.,* 324, 379, 1992.
112. Grillner, S., Control of locomotion in bipeds, tetrapods and fish, in *Handbook of Physiology,* Section 1: *The Nervous System,* Vol. II: *Motor Control,* Part 2, Brooks, V.P., Ed., American Physiological Society, Bethesda, MD, 1981, 1179.

2 Comparative Study of the Primate Retina

Luiz Carlos de Lima Silveira

CONTENTS

2.1 INTRODUCTION

The vertebrate retina is organized according a common general plan. This basic idea was already present in the work of leading investigators of the 18th century and especially in the work of Spanish Nobel laureate Santiago Ramón y Cajal.[1] Thus, the primate retina has in common with the retina of most vertebrates several important aspects.[2] The similarities grow further when comparing primates with other mammals. The retina of all mammals is organized in similar layers and possesses the same gamut of neurons, which are distributed throughout the retinal extension following the same basic rules.[3] One has to go deep into details of retinal morphology and physiology first to find and then to understand the differences in retinal organization among mammals. A considerable knowledge of retinal structure and function is necessary to grasp the subtleties of the primate retina, and it seems ironic that trichromacy, a feature reserved to primates among all mammals, depends on such a minute structural change in the photoreceptor array that it can only be anatomically demonstrated using the modern, very sophisticated adaptive optics retinal imaging.[4]

In this chapter the main aspects of the retinal anatomy and physiology of primates are reviewed, emphasizing those topics that can be related to adaptation to their

0-8493-1243-4/04/$0.00+$1.50

diurnal or nocturnal behavior. The findings supporting the present view of primate retinal organization were obtained by careful observation followed by quantitative analysis of retinal flat mounts and retinal sections stained with classical stains, such as Nissl, Golgi, or Gros-Schultze, or modern immunocytochemistry and neurotracing. For some species, there is also information available about the electrophysiology of retinal neurons, which was obtained by *in vivo* or *in vitro* retinal recording.

2.2 PHOTORECEPTORS, POSTRECEPTORAL CELLS, AND COLOR VISION

2.2.1 PHOTORECEPTORS

One of the most interesting results of comparative studies about the primate visual system was the discovery by John Mollon, James Bowmaker, and Gerald Jacobs that platyrrhines, in contrast to catarrhines, have several normal color vision phenotypes.[5] Together with a chromosome 7 gene that codes the S-cone opsin, several diurnal platyrrhines possess a single locus in the X chromosome that codes for M- or L-cone opsins. However, due to the existence of a polyallelism of the X chromosome opsin gene, there are several possible color vision phenotypes. All males and some females are dichromats, whereas the majority of females are trichromats. The number of di- and trichromatic phenotypes, as well as the proportion of females that are di- or trichromats, depends on the number of alleles present in the population. In some platyrrhines, such as *Ateles*, there are only two alleles; in others such as *Cebus*, *Saimiri*, *Callithrix*, and *Saguinus*, there are three alleles; whereas in some others, such as *Callicebus*, there are up to five alleles (reviewed by Jacobs[6]). There is at least one exception to this pattern among diurnal platyrrhines. The *Alouatta* has color vision similar to catarrhines, with two loci in the X chromosome coding M and L opsins, respectively, and a single allele for each locus.[6,7] Another exception is the nocturnal, monochromatic *Aotus*. It lacks a functioning gene on chromosome 7 and has a single allele for the X chromosome, coding for a M-cone opsin.[6,8]

Recently, it was shown that prosimians might be as diversified as platyrrhines in terms of color vision. Some species are monochromats,[9,10] others dichromats,[11] and still others have a mixed population of di- and trichromatic individuals due to X chromosome–linked polyallelism.[12]

The discovery of different kinds of color vision among platyrrhines and, more recently, prosimians renewed interest in the retinal anatomy and physiology of primates in general. The distribution of cones and rods in the retina of several platyrrhines, catarrhines, and prosimians has now been published. It seems that the photoreceptor topography of many primates can be described as consisting of three basic patterns, which are illustrated in Figure 2.1, although there are striking exceptions such as the *Tarsius*.[13] Two of these patterns are present in diurnal primates, both catarrhines and platyrrhines, independently of their color vision phenotype (Figure 2.1A and 2.1B). Primates with small eyes, such as *Callithrix*, exhibit a low rod-to-cone ratio (Figure 2.1A), whereas those with medium to large eyes, such as *Cebus*, exhibit a high rod-to-cone ratio (Figure 2.1B). Finally, nocturnal primates, such as *Aotus*, exhibit a very high rod-to-cone ratio (Figure 2.1C) and are devoid

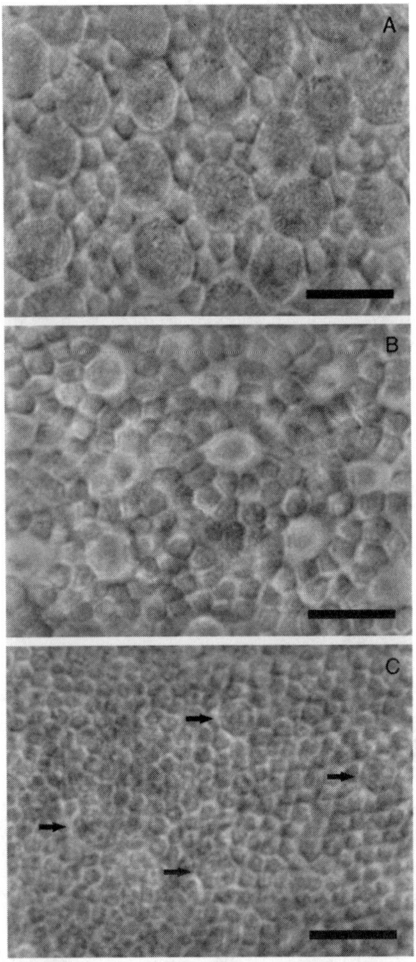

FIGURE 2.1 Photoreceptor mosaics. Representative platyrrhines were chosen to illustrate the three common patterns of photoreceptor mosaics that are found among primates.[110] (A) The retina of *Callithrix jacchus jacchus*, a diurnal platyrrhine with small eyes, has a low rod-to-cone ratio. Eccentricity: 6.7 mm ventral. (B) The retina of *Cebus apella*, a diurnal platyrrhine with large eyes, has a high rod-to-cone density. Eccentricity: 9.9 mm dorsal. (C) The retina of *Aotus trivirgatus*, a nocturnal platyrrhine with large eyes, has a very high rod-to-cone density. Eccentricity: 7.9 mm dorsal to the fovea. Arrows point to cones. Scale bars = 10 μm.

of well-developed fovea. Table 2.1 summarizes the data from the literature about cone and rod peak density and total counts for several primates.

2.2.2 Horizontal Cells

The horizontal, bipolar, and amacrine cells of primates have been studied with a variety of methods that label isolated neurons, revealing fine details of their

TABLE 2.1
Photoreceptor Total Count and Density Peak for Different Primate Retinas

Species	Lifestyle	Ref.	Retinal Area (mm^2)	Cone Total Count ($\times 10^3$ cones)	Rod Total Count ($\times 10^3$ rods)	Peak Cone Density (cones/mm^2)	Peak Rod Density (rods/mm^2)
Homo sapiens	Diurnal	58	1,012	4,600	—	208,300	—
Papio anubis	Diurnal	111	—	—	—	385,000	—
Macaca mulatta	Diurnal	55	670	3,100	—	140,500	—
		112	750	3,200	60,700	140,000	180,000
Macaca fascicularis	Diurnal	55	—	—	—	100,400	—
		60	—	—	—	250,000	—
Macaca nemestrina	Diurnal	113	800	3,100	60,100	210,000	177,000
Macaca irus	Diurnal	111	—	—	—	179,000	—
Cercopithecus aethiops	Diurnal	55	—	—	—	191,400	—
Alouatta caraya	Diurnal	110	446	3,600	58,700	394,000	209,200
Aotus trivirgatus	Nocturnal	114	—	—	—	7,300	387,000
		9	—	—	—	6,800	325,000
		110	569	2,300	140,600	17,100	388,000
Cebus apella	Diurnal	110	529	4,500	51,900	164,000	175,000
		115	602	3,800	48,200	169,100	117,200
Saimiri sciureus	Diurnal	56	—	—	—	62,000	—
Saimiri ustius	Diurnal	110	352	3,400	33,600	138,000	138,000
Saguinus midas niger	Diurnal	110	223	3,700	11,200	150,400	87,600
Callithrix jacchus jacchus	Diurnal	116	210	2,800	—	190,700	—
		66	—	—	—	211,700	—
		110	204	3,700	10,700	132,800	79,100
Tarsius spectrum	Nocturnal	13	476	—	—	12,900	—
Galago garnetti	Nocturnal	9	—	—	—	8,000	450,000
Microcebus murinus	Nocturnal	117	—	—	—	8,000	850,000

Note: Cone and rod total count rounded to 10,000 cells. Cone and rod peak density rounded to 100 cells/mm^2.

morphology, such as the method of Golgi, extracellular deposits of carbocyanine (DiI), or intracellular injection of Lucifer yellow, horseradish peroxidase (HRP), or Neurobiotin. They have also been studied with immunocytochemistry, which is well suited to quantitative analysis of density distribution.

All diurnal catarrhines and platyrrhines so far studied have the same classes of retinal elements in the inner nuclear layer. Using morphological criteria, different authors have described two, H1 and H2, or three, H1, H2, and H3, horizontal cell classes in diurnal anthropoids (*Homo*,[14–16] *Macaca*,[17–22] *Cebus*,[23] *Saimiri*,[20] *Callithrix*[24]). H1/H3 cells contact cones with their dendritic clusters and rods with their axon terminals, whereas H2 cell dendrites and axons make synapses only with cones. There have been very few studies about horizontal cell physiology in primates, all limited to the genus *Macaca*.[25–28] In contrast to lower vertebrates, it has been demonstrated that all classes of primate horizontal cells are non-opponent cells.[25,26] However, H1 and H2 differ in their cone specificity. H1 cells give a sustained hyperpolarizing response to M-cone or L-cone isolating stimuli, but no response to S-cone isolating stimuli, whereas H2 cells respond to all three kinds of stimuli and give a large response to S-cone isolating stimuli.[26]

The horizontal cells were also studied in the nocturnal platyrrhine *Aotus*.[29,30] Recently, using the same methodology, the morphology and size of *Aotus* horizontal cells were compared with those of the diurnal platyrrhine *Cebus* (Figure 2.2B through 2.2F).[23,30] After perfusion with fixative, the retina was dissected and small crystals of DiI were inserted into the retina at different locations. Neuronal labeling was visible after 1 to 3 weeks. The labeling was then photoconverted to a stable product using diaminobenzidine (DAB) as chromogen. The most commonly stained horizontal cells in the *Aotus* retina were classified as H1/H3 because they were qualitatively similar to *Cebus* H1/H3 cells and to H1/H3 cells of other diurnal anthropoids, although there were important quantitative differences regarding cell size, distance between dendritic clusters, and density of axon terminal knobs.[30] All these differences can be explained by the differences in cone and rod density in the retina of the two platyrrhines (Figure 2.2B through 2.2F). Other, rarely stained *Aotus* horizontal cells were tentatively classified as H2.[30] Their lower frequency of occurrence and distinct morphology when compared with *Cebus* H2 cells may be due to the absence of S cones in the *Aotus*.[9]

2.2.3 Two Pathways for Color Vision

The presence of three cone classes in the retina does not guarantee the joy of trichromacy. In addition to the correct number of cone classes, the retina should also have the appropriate postreceptoral mechanisms to process the cone signals. It has been proposed that primate color vision evolved in two steps.[31] Ancient mammals were nocturnal, possessed a single M/L cone, and thus were monochromats. The ancient form of color vision appeared when mammals acquired S cones, initially to enlarge their spectral sensitivity, and then developed the neural circuitry for blue-yellow color opponency.[31] This circuit comprises the appropriate synapses between S cones, H2 horizontal cells, S-cone bipolar cells, and small bistratified cells. All the neural elements of this circuit have been found both in diurnal catarrhines[32–37]

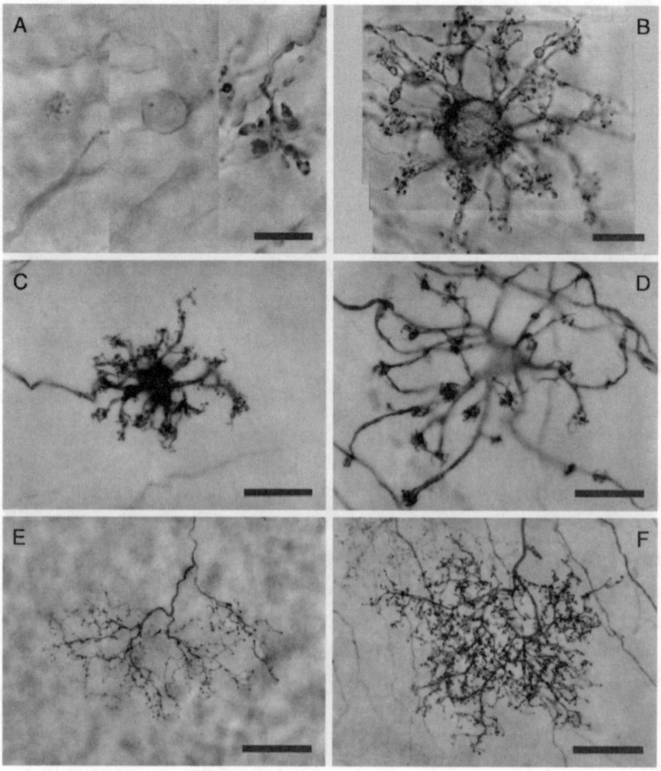

FIGURE 2.2 Photomicrographs of bipolar and horizontal cells of nocturnal and diurnal platyrrhines, *Aotus trivirgatus*[30,45] and *Cebus apella*,[23] respectively. The cells are all viewed from the internal retinal side. They were stained by postmortem transport of DiI followed by dye photoconversion to a stable DAB product. (A) Single-cone midget bipolar cell of the *Aotus* retina, 0.2 mm ventral to the fovea. From left to right three levels of focus show the single dendritic cluster (left panel), the cell body (middle panel), and the axon terminal (right panel). The axon terminal of this central midget bipolar cell matches the dendritic field of P ganglion cells of the same retinal region. Scale bar = 10 μm. (B) *Cebus* H1 horizontal cell, 3.8 mm temporal. Scale bar = 10 μm. (C) *Cebus* H1 horizontal cell, 9.6 mm dorsal. Scale bar = 25 μm. (D) *Aotus* H1 horizontal cell, 12.6 mm temporal. Scale bar = 25 μm. *Aotus* H1 cells are larger and have sparser dendritic clusters than *Cebus* H1 cells at similar eccentricities. (E) *Cebus* H1 axon terminal, 12 mm dorsal. Scale bar = 50 μm. (From dos Reis, J.W.L. et al., *J. Comp. Neurol.*, 443, 105, 2002. With permission of John Wiley & Sons.) (F) *Aotus* H1 axon terminal, 12.6 mm temporal. Scale bar = 50 μm. *Aotus* H1 horizontal cells have dense axon terminal arborizations, reflecting the high rod density in the retina of this primate.

and platyrrhines.[38–42] It is not known if nocturnal primates without S cones, such as the *Aotus* and some prosimians, also are devoid of other elements of the ancient color vision pathway, but there are indications that these neurons are either absent or very rare.[30,42]

The primates acquired full trichromacy after a mutation in the X chromosome that enabled them to express two M/L opsins in the retina together with the S opsin.

It is hypothesized that by that time the central region of the primate retina already had the so-called private lines, connecting a single M/L cone to a single midget bipolar cell, which in turn contacts a single P ganglion cell. This retinal circuit had evolved to fulfill the needs of spatiotemporal achromatic vision before the red-green color opponency had appeared. Thus, red-green color opponency and trichromacy were consequences of a lucky combination of two independent events.[31]

The one-to-one connection between M/L cones, midget bipolar, and P ganglion cells, which constitute the basis for red-green color opponency of trichromatic primates was first described in catarrhines.[2,43] Thus, it is of considerable importance to address the question of whether dichromatic platyrrhines or the monochromatic *Aotus* also have such a circuit, despite lacking red-green color-opponency. To answer this question, bipolar cells were labeled with DiI retinal implants and ganglion cells were labeled by retrograde transport of Biocytin.[44,45] It was found that the central retina of dichromatic *Cebus* and monochromatic *Aotus* preferentially had single-cone MB bipolar cells (Figure 2.2A), and that these cells had axon terminals whose sizes are in the range of the dendritic tree sizes of central P ganglion cells. These findings suggested that one-to-one connections between M/L cones, midget bipolar cells, and P ganglion cells are a feature of the P pathway shared by diurnal catarrhines, diurnal platyrrhines, and nocturnal platyrrhines, independently of the presence of one or two M/L-cone photopigments in their retina. They are consistent with the hypothesis that the P pathway evolved for achromatic vision before red-green color opponency appeared.[31]

2.3 GANGLION CELLS

The primate retina, similar to the retina of other vertebrates, has several classes of ganglion cells, connected with distinct sets of bipolar and amacrine cells, which send axons to specific targets in the mesencephalon and diencephalon. The ganglion cell axons convey the result of retinal processing to several subcortical centers and cortical visual areas for further analysis. There is considerable interest in characterizing the morphology and physiology of those ganglion cells that connect the retina to the lateral geniculate nucleus (LGN), and the thalamic relay neurons that connect the LGN layers to the primary visual cortex, because the retino-geniculate-striate pathway represents major information channels for the perception of color, movement, and recognition of spatial patterns in primates.

Placed on the origin of the retino-geniculo-striate circuits, there are ganglion cells with distinct morphology and physiology, whose axons synapse in specific layers of the LGN. The M cells send axons to two ventral magnocellular layers; the P cells send axons to four dorsal parvocellular layers; and a heterogeneous group of K cells synapse in the intercalated layers of the LGN.[46] The majority, if not all these cell classes, are more densely packed in the central retinal area, and progressively less frequent toward the retinal periphery.[47–50]

2.3.1 Ganglion Cell Distribution

Stone[3] proposed and extensively discussed the idea that, in the mammalian retina, the neurons are distributed along the retinal surface following a two-axis topographical pattern. The horizontal axis is dominated by the visual streak, a specialization devoted to processing information coming from the horizon.[3,51] The vertical axis is dominated by the nasotemporal division, a specialization related to stereopsis.[3] The *fovea centralis* of primates and the *area centralis* of cats and other mammals are located at the junction of those two axes.[3] In different mammals, one of the two axes may become a dominant feature of the retinal topography. In mammals with eyes located laterally in the head, such as the rabbit and the agouti, the visual streak is the most salient retinal feature, and the ganglion cell density peaks in the middle of this structure, at the projection of the optical axis.[52,53] There is a less prominent peak of large ganglion cells in the nasotemporal division, in the projection of the visual axis.[54] In animals with frontalized eyes, such as primates and felids, the nasotemporal division is more important, and the *fovea centralis* or *area centralis* is the most prominent retinal feature, located at the junction of the vertical and horizontal axes.[3]

The ganglion cell topography has been studied counting cell bodies in preparations usually stained by the method of Nissl. Additional criteria were developed to distinguish ganglion cells from other cell bodies of the ganglion cell layer, such as retrograde labeling of ganglion cells, immunocytochemical staining of displaced amacrine cells and astrocytes, and selective counting of displaced amacrine cells and neuroglia after selective degeneration of ganglion cells. In the majority of recent studies, the ganglion cells have been counted in retinal flat mounts, focusing through the layers of the foveal slope (e.g., Perry and Cowey[55]), although in some works cell densities have been measured in retinal sections followed by stereological corrections (e.g., Rolls and Cowey[56]). By using these methods, several species of primates were studied, including catarrhines (*Homo*,[57,58] *Papio*,[59] *Macaca*,[55–57,60] *Cercopithecus*[61]), platyrrhines (*Saimiri*,[56,57] *Cebus*,[62] *Aotus*,[63–65] *Callithrix*[66]), and prosimians (*Galago*[57,67]). For interspecies comparisons to be made, it is necessary to use the retinal magnification factor to express retinal eccentricity in angular distance. This scale conversion allows direct comparison between animals with different eye sizes, such as *Cebus* and *Callithrix*, or different optical design, such as *Cebus* and *Aotus*.[68]

In all primates, apart from small quantitative differences found in different species, the retinal ganglion cell topography can be understood as occurring in two basic patterns, for diurnal and nocturnal species. In both diurnal and nocturnal primates, the ganglion cells are distributed following a well-marked radial pattern, in such a way that density peaks in the central retina, and then steeply decreases toward the retinal periphery. The ganglion cell density is not very different in the retinal periphery of diurnal and nocturnal primates with large or small eyes. However, large differences exist in the degree of central specialization. All diurnal genera exhibit a well-developed fovea, with the ganglion cell and inner nuclear layers displaced from the center. In the foveal slope, the ganglion cell density reaches 30,000 to 60,000 cells/mm^2, depending on the species being considered.[55,58–62,66] The peak of ganglion cell sampling density is much higher, and must be calculated taking into account the displacement of retinal elements away from the foveola. It was estimated to be 668,250 cells/mm^2 in *M. fascicularis*[60] and 545,600 cells/mm^2 in *Callithrix jacchus*.[66] On the other hand, the nocturnal genera exhibit either a rudimentary fovea or an area centralis, and the ganglion cell density peak is only one third to one fifth of that found in diurnal genera, reaching 10,000 to 20,000 cells/mm^2.[64,65,67]

As mentioned, the ganglion cell topography in primates and other mammals is dominated by the nasotemporal division, the vertical axis of the Stone model, whereas the visual streak, the horizontal axis of the Stone model, is obscured by the strong radial distribution of ganglion cells. However, the visual streak can still be demonstrated by careful comparison of ganglion cell density at the four retinal quadrants. Ganglion cell density is roughly similar in the temporal, dorsal, and ventral quadrants, at equal distances from the fovea, but is higher in the nasal quadrant. Consequently, the isodensity contours are elongated along the horizontal axis, toward the nasal periphery, forming a rudimentary visual streak. The nasotemporal asymmetry and the visual streak are more marked in ground-dwelling, foraging, secondarily terrestrial vs. more arboreal species (e.g., *Papio* vs. *Macaca*)[59] and in diurnal vs. nocturnal species (e.g., *Cebus* vs. *Aotus*).[65]

Thus, the ganglion cell topographies of diurnal and nocturnal primates mainly differ in the degree of their retinal specializations, the diurnal species exhibiting more-developed central specialization and visual streak, whereas nocturnal species have a relatively more flattened ganglion cell density distribution. A similar difference between diurnal and nocturnal species of other mammalian orders has been described.[69] Thus, it seems that an important adaptation to nocturnal life consists of the sacrifice of detailed sampling of the central visual field or the horizon to the benefit of a smaller number of cells with large receptive fields, which are more efficient light collectors.

Figure 2.3 illustrates the difference in ganglion cell distribution between two closely related primates, the diurnal capuchin monkey (*C. apella*) and the nocturnal owl monkey (*A. trivirgatus*). The data from the two species were collected using the same methods, making a direct comparison easier: ganglion cells were counted in retinal flat-mounts stained with cresyl violet, and the criteria to distinguish ganglion cells from displaced amacrine cells were established by retrograde labeling of ganglion cells from HRP deposits in the optic nerve.[62,65] The ganglion cell density of *Cebus* retina peaks 0.5 mm nasal from the foveola at about 49,300 cells/mm^2 (between 40,300 and 56,900 cells/mm^2 in a sample of six retinae) (Figure 2.3A and 2.3B).[62] This value is comparable to those reported for other diurnal primates, both platyrrhines and catarrhines (Table 2.2). The nocturnal *Aotus* exhibits lower values for retinal ganglion cell density than *Cebus* at the central retinal region (Figure 2.3A and 2.3B). In *Aotus*, similar to *Cebus*, the highest ganglion cell density is still higher in the central retinal region in comparison to the retinal periphery, but the exact location of the density peak varies in different animals.[65] In a sample of four retinae, three of them had a rudimentary fovea, and the ganglion cell density reached a maximum of 15,300, 16,000, and 16,000 cells/mm^2, located at 0.25 mm temporal, 0.25 mm nasal, and 0.25 mm ventral to the fovea, respectively, whereas the fourth retina had an area centralis with a peak of 18,000 cells/mm^2.[65] The mean value for these four retinae peaked at 15,000 cells/mm^2 located at 0.25 mm dorsal to the fovea or area centralis.[65] Such low ganglion cell density in the central retinal region is comparable to the values reported for other nocturnal primates, such as the prosimian *Galago*, 11,300 to 12,200 cells/mm^2.[67]

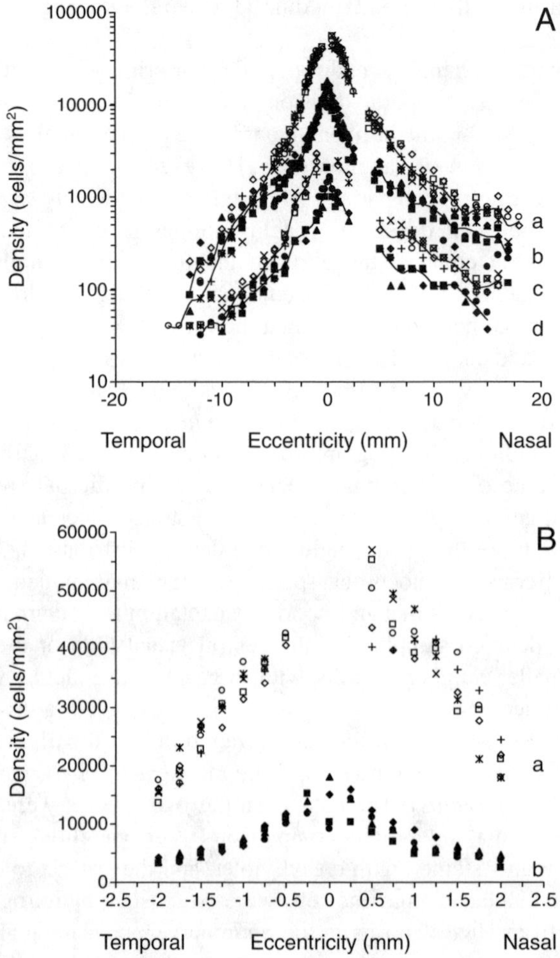

FIGURE 2.3 Ganglion cell density in diurnal and nocturnal platyrrhines, *Cebus apella* and *Aotus trivirgatus*, respectively. (A) Total ganglion cell density as a function of retinal eccentricity for *Cebus* (a) and *Aotus* (b) retinae. M ganglion cell density as a function of retinal eccentricity for *Cebus* (c) and *Aotus* (d) retinae. Ganglion cells were counted in retinal whole mounts stained using the method of Nissl.[62,65] M cells were counted in retinal whole mounts stained using the method of Gros-Schultze.[50] (B) Total ganglion cell density as a function of retinal eccentricity for *Cebus* (a) and *Aotus* (b) retinae. Only the values for the central 2 mm of the horizontal meridian are shown to illustrate the rapid changes in the central retinal region. In each plot, the different symbols represent values obtained from different animals.

In primates, the central few degrees of the visual field are overrepresented in the primary visual cortex.[70,71] An important question is whether this is due to an increased ganglion cell population subserving the central degrees of visual field, as described in the precedent paragraphs, or whether the visual cortex is additionally specialized for central vision. How is the retinal ganglion cell distribution related to the cortical topography? The answer to this controversial issue is complicated by

TABLE 2.2
Ganglion Cell Total Count and Density Peak for Different Primate Retinas

Species	Lifestyle	Ref.	Retinal Area (mm^2)	Ganglion Cell Total Count	Peak Ganglion Cell Density (cells/mm^2)
Homo sapiens	Diurnal	58	1,012	1,070,000	35,100
Papio anubis	Diurnal	59	793	1,480,000	58,000
Macaca mulatta	Diurnal	56	—	—	36,000
		55	670	1,500,000	33,000
Macaca fascicularis	Diurnal	60	—	—	48,000
Cercopithecus aethiops sabeus	Diurnal	61	707	1,230,000	43,200
Aotus trivirgatus	Nocturnal	63	—	—	19,600
		64	—	—	11,300
		65	684	480,000	15,000
Cebus apella	Diurnal	62	615	1,400,000	49,300
Saimiri sciureus	Diurnal	56	—	—	36,600
		57	345	—	—
Callithrix jacchus jacchus	Diurnal	66	—	—	55,000
Galago crassicaudatus	Nocturnal	67	344	750,000	11,300
Galago senegalensis	Nocturnal	67	332	520,000	12,200
		52	319	620,000	—
Microcebus murinus	Nocturnal	117	—	—	28,000

Note: Ganglion cell total count rounded to 10,000 cells. Ganglion cell peak density rounded to 100 ganglion cells/mm^2.

at least two major problems. First, the retinal topography in the central visual field is difficult to determine because of the displacement of cone pedicles, bipolar cells, and retinal ganglion cells away from the foveola. Second, the convoluted anatomy of the primate visual cortex also makes it difficult to estimate the cortical magnification factor with the appropriate degree of precision. Using different methodologies, some authors have advocated a precise relationship between retinal ganglion cell topography and the topography of the primary visual cortex,[60,64] whereas others have found that the central visual field is even more overrepresented in the primary visual cortex than in the retinal ganglion cell layer.[55,62,65,72] The *Aotus* is a good model in which to explore this subject, because its central retina is devoid of a well-developed fovea and its primary visual cortex is relatively flat, in comparison with its diurnal relatives. In this animal, the most recent studies point to an overrepresentation of the central retina in the striate cortex.[65] A more direct approach to this question was performed using WGA-HRP retrograde transneuronal labeling from cortex to retina to obtain directly the number of ganglion cells projecting to marked areas of primary visual cortex.[73] This method has been used previously only in *M. mulatta*. Results from macaques indicated that the central retina has a preferential representation in the primary visual cortex.[73]

The mismatch between the representation of the visual field in the retinal ganglion cell layer and the primary visual cortex of primates might be related to the enormous retinocortical mismatch observed in mammals with lateralized eyes.[52,53] In these animals, there is a very large overrepresentation of the "large-cell node," located in the intersection of the temporal end of the visual streak and the nasotemporal division, a region that might be the equivalent of the fovea or area centralis of animals with more frontalized eyes, such as the primates. The explanation for the retinocortical mismatch might be the presence of additional cortical wiring to process information coming from the visual axis, including the generation of stereopsis. In accordance with this, it has been shown that the nasotemporal density gradient of layer 4 nonpyramidal neurons of the rabbit primary visual cortex matches the nasotemporal variation in the cortical magnification factor.[74]

2.3.2 The M and P Ganglion Cell Classes

2.3.2.1 The Morphology of M and P Cells

The main ganglion cell classes are readily recognizable in retinal sections or flat-mounts stained with a variety of methods, such as the classical methods of Ehrlich, Golgi, and Gros-Schultze, as well as with neurotracer retrograde transport and dye intracellular injection. The M and P cells have been identified in all primates so far studied, including humans,[15,75–77] other catarrhines (*Pan*,[2] *Papio*,[78] *Macaca*[2,43,48,49,78–80]), diurnal platyrrhines (*Saimiri*,[81] *Cebus*,[50,80] *Callithrix*[38,83–85]), nocturnal platyrrhines (*Aotus*[50,82,85,86]), and prosimians (*Galago*[87]).

In all primates, M cells have large cell bodies, thick axons, and large dendritic trees with a radial branching pattern, whereas P cells have small cell bodies, thin axons, and small dendritic trees with a bushier and denser branching pattern (Figure 2.4). The M and P cells occur in two morphological varieties, one ramifying in the outer half of the inner plexiform layer (IPL), the outer cells, and the other ramifying in the inner half, the inner cells.[78,80] The outer cells correspond to the off-center and the inner cells to the on-center varieties. Every point of the photoreceptor matrix is connected with at least one inner and one outer ganglion cell of the two cell classes M and P.

Figure 2.4 illustrates the morphology of M and P cells from diurnal and nocturnal platyrrhines, *Cebus* and *Aotus*, respectively. The cells were retrogradely labeled by placing pellets of Biocytin within a small longitudinal cut in the optic nerve 1 to 3 mm behind the eyeball. After 18 to 48 h, the animal was given a lethal dose of barbiturate and perfused with fixative. After perfusion, the eye was removed; the retina dissected and incubated in avidin-peroxidase solution (ABC Vectastain) for 12 to 48 h, and then reacted for peroxidase histochemistry using DAB as the chromogen.[82,88] Immersing the retina in 0.1% osmium tetroxide solution intensified the reaction product.

In the *Cebus* retina, M and P cell morphology changes from central to peripheral retina in the same way as observed in other diurnal catarrhines and platyrrhines: very small P cells and small M cells are observed around the fovea; with increasing eccentricity the cell bodies, and dendritic fields increase in size but both cell classes maintain their distinctive dendritic branching pattern.[84] Dendritic field diameter increases less steeply along the nasal retina, so that at comparable eccentricities

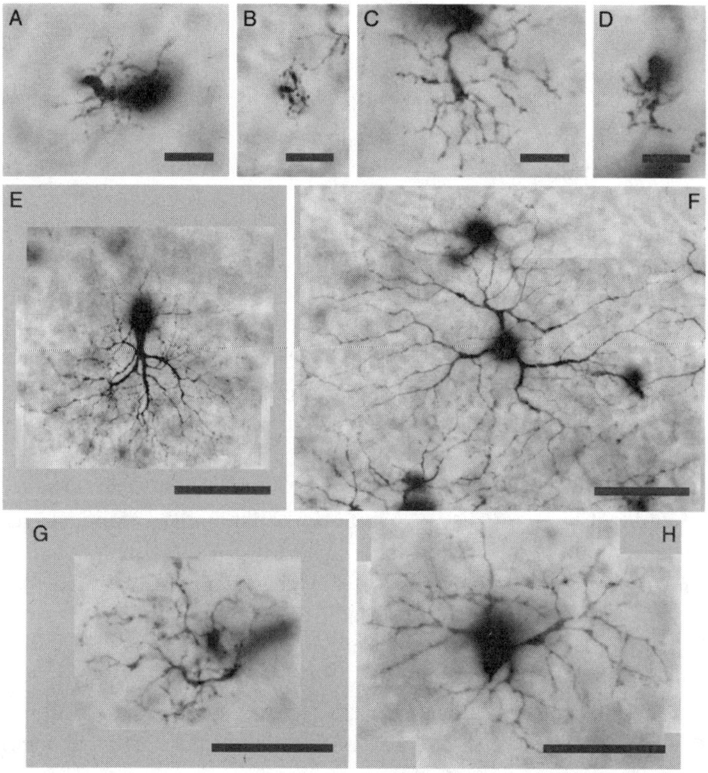

FIGURE 2.4 Photomicrographs of M and P ganglion cells of diurnal and nocturnal platyrrhines, *Cebus apella*[84] and *Aotus trivirgatus*,[86] respectively. The cells are all viewed from the internal retinal side and they were selected to compare the two primates at similar eccentricities. They were stained by retrograde transport of Biocytin. (A) *Cebus* M-off cell, 0.3 mm dorsal to the fovea. (B) *Cebus* P-on cell, 0.6 mm temporal. (C) *Aotus* M-off cell, 0.2 mm dorsal. (D) *Aotus* P-off cell 0.2 mm dorsal. (E) *Cebus* M-off, 4 mm temporal. (F) *Aotus* M-off, 4 mm temporal. (G) *Cebus* P-off, 4 mm temporal. (H) *Aotus* P-off, 4.3 mm temporal. *Aotus* M and P cells are larger than their *Cebus* counterparts at similar eccentricities, to compensate for a lower cone density (see Table 2.3).

nasal M cells were 1.1- to 2.5-fold smaller than temporal ones, and nasal P cells were 1.1- to 1.6-fold smaller than their temporal counterparts. The nasotemporal asymmetry of the dendritic field size was significant for retinal eccentricities greater than 2 mm, being very conspicuous for M cells at 5 mm or more from the fovea. M cell dendritic field diameter ranged from 28 μm at 0.5 mm to 318 μm at 11 mm of temporal eccentricity, and from 25 μm at 0.5 mm to 227 μm at 11 mm of nasal eccentricity, whereas P cell dendritic field diameter ranged from 8 μm at 0.5 mm to 114 μm at 11 mm of temporal eccentricity, and from 7 μm at 0.5 mm to 81 μm at 13.5 mm of nasal eccentricity. M cell dendritic field sizes are larger than those of P cells at any given eccentricity along both temporal and nasal retinal regions. The M cell dendritic fields are 3.5-fold larger than P cell dendritic fields in the central retinal region. This difference decreases toward the retinal periphery to a minimum of 2.5-fold.

As in the *Cebus* and all other primates so far studied, *Aotus* M cells have larger dendritic fields and cell bodies than P cells both in the central and peripheral retinal regions. Near the fovea, for eccentricities lower than 0.5 mm, the mean sizes for M cell and P cell dendritic fields are 36 and 14 μm, respectively, whereas in the retinal periphery, for eccentricities higher than 10 mm, the mean sizes are 340 and 136 μm in the nasal quadrant and 401 and 178 μm in the temporal quadrant. The difference between these two cell classes remains more or less constant along the horizontal meridian, with M cell dendritic fields about 2.3 to 2.7 times larger than those of P cells. The nasotemporal asymmetry of M and P dendritic fields in the *Aotus* retina is less pronounced than that observed in the *Cebus* retina, and it attains a significant level only in the retinal periphery, at 10 mm from the fovea.[86] This is consistent with the fact that the visual streak is less developed in nocturnal than in diurnal primates.

In the central retina of both *Cebus* and *Aotus*, the dendritic tree sizes of P cells as far as 2 mm from the fovea are in the same range as the axon terminal sizes of single-cone midget bipolar cells.[44,45] Moreover, the majority of midget bipolar cells in the central retinal region have a single dendritic cluster, which contacts a single cone pedicle.[44,45] These findings support the view that, as in diurnal catarrhines, central P cells of diurnal and nocturnal platyrrhines receive input from single midget bipolar cells, which, in turn, receive input from single cones.[44,45] This is consistent with the idea that a P pathway with one-to-one connectivity was present in the anthropoid ancestor before the divergence between catarrhines and platyrrhines.[44] P cells with similar morphology have also been found in the central retina of another diurnal platyrrhine, the *Callithrix*.[38,85]

The *Aotus* M and P cells are larger than those found in the same eccentricities in the *Cebus* retina. The dendritic tree area of *Aotus* M cells is about 3.9 times that of *Cebus* M cells at 0.5 mm from the fovea, the ratio decreasing toward the retinal periphery to 1.9 at 12 mm of eccentricity. *Aotus* P cells are larger than *Cebus* P cells by a factor of 5.9 at 0.5 mm from the fovea and 3.5 in the retinal periphery. The size difference between *Aotus* and *Cebus* M and P cells can be related to the relative density of cones and rods in the retina of the two primates.[86] Table 2.3 shows the cone and rod convergence to M and P cells as a function of retinal eccentricity. Although there are large differences in the sizes of M and P dendritic fields between *Cebus* and *Aotus*, the cone convergence to each ganglion cell class is of the same magnitude in the two primates.[86] This finding indicates that during development, ganglion cells from different primates adjust their dendritic size to connect to a similar number of cones.[83,86] Consistent with this hypothesis is the fact that ganglion cells and cones are generated during the first phase of retinal neurogenesis, whereas rods appear later in the second phase.[89] The "diurnal pattern" of *Aotus* cone convergence gives further support to this hypothesis.[86]

2.3.2.2 The Number of M and P Cells

M cells can be selectively stained using the Silveira–Perry protocol for the classic Gros-Schultze neurofibrillar stain[90] or antibodies against $GABA_A$-receptors.[49] With

TABLE 2.3
Photoreceptor Convergence of M and P Cells for Diurnal (*Cebus*) and Nocturnal (*Aotus*) Platyrrhines[86]

	Cone Convergence				Rod Convergence			
	M Cells		P Cells		M Cells		P Cells	
Ecc[a]	*Cebus*	*Aotus*	*Cebus*	*Aotus*	*Cebus*	*Aotus*	*Cebus*	*Aotus*
–9	318	302	65	56	4,301	16,759	877	3,125
–5	107	129	14	14	1,336	9,081	173	992
–4	74	92	7	13	840	6,936	84	975
–3	55	96	6	13	542	6,064	61	819
–2	24	49	2	6.7	183	2,933	16	403
–1	20	47	2	3	84	2,077	7	133
–0.5	24	25	2	2	—	935	—	86
0.5	22	18	1	2	—	652	—	59
1	20	37	1	3	75	1,561	4	107
2	29	33	3	4	197	2,777	17	369
4	61	87	6	13	715	4,931	64	726
5	91	136	6	17	985	9,751	68	1,240
10	220	359	24	44	2,925	20,688	318	2,519
14	320	417	47	62	3,335	16,184	485	2,394

[a] Negative and positive values of eccentricity (Ecc) represent temporal and nasal distance from fovea, respectively.

these methods, it is possible to estimate the density distribution of M cells. In addition, direct counting or statistical analysis of their mosaics can calculate M and P cell densities after retrograde labeling or intracellular injection of large numbers of such cells.[76,77,80,84] By using these methods, it was established that M and P cells comprise the majority of ganglion cells of the primate retina. In *Macaca*, using mosaics stained by retrograde transport of HRP or stained with neurofibrillar stain, it was estimated that M and P cells correspond, roughly, to 10% and 80% of the total ganglion cell population.[48,80] For platyrrhines, the integration of isodensity contours obtained from retinal whole-mounts that had been stained with neurofibrillar stain, resulted in 140,300 and 74,000 M cells for the *Cebus* and *Aotus* retinae, respectively, corresponding to about 10% and 15% of the total retinal ganglion cell population in these primates.[50] However, there are regional differences in the retina in the proportion of these two cell classes and these differences might be species specific.[48,50,84] Both cell classes increase in density toward the fovea, but the proportion of M cells increases in the nasal quadrant in diurnal catarrhines and platyrrhines such as *Macaca*[48] and *Cebus,*[50] and in less proportion in the nocturnal *Aotus*.[50] Figure 2.3A and 2.3B illustrates how M cell density changes with eccentricity for the retina of *Cebus* and *Aotus*.[50]

2.3.2.3 The Physiology of M and P Cells

The receptive field properties of M and P ganglion cells have been studied by stimulating the eye with different visual stimuli and recording cell activity in the retina,[32] optic nerve and tract,[91] or LGN.[92] In the LGN, the ganglion cell activity can be studied by recording the postsynaptic potentials in relay neurons, which bear a direct relationship with the ganglion cell input.[92] In addition, the LGN relay neurons located in the magnocellular and parvocellular layers respond similarly to their ganglion cell afferents, except that their firing rate, at least in anesthetized animals, is lower than that of ganglion cells. Thus, many questions about the properties of the M and P pathways have been answered by recording from their thalamic relay neurons.

Only recently, it has become possible to record M and P cells from *in vitro* retinal preparations and then label the same cells by intracellular injection of neurotracers.[37] With this technique, it was directly demonstrated for the first time that M cells are the phasic, broadband cells, and that P cells are the tonic, red-green color opponent cells, previously described by electrophysiologists.[32,33] The direct correlation between morphology and physiology has so far been achieved only for *Macaca*, but it is largely assumed that it also holds for other primates, both anthropoids and prosimians.

The responses of M and P cells to visual stimuli differ in several important aspects. Apart from their different responses to chromatic contrast (see below), the M and P cells respond quite differently to achromatic patterns, suggesting that they have distinct, complementary roles in achromatic vision.[93] The M-cell response is more localized in the time domain than the P-cell response. M cells are phasic and P cells are tonic.[32] The P-cell response is more localized in the space domain than the M-cell response, a consequence of the smaller receptive fields of P cells in comparison with M cells.[33,94] As the M- and P-cell responses are linear in great measure, and as predicted by the Fourier theory for linear systems, the M-cell responses are more localized in the spatial frequency domain whereas the P-cell responses are more localized in the temporal frequency domain.[93] In addition, there are also important differences in the way M and P cells respond to simultaneous or successive contrast: the M cells are more sensitive to achromatic contrast and their response saturates at lower contrast levels than P cells.[92,95,96]

However, the more striking characteristic that distinguishes M and P cells is their response to color contrast. M cells receive additive inputs of L and M cones both in the center and surround regions of their receptive fields,[32,33] and their spectral sensitivity is very similar to the human photopic luminosity curve.[97] On the other hand, P cells have chromatic opponent inputs from L and M cones to the center and surround of their receptive fields, and their spectral sensitivity is such that they are excited by some light wavelengths and inhibited by others. These findings led to the hypothesis that P cells might be primarily involved in chromatic vision, providing the mechanism for visual performance in psychophysical tasks, such as detection of chromatic modulation.[95,98] On the other hand, M cells would form the basis for visual performance in achromatic tasks, such as heterochromatic flicker photometry,[99] detection of luminance modulation,[95,98] minimally distinct border,[100] and hyperacuity.[101]

Platyrrhines exhibit a variety of color vision phenotypes and thus are good models to test current hypotheses on the role of each visual pathway in the processing of color information.[6,102] Some of these questions were addressed by recording the responses of M and P cells at the retinal ganglion cell layer of dichromatic and trichromatic *Cebus*,[103] as well as the responses of relay neurons at the LGN magnocellular and parvocellular layers of dichromatic and trichromatic *Callithrix*.[104] Using a combination of sine wave stimuli with different colors, amplitudes, and phases, it was possible to determine by single unit recording of M and P cells, which opsins were present in every animal, establishing its color vision phenotype. The electrophysiological results were confirmed by genetic analysis of DNA isolated from liver tissue of the same individuals. Then, the M- and P-cell responses to chromatic and achromatic contrast were studied in the temporal domain. The M and P cells of dichromatic and trichromatic platyrrhines were very similar to those found in catarrhines (all trichromats), with the exception that P cells were color-blind.[103,104] The recordings made so far in the retinal ganglion cell layer of *Aotus*[105] and LGN of *Aotus*[106–108] and *Galago*[109] have shown that these nocturnal, monochromatic primates also have M and P cells that share many physiological characteristics in common with the M and P cells of diurnal primates.

Thus, there is anatomical and physiological evidence indicating that both cell classes, M and P, are a general feature of the primate visual system with primary functions that are not related to the color vision phenotypes. This is consistent with the hypothesis that the P pathway evolved before red-green color opponency and thus was primarily involved in achromatic vision.[31]

ACKNOWLEDGMENTS

The author is a CNPq research fellow, supported by CAPES, CNPq, and FINEP, and thanks Dr. José Augusto P. C. Muniz, Head of the *Centro Nacional de Primatas* (Ananindeua, State of Pará, Brazil) for his support of primate research, and Cézar Akiyoshi Saito for research assistance.

REFERENCES

1. Ramón y Cajal, S., *Textura del Sistema nervioso del hombre y de los vertebrados*, 1904, English translation: *Histology of the Nervous System of Man and Vertebrates*, Swanson, N., and Swanson, L.W., Translators, Oxford University Press, New York, 1995.
2. Polyak, S., *The Retina*, University of Chicago Press, Chicago, 1941.
3. Stone, J., *Parallel Processing in the Visual System — The Classification of Retinal Ganglion Cells and Its Impact on the Neurobiology of Vision*, Plenum Press, New York, 1983.
4. Roorda, A., Metha, A.B., Lennie, P., and Williams, D.R., Packing arrangement of the three cone classes in primate retina, *Vision Res.*, 41, 1291, 2001.
5. Mollon, J.D., Bowmaker, J.K., and Jacobs, G.H., Variations of colour vision in a New World primate can be explained by polymorphism of retinal photopigments, *Proc. R. Soc. Lond. B*, 222, 373, 1984.
6. Jacobs, G.H., Photopigments and seeing — lessons from natural experiments, *Invest. Ophthalmol. Vis. Sci.*, 39, 2205, 1998.
7. Jacobs, G.H., Neitz, M., Deegan, J.F., II, and Neitz, J.A., Trichromatic colour vision in New World monkeys, *Nature*, 382, 156, 1996.

8. Jacobs, G.H., Deegan, J.F., II, Neitz, J.A., Crognale, M.A., and Neitz, M., Photopigments and color vision in the nocturnal monkey, *Aotus*, *Vision Res.*, 33, 1773, 1993.
9. Wikler, K.C. and Rakic, P., Distribution of photoreceptor subtypes in the retina of diurnal and nocturnal primates, *J. Neurosci.*, 10, 3390, 1990.
10. Deegan, J.F., II and Jacobs, G.H., Spectral sensitivity and photopigments of a nocturnal prosimian, the bushybaby (*Otolemur crassicaudatus*), *Am. J. Primatol.*, 40, 55, 1996.
11. Jacobs, G.H. and Deegan, J.F., II, Photopigments underlying color vision in ringtail lemurs (*Lemur catta*) and brown lemurs (*Eulemur fulvus*), *Am. J. Primatol.*, 30, 243, 1993.
12. Tan, Y. and Li, W.H., Trichromatic vision in prosimians, *Nature*, 402, 36, 1999.
13. Hendrickson, A., Djajadi, H.R., Nakamura, L., Possin, D.E., and Sajuthi, D., Nocturnal tarsier retina has both short and long/medium-wavelength cones in an unusual topography, *J. Comp. Neurol.*, 424, 718, 2000.
14. Ahnelt, P. and Kolb, H., Horizontal cells and cone photoreceptors in primate retina: a Golgi-electron microscopic study of spectral connectivity, *J. Comp. Neurol.*, 343, 406, 1994.
15. Kolb, H., Linberg, K., and Fisher, S.K., Neurons of the human retina: a Golgi study, *J. Comp. Neurol.*, 318, 147, 1992.
16. Kolb, H., Fernández, E., Schouten, J., Ahnelt, P., Linberg, K.A., and Fisher, S.K., Are there three types of horizontal cell in the human retina? *J. Comp. Neurol.*, 343, 370, 1994.
17. Kolb, H., Mariani, A., and Gallego, A., A second type of horizontal cell in the monkey retina, *J. Comp. Neurol.*, 189, 31, 1980.
18. Boycott, B.B., Hopkins, J.M., and Sperling, H.G., Cone connections of the horizontal cells of the rhesus monkey's retina, *Proc. R. Soc. Lond. B*, 229, 345, 1987.
19. Wässle, H., Boycott, B.B., and Röhrenbeck, J., Horizontal cells in the monkey retina: cone connections and dendritic network, *Eur. J. Neurosci.*, 1, 421, 1989.
20. Ahnelt, P. and Kolb, H., Horizontal cells and cone photoreceptors in primate retina: a Golgi-light microscopy study of spectral connectivity, *J. Comp. Neurol.*, 343, 387, 1994.
21. Goodchild, A.K., Chan, T.L., and Grünert, U., Horizontal cell connections with short-wavelength sensitive cones in macaque monkey retina, *Vis. Neurosci.*, 13, 833, 1996.
22. Wässle, H., Dacey, D.M., Haun, T., Haverkamp, S., Grünert, U., and Boycott, B.B., The mosaic of horizontal cells in the macaque monkey retina: with a comment on biplexiform ganglion cells, *Vis. Neurosci.*, 17, 591, 2000.
23. dos Reis, J.W.L., de Carvalho, W.A., Saito, C.A., and Silveira, L.C.L., Morphology of horizontal cells in the retina of the capuchin monkey, *Cebus apella*: how many horizontal cell classes are found in dichromatic primates? *J. Comp. Neurol.*, 443, 105, 2002.
24. Chan, T.L., Goodchild, A.K., and Martin, P.R., The morphology and distribution of horizontal cells in the retina of a New World monkey, the marmoset *Callithrix jacchus*: a comparison with macaque monkey, *Vis. Neurosci.*, 14, 125, 1997.
25. Dacheux, R.F. and Raviola, E., Physiology of H1 horizontal cells in the primate retina, *Proc. R. Soc. Lond. B*, 239, 213, 1990.
26. Dacey, D.M., Lee, B.B., Stafford, D.K., Pokorny, J., and Smith, V.C., Horizontal cells of the primate retina: cone specificity without spectral opponency, *Science*, 271, 656, 1996.
27. Lee, B.B., Dacey, D.M., Smith, V.C., and Pokorny, J., Horizontal cells reveal cone type-specific adaptation in primate retina, *Proc. Natl. Acad. Sci. U.S.A.*, 96, 14611, 1999.

28. Verweij, J., Dacey, D.M., Peterson, B.B., and Buck, S.L., Sensitivity and dynamics of rod signals in H1 horizontal cells of the macaque monkey retina, *Vision Res.*, 39, 3662, 1999.
29. Ogden, T.E., The morphology of retinal neurons of the owl monkey *Aotus*, *J. Comp. Neurol.*, 153, 399, 1974.
30. dos Santos, S.N., dos Reis, J.W.L., Kremers, J., and Silveira, L.C.L., Morphology and size of owl monkey horizontal cells, *Invest. Ophthalmol. Vis. Sci.*, 41, S944, 2000.
31. Mollon, J.D., Uses and evolutionary origins of primate colour vision, in *Vision and Visual Dysfunction,* Vol. 2, *Evolution of the Eye and Visual System*, Cronly-Dillon, J.R. and Gregory, R.L., Eds., Macmillan Press, London, 1991, 306.
32. Gouras, P., Identification of cone mechanisms in monkey ganglion cells, *J. Physiol.* (London), 199, 533, 1968.
33. de Monasterio, F.M. and Gouras, P., Functional properties of ganglion cells of the rhesus monkey retina, *J. Physiol.* (London), 251, 167, 1975.
34. Mariani, A.P., Bipolar cells in monkey retina selective for cones likely to be blue-sensitive, *Nature*, 308, 184, 1984.
35. Kouyama, N. and Marshak, D.W., Bipolar cells specific for blue cones in the macaque retina, *J. Neurosci.*, 12, 1233, 1992.
36. Dacey, D.M., Morphology of a small-field bistratified ganglion cell type in the macaque and human retina, *Vis. Neurosci.*, 10, 1081, 1993.
37. Dacey, D.M. and Lee, B.B., The "blue-on" opponent pathway in primate retina originates from a distinct bistratified ganglion cell type, *Nature*, 367, 731, 1994.
38. Ghosh, K.K., Goodchild, A.K., Sefton, A.E., and Martin, P.R., The morphology of retinal ganglion cells in the New World marmoset monkey *Callithrix jacchus*, *J. Comp. Neurol.*, 366, 76, 1996.
39. Ghosh, K.K., Martin, P.R., and Grünert, U., Morphological analysis of the blue cone pathway in the retina of a New World monkey, the marmoset *Callithrix jacchus*, *J. Comp. Neurol.*, 379, 211, 1997.
40. Ghosh, K.K. and Grünert, U., Synaptic input to small bistratified (blue-on) ganglion cells in the retina of a New World monkey, the marmoset *Callithrix jacchus*, *J. Comp. Neurol.*, 413, 417, 1999.
41. Luo, X., Ghosh, K.K., Martin, P.R., and Grünert, U., Analysis of two types of cone bipolar cells in the retina of a New World monkey, the marmoset, *Callithrix jacchus*, *Vis. Neurosci.*, 16, 707, 1999.
42. Silveira, L.C.L., Lee, B.B., Yamada, E.S., Kremers, J., Hunt, D.M., Martin, P.R., and Gomes, F.L., Ganglion cells of a short wavelength sensitive cone pathway in New World monkeys: morphology and physiology, *Vis. Neurosci.*, 16, 333, 1999.
43. Boycott, B.B. and Dowling, J.E., Organization of the primate retina: light microscopy, *Philos. Trans. R. Soc. Lond. B*, 255, 109, 1969.
44. Silveira, L.C.L., Lee, B.B., Yamada, E.S., Kremers, J., and Hunt, D.M., Post-receptoral mechanisms of colour vision in new world primates, *Vision Res.*, 38, 3329, 1998.
45. Silveira, L.C.L., dos Santos, S.N., and Kremers, J., The presence of single-cone midget bipolar cells in the retina of the owl-monkey, *Invest. Ophthalmol. Vis. Sci.*, 42, S375, 2001.
46. Casagrande, V.A., A third parallel visual pathway to primate area V1, *TINS*, 17, 305, 1994.
47. Perry, V.H. and Silveira, L.C.L., Functional lamination in the ganglion cell layer of the macaque's retina, *Neuroscience*, 25, 217, 1988.
48. Silveira, L.C.L. and Perry, V.H., The topography of magnocellular projecting ganglion cells (M-ganglion cells) in the primate retina, *Neuroscience*, 40, 217, 1991.

49. Grünert, U., Greferath, U., Boycott, B.B., and Wässle, H., Parasol (Pα) ganglion-cells of the primate fovea: immunocytochemical staining with antibodies against $GABA_A$-receptors, *Vision Res.*, 33, 1, 1993.
50. Lima, S.M.A., Silveira, L.C.L., and Perry, V.H., Distribution of M retinal ganglion cells in diurnal and nocturnal New-World monkeys, *J. Comp. Neurol.*, 368, 538, 1996.
51. Hughes, A., The topography of vision in mammals of contrasting life style: comparative optics and retinal organization, in *Handbook of Sensory Physiology,* Vol. VII/5: *The Visual System in Vertebrates*, Crescitelli, F., Ed., Springer-Verlag, Berlin, 1977, 614.
52. Picanço-Diniz, C.W., Silveira, L.C.L., de Carvalho, M.S.P., and Oswaldo-Cruz, E., Contralateral visual field representation in area 17 of the cerebral cortex of the agouti: a comparison between cortical magnification factor and retinal ganglion cell distribution, *Neuroscience*, 44, 325, 1991.
53. Picanço-Diniz, C.W., Silveira, L.C.L., and Oswaldo-Cruz, E., A comparative survey of magnification factors and retinal ganglion cell topography of lateral-eyed mammals, in *The Visual System from Genesis to Maturity*, Lent, R., Ed., Birkhauser, Boston, 1992, 188.
54. Provis, J.M., The distribution and size of ganglion cells in the retina of the pigmented rabbit: a quantitative analysis, *J. Comp. Neurol.*, 185, 121, 1979.
55. Perry, V.H. and Cowey, A., The ganglion cell and cone distributions in the monkey's retina: implication for the central magnification factors, *Vision Res.*, 25, 1795, 1985.
56. Rolls, E.T. and Cowey, A., Topography of the retina and striate cortex and its relationship to visual acuity in rhesus monkeys and squirrel monkeys, *Exp. Brain Res.*, 10, 298, 1970.
57. Stone, J. and Johnston, E., The topography of primate retina: a study of the human, bushbaby, and New and Old World monkeys, *J. Comp. Neurol.*, 196, 205, 1981.
58. Curcio, C.A. and Allen, K.A., Topography of ganglion cells in human retina, *J. Comp. Neurol.*, 300, 5, 1990.
59. Fischer, Q.S. and Kirby, M.A., Number and distribution of retinal ganglion cells in *anubis* baboons (*Papio anubis*), *Brain Behav. Evol.*, 37, 189, 1991.
60. Wässle, H., Grünert, U., Röhrenbeck, J., and Boycott, B.B., Retinal ganglion cell density and cortical magnification factor in the primate, *Vision Res.*, 11, 1897, 1990.
61. Herbin, M., Boire, D., and Ptito, M., Size and distribution of retinal ganglion cells in the St. Kitts green monkey (*Cercopithecus aethiops sabeus*), *J. Comp. Neurol.*, 383, 459, 1997.
62. Silveira, L.C.L., Picanço-Diniz, C.W., Sampaio, L.F.S., and Oswaldo-Cruz, E., Retinal ganglion cell distribution in the *Cebus* monkey: a comparison with the cortical magnification factors, *Vision Res.*, 29, 1471, 1989.
63. Jones, A.E., The retinal structure of (*Aotus trivirgatus*) the owl monkey, *J. Comp. Neurol.*, 125, 19, 1965.
64. Webb, S.V. and Kaas, J.H., The sizes and distribution of ganglion cells in the retina of the owl monkey, *Aotus trivirgatus*, *Vision Res.*, 16, 1247, 1976.
65. Silveira, L.C.L., Perry, V.H., and Yamada, E.S., The retinal ganglion cell distribution and the representation of the visual field in area 17 of the owl monkey, *Aotus trivirgatus*, *Vis. Neurosci.*, 10, 887, 1993.
66. Wilder, H.D., Grünert, U., Lee, B.B., and Martin, P.R., Topography of ganglion cells and photoreceptors in the retina of a New World monkey: the marmoset *Callithrix jacchus*, *Vis. Neurosci.*, 13, 335, 1996.
67. DeBruyn, E.J., Wise, V.L., and Casagrande, V.A., The size and topographic arrangement of retinal ganglion cells in the *Galago*, *Vision Res.*, 20, 315, 1980.

68. Yamada, E.S., Silveira, L.C.L., Gomes, F.L., and Lee, B.B., The retinal ganglion cell classes of New World primates, *Rev. Bras. Biol.*, 56(Suppl. 1), 381, 1996.
69. Silveira, L.C.L., Picanço-Diniz, C.W., and Oswaldo-Cruz, E., The distribution and size of ganglion cells in the retinae of large Amazon rodents, *Vis. Neurosci.*, 2, 221, 1989.
70. Allman, J.M. and Kaas, J.H., Representation of the visual field in striate and adjoining cortex of the owl monkey *Aotus trivirgatus*, *Brain Res.*, 35, 89, 1971.
71. van Essen, D.C., Newsome, W.T., and Maunsell, J.H.R., The visual field representation in striate cortex of the macaque monkey: asymmetries, anisotropies and individual variability, *Vision Res.*, 24, 429, 1984.
72. Myerson, J., Manis, P.B., Miezin, F.M., and Allman, J.M., Magnification in striate cortex and retinal ganglion cell layer of owl monkey: a quantitative comparison, *Science* (New York), 192, 855, 1977.
73. Azzopardi, P. and Cowey, A., Preferential representation of the fovea in the primary visual cortex, *Nature*, 361, 719, 1993.
74. Choudhury, B.P., Distribution of neurons in the pigmented rabbit's visual cortex, *Exp. Neurol.*, 101, 458, 1988.
75. Rodieck, R.W., Binmoeller, K.F., and Dineen, J., Parasol and midget ganglion cells of the human retina, *J. Comp. Neurol.*, 233, 115, 1985.
76. Dacey, D.M. and Petersen, M.R., Dendritic field size and morphology of midget and parasol ganglion cells of the human retina, *Proc. Natl. Acad. Sci. U.S.A.*, 89, 9666, 1992.
77. Dacey, D.M., The mosaic of midget ganglion cells in the human retina, *J. Neurosci.*, 13, 5334, 1993.
78. Watanabe, M. and Rodieck, R.W., Parasol and midget ganglion cells, *J. Comp. Neurol.*, 289, 434, 1989.
79. Leventhal, A.G., Rodieck, R.W., and Dreher, B., Retinal ganglion cell classes in the Old World monkey: morphology and central projections, *Science*, 213, 1139, 1981.
80. Perry, V.H., Oehler, R., and Cowey, A., Retinal ganglion cells that project to the dorsal lateral geniculate nucleus in the macaque monkey, *Neuroscience*, 12, 1101, 1984.
81. Leventhal, A.G., Ault, S.J., Vitek, D.J., and Shou, T., Extrinsic determinants of retinal ganglion cell development in primates, *J. Comp. Neurol.*, 286, 170, 1989.
82. Silveira, L.C.L., Yamada, E.S., Perry, V.H., and Picanço-Diniz, C.W., M and P retinal ganglion cells of diurnal and nocturnal New World monkeys, *NeuroReport*, 5, 2077, 1994.
83. Goodchild, A.K., Ghosh, K.K., and Martin, P.R., A comparison of photoreceptor spatial density and ganglion cell morphology in the retina of human, macaque monkey, cat, and the marmoset *Callithrix jacchus*, *J. Comp. Neurol.*, 366, 55, 1996.
84. Yamada, E.S., Silveira, L.C.L., and Perry, V.H., Morphology, dendritic field size, somal size, density and coverage of M and P retinal ganglion cells of dichromatic *Cebus* monkeys, *Vis. Neurosci.*, 13, 1011, 1996.
85. Yamada, E.S., Silveira, L.C.L., Gomes, F.L., and Lee, B.B., The retinal ganglion cell classes of New World primates, *Rev. Bras. Biol.*, 56(Suppl. 1), 381, 1996.
86. Yamada, E.S., Silveira, L.C.L., Perry, V.H., and Franco, E.C.S., Morphology and dendritic field size of M and P retinal ganglion cells of the owl monkey, *Vision Res.*, 41, 119, 2001.
87. Yamada, E.S., Marshak, D.W., Silveira, L.C.L., and Casagrande, V.A., Morphology of P and M retinal ganglion cells of the bush baby, *Vision Res.*, 38, 3345, 1998.
88. Picanço-Diniz, C.W., Silveira, L.C.L., Yamada, E.S., and Martin, K.A.C., Biocytin as retrograde tracer in mammal visual system, *Braz. J. Med. Biol. Res.*, 25, 62, 1992.

89. La Vail, M.M., Rapaport, D.H., and Rakic, P., Cytogenesis in the monkey retina, *J. Comp. Neurol.*, 309, 86, 1991.
90. Silveira, L.C.L. and Perry, V.H., A neurofibrillar staining method for retina and skin: a simple modification for improved staining and reliability, *J. Neurosci. Methods*, 33, 11, 1990.
91. Hubel, D.H. and Wiesel, T.N., Receptive fields of optic nerve fibres in the spider monkey, *J. Physiol.* (London), 154, 572, 1960.
92. Kaplan, E. and Shapley, R.M., The primate retina contains two types of ganglion cells, with high and low contrast sensitivity, *Proc. Natl. Acad. Sci. U.S.A.*, 83, 2755, 1986.
93. Silveira, L.C.L. and de Mello, H.D., Jr., Parallel pathways of the primate vision: sampling of the information in the Fourier space by M and P cells, in *Development and Organization of the Retina: From Molecules to Function,* NATO ASI Series, Series A: *Life Sciences,* Vol. 299, Chalupa, L.M. and Finlay, B.L., Eds., Plenum Press, New York, 1998, 173.
94. Croner, L.J. and Kaplan, E., Receptive fields of P and M ganglion cells across the primate retina, *Vision Res.*, 35, 7, 1995.
95. Kremers, J., Lee, B.B., Pokorny, J., and Smith, V.C., Responses of macaque ganglion cells and human observers to compound periodic waveforms, *Vision Res.*, 33, 1997, 1993.
96. Lee, B.B., Pokorny, J., Smith, V.C., and Kremers, J., Responses to pulses and sinusoids in macaque ganglion cells, *Vision Res.*, 34, 3081, 1994.
97. Lee, B.B., Martin, P.R., and Valberg, A., The physiological basis of heterochromatic flicker photometry demonstrated in the ganglion cells of the macaque retina, *J. Physiol.*, 404, 323, 1988.
98. Lee, B.B., Pokorny, J., Smith, V.C., Martin, P.R., and Valberg, A., Luminance and chromatic modulation sensitivity of macaque ganglion cells and human observers, *J. Opt. Soc. Am. A*, 7, 2223, 1990.
99. Smith, V.C., Lee, B.B., Pokorny, J., Martin, P.R., and Valberg, A., Responses of macaque ganglion cells to the relative phase of heterochromatically modulated lights, *J. Physiol.*, 458, 191, 1992.
100. Kaiser, P.K., Lee, B.B., Martin, P.R., and Valberg, A., The physiological basis of the minimally distinct border demonstrated in the ganglion cells of the macaque retina, *J. Physiol.*, 422, 153, 1990.
101. Lee, B.B., Wehrhahn, C., Westheimer, G., and Kremers, J., Macaque ganglion cell responses to stimuli that elicit hyperacuity in man: detection of small displacements, *J. Neurosci.*, 13, 1001, 1993.
102. Kremers, J., Silveira, L.C.L., Yamada, E.S., and Lee, B.B., The ecology and evolution of primate colour vision, in *Color Vision: From Molecular Genetics to Perception*, Gegenfurtner, K.R. and Sharpe, L.T., Eds., Cambridge University Press, Cambridge, U.K., 1999, 123.
103. Lee, B.B., Silveira, L.C.L., Yamada, E.S., Hunt, D.M., Kremers, J., Martin, P.R., Troy, J.B., and Silva-Filho, M., Visual responses of ganglion cells of a New World primate, the capuchin monkey, *Cebus apella*, *J. Physiol.* (London), 528(3), 573, 2000.
104. Yeh, T., Lee, B.B., Kremers, J., Cowing, J.A., Hunt, D.M., Martin, P.R., and Troy, J.B., Visual responses in the lateral geniculate nucleus of dichromatic and trichromatic marmosets (*Callithrix jacchus*), *J. Neurosci.*, 15, 7892, 1995.
105. Silveira, L.C.L., Lee, B.B., Kremers, J., Silva-Filho, M., Saito, C.A., and Kilavik, B.E., Receptor inputs and temporal dynamics of owl monkey ganglion cells, *Invest. Ophthalmol. Vis. Sci.*, 41, S937, 2000.

106. Sherman, S.M., Wilson, J.R., Kaas, J.H., and Webb, S.V., X- and Y-cells in the dorsal lateral geniculate nucleus of the owl monkey (*Aotus trivirgatus*), *Science*, 192, 475, 1976.
107. Usrey, W.M. and Reid, R.C., Visual physiology of the lateral geniculate nucleus in two species of New World monkey: *Saimiri sciureus* and *Aotus trivirgatis*, *J. Physiol.*, 15, 755, 2000.
108. Kilavik, B.E., Kremers, J., and Silveira, L.C.L., Receptive field spatial properties of lateral geniculate nucleus in the owl monkey (*Aotus azarae*), *Invest. Ophthalmol. Vis. Sci.*, 42, S406, 2001.
109. Norton, T.T., Casagrande, V.A., Irvin, G.E., Sesma, M.A., and Petry, H.M., Contrast-sensitivity functions of W-, X-, and Y-like relay cells in the lateral geniculate nucleus of bush baby, *Galago crassicaudatus*, *J. Neurophysiol.*, 59, 1639, 1988.
110. Franco, E.C.S., Finlay, B.L., Silveira, L.C.L., Yamada, E.S., and Crowley, J.C., Conservation of absolute foveal area in New World primates: a constraint on eye size and conformation, *Brain Behav. Evol.*, 56, 276, 2000.
111. Krebs, W. and Krebs, I.P., Quantitative morphology of the central fovea in the primate retina, *Am. J. Anat.*, 184, 225, 1989.
112. Wikler, K.C., Williams, R.W., and Rakic, P., Photoreceptor mosaic: number and distribution of rods and cones in the rhesus monkey retina, *J. Comp. Neurol.*, 297, 499, 1990.
113. Packer, O., Hendrickson, A.E., and Curcio, C.A., Photoreceptor topography of the retina in the adult pigtail macaque (*Macaca nemestrina*), *J. Comp. Neurol.*, 288, 165, 1989.
114. Ogden, T., The receptor mosaic of *Aotus trivirgatus*: distribution of rods and cones, *J. Comp. Neurol.*, 163, 193, 1975.
115. Andrade-da-Costa, B.L.S. and Hokoç, J.N., Photoreceptor topography of the retina in the New World monkey *Cebus apella*, *Vision Res.*, 40, 2395, 2000.
116. Troilo, D., Howland, H.C., and Judge, S.J., Visual optics and retinal cone topography in the common marmoset (*Callithrix jacchus*), *Vision Res.*, 33, 1301, 1993.
117. Dkhissi-Benyahya, O., Szel, A., Degrip, W.J., and Cooper, H.M., Short and mid-wavelength cone distribution in a nocturnal strepsirrhine primate (*Microcebus murinus*), *J. Comp. Neurol.*, 438, 490, 2001.

3 The Pulvinar Complex

Iwona Stepniewska

CONTENTS

3.1 INTRODUCTION

During the course of evolution of the human brain, there has been a relative enlargement of the association cortex and thalamus.[1] The pulvinar nuclear complex in the posterior thalamus is one of the regions that contributed most to the enlargement of the thalamus. Not identifiable in rodents and other small mammals,[2,3] the pulvinar became proportionally larger in advanced mammals, accounting for a quarter of the total mass in primates, and reaching its greatest dimensions in the human brain.[3] Thus, the evolution of the pulvinar appears to parallel the evolution of higher brain function.[2,4,5]

In primates, the pulvinar complex is the largest component of the subcortical visual system. With input from the retina and superior colliculus, and reciprocal connections to vast areas of neocortex, the pulvinar occupies a position in the extrastriate visual system that is analogous to the position of the lateral geniculate

0-8493-1243-4/04/$0.00+$1.50

nucleus in the primary visual pathway, but deals with higher-order visual and visuomotor processing. Although the involvement of the pulvinar complex in visual perception is well supported by a variety of evidence, its functional divisions and their specific roles remain relatively enigmatic.[1,6–9] However, it is well known that the pulvinar is not structurally or functionally homogeneous. Based on topographic and architectonic criteria, the pulvinar complex of the dorsal thalamus traditionally has been divided into four main divisions or nuclei: the inferior (visual), lateral (visual), medial (visual, multisensory), and anterior (somatosensory) divisions.[10,11] In recent years the connections of these divisions have become better known, and further subdivisions have been defined by patterns of inputs and outputs. New and more exact studies continue to delineate more subdivisions combining histochemical techniques and studies of connections.[8,12,13]

This chapter focuses on the three major divisions of the pulvinar (inferior, lateral, and medial) that are associated with processing of visual information. The anatomical and physiological data are discussed in the context of newly proposed architectonic subdivisions that are likely to be of functional significance. The material presented here is largely based on research carried out in simian primates, mostly macaque monkeys, which are considered to be the best available animal model for the human visual system.

3.2 ARCHITECTONIC SUBDIVISIONS OF THE PULVINAR COMPLEX

Early architectonic studies distinguished medial (PM), lateral (PL), and inferior (PI) "nuclei" as subdivisions of the pulvinar. Because the general morphology of basic types of neurons is similar through the pulvinar, classically defined subdivisions could only be recognized on the basis of cell density and the locations of fiber bundles. In an early cytoarchitectonic parcellation,[11] two dorsally located medial and lateral subdivisions were separated from the inferior pulvinar by the brachium of the superior colliculus (BrSC; Figure 3.1A). This parcellation differed somewhat from the portrayals based on myeloarchitecture where the presence of horizontally oriented fiber bundles suggested that the lateral pulvinar extends ventrally to the inferior margin of the pulvinar[10] (Figure 3.1B).

The classical architectonic subdivisions of the pulvinar do not correspond, however, to the subdivisions described based on electrophysiology[14–16] or connectional patterns.[17–19] Thus, the cyto- and myeloarchitectonic pulvinar components do not seem to represent the full complexity of nuclear organization. The advent of recent histo- and immunocytochemical techniques created the opportunity to reevaluate the architecture of the pulvinar. Compared with the traditional architectonic approach, neurochemical methods offer an index that more directly reflects functional subdivisions as they target molecular components (e.g., calcium-binding proteins) or specific transmitter systems (e.g., acetylcholinesterase, AChE).

The most significant advance in our understanding of the pulvinar has come from the recent finding in monkeys that the pulvinar consists of up to nine separate chemically distinct subdivisions that can be reliably distinguished by neurochemical stainings.[8,12,13,20–22] Although all major divisions of pulvinar probably have

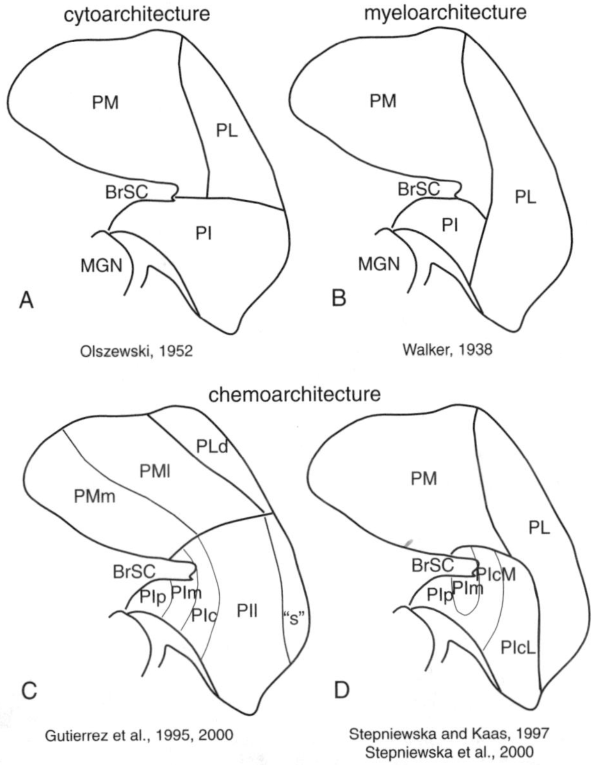

FIGURE 3.1 Comparison of the traditional parcellation of the macaque pulvinar based on distribution of Nissl substance (A) and myelinated fibers (B) with chemoarchitectonic subdivisions (C, D) based mostly on the distribution of calbindin. Lines represent borders between major divisions: medial, lateral, and inferior, and boundaries of subdivisions identified within inferior and medial pulvinar. BrSC = brachium of the superior colliculus; MGN = medial geniculate nucleus; PI = inferior pulvinar; PIc = central PI subdivision; PIcL = central lateral PI subdivision; PIcM = central medial PI subdivision; PIL = lateral PI subdivision; "s" = shell of PIL; PIm = medial PI subdivision; PIp = posterior PI subdivision; PL = lateral pulvinar; PLd = dorsal PL division; PM = medial pulvinar; PMm and PMl = medial and lateral PM divisions.

functionally significant subdivisions, the most compelling evidence has been for the inferior pulvinar, where several nuclei or subnuclei have been distinguished on the basis of architectonic distinctions and differences in connections.

3.2.1 Inferior Pulvinar

In an early effort based on Nissl and myelin staining, Lin and Kaas[17] distinguished three subdivisions of the inferior pulvinar in the owl monkey: a small posterior nucleus (PIp), a larger medial nucleus (PIm), and a large central nucleus (PIc). Most recently, neurochemical stainings have been used to parcellate the PI complex in a range of primate species, including humans.[8,12,13,20,21,23] These studies reaffirmed the identities of three subdivisions identified by Lin and Kaas[17] and revealed additional PI components (subunits). The inferior pulvinar of monkeys has been subdivided into four or five regions on the basis of differences in expression of the calcium binding protein calbindin (Cb), cytochrome oxidase (CO), AChE, and the SMI-32 antibody for nonphosphorylated neurofilament protein, as well as Cat-301 and *Wisteria floribunda* agglutinin binding.[12,13,20] There are two current neurochemical schemes of the primate inferior pulvinar (Figure 3.1C and 3.1D). One of them recognizes five subdivisions that encompass PI in macaque monkey — posterior (PIp), medial (PIm), central (PIc), lateral (PIL), and a "shell" region of PIL, PIL-s[12,20] (Figure 3.1C). Four of these subdivisions, although renamed, correspond roughly to subdivisions identified earlier in squirrel monkey.[8] As pictured, the PI in both species extends dorsally above the brachium of the superior colliculus on the most ventral parts of the lateral and medial pulvinar, laterally incorporating a ventral portion of the traditional lateral pulvinar (PL).

Using similar techniques, Stepniewska and Kaas[13] followed a more traditional pulvinar parcellation and proposed somewhat different immunocytochemical subdivisions for PI. In owl, squirrel, and macaque monkeys, they recognized four subdivisions within inferior pulvinar — PIp, PIm, and PIc, which was divided further into medial and lateral nuclei (PIcM and PIcL) (Figure 3.1D). PIm and PIcM (roughly corresponding to PIc of Gutierrez et al.[12]) of this portrayal extend above the BrSC, but PIcL is restricted to the lateral portion of traditional PI. Consequently, the region forming the lateral border of PIc with horizontally oriented fiber bundles is identified as the part of the lateral pulvinar (PL). Thus, PIcL of Stepniewska and Kaas[13] overlaps much, but not all, of PIL in macaque identified by Gutierrez et al.[12] while being somewhat larger than PIL in squirrel monkey proposed by Cusick et al.[8] The boundary of PIcL and PL is uncertain, and PIcL may contain additional subregions, as suggested by shell region "s" of Gutierrez et al.[12] Such remaining uncertainties need to be resolved.

In spite of some differences in nomenclature used, the pattern revealed by neurochemical stainings, especially the calbindin antibody, is similar in all primates studied. Thus, PIp stains darkly for Cb and lightly for AChE and CO, whereas PIm stains lightly for Cb and darkly in a patchy manner for AChE, CO, and Cat-301 (but see also figure 9.B in Reference 13 for Cat-301 pattern in squirrel monkey). In contrast, the region defined as PIc[8,12] or PIcM[13] stains darkly for Cb and can be distinguished from the lighter-stained, more lateral PIcL (PIL). Two of these subdivisions, the Cb-dark posterior nucleus (PIp) and Cb-light medial nucleus (PIm), strictly correspond to those previously described by Lin and Kaas.[17,24] However, the third central division PIc[17] contains obvious medial and lateral subdivisions. These subdivisions are apparent in New World and Old World monkeys, and they were first recognized by Cusick and co-workers,[8,12] who designated them as PIc and PIL. It has to be emphasized that the lateral part of PIL encroaches on the pulvinar region traditionally considered as PL and the medial part occupies lateral PIc of Lin and Kaas.[17] Stepniewska and Kaas[13] preferred to name the two subdivisions of PI, PIcM and PIcL to indicate that they derived from the original PIc of Lin and Kaas.[17] This terminology has the advantage of recognizing the newly revealed subdivisions, while preserving the original meaning of PIc and correcting errors of identification.

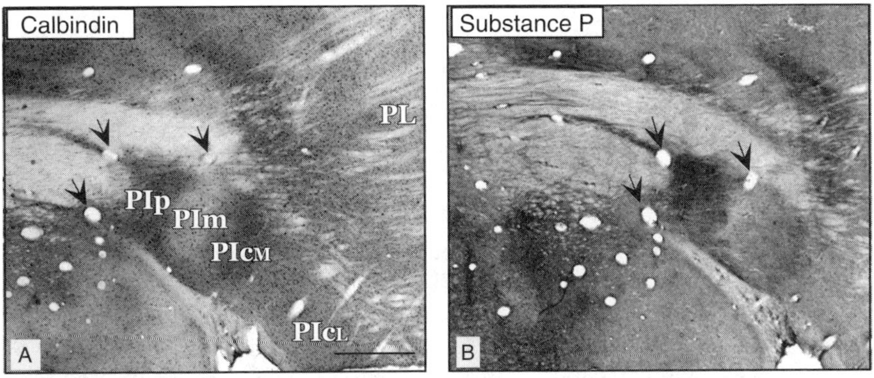

FIGURE 3.2 Photomicrographs showing architectonic subdivisions of the macaque inferior pulvinar in calbindin (A) and substance P (B) preparations. PIp and PIcM show similar dark patterns in both stainings, whereas PIp and PIcL are lightly stained in these preparations. The ventral stripe connecting PIp and PIcM is also darker than PIm. Arrows indicate the same blood vessels. Scale bar = 1 mm. Conventions as in Figure 3.1. (From Stepniewska, I. et al., *Vis. Neurosci.*, 17, 529, 2000. With permission.)

The most compelling evidence for the existence of two subdivisions of the original PIc comes from Cb preparations, in which the medial portion is darker than the rest of original PIc and contrasts clearly with the medially adjacent light PIm[8,12,13] (Figure 3.2A). Substance P (SP) distribution further demonstrates the existence of two PIc subdivisions, as substance P (possibly reflecting a class of superior colliculus inputs) is densely expressed in PIcM, but not in PIcL[25] (Figure 3.2B). The SP staining pattern is similar in Old World and New World monkeys although SP is more densely expressed in Old World monkeys.

Darkly expressed Cb and SP distributions extend ventrally in PIp and medial PIc beyond their previously established borders forming a bridge between these subdivisions.[25] Thus, PIm stands out as a light structure surrounded laterally, ventrally, and medially by the dark pattern of adjacent nuclei (Figure 3.2). The arrangement of myelinated fibers, as well as the distribution of SC inputs, also indicates the existence of a connecting PIp–PIcM bridge.[25] The dense zone of labeled SC terminations in PIcM and PIp often fuses ventrally, suggesting that PIm does not extend to the ventral margin of the pulvinar.

In conclusion, there is an agreement between two current neurochemical schemes of PI relative to the borders of PIp, PIm, and of the darker adjacent region named PIc[12,20,26] or PIcM.[13,21,27] The major controversy is related to the subdivision of the ventrolateral portion of the pulvinar. Cusick and colleagues,[8,12,20] based on a similar pattern of calbindin staining observed in the lateral portion of PI and in the ventral portion of PL, and by the fact that the V1 projection zone extends dorsal to the BrSC, consider this region to be a single subdivision named lateral PI (PIL). Stepniewska and Kaas[13] interpreted the lateral region of PI with slightly reduced staining in several types of preparation as a ventral extension of PL rather than a subdivision of the inferior pulvinar. In support of the Stepniewska and Kaas[13] pulvinar parcellation, most recently Adams et al.[21] used the pattern of striate (V1) and extrastriate

(V2, V4, and MT) projections to the pulvinar to argue that ventral PL should not be included as part of PI, in spite of the fact that PIcL and ventral PL look similar neurochemically. Uncertainties remain, and further studies are probably needed to delimit the inferior pulvinar dorsolaterally.

3.2.2 Lateral and Medial Pulvinar

Inferior pulvinar is capped by lateral (PL) and medial (PM) subdivisions that are often considered together as the superior pulvinar, especially in New World monkeys, where lateral pulvinar occupies a largely dorsal position relative to the inferior pulvinar.[14,28] Whereas the borders of PM seem well established and quite consistent in all studied primates, the full extent of the lateral pulvinar is less certain.[8,12,13,22]

In humans and Old World monkeys the *lateral pulvinar* is greatly expanded, so it extends ventrally to the edge of the pulvinar complex and caudally to form the caudolateral border of the thalamus. Some architectonic similarities of the ventral region of PL and lateral PI make the border between the two subdivisions quite difficult to determine (as noted above).

The main architectonic criterion for distinguishing PL from the rest of pulvinar has been the presence of mediolaterally directed fiber bundles in PL, and their absence in medial and inferior pulvinar.[28] In comparison with the inferior pulvinar, the cytoarchitecture of PL is more homogeneous although some differences attributed to uneven distribution of myelinated fibers between lateral and medial portions of PL have been reported.[13,28,29] Thus, in the lateral portion of PL neurons are arranged in horizontal rows distributed among numerous fiber bundles, whereas the medial part of the lateral pulvinar is less disrupted by fibers and has more uniform distribution of cells. Because the defining fiber bundles are thought to be fibers of passage, the lateral pulvinar was not defined by features that have functional significance for the pulvinar. Moreover, the connectional differences within classical PL make it appear unlikely that PL is a single coherent thalamic nucleus.

Recently calbindin immunocytochemistry has been used to define pulvinar compartments. PL contains numerous large and very darkly stained neurons on a background of a lighter-stained neuropil. The distribution differs somewhat lateromedially, indicating the existence of two medial (PLm) and lateral (PLl) subdivisions (see Fig. 10 in Stepniewska and Kaas[13]) as suggested by the myeloarchitecture of PL.[10,28] The distinctive group of quite large Cb-dark neurons, situated in medial PL close to PM border, may correspond to the electrophysiologically defined dorsomedial component of the pulvinar, Pdm.[30] A different interpretation of the PL was suggested by Gutierrez et al.,[12] who emphasized the similar pattern of Cb staining in the ventral portion of PL and the lateral portion of PI and proposed this region as a single entity designated as the lateral subdivision of the inferior pulvinar (PIL) (Figure 3.1C). According to this schema, the lateral pulvinar is much smaller than originally described[10,11,29] and restricted to the dorsal portion of traditional PL. Most recently, this dorsal PL region (PLd) (Figure 3.1C) was defined on the basis of histochemical criteria as a distinct, AChE- and parvalbumin-dense, and calbindin-light structure.[22] This PLd may correspond to a caudal extension of the lateral posterior (LP) nucleus[31–33] rather than PL of other portrayals.

The *medial pulvinar* can be distinguished from the lateral pulvinar by the lack of fiber bundles that are numerous in PL. This makes the appearance of PM more homogeneous. PM also has a higher cell density and its cells are more rounded than those in PL. The border between PM and PL is sharply marked in neurochemical preparations, especially Cb.[22] Although the medial pulvinar-like PL contains many small Cb-immunoreactive cells and a few scattered large neurons, it can be easily distinguished from PL by the more intensely stained neuropil. Moreover, in AChE and parvalbumin preparations, PM is lighter than PL (PLd). In addition, there is evidence for lateral (PMl) and medial (PMm) histochemical divisions of PM (Figure 3.1C). The PMm division has lighter AChE and parvalbumin staining of the neuropil. PMm also contains a distinctive central oval region (PMmc) that stains darker for AChE and parvalbumin than the surrounding PM region. The dense AChE staining of this region contrasts to the rest of PM as described earlier by Cavada et al.[34]

In summary, the lateral and medial pulvinar, although less differentiated than the inferior pulvinar, undoubtedly contain subdivisions as well. Although subdivisions have been proposed, their boundaries and possibly other basic units of organization need to be better demarcated.

3.3 CONNECTIONS

The pulvinar, together with the lateral geniculate nucleus (LGN), are the main thalamic centers of the primate visual system. However, unlike the LGN, pulvinar receives relatively few primary ascending afferents from the retina and other visual centers in the brain stem (e.g., superior colliculus), but it is strongly associated with the neocortex. All three major pulvinar divisions are differentially associated with the cortical structures of early visual processing (primary visual cortex), low- to medium-level extrastriate visual cortices (occipital areas), to higher cortices (parietal, frontal, cingulate cortex). The inferior pulvinar (PI) and the ventral part of lateral pulvinar (PL) receive major cortical inputs from striate and the adjoining extrastriate cortices, whereas the dorsal part of the lateral and medial pulvinar receive input from higher-level processing cortices such as the posterior parietal and parts of frontal cortex. All three major subdivisions receive also subcortical input from the superior colliculus. Inputs and outputs associated with the pulvinar are topographically organized. The major connections of pulvinar are summarized in Figure 3.3.

3.3.1 Connections from Retina

Data concerning the retinal input to the pulvinar are somewhat ambiguous. Several studies have provided evidence for direct retinal input to the pulvinar of macaque monkeys, but there is not complete agreement about where the retinal inputs terminate. Campos-Ortega et al.[35] first suggested the existence of a direct retinal projection to the pulvinar nucleus based on the observation of a few patches of degenerating retinal terminals in the ipsi- and contralateral pulvinar of baboons. Subsequent reports initially questioned the existence of direct retino-pulvinar projections in macaques,[36,37] but later studies provided evidence for this projection in several macaque species,[38–40] where retinal input was shown to end in contralateral

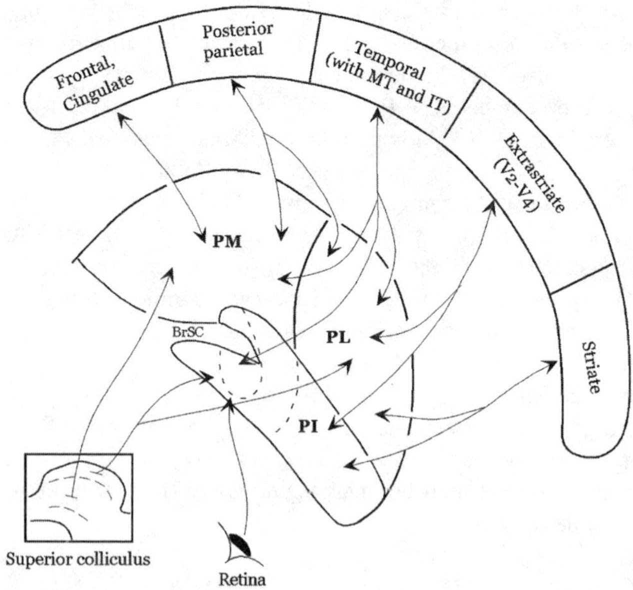

FIGURE 3.3 The major cortical and subcortical connections of the medial, lateral, and inferior pulvinar divisions. Cortical connections are reciprocal (double-headed arrows) although the degree of overlap of such connections with the various nuclei is still a matter of debate. PI and ventral PL are mostly connected to striate and near extrastriate cortices, whereas the dorsal PL and PM are connected to higher cortices (parietal, frontal, cingulate). There is an overall trend in the organization of connections between pulvinar and cortex so that ventrolateral portions of the pulvinar are connected to striate and near extastriate cortices, and more dorsomedial portions are connected to more associative cortical areas such as posterior parietal and frontal cortex. All pulvinar divisions (but not all parts) receive inputs from the superior colliculus, with PI and PL receiving inputs from superficial layers and PM from the deeper layers. Some direct retinal projections terminate in PI (retinal input to PM is uncertain). Dashed lines mark boundaries of chemoarchitectonic subdivisions within PI. MT = middle temporal area; IT = inferotemporal cortex. Other conventions as in Figure 3.1.

parts of the inferior pulvinar. The retinorecipient zone of PI was found to occupy roughly the part of PI adjacent to both medial geniculate nucleus (MGN) and the brachium of the superior colliculus (BrSC). Some results also suggested that the medial pulvinar (PM) receives retinal input,[38] although this was not confirmed in other studies.[40,41] More recently, retinal projections to the inferior pulvinar were related to pulvinar subdivisions revealed by the antibody against calbindin-D28K.[42] These results confirmed the presence of a direct retinal projection to the inferior (PI), but not medial pulvinar (PM) in macaque monkeys (described by Itaya and Van Hoesen[38]). Generally the relative location and extent of terminals appear quite similar to previous descriptions of this retino-thalamic pathway.[35,38–40] Moreover, the study revealed that the major retinal input was to the medial PI subdivision (PIm), with some involvement of the adjacent posterior and central subdivisions. These results suggest that PIm, which receives input from the retina and sends axons to visual area MT, could relay visual information from the retina to MT.[42] Possibly this explains the activation of area MT neurons after V1 lesions[43,44] (however, see Collins et al.[45]).

Recent studies based on iontophoretic injections of WGA-HRP into the inferior pulvinar confirmed the existence of direct retino-pulvinar projections and revealed the ganglion cells that are the source of these projections.[46] All three major ganglion cell types, with the gamma cells the most numerous, were found in mostly nonfoveal locations in the contralateral retina. The presence of beta (midget) cells, known to mediate color vision, was surprising, as these cells were thought to project solely to parvocellular layers of LGN. This beta cell pathway might mediate the wavelength (color) discrimination that has been demonstrated in monkeys and in human patients within the field defects caused by destruction of part of the striate cortex (for review, see Stoering and Cowey[47,48]).

An indirect retinal pathway to the pulvinar involves the superior colliculus. Recent connectional studies have shown that projections from retino-recipient layers of the SC avoid PIm and terminate in the adjacent PIp and PIcM subdivisions.[25] This pattern of connections suggests that direct and indirect retinal pathways to the inferior pulvinar have different functional roles.

In summary, there is evidence for retinal input to the inferior pulvinar, but the existence of retinal fibers within the medial pulvinar is uncertain.

3.3.2 Connections from the Superior Colliculus

The existence of input to the pulvinar from the superior colliculus (SC) is well documented. Superior colliculus projections to pulvinar in primates were established in early anterograde tracing studies (degeneration or autoradiography) in squirrel monkeys,[49] owl monkeys,[17] macaque monkeys,[41,50,51] and prosimian galagos.[52] These projections originate in neurons situated in the ventral half of the superficial gray layer (SGS) of the SC,[37,41,53,54] relay visual information, and terminate densely in the inferior pulvinar,[17,49,50,52,55] and sparsely in the lateral pulvinar.[25,41,51,56,57] These projections are predominantly ipsilateral, travel through the brachium of the superior colliculus, and are retinotopically organized. Most studies indicated that the terminations of collicular neurons in the pulvinar complex of macaque and squirrel monkeys are unevenly distributed with higher concentrations in the medial portion of the inferior pulvinar,[49,50,58] a location that appears to correspond to the neurochemical PIm subdivision.[8,12,13] In macaques, SC terminals have also been related to regions with a high concentration of AChE.[59] However, in owl monkeys, where label was related to myeloarchitectonic subdivisions of PI, injections of tritiated leucine into the SC resulted in dense terminal label in posterior and central parts of the inferior pulvinar, adjacent to darkly myelinated PIm, but not in PIm.[17]

Recent studies in New World marmoset, owl, and squirrel monkeys as well as Old World macaque monkeys extended these observations and demonstrated more precisely the pattern of colliculopulvinar connections by relating them to immunocytochemically defined subdivisions of PI.[25] According to this pattern, SC input is distributed mostly to the posterior (PIp) and central medial (PIcM) subdivisions of the inferior pulvinar, much as Lin and Kaas[17] had described earlier in owl monkeys. Some sparse collicular projections are also directed to lateral PIc (PIcL) and the lateral pulvinar (PL). The medial subdivision (PIm), which sends a massive input to area MT, does not receive a significant input from the SC, although biochemical features of PIm that are characteristic of relay nuclei (dense CO and parvalbumin staining) suggest that it could function as a nucleus relaying information from SC to area MT.[8]

In macaque monkeys, a projection from the SC to the principal tectorecipient zone of the posterior thalamus is identified by substance P immunoreactivity.[60] Pulvinar subdivisions receiving projections from the SC are also immunoreactive for Cb, but Cb is not uniquely expressed in SC projection neurons.[61] Thus, substance P serves as a specific marker of colliculothalamic terminations, although its functional role in these projections is unknown (see Wall and Fitzgerald,[62] Hutsler and Chalupa[61]).

Neurons in the deeper layers of the superior colliculus project outside the PI, to PL and PM. In both squirrel monkeys[54,56] and macaques,[51] the superior colliculus clearly projects to PL although these projections are not always obvious.[58] Some label was apparent in the PM after injections in the superior colliculus in both squirrel and macaque monkeys.[51,56] However, PM is usually considered a target of only the deeper layers of the superior colliculus (see Kaas and Huerta[28]). The projection system originating in the deeper SC layers appears to connect also to the functionally defined dorsomedial area of pulvinar, Pdm.[30,41]

With the exception of the superior colliculus, little is known about other subcortical regions that project to the pulvinar. Some early studies described afferent projections from brain stem nuclei such as the nucleus of the optic tract and the lateral terminal nucleus,[41,50,51] as well as the thalamic reticular nucleus, claustrum,[37] and amygdala.[63]

3.3.3 Cortical Connections

The primate pulvinar connects with the entire array of known visual areas. Most of these connections are reciprocal and predominantly ipsilateral. Each major subdivision of the pulvinar connects with multiple but different sets of cortical areas. The inferior pulvinar and the adjacent ventral portions of the lateral pulvinar, which show well-defined receptive fields and a discernible retinotopic organization,[16,30] relate to retinotopically organized striate (V1) and extrastriate cortex, (V2, V3, V4)[18,19,64–66] as well as areas in the superior temporal sulcus (MT, MST, FST) and inferotemporal visual cortex (TEO and TE).[8,19,33,64,67–72]

In contrast, most of the lateral and medial pulvinar (PL and PM) do not seem to be retinotopically organized and their cortical connections are more widespread than PI. They do not have connections with striate cortex V1 or area V2,[18,67,73] but connect with certain extrastriate visual areas including V4, inferotemporal, and higher-order areas such as posterior parietal cortex.[33,70,74–77] Moreover, PM interconnects with auditory and somatosensory cortical areas,[78–81] as well as with nonmodality-specific higher-order cortex such as insular, superior temporal polysensory, temporopolar, prefrontal, parahippocampal, and cingulate cortex.[63,82,83] The topography of PM cortical connections suggests that specific subregions of PM may be predominantly visual, whereas other subregions may be dominated by other modalities of sensation or participate in distributed, hemisphere-wide circuits that integrate different types of sensory information.[22,63]

In summary, the pattern of cortical connections supports the hypothesis that the inferior pulvinar and adjacent regions of lateral pulvinar are involved in the early stages of visual processing, whereas the medial pulvinar and most of lateral pulvinar that are less tied to early visual centers are involved in higher stages of processing.

3.3.3.1 Relation to Functional Streams of Visual Processing

The extrastriate areas have been conceptualized as comprising distinct dorsal and ventral streams of visual information processing. It has been suggested that these visual pathways have different overall patterns of corticothalamic connectivity, particularly with the pulvinar complex.[76] A more specific recent proposal is that the two pathways through the prestriate cortex are replicated within dorsal and ventral pulvinar domains.[84] The strong connections of inferior pulvinar and ventral region of lateral pulvinar to ventral extrastriate areas such as V4, and especially to inferotemporal cortex, associates the inferior pulvinar with the ventral processing stream for object recognition.[68] On the other hand, the dorsal pulvinar densely connects to posterior parietal cortex and may relate to the dorsal cortical visual stream dealing with visuomotor or visuospatial functions.[33,65,70,74,83,85–88]

Although the dorsal and ventral pulvinar show different cortical connections, they cannot be considered simply ventral or dorsal stream components. For example, PI associated with ventral stream processing also has connections with motion-related areas (MT, MST, FST), and the dorsal pulvinar, especially PM, has well-established connections with inferotemporal cortex (IT), which is a part of the ventral stream. Moreover, V1 and V2, the sources of both streams, connect to the ventral pulvinar only.[18,66,67,89]

Thus, on one hand, the different major connections of ventral and dorsal pulvinar components may reflect dichotomous contributions for visual perception vs. action in space[90] for the inferior and dorsal pulvinar, respectively. On the other hand, however, the lack of absolute separation of visual cortical connections between dorsal and ventral pulvinar suggests a possible role for the pulvinar in integrating information from the dorsal and ventral processing streams. In a recently proposed neurobiological model of cortical information processing, the pulvinar was assigned a key role in the integration of dorsal and ventral stream activities.[91] Indeed, a visual zone in the lateral part of PM (PMl) has connections with areas that belong to both streams and has neurons sensitive to both color/pattern and spatial stimulus properties.[92] This zone might be a candidate site for dorsal/ventral stream integration.

3.3.3.2 Laminar Distribution

The pulvinocortical and corticopulvinar systems have been shown to be generally reciprocal, but specific details are still unknown. For example, corticopulvinar connections of temporal lobe cortex may be more widespread than pulvinocortical connections,[72] and pulvinocortical connections to area V2 are stronger than those to V1.[73]

Pulvinocortical connections are known to terminate in layer 1 in area V1, and in several layers (1, 3, 4, and 5) in other cortical areas, with some variability according to area, species, and study.[82,89,93–97] The postsynaptic targets of pulvinocortical terminations are not known, but they likely contact parts of apical and basal dendrites of different pyramidal cell populations.

Corticopulvinar connections from visual areas originate from neurons in layer 5,[75,82,97] although from some cortical areas such as posterior cingulate and parahippocampal (TF and TH, Bonin and Bailey[98]) areas, they may originate from neurons in both layers 5 and 6.[75,82] In the prefrontal cortex, neurons forming connections with the pulvinar are exclusively confined to layer 6.[63]

3.3.3.3 Microcircuitry of Pulvinocortical Connections

Less is known about microcircuitry of pulvinocortical connections, and how they function as part of the corticopulvinar-cortical loop. Like other thalamic nuclei, the pulvinar consists of both projection neurons and local circuit neurons.[29,99–102] Most projection neurons are larger than local circuit neurons and only relatively small differences in morphology of neurons were observed between the pulvinar subdivisions. Although the form and shape of projection neurons are almost identical in all pulvinar subdivisions,[101,102] a striking element of inferior and lateral visual subdivisions is the presence of very large projection neurons. However, the functional attributes of medial, lateral, and inferior pulvinar likely depend on differences in connections rather than the limited cytoarchitectural differences noted between them.[99]

Golgi studies have described corticopulvinar terminations as classical type 1 small terminals,[29,101,103] in contrast to larger (type 2) endings associated with specific sensory inputs. Two populations of corticopulvinar terminals have been demonstrated in PI and PL.[104] One type corresponds to small stalked (Golgi type 1) terminations and the second, less frequent type consists of large beaded terminations (Golgi type 2). Spiny type 1 and beaded type 2 are also called E and R from the shape of their terminal specialization and arbor.[75] Terminations of axons originating from V1 and V2 are both R and E types, but terminations of axons from more anterior cortical areas are almost exclusively E type.[75,105] Because of their high resolution, new tracing techniques phaseolus vulgaris leucoagglutinin (PHA-L) and biotinylated dextran-amine (BDA) further reveal that the different types of terminations are mixed rather than preferentially distributed within particular terminal foci. This lack of separation of different types of terminations in specific neurochemical pulvinar subdivisions[104,105] indicates an interaction of these two subsystems rather than their segregation. This may be important for pulvinar functional properties.

3.4 RETINOTOPIC AND CONNECTIONAL DOMAINS

The retinotopic organization of the pulvinar was studied using electrophysiological techniques in New World owl[14] and *Cebus*[15] monkeys, and Old World macaque[16] monkeys. The early mapping studies in owl monkeys revealed a single map of the inferior pulvinar (PI) that contained the representation of central vision dorsally and rostrally, and the more peripheral visual field that is represented ventrally.[14] In another New World monkey, *Cebus*, two visual representations have been mapped in the pulvinar, but only one occupied PI as defined in owl monkeys.[15] Another map located in the lateral pulvinar was less clear. The retinotopic organization of the pulvinar is best known from the electrophysiological mapping in Old World macaque monkeys,[16,30] where two complete retinotopic maps of the contralateral hemifield have been found in the region inferior to the brachium of the superior colliculus. The medial region called "PI" corresponds to traditional architectonic PI and medial portions of adjacent PL and the lateral mapped region, termed "PL," partially surrounds PI and occupies the ventrolateral portion of the adjoining architectonic PL. The two maps adjoin along a common representation of the vertical meridian whose location can be correlated with a prominent feature of the pulvinar myeloarchitecture. A representation of the horizontal meridian crosses both maps diagonally and splits into dorsal and ventral parts along the outer margin of the PL map (i.e., along the lateral edge of PL). Thus, the lower visual field is represented dorsally and the upper field is represented ventrally in both maps. Dorsal and medial to the PI and PL maps, a separate visually responsive dorsomedial region of lateral pulvinar (Pdm) with a crude visuotopic organization has also been identified.[16,30,106,107] Some neurons in a separate region at the caudal pole of the pulvinar (not involved in Pdm) also respond to stimulation of wide areas of the visual field but show little or no retinotopic organization.[108]

Crucial evidence for determining how many separate visual representations may exist in the pulvinar could come from determining the number of separate topographic projection zones from individual cortical areas. Connection patterns of the large, retinotopically organized striate cortex have been especially useful in the identification of visual representations in the pulvinar. Early anatomical studies of striate cortex connections in monkeys similar to physiological results suggested the existence of one retinotopic representation within much of the PI and another in adjacent PL.[67,109–113] Later, Ungerleider et al.[18,19] provided more extensive evidence from V1 and middle temporal area (MT) projections for three topographical representations of the contralateral hemifield in the pulvinar, P1, P2, and P3. Connected to V1, P1 and P2 fields correspond to the "PI" and "PL" visuotopic maps described by Bender;[16] the P3 field (not identified in recordings) with MT connections is located posteromedially in the inferior pulvinar and includes small ventral portions of lateral and medial pulvinar, located dorsal to the brachium of the superior colliculus.

Some aspects of retinotopy of the pulvinar were also indicated by its pattern of connections with other visual areas and the superior colliculus.[50,58,109] The P1 and P2 fields have connections with V1, V2, V4, and MT[18,19,64,84,89,109,111,113–115] and both receive inputs from the superficial layers of the superior colliculus.[41,50,51,58] The P3 field has connections with areas in the superior temporal sulcus (MT, MST, and FST).[19,65,114,115] Pdm is interconnected with posterior parietal (area 7), MT, and PO cortical areas[32,93] and receives input from both the superficial and intermediate superior colliculus.[16,30,41]

In summary, the results of electrophysiological and early anatomical studies indicated a visuotopic map of the entire contralateral field within the PI and adjacent PL region. Central vision is represented laterally and posteriorly, with the fovea represented at the caudal pole of the nucleus. Conversely, far peripheral vision is represented medially and anteriorly, in the region adjacent to the MGN. Both anatomical and electrophysiological data also indicate the existence of second visual map in the lateral part of PL. According to Bender,[16] the visuotopic map in PL encompasses the whole contralateral hemifield. However anatomical data indicate that in the P2 map, unlike the P1 map, striate projections are limited to the representation of the central visual field only.[18] The lower visual field is represented dorsally in each mapped region, and the upper visual field ventrally. In contrast to P1 and P2, P3 does not seem to have a well-defined retinotopic map, although it has yet to be mapped electrophysiologically. Similarly, the electrophysiologically defined Pdm region has a poor retinotopic arrangement.

3.4.1 Correspondence to Neurochemical Pulvinar Subdivisions

The existence of multiple patches of connections does not itself indicate separate subdivisions of the pulvinar, but their correlation with architectonic features can help define valid subdivisions. Most early anatomical studies supported the idea of separate functional fields in the pulvinar, but they did not provide a way to relate these fields to architectonic features of the pulvinar, as none of the described visual fields (P1 to P3) corresponds precisely to the traditional cytoarchitectonic divisions. The subdivisions of the pulvinar defined by immunocytochemistry, especially Cb immunoreactivity, offer a new scheme that could be related to functional organization of the pulvinar. However, the exact contribution of these Cb-immunoreactive cells to visual function remains to be determined.

Most recent studies used the new neurochemical compartments of pulvinar as a framework for locating connections.[8,21,25–27,34,60,66] Correlation of histochemical subdivisions of the pulvinar with connections of striate and extrastriate visual cortex in squirrel,[8] *Cebus*,[26] and macaque monkeys[21,66] indicated that divisions of PI connect differentially with these cortical areas. Thus, V1 and V2 areas were found to have the strongest connections with the most lateral PI subdivisions PIcL (PIL),[21,66] MT as described earlier by Lin and Kaas[17] was the main target of PIm,[8,21] and cortex directly caudal to MT (DLr) had the strongest connections with the medial PIc (PIcM).[8,71] Judging from the retinotopic organization of these areas, some degree of precision in the retinotopy similar to that previously described was suggested within the largest PI subdivision, PIc. Because of their smaller sizes, the retinotopic organization of PIcM and PIm are difficult to assess. The retinotopic organization of PIm is best known from its connection pattern with areas MT and V1 (macaques,[18,19,87,116] owl monkeys,[17,24] squirrel monkeys,[8] marmosets[117,118]). Central vision seems to be represented in its caudal half, with the lower field representation dorsal to the upper field. More peripheral representations appeared to lie mainly in rostral PIm.[8] Connection patterns with the superior colliculus provide additional evidence for retinotopic organization of functional subdivisions of PI in owl[17,25] and macaque monkeys.[60] The topography of SC-pulvinar projections supports the view that central vision is represented dorsally and peripheral vision ventrally for most or all divisions of PI.

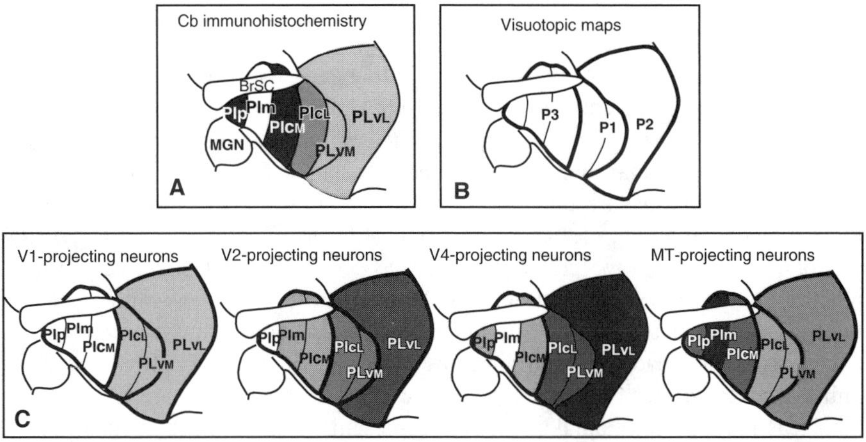

FIGURE 3.4 Relationship of neurochemical pulvinar subdivisions (A) to visuotopic maps (B) and cortical connections with areas V1, V2, V4, and V5/MT (C). Lateral PI (PIcL) and medial PL (PLvm) subdivisions form the P1 map, lateral PL (PLvl) forms the P2 map, and the rest of the pulvinar, posterior (PIp), medial (PIm), and the central medial (PIcM) nuclei, forms P3 map. Different shades of gray in A represent different densities of Cb-positive neurons. Different shades of gray in C indicate the density of connections. Other conventions as in Figure 3.1. (From Adams, M.M. et al., *J. Comp. Neurol.,* 419, 377–393, 2000. With permission.)

Most recently there have been attempts to unify histochemical, connectional, and physiological aspects of pulvinar organization. Probably the most complex and informative is a recent study of Adams et al.,[21] demonstrating functional fields in the pulvinar of macaque monkeys based on estimates of previously described visual field maps and connectional experiments, as well as Cb immunochemistry. In this study, the locations of V1-, V2-, V4-, and MT-projecting neurons were related to the neurochemical pulvinar subdivisions and to three visuotopic fields P1, P2, and P3. This relationship is illustrated in Figure 3.4. The P1 field in rostrolateral PI corresponds to the architectonic PIcL subdivision and extends into adjacent ventromedial portions of PL (PLvm). The P2 field, partially surrounding P1, is located entirely in the ventrolateral portion of PL (PLv), and the P3 field corresponds to posteromedial PI including the PIp, PIm, and PIcM subdivisions. Therefore, calbindin immunostaining, as noted by the authors, marks P3 borders, but cannot be used as a marker for P1 and P2.

In agreement with previous studies, neurons projecting to V1 were found almost exclusively in PIcL (P1), and less frequently in PL (P2). Neurons projecting to V2 overlapped those in PIcL and PL, and there was an additional projection from PIcM and less frequently PIm to V2. Neurons projecting to V4 were observed in PIcL (P1) and PL (P2) and in PIcM and PIp (P3). No cells projecting to V4 were found in PIm (P3), which was very densely connected to area MT. The differential pattern of cortical connectivity of the calbindin-rich and calbindin-poor regions of P3 suggests that Cb may further delineate functional subdivisions in pulvinar. However, multiple patches of projections not only can reflect the presence of separate visuotopically organized subdivision, but also could represent separate modules within a single subdivision.[12]

Similar studies in the New World *Cebus* monkey also revealed three (P1, P2, P3) neurochemically differentiated and visuotopically organized pulvinar regions.[26] As in macaques, P1 and P3 were located in the centrolateral and medial portions of PI. However, the P2 map within the lateral pulvinar of *Cebus* lies entirely dorsal to, rather than partially surrounding the inferior pulvinar. This location is consistent with the results of earlier electrophysiological studies.[15] The locations and orientations of visual maps in the pulvinar that have been described in *Cebus* monkeys[15] are similar to those described in the New World marmoset monkey on the basis of connections with visual areas.[118] A comparison of the representations of visual field in the pulvinar of macaques and marmosets indicates that these animals have the similar visuotopic maps, but their maps are rotated about 90° in relation to one another.[118] Thus, the horizontal meridian is represented ventrodorsally in marmosets, whereas in rhesus monkeys it is represented mediolaterally.

The anatomical studies cited above demonstrated local topographies of cortical connections in the pulvinar. In the search for a global correspondence between connectional and retinal zones (topographies) with the new architectonic pulvinar subdivisions, a model of pulvinar organization was developed.[119] The model considers the cortical connections, the visual zones, and the neurochemical subdivisions of the pulvinar in the horizontal plane. The proposed general topography of pulvinocortical connections follows a rostrolateral to caudomedial shift of the projection zones of successively more rostral cortical areas. For P1 and P2 maps (involving PIcL and ventral PL) in the ventral pulvinar, the projection zones of successively more distant extrastriate cortex with respect to that of V1 along occipitotemporal cortical gradient shift caudomedially. Thus, projections from most occipital area (V1) occupy the rostrolateral pole of lines of isorepresentations in pulvinar, and projections from successive areas (e.g., V4, TEO, and TE) are connected to overlapping, but progressively more caudomedial zones of ventral pulvinar.[33,72,76,84] For the dorsal pulvinar, rostrolateral to caudomedial connectional lines correspond to a dorsal cortical gradient running from medial occipitoparietal to anterior superiotemporal areas.[22,63,87,116] Much of this territory is association cortex (visuosomatosensory parietal or visuoauditory areas in STS) known to be related to the dorsal visual stream of processing. According to the model, the similar connections of dorsal pulvinar and the P3 visual map in the ventral pulvinar with areas of dorsal stream within the STS[65] establish connectional continuity of P3 with the dorsal pulvinar. Thus, the two dorsal and ventral cortical pathways through prestriate cortex may be replicated in the pulvinar, within P1/P2 and P3/dorsal pulvinar domains.[119]

Although this model may indicate some regularities in the visual pulvinar topographies, it certainly oversimplifies the complex nature of pulvinar organization, especially in light of recent histochemical data demonstrating that the patterns of pulvino-cortical connections are more specific than previously thought.

In summary, most recent anatomical and architectonic studies revealed several retinotopic zones (with the possibility of additional zones connected with striate and "near" extrastriate visual areas) within the inferior pulvinar, suggesting that the electrophysiological evidence for visual subdivisions of the inferior pulvinar is incomplete. A single electrophysiological map in the inferior pulvinar may result from recording mainly from its largest subdivision PIc, as suggested by Kaas and Huerta.[28] The few recordings from other parts of PI would not necessarily provide compelling evidence for additional representations, because the vertical approach of electrodes in such studies would not easily reveal separate but parallel representations of the visual field and clear reversals of retinotopic patterns. Additional data are needed to assess directly the potential correspondence between the neurochemical subdivisions and the visuotopic organization of the inferior pulvinar.

3.5 FUNCTIONS

3.5.1 Properties of Pulvinar Neurons

Properties of pulvinar neurons have been studied in Old World macaque[16,30,92,108] and New World *Cebus,*[120] owl,[14] and squirrel[121] monkeys, and have been described as qualitatively similar in all species. Neurons in the inferior and lateral pulvinar have visual properties with receptive fields arranged in precise retinotopic maps of the visual field.[16,18,122] Visual responses of neurons in two visuotopic maps are very similar to each other.[30] Thus, generally these neurons are binocular and respond to visual stimuli presented through either eye.[14,16,120,123] Single and multiunit recording studies showed that PI and PL neurons have well-defined receptive fields with sizes positively correlated with their eccentricity, and most neurons are sensitive to moving stimuli.[30,108,120,121,123] These neurons also may be characterized by sensitivity to stimulus orientation, directional selectivity for motion, or both orientation and direction selectivity; however, the majority of cells are either broadly tuned or nonselective for these attributes.[30] In most cases the properties of neurons are not exceptional for a particular pulvinar subdivision, and a concentration of neurons of a particular type is not clearly characteristic of any subdivision.

The results of experiments of single unit recording during visual behavior[30] suggest that there are functional differences between inferior and lateral pulvinar and dorsomedial pulvinar region (Pdm) and that they most likely contribute to different aspects of visual behavior. Whereas PI and PL have a simple retinotopic organization, Pdm has a poor retinotopic arrangement.[16,30] Neurons in Pdm have many physiological properties that distinguish them from those in PI and PL.[30] Thus, their visual receptive fields are larger, their discharge latencies are longer, and the cells that discharge after saccadic eye movements are more common in Pdm than in PI and PL.[106] A recent study[92] has shown that neurons in lateral PM and PL (involving Pdm) also display preferences for pattern and color stimuli. In fact, these neurons encode both form/color and saccade-related information. In addition to these visual responses, there are nonvisual influences (e.g., auditory) that distinguish Pdm.[120,124] Physiological studies demonstrated that PM neurons have visual, auditory, somatosensory, and oculomotor responses.[92,125–127]

An additional population of neurons with visual responses was found in the caudal part of the lateral pulvinar of macaque called PLγ.[108] Some properties of cells in this pulvinar region caudal to Pdm are quite similar to Pdm. Thus, both have cells with large receptive fields and neither area has a precise retinotopic organization.

However, PLγ in contrast to Pdm frequently has ipsi- and bilateral receptive fields. Felsten et al.[128] described a few cells in Plγ and other parts of the pulvinar that are selective for color stimuli. The observation that Plγ has different connections and physiological properties suggests that it may represent an additional subdivision of the pulvinar.

3.5.2 Role of Corticopulvinar-Cortical Connections

The results of visual cortex and superior colliculus lesions in monkeys have shown that almost all pulvinar neurons are critically dependent on descending input from the cortex.[129] Thus, it seems unlikely that monkey pulvinar can serve primarily as a thalamic relay nucleus in a second visual system, independent of the geniculostriate pathway. Instead, the pulvinar may provide the cortex with another route in addition to the corticocortical pathways, by which it can influence visual processing within the vast expanse of prestriate cortex.

Corticopulvinar-cortical circuitry is acknowledged as an alternative source for intercommunication between visual cortical areas.[18,89,130] There is some evidence that cortical areas that communicate directly are likely to have overlapping connections in the pulvinar, and pulvinar connection fields of areas that are not interconnected mostly avoid each other (separate or interdigitating).[22,32,64] Thus, corticopulvinar-cortical connections tend to retinotopically match direct corticocortical connections. However, given the obvious differences in anatomy and physiology of indirect relays, the idea of pulvinar circuitry as an alternative to cortical circuitry is doubtful. Also, the pattern of corticopulvinar-cortical connections does not seem to compensate for "deficiencies" in the corticocortical network. Then, what is the significance of corticopulvinar-cortical pathways?

Supposedly, corticopulvinar-cortical linkages enable functions that are not an exact duplication of direct corticocortical information transfer, but may play a more regulatory, or coordinating role in cortical signal processing.[84,119] Parts of the pulvinar receiving input from the tectum and from the retina also receive primary afferents from the cortex. Thus, if the one visual area sends afferents to the pulvinar and through the pulvinar to other visual cortical areas, it is likely to provide a modulated signal of major importance to the secondary areas. From this perspective a corticopulvinar-cortical loop is less an alternative channel for stimulus-specific information, but more a coordinator of activity among groups of cortical areas that are locally interconnected.

The pulvinar may also provide a more secure, indirect pathway for long-range cortical communication. It is known that the resonant interaction between the thalamus and cortex acts to synchronize neural firing so both structures may operate as a unified oscillatory system.[131,132] Rhythmicity of thalamic neurons enables the corticopulvinar-cortical link to facilitate long-range synchronization of cortical activity and, in consequence, strengthened neural communication.[133,134]

The component of corticocortical processing involving the transthalamic route must have functional consequences. Whatever the nature of messages that the pulvinar region sends to the cortex, they must play some role in defining the function of the recipient cortical areas.

3.5.3 Role of Pulvinar in Behavior

Despite anatomical and physiological evidence of the affiliation of the pulvinar with the visual system, surprisingly little is known about its functions. The poor understanding of pulvinar functions is partly because most lesion studies of the pulvinar have failed to produce clear and pronounced functional deficits.[135–137]

As described earlier, the pulvinar receives major afferents from the visual cortex and superior colliculus. Visual cortex is related to detailed analysis for visual perception, whereas the superior colliculus is involved in the use of visual stimuli as targets for orienting movements. As early physiological studies have shown that visual responses in pulvinar are dependent on striate cortex and not the colliculus,[16,129] the significance of the contribution of collicular pathway to the pulvinar has been uncertain. Now, there is growing evidence for modulation of activity of pulvinar neurons by eye movements, an effect that may be mediated by collicular inputs. Saccade-related activity has been reported in the inferior pulvinar (PI) of macaques.[30,106,138] Strong visual responses of pulvinar neurons to moving stimuli during fixation are suppressed during eye movements.[127,139] All these data and the fact that pulvinar lesions may produce an abnormal pattern of eye movements[140,141] suggest that the pulvinar may be involved in the integration of visual and oculomotor information.

Recent electrophysiological studies pointed to a role for the pulvinar in visual attention.[107] Monkeys and humans with pulvinar damage can no longer determine what is important in a visual scene, that is, what is visually salient.[142,143] As a part of attentional mechanisms, the pulvinar has been proposed to contribute to generation of visual salience by both suppressing information that is irrelevant and facilitating information that has behavioral importance.[6,7,30,107,144] Cortical connections, functional characteristics of pulvinar neurons, and neurophysiological studies in monkeys and positron emission tomography (PET) studies in humans support a role for the pulvinar in visual attention, including attentional filtering in the presence of visual distractors[145] and selective attention to stimulus shape, speed, and color.[146]

The differing histochemical features of major pulvinar divisions, as well as evidence for differences in connections and properties of their neurons, all support the contention that each division may have a functionally distinct role. The properties of neurons in PI and PL suggest that these parts deal with the visual effects of eye movements.[106] Because of its topographical projections to striate and early extrastriate cortex the retinotopically organized inferior pulvinar and the adjoining ventral portion of the lateral pulvinar may be involved in focusing attention selectively on the form or color of a visual stimulus. Thus, ventral pulvinar may play a role in processes such as visual fixation, visual pattern discrimination, and attending to salient stimuli.[107,125,147]

In contrast, the dorsal portions of the lateral pulvinar and medial pulvinar that have no clear retinotopic organization and contain neurons that are directionally sensitive seem to have a different functional role in processes such as visuospatial (directed) attention and spatially directed reaching.[30,92,125] Their connections with posterior parietal and frontal cortices and intermediate layers of the superior colliculus, as well as recent PET and MRI studies,[148–150] further suggest that the dorsal region of pulvinar (especially PM) is likely to be related to directed attention and visual salience. Moreover, the medial pulvinar connected with frontal cortex and amygdala may play a role in memory function, which may be secondary to its role in controlling or regulating attention.[151] The differential connectivity of the PM/PL and PI nuclei with the dorsal and ventral extrastriate regions may also play a role in this functional differentiation of visual processes.

Involvement of the pulvinar in the control of eye movements, the selection of salient stimuli, and the modulation of attention, suggests that the pulvinar may act to gate incoming information to the cortex.[6,141] The most recent model of visual attention pointed out that the role of pulvinar neurons in the corticopulvinar-cortical pathway may be to dynamically modify the strength of intracortical connections.[91] These modified connections are used to route information through higher cortical areas. The model proposes that the pulvinar provides control signals for routing attended information through the cortex.

3.6 SUMMARY

The mammalian thalamus and cortex evolved together, and they are extensively interconnected. Basically, all the cortex relies heavily on the thalamus for its input, and without each other, neither the thalamus nor cortex would function. Understanding the role that the pulvinar plays in the processing of visual information is critical to understanding cortical function.

The pulvinar complex has been classically divided into four subdivisions, with three of them involved in visual processes. The inferior and lateral pulvinar are closely connected to primary visual centers (retina, striate cortex, superficial layers of the superior colliculus), whereas the medial pulvinar is related more to association cortex (parietal and frontal cortex). Except their affiliation with the visual system, little has been known about these pulvinar components. Our understanding of the functional organization of the pulvinar has increased dramatically during the last decade. Progress is mostly due to morphological studies combining axonal tracing with immunohistochemistry. New histo- and immunohistochemical techniques have been used to divide the major pulvinar divisions into smaller units, showing that the fundamental organization of the pulvinar is more complex than previously thought. At present, nine subdivisions have been identified within the pulvinar. These subdivisions, defined by appearance and connections, cross the boundaries of the originally defined pulvinar divisions. The different histochemical features of proposed nuclei, as well as evidence for differences in connections and properties of neurons, all support the contention that each subdivision has a functionally distinct role. More subdivisions imply more channels of processing, and further differentiation and specialization within channels. The complexity of the connections and patterns of neurochemical activity within the pulvinar subdivisions suggest that functionally and behaviorally meaningful components will be found within even smaller regions of each subdivision.

The pulvinar is not a simple passive relay of information to the cortex, but instead is involved in many dynamic processes that significantly alter the nature of the information relayed to cortex. Controlled by a complex array of modulatory and activatory inputs (cortical and subcortical), the pulvinar may play a crucial role in modifying the messages that pass from one cortical area to another and in coordinating corticocortical communication. A coordinating or regulatory action upon the cortex is consistent with a role in visual attention, which has been a recurrent theme in functional hypotheses regarding the pulvinar.

ACKNOWLEDGMENTS

The author thanks Dr. Soumya Iyengar for helpful suggestions and valuable comments on this manuscript.

REFERENCES

1. Chalupa, L.M., The visual function of pulvinar, in *The Neural Basis of Visual Function*, A.G. Leventhal, Ed., Macmillan Press, London, 1991, 140–159.
2. Rakic, P., Embryonic development of the pulvinar-LP complex in man, in *The Pulvinar-LP complex,* M. Riklan, I.S. Cooper, and P. Rakic, Eds., Charles C Thomas, Springfield, IL, 1974, 3–30.
3. Browne, B. and Simmons, R.M., Quantitative studies of the evolution of the thalamus in primates, *J. Hirnforsch.*, 25, 261, 1984.
4. Harting, J.K., Glendenning, K.K., Diamond, I.T., and Hall, W.C., Evolution of the primate visual system: anterograde degeneration studies of the tecto-pulvinar system, *Am. J. Phys. Anthropol.*, 38, 383–392, 1973.
5. Chalupa, L.M., A review of cat and monkey studies implicating the pulvinar in visual function, *Behav. Biol.,* 20, 149–167, 1977.
6. Desimone, R.M., Wessinger, L., Thomas, L., and Schneider, W., Attentional control of visual perception: cortical and subcortical mechanisms, *Cold Spring Harbor Symp. Quant. Biol.*, 55, 963–971, 1990.
7. Posner, M.I. and Petersen, S.E., The attention system of the human brain, *Annu. Rev. Neurosci.*, 13, 25–42, 1990.
8. Cusick, C.G., Scripter, J.L., Darensbourg, J.G., and Weber, J.T., Chemoarchitectonic subdivisions of the visual pulvinar in monkeys and their connectional relations with the middle temporal and rostral dorsolateral visual areas, MT and DLr, *J. Comp. Neurol.,* 336, 1–30, 1993.
9. Robinson, D.L. and Cowie, R.J., The primate pulvinar: structural, functional and behavioral components of visual salience, in *Thalamus: Organization and Functions*, M. Steriade, E.G. Jones, and D.A. McCormick, Eds., Elsevier, Amsterdam, 1997, 53–91.
10. Walker, A.E., *The Primate Thalamus*, University of Chicago Press, Chicago, 1938.
11. Olszewski, P.J., *The Thalamus of the* Macaca mulatta, Karger, Basel, 1952.
12. Gutierrez, C., Yaun, A., and Cusick, C.G., Neurochemical subdivisions of the inferior pulvinar in macaque monkeys, *J. Comp. Neurol.*, 363, 545–562, 1995.
13. Stepniewska, I. and Kaas, J.H., Architectonic subdivisions of the inferior pulvinar in New World and Old World monkeys, *Vis. Neurosci.,* 14, 1043–1060, 1997.
14. Allman, J.M., Kaas, J.H., Lane, R.H., and Miezin, F.M., A representation of the visual field in the inferior nucleus of the pulvinar in the owl monkey (*Aotus trivirgatus*), *Brain Res.*, 40, 291–302, 1972.

15. Gattas, R., Oswaldo-Cruze, E., and Sousa, A.P.B., Visuotopic organization of the *Cebus* pulvinar: A double representation of the contralateral hemifield, *Brain Res.*, 152, 1, 1978.
16. Bender, D.B., Retinotopic organization of macaque pulvinar, *J. Neurophysiol.,* 46, 672–693, 1981.
17. Lin, C.-S. and Kaas, J.H., The inferior pulvinar complex in owl monkeys: architectonic subdivisions and patterns of input from the superior colliculus and subdivisions of visual cortex, *J. Comp. Neurol.*, 187, 655–678, 1979.
18. Ungerleider, L.G., Galkin, T.W., and Mishkin, M., Visuotopic organization of projections from striate cortex to inferior and lateral pulvinar in rhesus monkey, *J. Comp. Neurol.,* 217, 137–157, 1983.
19. Ungerleider, L.G., Desimone, R., Galkin, T.W., and Mishkin, M., Subcortical projections of area MT in the macaque, *J. Comp. Neurol.*, 223, 368–386, 1984.
20. Gray, D., Gutierrez, C., and Cusick, C.G., Neurochemical organization of inferior pulvinar complex in squirrel monkeys and macaques revealed by acetylcholinesterase histochemistry, calbindin and Cat-301 immunostaining, and *Wisteria floribunda* agglutinin binding, *J. Comp. Neurol.,* 409, 452–468, 1999.
21. Adams, M.M., Hof, P.R., Gattas, R., Webster, M.J., and Ungerleider, L.G., Visual cortical projections and chemoarchitecture of macaque monkey pulvinar, *J. Comp. Neurol.*, 419, 377–393, 2000.
22. Gutierrez, C., Cola, M.G., and Cusick, C., Neurochemical and connectional organization of the dorsal pulvinar complex in monkeys, *J. Comp. Neurol.,* 419, 61–86, 2000.
23. Cola, M.G., Gray, D.N., Seltzer, B., and Cusick, C.G., Human thalamus: neurochemical mapping of inferior pulvinar complex, *NeuroReport,* 12, 3733–3738, 1999.
24. Lin, C.S. and Kaas, J.H., Projections from the medial nucleus of the inferior pulvinar complex to the middle temporal area of the visual cortex, *Neuroscience*, 5, 2219–2228, 1980.
25. Stepniewska, I., Qi, H.-X., and Kaas, J.H., Do superior colliculus projection zones in the inferior pulvinar project to MT in primates? *Eur. J. Neurosci.*, 11, 469–480, 1999.
26. Soarez, J.G.M., Gattass, R., Souza, A.P.B., Rosa, M.G.P., Fiorani, M., Jr., and Brandao, B.L., Connectional and neurochemical subdivisions of the pulvinar in *Cebus* monkeys, *Vis. Neurosci.*, 18, 25–41, 2001.
27. Beck, P.D. and Kaas, J.H., Thalamic connections with the dorsomedial visual cortical area (DM), *J. Comp. Neurol.*, 396, 381–398, 1998.
28. Kaas, J.H. and Huerta, M.F., The subcortical visual system of primates, in *Comparative Primate Biology*: *Neurosciences*, H.P. Steklis, Ed., Alan R. Liss, New York, 1988, 327–391.
29. Jones, E.G., *The Thalamus*, Plenum Press, New York, 1985.
30. Petersen, S.E., Robinson, D.L., and Keys, W., Pulvinar nuclei of the behaving rhesus monkey: visual responses and their modulation, *J. Neurophysiol.*, 54, 867–886, 1985.
31. Trojanowski, J.Q. and Jacobson, S., Medial pulvinar afferents to frontal eye field in rhesus monkey demonstrated by horseradish peroxidase, *Brain Res.*, 80, 395–411, 1974.
32. Baleydier, C. and Mauguiere, F., Network organization of connectivity between parietal area 7, posterior cingulate cortex and medial pulvinar nucleus: a double fluorescent tracer study in monkey, *Exp. Brain Res.*, 66, 385–393, 1987.
33. Baleydier, C. and Morel, A., Segregated thalamocortical pathways to inferior parietal and inferotemporal cortex in macaque monkey, *Vis. Neurosci.*, 8, 391–406, 1992.

34. Cavada, C., Company, T., Hernandez-Gonzales, A., and Reinoso-Suarez, F., Acetylcholinesterase histochemistry in the macaque thalamus reveals territories selectively connected to frontal, parietal and temporal association cortices, *J. Chem. Neuroanat.*, 8, 245–257, 1995.
35. Campos-Ortega, J.A., Hayhow, W.R., and Cluver, P.F.V., A note on the problem of retinal projections to the inferior pulvinar nucleus of primates, *Brain Res.*, 22, 126–130, 1970.
36. Hendrickson, A., Wilson, M.E., and Toyne, M.J., The distribution of optic nerve fibers in *Macaca mulatta*, *Brain Res.*, 23, 425—427, 1970.
37. Trojanowski, J.O. and Jacobson, S., Peroxidase labeled subcortical afferents to pulvinar in rhesus monkey, *Brain Res.*, 97, 144–150, 1975.
38. Itaya, S.K. and Van Hoesen, G.W., Retinal projections to the inferior and medial pulvinar nuclei in the old-world monkey, *Brain. Res.*, 269, 223–230, 1983.
39. Mizuno, N., Itoh, K., Uchida, K., Uemura-Sumi, M., and Matsushima, R., A retino-pulvinar projection in the macaque monkey as visualized by the use of anterograde transport of horseradish peroxidase, *Neurosci. Lett.*, 30, 199–203, 1982.
40. Nakagawa, S. and Tanaka, S., Retinal projections to the pulvinar nucleus of the macaque monkey: a reinvestigation using autoradiography, *Exp. Brain Res.*, 57, 151–157, 1984.
41. Benevento, L.A. and Standage, G.P., The organization of projections of the retinorecipient and nonretinorecipient nuclei of the pretectal complex and layers of the superior colliculus to the lateral pulvinar and medial pulvinar in the macaque monkey, *J. Comp. Neurol.*, 217, 307–336, 1983.
42. O'Brien, B.J., Abel, P.L., and Olavarria, J.F., The retinal input to calbindin-D28k-defined subdivisions in macaque inferior pulvinar, *Neurosci. Lett.*, 312, 145–148, 2001.
43. Rodman, H.R., Gross, C.G., and Albrigt, T.D., Afferent basis of visual response properties in area MT of the macaque. I. Effects of striate cortex removal, *J. Neurosci.*, 9, 2033–2050, 1989.
44. Rodman, H.R, Gross, C.G., and Albrigt, T.D., Afferent basis of visual response properties in area MT of the macaque. Effects of superior colliculus removal, *J. Neurosci.*, 10, 1154–1164, 1990.
45. Collins, C.E., Lyon, D.C., and Kaas, J.H., Responses of neurons in MT after long-standing lesions of V1 in adult New World monkeys, *J. Neurosci.*, 23, 2251–2264, 2003.
46. Cowey, A., Stoerig, P., and Bannister, M., Retinal ganglion cells labelled from the pulvinar nucleus in macaque monkeys, *Neuroscience*, 61, 691–705, 1994.
47. Stoering, P. and Cowey, A., Increment-threshold spectral sensitivity in blindsight: evidence for colour opponency, *Brain*, 114, 1487–1512, 1991.
48. Stoering, P. and Cowey, A., Wavelength discrimination in blindsight, *Brain*, 115, 425–444, 1992.
49. Mathers, L.H., Tectal projections to the posterior thalamus of the squirrel monkey, *Brain Res.*, 35, 295–298, 1971.
50. Benevento, L.A. and Fallon, J.H., The ascending projections of the superior colliculus in the rhesus monkey (*Macaca mulatta*), *J. Comp. Neurol.*, 160, 339–362, 1975.
51. Harting, J.K., Huerta, M.F., Frankfurter, A.J., Strominger, N.L., and Royce, G.J., Ascending pathways from the monkey superior colliculus: an autoradiographic analysis, *J. Comp. Neurol.*, 192, 853–882, 1980.
52. Glendenning, K.K., Hall, J.A., Diamond, I.T., and Hall, W.C., The pulvinar nucleus of *Galago senegalensis*, *J. Comp. Neurol.*, 161, 419–458, 1975.

53. Raczkowski, D. and Diamond, I.T., Cells of origin of several afferent pathways from the superior colliculus in *Galago senegalensis*, *Brain Res.*, 146, 351–357, 1978.
54. Huerta, M.F. and Harting, J.K., Sublamination within the superficial gray layer of the squirrel monkey: an analysis of the tectopulvinar projection using anterograde and retrograde transport methods, *Brain Res.*, 261, 119–126, 1983.
55. Molotchnikoff, S., Casanova, C., and Cerat, A., The consequences of the superior colliculus output on lateral geniculate and pulvinar responses, *Prog. Brain Res.*, 75, 67–74, 1988.
56. Harting, J.K., Casagrande, V.A., and Weber, J.T., The projection of the primate superior colliculus upon the dorsal lateral geniculate nucleus: autoradiographic demonstration of interlaminar distribution of tectogeniculate axons, *Brain Res.*, 150, 593–599, 1978.
57. Huerta, M.F. and Harting, J.K., The mammalian superior colliculus: studies of its morphology and connections, in *Comparative Neurology of the Optic Tectum*, H. Vanegas, Ed., Plenum Press, New York, 1984, 687–773.
58. Partlow, G.D., Colonnier, M., and Szabo, J., Thalamic projections of the superior colliculus in the rhesus monkey, *Macaca mulatta*, a light and electron microscopy study, *J. Comp. Neurol.*, 171, 285–318, 1977.
59. Lysakowski, A., Standage, G.P., and Benevento, L.A., Histochemical and architectonic differentiation of zones of pretectal and collicular inputs to the pulvinar and lateral geniculate nucleus in the macaque, *J. Comp. Neurol.*, 250, 431–448, 1986.
60. Stepniewska, I., Qi, H.-X., and Kaas, J.H., Projections of the superior colliculus to subdivisions of the inferior pulvinar in New World and Old World monkeys, *Vis. Neurosci.*, 17, 529–549, 2000.
61. Hutsler, J.J. and Chalupa, L.M., Substance P immunoreactivity identifies a projection from the cat's superior colliculus to the principal tectorecipient zone of the lateral posterior nucleus, *J. Comp. Neurol.*, 312, 379–390, 1991.
62. Wall, P.D. and Fitzgerald, M., If substance P fulfills the criteria as a neurotransmitter in somatosensory afferents, what might be its function? *Ciba Found. Symp.*, 91, 249–266, 1982.
63. Romanski, L.M., Giguere, M., Bates, J.F., and Goldman-Rakic, P.S., Topographic organization of medial pulvinar connections with the prefrontal cortex in the rhesus monkey, *J. Comp Neurol.*, 379, 313–331, 1997.
64. Benevento, L.A. and Davis, B., Topographical projections of the prestriate cortex to the pulvinar nuclei in the macaque monkey: an autoradiographic study, *Exp. Brain Res.*, 30, 405–424, 1977.
65. Boussaoud, D., Desimone, R., and Ungerleider, L.G., Subcortical connections of visual areas MST and FST in macaques, *Vis. Neurosci.*, 9:291–302, 1992.
66. Gutierrez, C. and Cusick, C.G., Area V1 in macaque monkeys projects to multiple histochemically defined subdivisions of the inferior pulvinar complex, *Brain Res.*, 765, 349–356, 1997.
67. Rezak, M. and Benevento, L.A., A comparison of the organization of the projections of the lateral geniculate nucleus, the inferior pulvinar and adjacent lateral pulvinar to primary visual cortex (area 17) in the macaque monkey, *Brain Res.*, 167, 19–40, 1979.
68. Ungerleider, L.G. and Mishkin, M., Two cortical visual systems, in *Analysis of Visual Behavior*, D.J. Ingle, M.A. Goodale, and R.J.W. Mansfield, Eds., MIT Press, Cambridge, MA, 1982, 549–586.
69. Felleman, D.J. and Van Essen, D.C., Distributed hierarchical processing in the primate cerebral cortex, *Cereb. Cortex*, 1, 1–47, 1991.

70. Baizer, J.S., Desimone, R., and Ungerleider, L.G., Comparison of subcortical connections of inferior temporal and posterior parietal cortex in monkeys, *Vis. Neurosci.*, 10, 59–72, 1993.
71. Steele, G.E. and Weller, R.E., Subcortical connections of subdivisions of inferior temporal cortex in squirrel monkeys, *Vis. Neurosci.*, 10, 563–583, 1993.
72. Webster, M.J., Bachevalier, J., and Ungerleider, L.G., Subcortical connections of inferior temporal areas TE and TEO in macaque monkeys, *J. Comp. Neurol.*, 335, 73–91, 1993.
73. Kennedy, H. and Bullier, J., A double-labeling investigation of the afferent connectivity to cortical areas V1 and V2 of the macaque monkey, *J. Neurosci.*, 5, 2815–2830, 1985.
74. Selemon, L.D. and Goldman-Rakic, P.S., Common cortical and subcortical targets of the dorsolateral prefrontal and posterior parietal cortices in rhesus monkey: evidence for a distributed neural network subserving spatially guided behavior, *J. Neurosci.*, 8, 4049–4068, 1988.
75. Rockland, K.S., Two types of corticopulvinar terminations: round (type 2) and elongate (type 1), *J. Comp Neurol.*, 368, 57–87, 1996.
76. Yeterian, E.H. and Pandya, D.N., Corticothalamic connections of extrastriate visual areas in rhesus monkeys, *J. Comp. Neurol.*, 378, 562–585, 1997.
77. Weller, R.E., Steele, G.E., and Kaas, J.H., Pulvinar and other subcortical connections of dorsolateral visual cortex in monkeys, *J. Comp. Neurol.*, 450, 215–240, 2002.
78. Burton, H. and Jones, E.G., The posterior thalamic region and its cortical projections in the New World and Old World monkeys, *J. Comp. Neurol.*, 168, 249–301, 1976.
79. Acuna, C., Cudeiro, J., Gonzales, F., Alonso, J.M., and Perez, R., Lateral-posterior and pulvinar reaching cells comparison with parietal area 5a: a study in behaving *Macaca nemestrina*, *Exp. Brain Res.*, 82, 158–166, 1990.
80. Pandya, D.N., Anatomy of the auditory cortex, *Rev. Neurol.*, 151, 486–494, 1995.
81. Hackett, T.A., Stepniewska, I., and Kaas, J.H., Thalamocortical connections of the auditory parabelt cortex in macaque monkeys, *J. Comp. Neurol.*, 400, 271–286, 1998.
82. Baleydier, C. and Mauguiere, F., Anatomical evidence for medial pulvinar connections with posterior cingulate cortex, the retrosplenial area, and the posterior parahippocampal gyrus in monkeys, *J. Comp. Neurol.*, 232, 219–228, 1985.
83. Barbas, H., Tenion, T.H., and Dermon, C.R., Diverse thalamic projections to the prefrontal cortex in the rhesus monkey, *J. Comp. Neurol.*, 313, 65–94, 1991.
84. Shipp, S., Corticopulvinar connections of areas V5, V4, and V3 in the macaque monkey: a dual model of retinal and cortical topographies, *J. Comp. Neurol.*, 439, 469–490, 2001.
85. Asanuma, C., Andersen, R.A., and Cowan, W.M., The thalamic relations of the caudal inferior parietal lobule and the lateral prefrontal cortex in monkeys: divergent cortical projections from cell clusters in the medial pulvinar nucleus, *J. Comp. Neurol.*, 241, 357–381, 1985.
86. Huerta, M.F., Krubitzer, L.A., and Kaas, J.H., The frontal eye field as defined by intracortical microstimulation in squirrel monkeys, owl monkeys, and macaque monkeys. I. Subcortical connections, *J. Comp. Neurol.*, 253, 415–439, 1986.
87. Yeterian, E.H. and Pandya, D.N., Thalamic connections of the cortex of the superior temporal sulcus in the rhesus monkey, *J. Comp. Neurol.*, 282, 80–97, 1989.
88. Hardy, S.G. and Lynch, J.C., The spatial distribution of pulvinar neurons that project to two subregions of the inferior parietal lobule in the macaque, *Cereb. Cortex*, 2, 217–230, 1992.

89. Benevento, L.A. and Rezak, M., The cortical projections of the inferior pulvinar and adjacent lateral pulvinar in rhesus monkey (*Macaca mulatta*): an autoradiographic study, *Brain Res.*, 108, 1–24, 1976.
90. Milner, A.D. and Goodale, M.A., Visual pathways to perception and action, *Prog. Brain Res.*, 95, 317–337, 1993.
91. Olshausen, B.A., Anderson, C.H., and Van Essen, D.C., A neurobiological model of visual attention and invariant pattern recognition based on dynamic routing of information, *J. Neurosci.*, 11, 4700–4719, 1993.
92. Benevento, L.A. and Port, J.D., Single neurons with both form/color differential responses and saccade-related responses in the nonretinotopic pulvinar of the behaving macaque monkey, *Vis. Neurosci.*, 12, 523–544, 1995.
93. Trojanowski, J.O. and Jacobson, S., Areal and laminar distribution of some pulvinar cortical efferents in rhesus monkey, *J. Comp. Neurol.*, 169, 371–392, 1976.
94. Ogren, M.P. and Hendrickson, A.E., The distribution of pulvinar terminals in area 17 and 18 of the monkey, *Brain Res.*, 137, 343–350, 1977.
95. Wong-Riley, M.T.T., Connections between the pulvinar nucleus and the prestriate cortex in the squirrel monkey as revealed by peroxidase histochemistry and autoradiography, *Brain Res.*, 134, 249–267, 1977.
96. Curcio, C.A. and Harting, J.K., Organization of pulvinar afferents to area 18 in the squirrel monkey: evidence for stripes, *Brain Res.*, 143, 155–161, 1978.
97. Lund, J.S., Hendrickson, A.E., Ogren, M.P., and Tobin, E.A., Anatomical organization of primate visual cortex area VII, *J. Comp. Neurol.*, 202, 19–45, 1981.
98. Bonin, G. and Bailey, P., *The Neocortex of* Macaca mulatta, University of Illinois Press, Urbana, 1947.
99. Lynch, J.C., Attallah, H., and Ma, T.P., Cytoarchitecture of the primate medial pulvinar nucleus, *Soc. Neurosci. Abstr.,* 20, 145, 1994.
100. Rafols, J.A., Attallah, H., and Ma, T.P., Networks of pulvinar local circuit neurons in primate, *Soc. Neurosci. Abstr.*, 20, 563, 1994.
101. Ogren, M.P. and Hendrickson, A.E., The structural organization of the inferior and lateral subdivisions of the *Macaca* monkey pulvinar, *J. Comp. Neurol.*, 188, 147–178, 1979.
102. Ma, T.P., Lynch, J.C., Donahoe, D.K., Attallah, H., and Rafols, J.A., Organization of the medial pulvinar nucleus in the macaque, *Anat. Rec.*, 250, 220–237, 1998.
103. Guillery, R.W., Patterns of synaptic interconnections in the dorsal lateral geniculate nucleus of the cat and monkey. A brief review, *Vision Res.*, 3, 211–227, 1971.
104. Rockland, K.S., Further evidence for two types of corticopulvinar neurons, *NeuroReport*, 5, 1865–1868, 1994.
105. Weller, R.E. and Davis, M.D., Corticopulvinar terminations from the caudal subdivision of the dorsolateral visual area in squirrel monkeys, *Soc. Neurosci. Abstr.*, 549.2, 2000.
106. Robinson, D.L., Petersen, S.E., and Keys, W., Saccade-related activity in the pulvinar nuclei of the behaving rhesus monkey, *Exp. Brain Res.*, 62, 625–634, 1986.
107. Robinson, D.L. and Petersen, S.E., The pulvinar and visual salience, *TINS*, 15, 127–132, 1992.
108. Benevento, L.A. and Miller, J., Visual responses of single neurons in the caudal lateral pulvinar of the macaque monkey, *J. Neurosci.*, 1, 1268–1278, 1981.
109. Campos-Ortega, J.A. and Hayhow, W.R., On the organization of the visual cortical projection to the pulvinar in *Macaca mulatta*, *Brain Behav. Evol.*, 6, 394–423, 1972.
110. Hollander, H., Projections form the striate cortex to the diencephalon in the squirrel monkey (*Saimiri sciureus*): a light microscopic radioautographic study following intracortical injection of ^{3}H-leucine, *J. Comp. Neurol.*, 155, 425–440, 1974.

111. Ogren, M.P. and Hendrickson, A.E., Pathways between striate cortex and subcortical regions in *Macaca mulatta* and *Saimiri sciureus*: evidence for reciprocal pulvinar connections, *Exp. Neurol.*, 53, 780–800, 1976.
112. Graham, J., Lin, C.-S., and Kaas, J.H., Subcortical projections of six visual cortical areas in the owl monkey, *Aotus trivirgatus*, *J. Comp. Neurol.*, 187, 557–580, 1979.
113. Graham, J., Some topographic connections of the striate cortex with subcortical structures in *Macaca fascicularis*, *Exp. Brain Res.*, 47, 1–14, 1982.
114. Maunsell, J.H.R. and Van Essen, D.C., The connections of the middle temporal visual area (MT) and their relationship to a cortical hierarchy in the macaque monkey, *J. Neurosci.*, 3, 2563–2586, 1983.
115. Standage, G.P. and Benevento, L.A., The organization of connections between the pulvinar and visual area MT in the macaque monkey, *Brain Res.*, 262, 288–294, 1983.
116. Yeterian, E.H. and Pandya, D.N., Corticothalamic connections of the superior temporal sulcus in rhesus monkeys, *Exp. Brain Res.*, 83, 268–284, 1991.
117. Spatz, W.B., Thalamic and other subcortical projections to area MT (visual area of superior temporal sulcus) in the marmoset *Callithrix jacchus*, *Brain Res.*, 99, 129–134, 1975.
118. Dick, A., Kaske, A., and Creutzfeld, O.D., Topographical and topological organization of the thalamocortical projection to the striate and prestriate cortex in the marmoset (*Callithrix jacchus*), *Exp. Brain Res.*, 844, 233–253, 1991.
119. Shipp, S., The functional logic of cortico-pulvinar connections. *Philos. Trans. R. Soc. London B*, 2002 (published online).
120. Gattas, R., Oswaldo-Cruz, E., and Sousa, A.P.B., Visual receptive fields of units in the pulvinar of cebus monkey, *Brain Res.*, 160, 413–430, 1979.
121. Mathers, L.H. and Rappisardi, S.C., Visual and somatosensory receptive fields of neurons in the squirrel monkey pulvinar, *Brain Res.*, 64, 65–83, 1973.
122. Robinson, D.L. and McClurkin, J.W., The visual superior colliculus and pulvinar, in *The Neurobiology of Saccadic Eye Movements. Review of Occulomotor Research*, R.H. Wurtz and M.E. Goldberg, Eds., Elsevier, Amsterdam, 1989, 337–360.
123. Bender, D.B., Receptive field properties of neurons in the macaque inferior pulvinar, *J. Neurophysiol.*, 48, 1–17, 1982.
124. Yirmiya, R. and Hocherman, S., Auditory- and movement-related neural activity interact in the pulvinar of the behaving rhesus monkey, *Brain Res.*, 402, 93–102, 1987.
125. Acuna, C., Gonzales, F., and Dominguez, R., Sensorimotor unit activity related to intention in the pulvinar of behaving *Cebus apella* monkeys, *Exp. Brain Res.*, 52, 411–422, 1983.
126. Cudeiro, J., Gonzales, F., Perez, R., Alonso, J.M., and Acuna, C., Does the pulvinar-LP complex contribute to motor programming? *Brain Res.*, 484, 367–370, 1989.
127. Robinson, D.L., McClurkin, J.W., Kertzman, C., and Petersen, S.E., Visual responses of pulvinar and collicular neurons during eye movements of awake, trained macaques, *J. Neurophysiol.*, 66, 485–496, 1991.
128. Felsten, G., Benevento, L.A., and Burman, D., Opponent-color responses in macaque extrageniculate visual pathways: the lateral pulvinar, *Brain Res.*, 288, 363–367, 1983.
129. Bender, D.B., Visual activation of neurons in the primate pulvinar depends on cortex but not colliculus, *Brain Res.*, 297, 258–226, 1983.
130. Sherman, S.M. and Guillery, R.W., Functional organization of thalamocortical relays, *J. Neurophysiol.*, 76, 1367–1395, 1996.
131. Steriade, M., Synchronized activities of coupled oscillators in the cerebral cortex and thalamus at different levels of vigilance, *Cereb. Cortex*, 7, 583–604, 1997.

132. Steriade, M., Corticothalamic resonance, states of vigilance and mentation, *Neuroscience*, 101, 243–276, 2000.
133. Miller, R., Cortico-thalamic interplay and the security of operation of neural assemblies and temporal chains in the cerebral cortex, *Biol. Cybern.*, 75, 263–275, 1996.
134. Miller, R., Wheels within wheels: circuits for integration of neural assemblies on small and large scales, in *Cortical Areas: Unity and Diversity*, R. Miller and A. Schultz, Eds., Taylor & Francis, London, 2002, 423–458.
135. Mishkin, M., Cortical visual areas and their interaction, in *Brain and Human Behavior*, A.G. Karczmar and J.C. Eccles, Eds., Springer, Berlin, 1972, 187–208.
136. Ungerleider, L.G. and Pribram, K.H., Inferotemporal versus combined pulvinar-prestriate lesions in the rhesus monkey: effects on color, object and pattern discrimination, *Neuropsychologia*, 15, 481–498, 1977.
137. Ungerleider, L.G., Ganz, L., and Pribram, K.H., Size constancy in rhesus monkeys: effects of pulvinar, prestriate and inferotemporal lesions, *Exp. Brain Res.*, 27, 251–269, 1977.
138. Robinson, D.L., McClurkin, J.W., and Kertzman, C., Orbital position and eye movement influences on visual responses in the pulvinar nuclei of the behaving macaque, *Exp. Brain Res.*, 82, 235–246, 1990.
139. Robinson, D.L. and Petersen, S.E., Response of pulvinar neurons to real and self induced stimulus movement, *Brain Res.*, 338, 392–394, 1985.
140. Ungerleider, L.G. and Christensen, C.A., Pulvinar lesions in monkeys produce abnormal eye movements during visual discrimination training, *Brain Res.*, 136, 189–196, 1977.
141. Ungerleider, L.G. and Christensen, C.A., Pulvinar lesions in monkeys produce abnormal scanning of a complex visual array, *Neuropsychologia*, 17, 493–50, 1979.
142. Ogren, M.P., Mateer, C., and Wyler, A.R., Alterations in visually related eye movements following left pulvinar damage in man, *Neuropsychologia*, 22, 187–196, 1984.
143. Zihl, J. and Von Cramon, D., The contribution of the "second" visual system to directed visual attention in man, *Brain*, 102, 835–856, 1979.
144. Robinson, D.L., Functional contribution of the primate pulvinar, in *Progress in Brain Research*, T.P. Hicks, S. Molotshcnikoff, and T. Ono, Eds., Elsevier Science Publishers, Amsterdam, 1993, 371–379.
145. LaBerge, D. and Buchsbaum, M.S., Positron emission tomographic measurements of pulvinar activity during an attention task, *J. Neurosci.*, 10, 613–619, 1990.
146. Corbetta, M., Miezin, F.M., Dobmeyer, S., Shulman, G.L., and Petersen, S.E., Selective and divided attention during visual discrimination of shape, color, and speed: functional anatomy by positron emission tomography, *J. Neurosci.*, 11, 2383–2402, 1991.
147. Chalupa, L.M., Coyle, R.S., and Lindsey, D.B., Effect of pulvinar lesions on visual pattern discrimination in monkeys, *J. Neurophysiol.*, 39, 354–369, 1976.
148. Andersen, R.A., The role of the inferior parietal lobule in spatial perception and visual-motor integration, in *The Handbook of Physiology, The Nervous System, Higher Functions of the Brain*, F. Plum, V.B. Mountcastle, and S.R. Geiger, Eds., American Physiological Society, Bethesda, MD, 1987, 483–518.
149. Mesulam, M.-M., Large scale neurocognitive networks and distributed processing for attention, language and memory, *Ann. Neurol.*, 28, 597–613, 1990.
150. Morecraft, R.J., Geula, C., and Mesulam, M.-M., Architecture of connectivity within a cingulo-fronto-parietal neurocognitive network for directed attention, *Arch. Neurol.*, 50, 279–284, 1993.
151. Salzmann, E., Attention and memory trials during neuronal recording from the primate pulvinar and posterior parietal cortex (area PG), *Behav. Brain Res.*, 67, 241–253, 1995.

4 Normal and Abnormal Development of the Neuronal Response Properties in Primate Visual Cortex

Yuzo M. Chino, Hua Bi, and Bin Zhang

CONTENTS

0-8493-1243-4/04/$0.00+$1.50

4.1 INTRODUCTION

Over the past two decades, we have witnessed important advances in our understanding of the common rules that govern visual system development. Many studies have concentrated on measuring visual capacities at different stages of normal development or after experiencing abnormal vision early in life and identifying the factors that limit visual performance. There is considerable debate regarding the nature of the neural factors that constrain vision development. However, a relatively clear picture is emerging, at least, for the primate visual brain. This chapter is not a comprehensive review on the subject, but rather presents a concise description of our current understanding of how neuronal responses mature in the primate visual brain and how this maturation can be influenced by visual experience early in life.

4.2 NORMAL DEVELOPMENT

Monkeys can "see" relatively well at birth.[1,2] The machinery that maintains the optics of the eyes at its optimal condition is functioning shortly after birth. Newborn monkeys have a pair of eyes with fine optical quality, clear optic media, and exhibit acceptable accommodation and normal pupil responses. The eye movements of newborn infants are crude but sufficient to support vision.[1,3–6] Infant monkeys and humans, however, are affected by nasotemporal asymmetry in smooth pursuit eye movement and visual tracking (OKN) under monocular viewing conditions.[7–11] Because normal infants typically view the world with both eyes open, it is not clear whether this asymmetry in oculomotor responses seriously limits an infant's vision.

4.2.1 Monocular Response Properties

Despite relatively good optical quality, the monocularly measured visual acuity of infant monkeys is very poor (about 1 to 2 cycles/degree) compared to that in adults (30 to 50 cycles/degree).[12,13] Although the shape of an infant monkey's contrast sensitivity function is qualitatively adultlike near birth,[14,15] the peak spatial frequency, peak contrast sensitivity, and spatial resolution continue to improve over the first 12 months of a monkey's life. Similarly, vernier acuity is very low soon after birth, much worse than grating acuity, but improves by a factor of 50 to 60 over the same time period.[12]

4.2.1.1 Lateral Geniculate Nucleus

In monkeys, the six neuronal layers of the lateral geniculate nucleus (LGN) are well segregated at birth,[16,17] and, thus, the laminar segregation does not appear to require visual experience. The receptive field properties of both parvocellular (P) and magnocellular (M) cells are adultlike at 2 or 4 days of age, although the peak firing rate, contrast sensitivity, and spatial resolution of both M and P cells are extremely poor[18,19] (Figure 4.1). The responsiveness and the spatial resolution of these units improve most rapidly during the first 8 weeks of life, and the adultlike responsiveness and receptive-field properties are attained by 8 months of age. The maturation

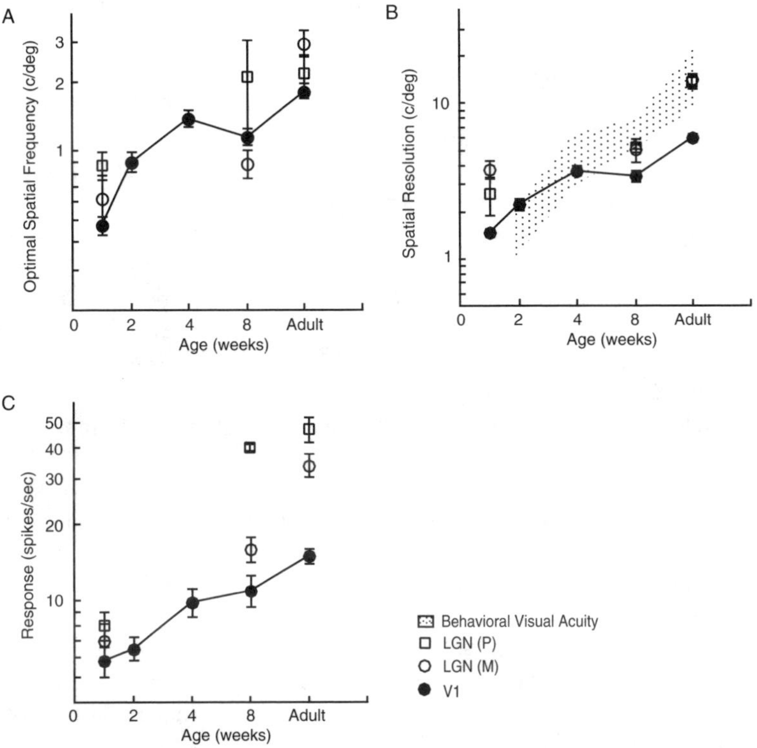

FIGURE 4.1 Normal development of monocular response of LGN and V1 neurons in macaque monkeys. (A) Mean (± SE) optimal spatial frequency of parvocellular (open squares) and magnocellular (open circles) LGN neurons (estimated from Reference 19) and V1 units.[31] (B) Mean (± SE) spatial resolution of LGN and V1 units. Conventions as in A. The dotted lines with shading indicate the range of behaviorally measured grating acuity in macaque monkeys (estimated from Reference 12). (C) Peak monocular response amplitude of LGN and V1 neurons. Conventions as in A and B.

of M and P cell temporal frequency tuning functions follows a similar time course. Much of the impoverished response properties of the LGN neurons during the first 8 weeks of life appears to reflect the known retinal immaturity, particularly that of photoreceptors.[4,20–22]

4.2.1.2 Striate Cortex (V1)

At birth, all six neuronal layers in V1 are present and its cytoarchitecture is qualitatively adultlike.[23,24] However, there are subtle immaturities in the anatomical organization of V1 at birth. The cytochrome oxidase (CO) activity of the input layer (IVC) is substantially reduced in newborn monkeys[23,25] and this subnormal CO activity persists at least for the first 4 weeks of life.[24] The LGN axon arbors are not fully developed in some primates,[26,27] and the density of dendritic spines of V1 neurons increases until 8 weeks of age, after which it decreases to the adult level.[28,29]

V1 neurons in newborn monkeys show qualitatively adultlike monocular response properties.[30] All simple, complex, and "hypercomplex" cells have orientation tuning that is qualitatively similar to that of adult monkeys. However, the quantitatively measured orientation and direction selectivity of V1 neurons in 6-day-old monkeys is lower than in adults.[31,32] Both the orientation tuning and directional bias of individual neurons improve rapidly during the first 4 weeks of life and the subsequent changes to the adult level are relatively small.

The shape and the bandwidth of the spatial frequency response functions of individual units are qualitatively adultlike in 6-day-old monkeys. However, the average optimal spatial frequency and the spatial resolution of V1 units are about two octaves lower than those in adults (e.g., the mean optimal frequency = 0.45 ± 0.02 cycle/degree and the spatial resolution = 1.3 ± 0.01 cycles/degree at 6 days of age).[31] Both the optimal spatial frequency and resolution rapidly improve to about 2 and 4 cycle/degree, respectively, by 4 weeks of age after which improvement is more gradual (Figure 4.1A and 4.1B).[31] The mean spatial resolution of V1 units is nearly identical to the behaviorally measured acuity for macaque monkeys at 1 and 4 weeks of age,[12] but far lower than the qualitatively determined resolution of LGN neurons[18] or the quantitatively determined high-frequency cutoffs for geniculate P or M cells.[19] These contrasting results in the LGN and V1 suggest that spatial and/or temporal imprecision in the convergence of afferent signals may have influenced the spatial frequency response characteristics of V1 units during the first 4 postnatal weeks. Regardless, the normal development of behaviorally measured visual acuity appears to parallel the relative time course of improvement in the spatial resolutions of V1 units.

The peak monocular response amplitude of both simple and complex cells is very low at 6 days of age, but it rapidly improves during the first 4 to 8 postnatal weeks (Figure 4.1C).[31] This increase in responsiveness of V1 units is very similar to the rapid improvement in the peak firing rate of the LGN units during the same period[19] and, as previously mentioned, largely reflects the maturation of the retina. The low CO activity in layer IVC during the first 4 weeks of life,[23–25] then, may be a consequence of this impoverished afferent input and/or the morphological immaturity of LGN axon arbors.

In addition to the subnormal feedforward connections in the striate cortex in newborn monkeys, the intrinsic long-range horizontal connections of primates are not well developed at birth.[33–35] In nonprimate species, the long-range connections are refined by activity-dependent processes shortly after birth, and this refinement process parallels the emergence of adultlike functional organization in the striate cortex.[36–38]

In mature cats and monkeys, the stimulation of these connections alone typically does not initiate spike activity in the unit under study; however, their activation strongly influences the response initiated by their feedforward projections.[39–41] The subthreshold influences can be excitatory or inhibitory depending on the functional status of target cells, the distance and alignment between the center and surround stimuli, and the nature of surround stimuli.[40–46]

Unlike in cats and ferrets, the structures of the long-range horizontal connections in monkey V1 appear to be adultlike as early as 10 days of age.[47] However, at even 5 weeks of age, some functional aspects of the long-range intrinsic connections in adults are absent in infant monkeys despite their adultlike structural organization.[48] Consistent with this observation, the influence of the surround stimulation on the response of the classical receptive field (CRF) of individual V1 neurons was reported to be subnormal in infant monkeys during the first 4 weeks of life.[34,35] These subnormal surround effects are presumably due to immaturity in the long-range horizontal connections in V1, although a lack of robust feedback signals from the extrastriate visual areas has not been ruled out.

Taken together, prior to 4 weeks of age, the monocular responses of V1 units are characterized by their very low optimal spatial frequencies, reduced directional biases, and overall unresponsiveness.[24,31,32] Interestingly, these subnormal monocular responses may be the primary cause of the aforementioned nasotemporal asymmetry in the eye movements of young infant monkeys under monocular viewing.[32] Because of the impoverished cortical signals, the cortical control over the brain stem nuclei is reduced or absent in neonates. This appears to enhance the innate tendency of the subcortical nuclei to drive the eyes more effectively in the temporal-to-nasal direction.[49–51]

4.2.1.3 Extrastriate Visual Areas

Although the role of the extrastriate visual areas in a variety of perceptual phenomena is attracting much attention, there have been few studies dealing with normal development of higher-order visual areas in monkeys.[52–54] Some cells in inferotemporal cortex (IT) are visually responsive as early as 6 weeks of age (the earliest age tested) but only 10% of units in infant monkeys younger than 4 months of age could be visually driven, compared with 90% in adults. Adultlike response properties emerge after 4 months of age; however, their responsiveness is less than one half of the adult. The response latencies of these IT cortex units in infant monkeys are longer and more variable than those in adult monkeys. Visually driven units are not found in the superior temporal polysensory area (STP) until 4 months of age. Both IT cortex and the STP regions in adult monkeys are several synapses away from V1 and are known to be involved in higher-order processing of visual information or visual cognitive functions (e.g., face recognition for IT cortex or analyses of complex visual motion and orientation functions for STP) (see References 55 through 57 for reviews). Because the monocular responses of V1 neurons are nearly adultlike by 4 to 8 weeks of age,[24,31,32] the poor responses in these extrastriate areas of anesthetized infant monkeys may result from immaturities in the functional connections between V1 and IT or STP.

4.2.2 Binocular Response Properties

The binocular visual capacities of monkeys are severely limited in neonates. Stereopsis, a highly sensitive measure of the functional status of binocular vision, is absent at birth, but suddenly emerges around 4 to 6 weeks of age.[58] Subsequently, stereoacuity steadily improves over the next 7 to 9 weeks until it reaches nearly the adult level around 13 weeks of age. A similar developmental time course for stereopsis is found in human infants,[59,60] although the 4 to 1 rule[1] must be applied (i.e., weeks in monkeys are roughly equivalent to months in humans). The neural factors that constrain binocular vision development in primates have not been as thoroughly investigated as the limits imposed by cortical neurons on the maturation of monocular vision.

4.2.2.1 Striate Cortex

Until about 3 weeks prior to birth, the LGN axon arbors for the two eyes completely overlap in the input layer (IVC) of V1.[61] It was long thought that the segregation of ocular dominance columns in layer IVC requires at least 6 weeks of normal visual experience,[62,63] and that a greater mixing of left- and right-eye activity occurs in layer IVC of 8-day-old or 3-week-old monkeys compared with older infant monkeys or adults. Based on these findings it was hypothesized that stereopsis is absent during the first 4 to 6 weeks of life because the intrinsic connections of V1 required for the development of binocular disparity sensitivity are not fully developed near birth.[64]

According to more recent studies, however, the ocular dominance columns in layer IVC of newborn monkeys that had no visual experience are adultlike.[23] Moreover, a normal proportion of V1 units are sensitive to interocular image disparity as early as 6 days of age, several weeks before the onset age for stereopsis in monkeys (Figure 4.2).[31,58] The ocular dominance distribution of cells in infant monkeys is also indistinguishable from that of adults.[31] Thus, binocular vision development in monkeys is not constrained by a paucity of disparity sensitive neurons in V1 or immaturity in ocular dominance columns.

Factors that may limit binocular vision near birth are not well agreed on. Abnormal eye movements[58] or immaturity in the extrastriate visual areas that support stereopsis (e.g., MT or MST)[58] are frequently mentioned as the primary

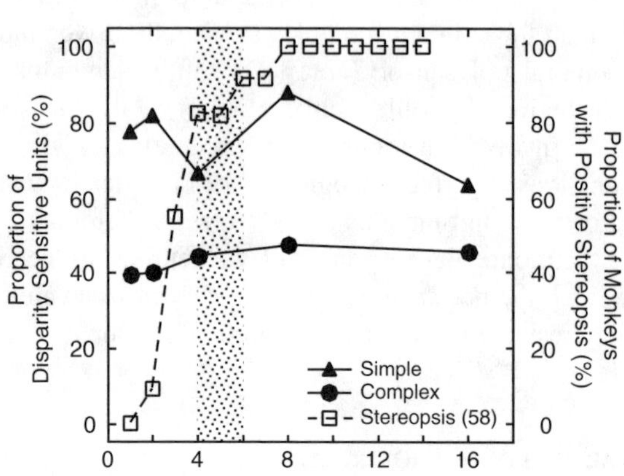

FIGURE 4.2 Normal development of binocular disparity sensitivity in simple (open circles and left vertical axis) and complex cells (filled circles) of monkey V1.[31] The development of stereopsis in monkeys (open squares and right vertical axis) is also plotted based on the data published by O'Dell and Boothe.[58] Vertical stippled area indicates the ages during which stereopsis emerges in normal monkeys.

constraining factors. However, experimental data to support either of these ideas are scarce or absent.

An alternative view is that the relative immaturity of the spatial properties (Figure 4.1A and 4.1B) and the overall unresponsiveness of existing disparity sensitive neurons in V1 (Figure 4.1C) may be the primary reasons stereopsis is absent during the first 4 weeks of life.[31] According to this hypothesis, the disparity-sensitive V1 neurons in neonates become capable of signaling the nature and magnitude of binocular retinal image disparity to higher-order cortical neurons only when a certain "threshold" binocular amplitude is exceeded or each cell's absolute firing rate per unit of angular disparity surpasses a certain level. According to this view, the rapid increase in responsiveness together with the improvement in spatial resolution of V1 during the first 4 to 6 weeks[31] is required for the emergence of stereopsis (compare Figure 4.1 and Figure 4.2).

Although the basic connections required for binocular vision are present in V1 at birth, the binocular excitatory connections in monkey V1 are not fully developed in infant monkeys prior to 4 weeks of age.[24] Specifically, V1 neurons prior to 8 weeks of age show a higher prevalence and greater magnitude of interocular cross-orientation suppression than in adult monkeys. This suppression to a pair of orthogonally oriented sine wave gratings decreases and reaches the adult level between 4 and 8 weeks of age. The major reason for the reported increase in interocular suppression can be found in the relative immaturity of the excitatory binocular connections in infant monkeys before 4 to 6 weeks of age. The excitatory connections from the two eyes, whether they are the feedforward receptive-field-generating projections or the intrinsic local and long-range horizontal connections, are relatively immature compared with those in adults.[31,32,48] However, the generally orientation-insensitive inhibitory connections (particularly the intrinsic long-range connections) function relatively more vigorously during early infancy.[34,35,65] Consequently, changing stimulus orientation from a neuron's preferred stimulus orientation may reduce the overall binocular excitatory drive without a concomitant reduction in binocular inhibitory signals. As a result, the binocular response rate becomes less than that of the dominant monocular response amplitude for the orthogonally oriented pair of sine wave gratings.[24,31]

4.2.2.2 Extrastriate Visual Areas

Immaturity in the extrastriate visual areas is often suggested as a major cause of the very limited binocular visual capacities in newborn monkeys. However, the binocular response properties of individual neurons have not been investigated at all in any of the higher-order visual areas.

4.3 ABNORMAL DEVELOPMENT

Chronically uncorrelated signals from the two eyes or binocular imbalance early in life can have devastating effects on binocular vision development. Since the pioneering work of Wiesel and Hubel[66] in cat area 17, animal models of binocular vision disorders have been employed to investigate how sensory deficits emerge in subjects that are reared with abnormal visual experience early in life. The majority of the previous studies in nonprimate species have concentrated on gathering the type of information that is necessary to understand the rules governing early visual system development. The most common form of abnormal visual experience was monocular form deprivation and, with some exceptions, ocular dominance tests (*monocular tests* of binocular vision) were employed to assess the functional alterations in the visual cortex (see References 5, 6, 37, and 67 through 71 for reviews).

Nonhuman primates are ideal animal models for investigating binocular vision disorders in humans, because the monkey visual system is closest to a human's and, as described in an earlier section, the relative time course of normal development parallels that in humans.[31,58,60] Moreover, psychophysical experiments in monkeys that measured their monocular and binocular visual capacities have demonstrated striking similarities between monkeys and humans (e.g., References 12, 14, 58, and 72 through 74).

Types of early discordant binocular visual experience that have been a focus of investigations include monocular form deprivation (associated with early ptosis or cataract), defocus (due to anisometropia or uncorrected refractive errors early in life), and ocular misalignment (caused by infantile strabismus). These abnormal conditions have been successfully simulated in monkey models and are known to result in binocular vision disorders that are similar to human's, including stereoblindness, chronic suppression, and, if untreated, amblyopia ("dull vision" or a reduction in visual sensitivity of one or both eyes) (for reviews, see References 1, 60, and 75 through 78).

This section is guided by the central theme that binocular vision deficiencies in primates can be explained to a large extent by experience-dependent alterations in the functional connections of the primary visual cortex (V1). However, comprehensive investigations of the higher-order visual areas may be necessary to appreciate fully the complex nature of the neural deficits associated with amblyopia.

4.3.1 Monocular Form Deprivation

Binocular imbalance early in life that results from monocular eyelid closure severely disrupts vision development in monkeys. The ocular dominance columns in layer IVC of V1 that are innervated by the deprived eye exhibit substantial shrinkage,[63,79,80] and there is a clear shift in the ocular dominance distribution of cortical units to favor the nondeprived eye.[63,79] Functionally, the visual sensitivity of the deprived eye is drastically reduced (deprivation amblyopia).[1,81–84] Not surprisingly, monocularly form-deprived monkeys lose virtually all binocular functions including the binocular summation of contrast sensitivity. Another notable alteration is the shrinkage of cell bodies of both P and M cells in the LGN laminae that are innervated by the deprived eye.[63,85–87] However, the monocular response properties of LGN neurons in monkeys are not affected by early monocular form deprivation.[79,88] This is in contrast to substantial morphological alterations of retinal ganglion cell axon arbors and neuronal responses in the LGN of nonprimate species following early deprivation or strabismus (e.g., References 67 and 89 through 92).

The functional organization of monkey V1 is very dynamic during the first 4 weeks of life,[31] and thus, binocular imbalance during this period is expected to have a far greater impact on visual system development than that during the later stages of development. The effects of early monocular form deprivation on the development of ocular dominance columns in monkey V1 are a case in point.[80] When the age of monocular eyelid closure has its onset at 1, 3, 5, 7, and 12 weeks of age, the most severe column shrinkage (32 to 44%) occurs in the infant monkeys with the earliest onset age. The degree of shrinkage becomes progressively smaller as the onset age is delayed (i.e., 15 to 20% for 3 weeks, 9 to 17% for 5 weeks, 5 to 7% for 7 weeks, and virtually no shrinkage for 12 weeks). Thus, V1 is most sensitive to monocular form deprivation during the first 4 weeks of life. Because the ocular dominance columns in monkeys, unlike in nonprimate species,[93] are largely adultlike at birth,[23,61] Horton and Hocking[80] suggested that the columnar shrinkage and expansion in their study are due primarily to retraction and sprouting of the afferent geniculo-striate fibers along the borders between columns. Clinically, these results suggest that the removal of congenital cataract and instatement of good optical quality should be accomplished at the earliest possible age.[60,80,94–97]

Interestingly, the plastic period for behaviorally measured binocular functions (e.g., binocular summation of contrast sensitivity) in monkeys continues beyond 2 years of age.[73,98] This is very different from the plastic period for alterations in ocular dominance columns upon which monocular form deprivation at 12 weeks of age has no effect.[80] The plastic period for binocular vision associated with monocular form deprivation also varies with the cortical sites (e.g., between layer IVC and layer II to III within V1) and between V1 and higher-order visual areas.[73,76,98] An emerging view is that the higher the level of cortical processing, the longer the plastic period appears to be. However, very few data are available to evaluate this important hypothesis.

4.3.2 Alternating Monocular Occlusion

Early monocular form deprivation immediately puts the deprived eye into a competitive disadvantage and typically leads to virtual blindness of that eye. Daily or semidaily alternation of monocular form deprivation between the two eyes (alternating monocular occlusion) creates a unique animal model for investigating the effects of extreme binocular imbalance on binocular vision development, without giving a lasting competitive advantage for either eye. Our preliminary study indicated that alternating occlusion for half the lighting cycle every day (i.e., 6 h) between 4 weeks and 4 months of age in infant monkeys leads to a severe loss of binocularly balanced V1 units and a proportional increase in "monocular" units.[99] An additional consequence is a large reduction in the sensitivity of V1 units to interocular spatial phase disparity, which is more severe in complex cells than in simple cells. The residual binocular interactions are minimal in that binocular facilitation is drastically reduced, while interocular suppression, a highly prevalent form of residual binocular interaction in strabismic monkeys,[100] is virtually absent. The retention of interocular suppression in monkeys reared with optically induced strabismus may be associated with the fact that both eyes have relatively clear images that cannot be fused, but neither eye has competitive advantage for weeks and months or even longer periods of time.

There are no interocular differences in the monocular response properties of V1 units such as the orientation/direction tuning and spatial frequency tuning in monkeys reared with early alternating occlusion. These spatial properties remain largely normal, a result consistent with behavioral observations in similarly raised monkeys that indicated no sign of amblyopia.[101]

If daily alternating monocular occlusion begins in monkeys at birth and continues until 6 months of age, a presumed loss of binocularity in visual cortical neurons leads to a variety of abnormal eye positions and ocular motility. The treated monkeys exhibit incomitant strabismus (esotropia or exotropia), saccadic disconjugacy (i.e., the amplitude of saccades in the occluded eye is less than that in the viewing eye),[102] and nasotemporal asymmetry in visual tracking under monocular viewing conditions (e.g., optokinetic nystagmus or OKN).

Because alternating occlusion that begins at 4 weeks of age does not cause any of these oculomotor anomalies,[99] the plastic period for the neural mechanisms that guide the ocular motility and alignment to the "normal" position after birth appears to be relatively short, i.e., between birth and 4 weeks of age.

4.3.3 Anisometropia

4.3.3.1 Unilateral Defocus

Anisometropia is a condition in which a significant difference in refractive error exists between the two eyes. In primates, which typically begin life with modest hyperopic refractive errors in both eyes, uncorrected anisometropia early in life leads to constant defocus in one eye. As a result, stereo-deficiencies, abnormal binocular summation, and/or amblyopia may develop.[78,84,100,103–105]

Anatomical changes in the geniculostriate pathway of monkeys reared with early unilateral atropinization[106] are generally similar to those described for monocularly form-deprived monkeys.[63,107] However, there are subtle differences between the two conditions. For example, in unilateral atropinization the shrinkage of LGN neurons is confined to P cells and the narrow bands of the marked alterations of CO activity in layer IVC are "associated with" the centers of the ocular dominance columns for the defocused eye. Surprisingly, the CO activity in these bands is higher than the alternate bands representing the nondefocused eye.[106]

Rather predictable changes occur in the visual cortex of monkeys reared with early unilateral defocus. The ocular dominance distribution of V1 units is marked by a substantial loss of binocularly balanced cells and a small bias that favors the nontreated eye.[100,104,108] According to the only quantitative study on binocular signal interactions in V1 of anisometropic monkeys,[100] many "monocular" neurons based on ocular dominance measures exhibit clear residual binocular interactions if the two eyes are stimulated together. However, the sensitivity of V1 neurons to interocular spatial phase disparity is substantially reduced in anisometropic monkeys. The average peak binocular response amplitude of V1 units is also lower in the treated monkeys. In contrast, the peak monocular amplitude is not different between the treated and normal monkeys. These subnormal binocular responses in V1 may explain the reduced binocular capacities of these monkeys.[100]

In normal mature monkeys, the disparity tuning of V1 neurons can be largely accounted for by a simple addition of signals from the two eyes. Thus, it is reasonable to expect that binocular phase tuning functions of V1 units in monkeys would depend on their monocular receptive field properties.[109] Constant defocus in one eye (anisometropia) during early development disrupts the excitatory and inhibitory spatial organization of the monocular receptive fields dominated by the affected eye, and the normally precise spatial organization of monocular signals is scrambled.[108,110] Consequently, monocular signals from the eye that experienced chronic early defocus may be weak or distorted and the resulting interocular imbalance in the monocular inputs is assumed to interfere with the development of normal binocular connections in the visual cortex.[110]

The mean optimal spatial frequency, spatial resolution, and contrast sensitivity of V1 neurons dominated by the "amblyopic" eye of atropinized monkeys[108] or lens-reared monkeys[100,104] are significantly lower in units driven by the treated eye. Thus, the neural correlate of the spatial contrast sensitivity loss that occurs with high prevalence among monkeys and humans with anisometropic amblyopia appears to exist at least in part at the site where excitatory binocular signals are first combined, i.e., in V1.

4.3.3.2 Alternating Defocus

To avoid complications from monocular deficits (e.g., amblyopia) in the treated eyes of monkeys reared with early unilateral defocus, defocus can be alternated daily between the two eyes. The effects of this manipulation on binocular vision development are very revealing. Daily alternation of monocular defocus in monkeys (by wearing 1.5, 3.0, and 6.0 diopter negative lenses) between 3 weeks and 9 months of age results in a spatial-frequency dependent loss of local stereopsis, which is generally proportional to the amount of defocus.[111] None of these monkeys developed amblyopia.

Normal binocular vision depends on the correct interocular matching of monocular signals. Monocular defocus creates interocularly conflicting signals that increase in severity with the spatial frequency of the stimuli. Thus, it is not surprising that the monkeys used in the behavioral study of Wensveen et al.[111] exhibited a spatial-frequency-dependent reduction in the sensitivity of individual neurons to interocular spatial phase disparity and that this reduction was generally proportional to the power of the defocusing lenses worn by the treated monkeys.[112] The monocular response tuning properties of these units did not show measurable abnormalities that favored one eye over the other, presumably because each eye received undisturbed vision on alternate days. However, the interocular matching of the monocular receptive field properties of V1 neurons in the experimental monkeys, particularly those that experienced –6D defocus, was not as tight as in normal monkeys. Together, these findings support the hypothesis that local stereopsis is spatial frequency dependent and that binocular disparities are processed by independent channels tuned to different spatial frequencies.[113–119] Moreover, the local disparity processing mechanisms in V1 can be independently compromised, according to their spatial frequency tuning, by early abnormal visual experience.

An emerging view on fine binocular vision is that V1 neurons alone may not be sufficient to support stereopsis and the higher-order visual neurons are critically important in generating a three-dimensional percept of the world.[119–123] However, our studies on the effects of alternating defocus[111,112] underscore the importance of the normal presence of disparity-sensitive V1 neurons for local stereopsis.

4.3.4 Strabismus

Strabismus is a misalignment of the visual axes that can result in double vision (diplopia) and confusion.[124] Untreated strabismus early in life can lead to deficiencies in stereoscopic vision, clinical suppression, and/or a reduction in the visual capacities of the deviating eye (strabismic amblyopia) (see References 1, 5, 6, and 78).

In monkey models of binocular vision anomalies, strabismus is simulated by surgical or optical methods that are applied shortly after birth. The nature of early abnormal visual experience and the consequences of these two rearing methods can be substantially different. In the surgical method,[125] the extraocular muscles of one eye are sectioned or tied to induce misalignment, thereby simulating a paralytic type of strabismus in humans, which is less common in humans than comitant strabismus (i.e., the angle of deviation does not change with gaze). In animals with noncomitant paralytic strabismus, the treated eye is immediately placed at a competitive disadvantage and the untreated eye is most likely to become the "fixating" eye. Eventually, the treated eye is likely to develop amblyopia due to uncontrolled defocus in that eye (e.g., References 81, 84, and 104).

The optical method that simulates comitant strabismus uses a pair of base-in prisms to cause binocular dissociation.[100,126–129] Neither eye becomes disadvantaged for a considerable period of time because the fixation is typically alternated between the two eyes. Consequently, prism-reared monkeys are less likely to develop amblyopia.[100] The morphological changes in the LGN that are common in surgical strabismus[125,130] are not found in the prism-reared monkeys.[127] Thus, it is important to know which method (surgical or optical) was used to simulate strabismus when the data from animal studies are evaluated.

4.3.4.1 Binocular Response Properties

Infantile or congenital strabismus in humans typically results in stereo-deficiencies and other binocular vision anomalies.[60] According to the majority of studies on binocular function that used a monkey model of strabismus, the neural factors that constrain binocular vision development appear to reside largely in the earliest stages of cortical processing.

4.3.4.1.1 Ocular Dominance

Early investigations of the effects of strabismus on cortical binocularity primarily employed ocular dominance measurements. In V1 of monkeys reared with either early surgical[125] or optical strabismus,[100,127–129,131] the proportion of units that can be excited by stimulation of either eye ("binocular cells") is drastically reduced. In monkeys with surgical strabismus, the ocular dominance distribution shifts to favor the nondeviating eye. Similar changes in ocular dominance distribution are reported in cortical area MT of macaque moneys reared with early strabismus induced by application of a toxin to the extraocular muscles,[132] which, like surgery, results in a noncomitant deviation. However, a much smaller loss of binocularly balanced cells was reported recently in area V2 of monkeys reared with surgical strabismus, provided that the monkeys did not develop severe amblyopia.[133] MT recordings showed similar results if the strabismus developed "naturally" as a result of binocular eyelid closure that lasted from 30 to 50 days after birth (Mustari, personal communication).

Ocular dominance measurements alone cannot explain many aspects of the binocular vision anomalies in strabismic monkeys. For example, suppressive binocular interactions, the most common form of residual binocular interactions in strabismic subjects, cannot be detected in cortical neurons by the measurement of ocular dominance. Also it is not possible to investigate subthreshold and often functionally significant binocular interactions with ocular dominance tests. Binocular signal interactions were directly investigated first in strabismic cats[92] and then in strabismic monkeys using dichoptic stimuli and quantitative analyses of the units' responses.[100]

4.3.4.1.2 Disparity Sensitivity

Infant monkeys reared with a pair of base-in prisms (a total of 27 prism diopters) behaviorally become stereo-deficient[131,134,135] and lack binocular summation in their spatial contrast sensitivity functions.[136] In V1 of these subjects, the proportion of binocular disparity sensitive units is drastically reduced, reflecting an increase in the proportion of neurons that have no sign of binocular interaction and respond only to monocular stimuli presented to one eye (truly monocular cells).[100] Thus, a fundamental neural requirement for fusion and stereopsis (i.e., a normal proportion of disparity detectors) is absent in V1 of these stereo-deficient monkeys.

4.3.4.1.3 Binocular Suppression

Humans with untreated early strabismus often develop one or more "adaptive" phenomena to eliminate the disturbing impact of diplopia and confusion (e.g., suppression, amblyopia, and/or anomalous correspondence).[124] Suppression is the most pervasive of these adaptive mechanisms. Prism-reared monkeys also exhibit binocular suppression that is similar to clinical or pathological suppression in strabismic humans.[137] Within V1, prism-reared monkeys have a higher-than-normal prevalence of binocular suppression, particularly in complex cells, and a clear loss of excitatory binocular interactions.[100] These studies clearly establish parallels between V1 neurophysiology and the behavioral deficits in strabismic monkeys. Interestingly, a CO study found clear evidence for robust binocular suppression immediately following a surgical induction of divergent strabismus in V1 of *adult* monkeys.[107]

4.3.4.1.4 Onset Age and Duration of Strabismus

Monkey V1 is most sensitive to monocular form deprivation during the first 4 weeks of life.[80] In contrast, the consequences of varying the onset age or duration of strabismic misalignment are very different.[128] Only 2 weeks of binocular dissociation by prism-rearing is sufficient to drastically reduce the disparity sensitivity of V1 neurons and increase the prevalence of binocular suppression. Surprisingly, the magnitude of these binocular deficits is greater if strabismus occurs between 4 and 6 weeks of age or between 6 and 8 weeks of age rather than before 4 weeks of age, e.g., between 2 and 4 weeks of age.[128,129] Complex cells are more severely affected than simple cells in all strabismic monkeys.

What is the latest onset age of strabismus that can cause substantial binocular and monocular vision deficits? Our preliminary data indicate that substantial response alterations occur in V1 and V2 even when surgical strabismus is induced at 6 months of age (roughly equivalent to 2 years of age in humans).[133] Because monocular form deprivation in monkeys that begins at 2 years of age disrupts binocular vision development[73] and both behavioral and neural changes are quite severe in strabismic monkeys with the onset age of 6 months, the effects of even later onset should be measured to determine the "end" of the plastic period for binocular functions in strabismic monkeys.

Strabismus that lasts longer than two weeks (e.g., 4 or 8 weeks) does not result in a greater loss of disparity sensitivity or a higher prevalence of binocular suppression if the onset age is kept constant at 4 weeks of age (Figure 4.3).[129] However, the duration of strabismus has a very different impact on the maturation of monocular receptive field properties (see Section 4.3.4.2).

What is the minimum duration of misalignment required to alter the normal development of binocular properties of V1 neurons in monkeys? According to our preliminary results, 7 days (equivalent to 4 weeks in humans) beginning at 4 weeks of age was sufficient to disrupt the development of binocular disparity sensitivity in V1 neurons, although the deficits were not as severe as those that occur after 2 weeks of misalignment. Three days of optical strabismus beginning at 4 weeks of age had virtually no effect on disparity sensitivity. However, after only 3 days of strabismus (12 days in human infants), the prevalence and magnitude of interocular suppression were higher than those in the age-matched normal monkeys. Thus, the first response alteration in V1 that can be detected following just days of ocular misalignment appears to be binocular suppression.

Does a period of "normal" visual experience after experimentally induced strabismus have any effects on the maintenance of binocular connections? Infant monkeys reared with either 4 or 8 weeks of optical strabismus were kept in a normal housing environment following the prism-rearing until maturity and behavioral testing was conducted to determine their monocular and binocular capacities before the recording experiments. A small but significantly higher degree of disparity sensitivity was found in V1 compared to those in similarly reared infant monkeys that did not have poststrabismus visual experience if the prism-rearing was terminated after 4 weeks of strabismus (i.e., at 8 weeks of age), but not after 8 weeks of strabismus (i.e., at 12 weeks of age).[129]

The findings in our experiments on the effects of the onset age and duration of strabismus point to a fundamental difference between how strabismus and form deprivation alter the early development of binocular connections in V1. Unlike monocular form deprivation, optical strabismus creates well-focused images that do not match between the two eyes. Consequently, the most sensitive period during which binocularly discordant signals affect the disparity sensitivity of V1 neurons is not immediately after birth but soon after the known onset age of stereopsis (i.e., 4 weeks of age). By this time of postnatal development, V1 neurons attain near adultlike monocular tuning properties and responsiveness.[24,31,32] As a result, small

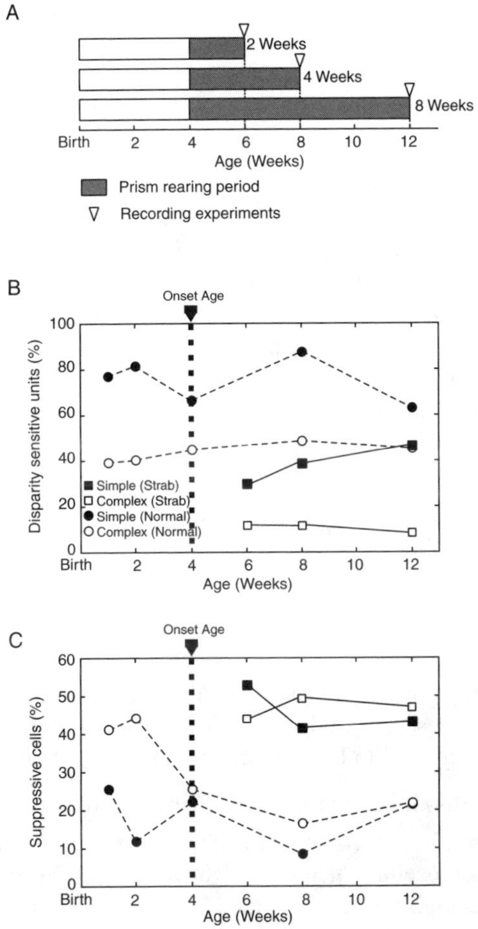

FIGURE 4.3 Effects of the duration of strabismus on the disparity sensitivity and the prevalence of interocular suppression of V1 neurons. (A) Rearing regimen. The onset of prism-rearing was kept constant at 4 weeks of age and the duration was varied for 2, 4, and 8 weeks. Open triangles indicate the ages at which the recording experiments were conducted. (B) The effects of strabismus duration on the disparity sensitivity of V1 neurons. Filled squares indicate simple cells and open squares indicate complex cells. Units were operationally defined as "disparity sensitive" if their binocular interaction index (BII) values were greater than or equal to 0.3.[109,157] The data for age-matched normal infant monkeys are also shown (filled circles, simple cells; open circles, complex cells). (C) The effects of strabismus duration on the prevalence of interocularly suppressive units (i.e., the peak binocular response amplitude/the dominant monocular response amplitude < 1.0).[31,109] Conventions as in B.

imbalances in the signals between the two eyes are more readily detected by binocular neurons and are more effective in disrupting the normal development of binocular response properties of V1 neurons. It is remarkable that during this critical time of development, only 7 days of strabismus is sufficient to cause significant damage to the binocular signal processing mechanisms in primate V1, and that periods of strabismus longer than 2 weeks result in little additional loss of disparity sensitivity. Another important difference is that strabismus causes substantial binocular suppression in V1 neurons soon after misalignment occurs, whereas, as described previously, monocular form deprivation rarely generates interocular suppression in cortical units.[112]

4.3.4.1.5 Clinical Significance

One of the most debated issues in vision care with respect to the management of infantile strabismus is at what age corrective measures (e.g., surgical alignment) should be taken to increase the odds of preserving fine binocular vision (e.g., References 138 through 140). Although data from animals studies must be interpreted with caution, our findings on the effects of strabismus onset and duration unambiguously support the hypothesis that alignment before the known onset age of stereopsis (4 to 6 months of age in humans and 4 to 6 weeks of age in monkeys) may be critically important for maintaining binocular sensory function and interocular alignment at later stages of development.[140]

4.3.4.2 Monocular Response Properties of LGN and Cortical Neurons

Humans who experience uncorrected, early-onset strabismus often exhibit amblyopia. Psychophysical experiments have identified two primary deficits in strabismic amblyopes, contrast sensitivity losses for mid to high spatial frequencies[141,142] and impaired precision of positional judgments.[143–145] More recently, higher-order visual functions were also reported to be abnormal in amblyopic humans.[146]

4.3.4.2.1 Spatial Frequency Tuning and Contrast Sensitivity

Early-onset strabismus has no effect on the development of the monocular response properties of LGN parvocellular neurons.[147] However, it is important to bear in mind that the effects of strabismus on magnocellular neurons have not been studied. Nevertheless, the earliest site in the central visual pathway where the monocular response properties of visual neurons are unambiguously affected by strabismus appears to be the primary visual cortex.

The neural mechanisms underlying strabismic amblyopia is a matter of considerable debate and conjecture. Most hypotheses are based largely on psychophysical findings in humans and neurophysiological data obtained in the cat visual cortex that typically lack behavioral documentation of amblyopia. Neurophysiological investigations in monkey models of amblyopia have been quite scarce (see References 76 and 78 for recent reviews). It is widely assumed that a relative paucity and/or a reduced contrast sensitivity of cortical neurons tuned to higher spatial frequencies are the primary factors responsible for the contrast sensitivity deficits in the amblyopic eye.[78,104,148] Thus far, however, only one study has reported a contrast sensitivity reduction in V1 units in monkeys reared with surgical strabismus, particularly those units tuned to higher spatial frequencies. Neurophysiological deficits generally paralleled the behavioral contrast sensitivity losses.[104] The neural basis for more complex anomalies such as crowding and position uncertainty,[143–145] deficits in contour integration,[149] abnormal topographic mapping of visual space or "jitter,"[150–152] and higher-order cognitive disorders[146] have not been investigated in amblyopic monkeys with neurophysiological methods.

Monkeys reared with optical strabismus that begins at 4 weeks of age typically do not develop amblyopia unless the duration of misalignment is very long.[100] In our investigation of the duration of strabismus,[129] the only monkeys that behaviorally became amblyopic were those monkeys that experienced 8 weeks of optical strabismus. Although this extended period of prism-rearing caused a behavioral loss of spatial contrast sensitivity for mid to high spatial frequencies and resulted in interocular acuity differences of about one octave,[137] V1 units of these monkeys failed to exhibit an obvious interocular difference in average spatial resolution or a reduction in the proportion of cells tuned to higher spatial frequencies for either eye (Figure 4.4). Also, there was no significant shift in ocular dominance distribution away from the amblyopic eye.[100,153,154] However, the same units, under dichoptic stimulation, exhibited robust interocular suppression that was greater in prevalence and magnitude for the units dominated by the behaviorally amblyopic eye and tuned to highest spatial frequencies. Behaviorally, these monkeys exhibited robust "clinical" suppression under dichoptic conditions.[137]

Early diplopia, unless accompanied by consistent and extended periods of defocus in one eye (e.g., more frequently found after early surgical strabismus and unilateral defocus), does not appear to affect dramatically the monocular spatial filter properties of V1 units. However, the binocular signal transmission characteristics of V1 neurons are altered such that units dominated by the amblyopic eye and/or tuned to high spatial frequencies are chronically suppressed during early development. The functional connections from these affected V1 units to extrastriate visual areas may therefore be altered, and as a result extrastriate neurons (e.g., V2 neurons) tuned to high spatial frequencies may be lost or become unresponsive to the amblyopic eye. Our preliminary data in V2 of strabismic monkeys that developed amblyopia are consistent with this hypothesis. In a sample of amblyopic monkeys (i.e., interocular differences in acuity were over one octave), the ocular dominance distribution of V2 units shifted substantially away from the amblyopic eye, whereas V1 units of the same monkey exhibited a weaker bias favoring the non-amblyopic eye (Figure 4.5). Instead, this monkey lost binocularly activated V1 units. Interocular comparisons of the spatial resolution of the units were not possible for this monkey, because very few units were dominated by the deviating eye. These results are in stark contrast to the data from monkeys showing less severe amblyopia. In mildly amblyopic monkeys (e.g., interocular acuity differences < 1.0 octave), the ocular dominance distribution was balanced between the two eyes in both V1 and V2 and the mean spatial resolution of V1 or V2 units dominated by the deviating eye did not differ from those dominated by the nondeviating eye or control units in normal monkeys.

4.3.4.2.2 Orientation/Direction Selectivity

Adult humans and monkeys with early-onset strabismus exhibit monocular nasotemporal asymmetries in pursuit eye movements and visual tracking, similar to those found in normal neonates.[132,155] Although there is debate regarding the neural basis of these oculomotor asymmetries in strabismic subjects, it is generally agreed that

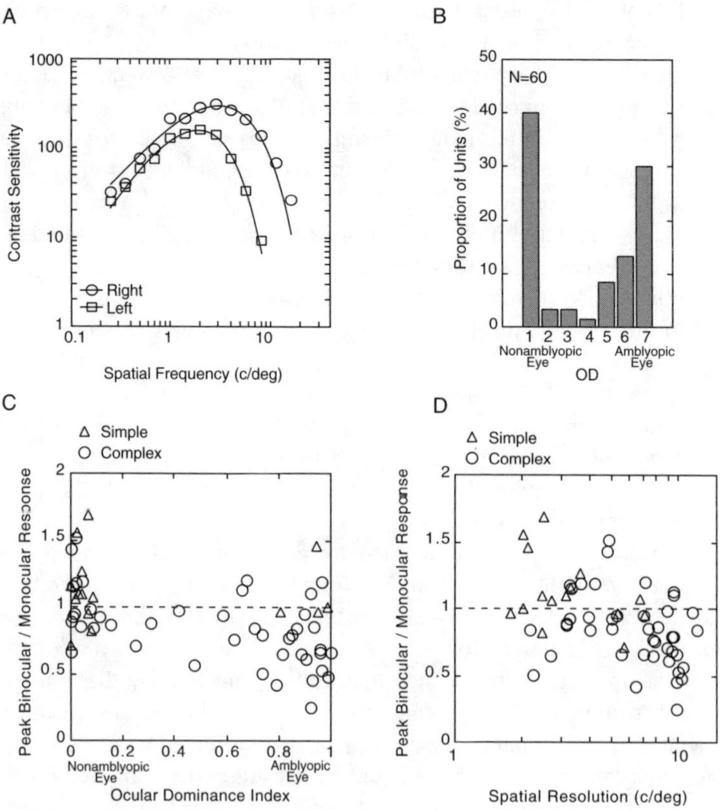

FIGURE 4.4 (A) Mean spatial contrast sensitivity functions of a monkey reared with prism-induced strabismus. Note that the right eye (open circles) became amblyopic. (B) Ocular dominance distribution of V1 units for the amblyopic monkey in A. Note that despite a severe loss of binocularly balanced cells, the ocular dominance distribution did not favor the non-amblyopic eye. (C) The peak binocular response amplitude of individual units as a function of their ocular dominance index values. Units with the peak binocular/monocular response ratios less than 1.0 are considered as those exhibiting interocular suppression. (D) The peak binocular/monocular ratios of individual units as a function of their spatial resolution. Note that the units with high spatial resolution are more likely to be binocularly suppressed (i.e., the binocular/monocular response ratios are less than 1.0).

these eye movement asymmetries reflect altered cortical inputs to the subcortical nuclei that control eye movements and visual tracking (e.g., the nucleus of optic tract, NOT). One hypothesis based on mVEP studies in primates states that the oculomotor asymmetry is a direct consequence of a similar nasotemporal asymmetry in visual motion processing in the early cortical visual areas (e.g., V1/V2).[11] However, in our single-unit studies of V1 and V2 in monkeys reared with either optical or surgical strabismus, we found no overabundance of units that preferred stimulus motion in the temporal to nasal direction. Instead, we found a significant reduction in overall directional bias and direction selectivity of V1 and V2 neurons associated

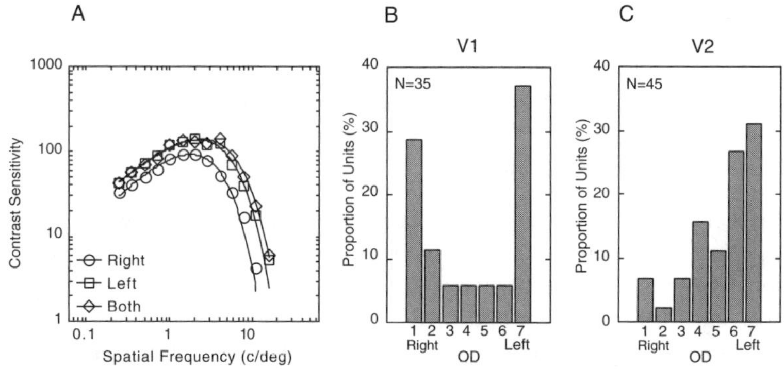

FIGURE 4.5 (A) Contrast sensitivity functions of a monkey reared with a surgical esotropia. Note that the right eye became amblyopic, and that virtually no binocular summation of contrast sensitivity was found for this monkey. (B) Ocular dominance distribution of V1 units for the amblyopic monkeys in A. Note that despite a clear loss of binocularly balanced units, the distribution did not favor the non-amblyopic (left) eye. (C) Ocular dominance distribution of V2 neurons in the same monkeys as in A and B. Note a drastic reduction of the units dominated by the amblyopic (right) eye.

with an increased interocular suppression.[154,156] The laminar analysis of all units indicated that the units in layer V, most of which project to the midbrain, also failed to show an obvious nasotemporal asymmetry and that their overall direction selectivity is significantly reduced as in the other layers. Thus, a recent hypothesis that only units in layer V are affected by early strabismus cannot be confirmed.[101] An alternative view is that the oculomotor asymmetries in strabismic monkeys may result from an interocular imbalance in the inputs from the MT/MST areas to the brain stem, rather than an alteration in cortical visual motion processing.[132] The data from our laboratory for both normal neonatal and strabismic monkeys indicate that impoverished cortical signals (reduced direction selectivity and/or sluggish responses) to the brain stem nuclei may be the primary reason for the observed oculomotor asymmetries.[32,156]

4.4 SUMMARY AND CONCLUSION

4.4.1 Normal Development

What constrains binocular vision in primates near birth? Previous studies have given us only a partial answer to this critical question in vision development. The functional connections that can support binocular vision are present in V1 and are operational at birth. However, the overall responsiveness and the monocular response properties of individual neurons are critically subnormal. During the first 4 weeks of life, neuronal responses rapidly mature to a qualitatively adult level just preceding the emergence of stereopsis and subsequently show a slow improvement until about 1 year of age. This substantial immaturity in the monocular response properties and responsiveness of V1 neurons during the first 4 postnatal weeks appears to limit fine binocular vision development by imposing a constraint on the neuronal functioning of higher-order visual areas, the neural sites that are strongly implicated for stereopsis in adults. Taken together, future studies may greatly benefit from focusing on the investigation of how the extrastriate visual areas mature after birth.

4.4.2 Abnormal Development

Many aspects of binocular anomalies resulting from abnormal visual experience early in life can be explained by response alterations of the neurons in V1. The developing binocular connections in V1 are extremely sensitive to binocularly discordant signals. Only 7 days of optically induced strabismus, beginning at the known onset age for stereopsis in monkeys (4 weeks of age), is sufficient to break down the sensitivity of V1 neurons to binocular disparity and only 3 days of misalignment can initiate abnormal binocular interactions (i.e., interocular suppression). In contrast to the critical period of normal development, the plastic period in monkeys during which the binocular response properties of V1 neurons can be altered by early visual experience appears to be very long. Surgical strabismus at 6 months of age (roughly equivalent to 2 years of age in humans) results in stereo-deficiencies and binocular neuronal response deficits that are as severe as those if the onset of strabismus is much earlier, e.g., 3 or 6 weeks of age. Monocular form deprivation at 2 years of age (8 years in humans) causes a severe loss of behaviorally measured binocular function. However, the "end" of plastic periods for binocular functions has not yet been determined for monkeys.

Amblyopia, "dull vision," is a far more complex phenomenon than traditionally thought. Although some impressive theories exist to explain the neural basis of amblyopia (e.g., "undersampling" vs. "topographic jitter"), a great majority of these theories have not been tested directly by neurophysiological experiments in monkey models of strabismic amblyopia. In V1 of severely amblyopic monkeys, neurons dominated by the amblyopic eye are less frequently encountered and typically show lower contrast sensitivity for mid- to high-spatial-frequency stimuli. These neuronal deficits may explain the well-documented contrast sensitivity loss in form deprivation amblyopia and severe cases of anisometropic or strabismic amblyopia. However, the neural constraints imposed by altered visual input that may lead to position uncertainty, poor vernier acuity, crowding, and deficits in contour integration or other higher-order cognitive functions in strabismic subjects have not been extensively investigated in monkeys. Systematic studies of the extrastriate visual areas in monkeys reared with abnormal binocular experience are necessary to understand the true nature of amblyopia, including the potential role of binocular suppression in its development.

ACKNOWLEDGMENTS

This work and the original experiments described here are supported by National Institutes of Health Grants, RO1 EY-08182 and RR-07146. The author thanks Ronald S. Harwerth for making comments on the earlier version of this chapter and for providing behavioral data, and Harold Bedell for his comments.

REFERENCES

1. Boothe, R.G., Dobson, V., and Teller, D.Y., Postnatal development of vision in human and nonhuman primates, *Annu. Rev. Neurosci.*, 8, 495–545, 1985.
2. Weisel, T.N., Postnatal development of the visual cortex and the influence of environment, *Nature,* 299, 583, 1982.
3. Fuchs, A.F. and Mustari, M.J., The optokinetic response in primates and its possible neuronal substrate, in Miles, S.A. and Wallman, J., Eds., *Visual Motion and Its Role in the Stabilization of Gaze,* Elsevier, New York, 1993, 343–369.
4. Jacobs, D.S. and Blakemore, C., Factors limiting the postnatal development of visual acuity in the monkey, *Vision Res.*, 28(8), 947–958, 1988.
5. Daw, N.W., *Visual Development*, Plenum Press, New York, 1995.
6. Howard, I.P. and Rogers, B.J., *Seeing in Depth,* Porteous I, Toronto, 2002.
7. Atkinson, J. and Braddick, O., Development of optokinetic nystagmus in infants: an indicator of cortical binocularity? in Fisher, D.F., Monty, R.A., and Senders, J.W., Eds., *Eye Movements: Cognition and Vision Perception,* Erlbaum, Hillsdale, NJ, 1981, 53–64.
8. Demer, J. and Von Noorden, G., Optokinetic asymmetry in esotropia, *J. Pediatr. Ophthalmol. Strabismus*, 25, 286–292, 1988.
9. Nocia, A.M., Garcia, H., Humphrey, R., Holmes, A., Hamer, R.D., and Orelbixler, D., Anomalous motion VEPs in infants and infantile esotropia, *Invest. Ophthalmol. Vis. Sci.*, 32, 436–439, 1991.
10. Teller, D.Y., Succop, A., and Mar, C., Infant eye movement asymmetries: stationary counterphase gratings elicit temporal-to-nasal optokinetic nystagmus in two-month-old infants under monocular test conditions, *Vision Res.*, 33(13), 1859–1864, 1993.
11. Brown, R.J., Wilson, J.R., Norcia, A.M., and Boothe, R.G., Development of directional motion symmetry in the monocular visually evoked potential of infant monkeys, *Vision Res.*, 38(9), 1253–1263, 1998.
12. Kiorpes, L., Development of vernier acuity and grating acuity in normally reared monkeys, *Vis. Neurosci.*, 9(3–4), 243–251, 1992.
13. Kiorpes, L. and Kiper, D.C., Development of contrast sensitivity across the visual field in macaque monkeys (*Macaca nemestrina*), *Exp. Brain Res.*, 36(2), 239–247, 1996.
14. Boothe, R.G., Kiorpes, L., Williams, R.A., and Teller, D.Y., Operant measurements of contrast sensitivity in infant macaque monkeys during normal development, *Exp. Brain Res.*, 28(3), 387–396, 1988.
15. Movshon, J.A. and Kiorpes, L., Analysis of the development of spatial contrast sensitivity in monkey and human infants, *J. Opt. Soc. Am. A,* 5(12), 2166–2172, 1988.
16. Rakic, P., Prenatal development of the visual system in the rhesus monkey, *Philos. Trans. R. Soc. Lond. B*, 278, 245–260, 1977.
17. Rakic, P., Genesis of the dorsal lateral geniculate nucleus in the rhesus monkey: site and time of origin, kinetics of proliferation, routes of migration and pattern of distribution of neurons, *J. Comp. Neurol.*, 176, 23–52, 1977.
18. Blakemore, C. and Vital-Durand, F., Organization and post-natal development of the monkey's lateral geniculate nucleus, *J. Physiol.*, 380, 453–491, 1986.
19. Hawken, M.J., Blakemore, C., and Morley, J.W., Development of contrast sensitivity and temporal-frequency selectivity in primate lateral geniculate nucleus, *Exp. Brain Res.*, 114(1), 86–98, 1997.
20. Hendrickson, A. and Kupfer, C., The histogenesis of the fovea in the macaque monkey, *Invest. Ophthalmol. Vis. Sci.*, 15(9), 746–756, 1976.

21. Packer, O., Hendrickson, A.E., and Curcio, C.A., Development redistribution of photoreceptors across the *Macaca nemestrina* (pigtail macaque) retina, *J. Comp. Neurol.*, 298(4), 472–493, 1990.
22. Banks, M.S. and Bennett, P.J., Optical and photoreceptor immaturities limit the spatial and chromatic vision of human neonates, *J. Opt. Soc. Am. A*, 5(12), 2059–2079, 1988.
23. Horton, J.C. and Hocking, D.R., An adult-like pattern of ocular dominance columns in striate cortex of new born monkeys prior to visual experience, *J. Neurosci.,* 16, 1791–1807, 1996.
24. Endo, M., Kaas, J.H., Jain, N., Smith, E.L., III, and Chino, Y., Binocular cross-orientation suppression in the primary visual cortex (V1) of infant rhesus monkeys, *Invest. Ophthalmol. Vis. Sci.*, 41(12), 4022–4031, 2000.
25. Kennedy, H., Bullier, J., and Dehay, C., Cytochrome oxidase activity in the striate cortex and lateral geniculate nucleus of the newborn and adult macaque monkey, *Exp. Brain Res.*, 61(1), 204–209, 1985.
26. Florence, S.L. and Casagrande, V.A., Development of geniculocortical axon arbors in a primate, *Vis. Neurosci.*, 5(3), 291–309, 1990.
27. Pospichal, M.W., Florence, S.L., and Kaas, J.H., The postnatal development of geniculocortical axon arbors in owl monkeys, *Vis. Neurosci.*, 11(1), 71–90, 1994.
28. Boothe, R.G., Greenough, W.T., Lund, J.S., and Wrege, K., A quantitative investigation of spine and dendrite development of neurons in visual cortex (area 17) of *Macaca nemestrina* monkeys, *J. Comp. Neurol.*, 186(3), 473–489, 1979.
29. Mates, S.L. and Lund, J.S., Neuronal composition and development in lamina 4C of monkey striate cortex, *J. Comp. Neurol.*, 221(1), 60–90, 1983.
30. Hubel, D.H. and Wiesel, T.N., Sequence regularity and geometry of orientation columns in the monkey striate cortex, *J. Comp. Neurol.*, 158(3), 267–293, 1974.
31. Chino, Y.M., Smith, E.L., III, Hatta, S., and Cheng, H., Postnatal development of binocular disparity sensitivity in neurons of the primate visual cortex, *J. Neurosci.*, 17(1), 296–307, 1997.
32. Hatta, S., Kumagami, T., Qian, J., Thornton, M., Smith, E.L., III, and Chino, Y.M., Nasotemporal directional bias of V1 neurons in young infant monkeys, *Invest. Ophthalmol. Vis. Sci.*, 39(12), 2259–2267, 1998.
33. Burkhalter, A., Bernardo, K.L., and Charles, V., Development of local circuits in human visual cortex, *J. Neurosci.*, 13(5), 1916–1931, 1993.
34. Movshon, J.A., Kiorpes, L., James, R., Cavanaugh, R., and Hawken, J., Receptive field properties and surround interactions in V1 neurons in infant macaque monkeys, *Soc. Neurosci. Abstr.,* 25, 1048, 1999.
35. Movshon, J.A., Kiorpes, L., Cavanaugh, J.R., and Hawken, M.J., Developmental reorganization of receptive field surrounds in V1 neurons in macaque monkeys, *Invest. Ophthalmol. Vis. Sci.* (Suppl.), 41(4), S333, 2000.
36. Katz, L.C. and Calloway, E.M., Development of local circuits in mammalian visual cortex, *Annu. Rev. Neurosci.,* 15, 31–56, 1992.
37. Katz, L.C. and Shatz, C.J., Synaptic activity and the construction of cortical circuits, *Science,* 274, 1133–1138, 1996.
38. Weiliky, M.K. and Katz, L., Functional mapping of horizontal connections in developing ferret visual cortex: experiments and modeling, *J. Neurosci.,* 14, 7291–7305, 1997.
39. Das, A. and Gilbert, C.D., Long-range horizontal connections and their role in cortical reorganization revealed by optical recording of cat primary visual cortex, *Nature,* 375(6534), 780–784, 1995.
40. Knierim, J.J. and Van Essen, D.C., Visual cortex: cartography, connectivity, and concurrent processing, *Curr. Opin. Neurobiol.*, 2(2), 150–155, 1992.

41. Walker, G.A., Ohzawa, I., and Freeman, R.D., Suppression outside the classical cortical receptive field, *Vis. Neurosci.*, 17(3), 369–379, 2000.
42. Gilbert, C.D., Das, A., Ito, M., Kapadia, M., and Westheimer, G., Spatial integration and cortical dynamics, *Proc. Natl. Acad. Sci. U.S.A.,* 93, 615–622, 1996.
43. Hirsch, J.A. and Gilbert, C.D., Synaptic physiology of horizontal connections in the cat's visual cortex, *J. Neurosci.*, 11(6), 1800–1809, 1991.
44. Lund, J.S., Wu, Q., Hadingham, P.T., and Levitt, J.B., Cells and circuits contributing to functional properties in area V1 of macaque monkey cerebral cortex: bases for neuroanatomically realistic models, *J. Anat.*, 187(3), 563–581, 1995.
45. Toth, L.J., Rao, S.C., Kim, D.S., Somers, D., and Sur, M., Subthreshold facilitation and suppression in primary visual cortex revealed by intrinsic signal imaging, *Proc. Natl. Acad. Sci. U.S.A.*, 93(18), 9869–9874, 1996.
46. Levitt, J.B. and Lund, J.S., Contrast dependence of contextual effects in primate visual cortex, *Nature*, 387(6628), 73–76, 1997.
47. Tychsen, L., Hanaway, K., and Burkhalter, A., Horizontal connections of blob and interblob compartments in neonatal monkey visual cortex, *Invest. Ophthalmol. Vis. Sci.* (Suppl.), 38, 621, 1997.
48. Levitt, J.B., Lund, J.S., and Yoshioka, T., Anatomical substrates for early stages in cortical processing of visual information in the macaque monkey, *Behav. Brain Res.*, 76(1–2), 5–19, 1996.
49. Hoffmann, K.P., Control of the optokinetic reflex by the nucleus of the optic tract in primates, *Prog. Brain Res.*, 80, 173–182, 1989.
50. Mustari, M.J. and Fuchs, A.F., Discharge patterns of neurons in the pretectal nucleus of the optic tract (NOT) in the behaving primate, *J. Neurophysiol.*, 64(1), 77–90, 1990.
51. Fuchs, A.F. and Mustari, M.J., The optokinetic response in primates and its possible neuronal substrate, in *Reviews of Oculomotor Research. Visual Motion and Its Role in Stabilization of the Gaze*, Vol. 5, Wallman, J. and Miles, F.A., Eds., Elsevier, Amsterdam, 1993, 343–369.
52. Rodman, H.R., Skelly, J.P., and Gross, C.G., Stimulus selectivity and state dependence of activity in inferior temporal cortex of infant monkeys, *Proc. Natl. Acad. Sci. U.S.A.*, 88(17), 7572–7575, 1991.
53. Rodman, H.R., Scalaidhe, S.P., and Gross, C.G., Response properties of neurons in temporal cortical visual areas of infant monkeys, *J. Neurophysiol.*, 70(3), 1115–1136, 1993.
54. Rodman, H.R. and Consuelos, M.J., Cortical projections to anterior inferior temporal cortex in infant macaque monkeys, *Vis. Neurosci.*, 11(1), 119–133, 1994.
55. Tanaka, K., Inferotemporal cortex and object vision, *Annu. Rev. Neurosci.*, 19, 109–139, 1996.
56. Tanaka, K., Mechanisms of visual object recognition: monkey and human studies, *Curr. Opin. Neurobiol.,* 7 (4), 523–529, 1997.
57. Tanaka, K., Mechanisms of visual object recognition studies in monkeys, *Spat. Vis.*, 13(2–3), 147–163, 2000.
58. O'Dell, C. and Boothe, R.G., The development of stereoacuity in infant rhesus monkeys, *Vision Res.*, 37(19), 2675–2684, 1997.
59. Birch, E.E., Gwiazda, J., and Held, R., Stereoacuity development for crossed and uncrossed disparities in human infants, *Exp. Brain Res.,* 22, 507–513, 1982.
60. Birch, E.E., Stereopsis in infants and its developmental relation to visual acuity, in *Early Development: Normal and Abnormal*, Simons, K., Ed., Oxford University Press, Oxford, U.K., 1993, 224–236.

61. Rakic, P., Prenatal genesis of connections subserving ocular dominance in the rhesus monkey, *Nature,* 261, 467–471, 1976.
62. Hubel, D.H., Wiesel, T.N., and LeVay, S., Plasticity of ocular dominance columns in monkey striate cortex, *Philos. Trans. R. Soc. Lond. B Biol. Sci.*, 278(961), 377–409, 1977.
63. LeVay, S., Wiesel, T.N., and Hubel, D.H., The development of ocular dominance columns in normal and visually deprived monkeys, *J. Comp. Neurol.,* 191, 1–51, 1980.
64. Held, R., Two stages in the development of binocular vision and eye alignment, in *Early Development, Normal and Abnormal,* Simons, K., Ed., Oxford University Press, New York, 1993, 250–257.
65. Shaw, C.L., Cameron, D., March, M., Cynader, B., Zielienski, B., and Henndrickson, A., Pre- and postnatal development of GABA receptors in *Macaca* monkey visual cortex, *J. Neurosci.*, 11, 3943–3959, 1991.
66. Wiesel, T.N. and Hubel, D.H., Comparison of the effects of unilateral and bilateral eye closure on cortical unit responses in kittens, *J. Neurophysiol.*, 28(6), 1029–1040, 1965.
67. Sherman, S.M. and Spear, P.D., Organization of visual pathways in normal and visually deprived cats, *Physiol. Rev.*, 62(2), 738–855, 1982.
68. Singer, W., Development and plasticity of cortical processing architectures, *Science*, 270(5237), 758–764, 1995.
69. Movshon, JA. and Kiorpes, L., The role of experience in visual development, in *Development of Sensory Systems in Mammals,* J.R. Coleman, Ed., John Wiley & Sons, New York, 1990, 156–202.
70. Movshon, J.A. and Kiorpes, L., Biological limits on visual development in primates, in *Early Development; Normal and Abnormal,* K. Simons, Ed., Oxford University Press, New York, 1993, 296–305.
71. Katz L.C. and Crowley, J.C., Development of cortical circuits: lessons from ocular dominance columns, *Nat. Rev. Neurosci.*, 3(1), 34–42, 2002.
72. Harwerth, R.S., Boltz, R.L., and Smith, E.L., III, Psychophysical evidence for sustained and transient channels in the monkey visual system, *Vision Res.*, 20(1), 15–22, 1980.
73. Harwerth, R.S., Smith, E.L., III, Crawford, M.L., and von Noorden, G.K., Behavioral studies of the sensitive periods of development of visual functions in monkeys, *Behav. Brain Res.*, 41(3), 179–198, 1990.
74. Boothe, R.G., Williams, R.A., Kiorpes, L., and Teller, D.Y., Development of contrast sensitivity in infant *Macaca nemestrina* monkeys, *Science*, 208(4449), 1290–1292, 1980.
75. Crawford, M.L., Harwerth, R.S., Chino, Y.M., and Smith, E.L., III, Binocularity in prism-reared monkeys, *Eye*, 10(2), 161–166, 1996.
76. Daw, N.W., Critical periods and amblyopia, *Arch. Ophthalmol.*, 116(4), 502–505, 1998.
77. Sengpiel, F. and Blakemore, C., The neural basis of suppression and amblyopia in strabismus, *Eye*, 10(2), 250–258, 1996.
78. Kiorpes, L. and McKee, S.P., Neural mechanisms underlying amblyopia, *Curr. Opin. Neurobiol.*, 9(4), 480–486, 1999.
79. Hubel, D.H., Wiesel, T.N., and LeVay, S., Plasticity of ocular dominance columns in monkey striate cortex, *Philos. Trans. R. Soc. Lond. B Biol. Sci.*, 278(961), 377–409, 1977.
80. Horton, J.C. and Hocking, D.R., Timing of the critical period for plasticity of ocular dominance columns in macaque striate cortex, *J. Neurosci.*, 17, 3684–3709, 1997.

81. von Noorden, G.K., Experimental amblyopia in monkeys. Further behavioral observations and clinical correlations, *Invest. Ophthalmol.*, 12(10), 721–726, 1973.
82. Hendrickson, A., Boles, J., and McLean, E.B., Visual acuity and behavior of monocularly deprived monkeys after retinal lesions, *Invest. Ophthalmol. Vis. Sci.*, 16(5), 469–473, 1977.
83. Harwerth, R.S., Crawford, M.L., Smith E.L., III, and Boltz, R.L., Behavioral studies of stimulus deprivation amblyopia in monkeys, *Exp. Brain Res.*, 21(6), 779–789, 1981.
84. Harwerth, R.S., Smith, E.L., III, Boltz, R.L., Crawford, M.L., and von Noorden, G.K., Behavioral studies on the effect of abnormal early visual experience in monkeys: spatial modulation sensitivity, *Exp. Brain Res.*, 23(12), 1501–1510, 1983.
85. Headon, M.P. and Powell, T.P., Cellular changes in the lateral geniculate nucleus of infant monkeys after suture of the eyelids, *J. Anat.*, 116(1), 135–145, 1973.
86. Vital-Durand, F., Garey, L., and Blakemore, C., Monocular and binocular deprivation in the money: morphological effects and their reversibility, *Brain Res.*, 158, 45–64, 1978.
87. von Noorden, G.K. and Crawford, M.L., Morphological and physiological changes in the monkey visual system after short-term lid suture, *Invest. Ophthalmol. Vis. Sci.*, 17(8), 762–768, 1978.
88. Blakemore, C. and Vital-Durand, F., Effects of visual deprivation on the development of the monkey's lateral geniculate nucleus, *J. Physiol.* 380, 493–511, 1986.
89. Sur, M., Humphrey, A.L., and Sherman, S.M., Monocular deprivation affects X- and Y-cell retinogeniculate terminations in cats, *Nature*, 300(5888), 183–185, 1982.
90. Garraghty, P.E. and Sur, M., Competitive interactions influencing the development of retinal axonal arbors in cat lateral geniculate nucleus, *Physiol. Rev.*, 73(3), 529–545, 1993.
91. Cheng, H., Chino, Y.M., Smith, E.L., III, Hamamoto, J., and Yoshida, K., Transfer characteristics of X LGN neurons in cats reared with early discordant binocular vision, *J. Neurophysiol.*, 74(6), 2558–2572, 1995.
92. Chino, Y.M., Smith, E.L., III, Yoshida, K., Cheng, H., and Hamamoto, J., Binocular interactions in striate cortical neurons of cats reared with discordant visual inputs, *J. Neurosci.*, 14(8), 5050–5067, 1994.
93. Issa, N.P., Trachtenberg, J.T., Chapman, B., Zahs, K.R., and Stryker, M.P., The critical period for ocular dominance plasticity in the ferret's visual cortex, *J. Neurosci.*, 19(16), 6965–6978, 1999.
94. Birch, E.E. and Stager, D.R., Prevalence of good visual acuity following surgery for congenital unilateral cataract, *Arch. Ophthalmol.*, 106(1), 40–43, 1988.
95. Drummond, G.T., Scott, W.E., and Keech, R.V., Management of monocular congenital cataracts, *Arch. Ophthalmol.*, 107(1), 45–51, 1989.
96. Wright, K.W., Matsumoto, E., and Edelman, P.M., Binocular fusion and stereopsis associated with early surgery for monocular congenital cataracts, *Arch. Ophthalmol.*, 110(11), 1607–1609, 1992.
97. Jeffrey, B.G., Birch, E.E., Stager, D.R., Jr., Stager, D.R., Sr., and Weakley, D.R., Jr., Early binocular visual experience may improve binocular sensory outcomes in children after surgery for congenital unilateral cataract, *J. AAPOS*, 5(4), 209–216, 2001.
98. Harwerth, R.S., Smith, E.L., III, Duncan, G.C., Crawford, M.L., and von Noorden, G.K., Multiple sensitive periods in the development of the primate visual system, *Science*, 232(4747), 235–238, 1986.
99. Zhang, B., Mori, T., Smith, E.L., III, and Chino, Y.M., Contrasting effects of early alternating occlusion and strabismus on the development of neurons in monkey visual cortex, *Invest. Ophthalmol. Vis. Sci.* (Suppl.), 42(4), S53, 2001.

100. Smith, E.L., III, Chino, Y.M., Ni, J., Cheng, H., Crawford, M.L.J., and Harwerth, R.S., Residual binocular interactions in the striate cortex of monkeys reared with abnormal binocular vision, *J. Neurophysiol.,* 78, 1353–1362, 1997.
101. Fu, L.N. and Boothe, R.G., A psychophysical measurement and analysis of motion perception in normal and binocularly deprived monkeys, *Invest. Ophthalmol. Vis. Sci.*, 42(11), 2547–2553, 2001.
102. Fu, L.N., Das, V.E., Mustari, M.J., and Tusa, R.J., Saccadic disconjugacy in monkeys with strabismus, *Soc. Neurosci. Abstr.,* 71, 42, 2001.
103. Kiorpes, L., Boothe, R.G., Hendrickson, A.E., Movshon, J.A., Eggers, H.M., and Gizzi, M.S., Effects of early unilateral blur on the macaque's visual system. I. Behavioral observations, *J. Neurosci.*, 7(5), 1318–1326, 1987.
104. Kiorpes, L., Kiper, D.C., O'Keefe, L.P., Cavanaugh, J.R., and Movshon, J.A., Neuronal correlates of amblyopia in the visual cortex of macaque monkeys with experimental strabismus and anisometropia, *J. Neurosci.*, 18(16), 6411–6424, 1998.
105. Smith, E.L., III, Harwerth, R.S., and Crawford, M.L., Spatial contrast sensitivity deficits in monkeys produced by optically induced anisometropia, *Invest. Ophthalmol. Vis. Sci.*, 26(3), 330–342, 1985.
106. Hendrickson, A.E., Movshon, J.A., Eggers, H.M., Gizzi, M.S., Boothe, R.G., and Kiorpes, L., Effects of early unilateral blur on the macaque's visual system. II. Anatomical observations, *J. Neurosci.*, 7(5), 1327–1339, 1987.
107. Horton, J.C. and Hocking, D.R., Effect of early monocular enucleation upon ocular dominance columns and cytochrome oxidase activity in monkey and human visual cortex, *Vis. Neurosci.*, 15(2), 289–303, 1998.
108. Movshon, J.A., Eggers, H.M., Gizzi, M.S., Hendrickson, A.E., Kiorpes, L., and Boothe, R.G., Effects of early unilateral blur on the macaque's visual system. III. Physiological observations, *J. Neurosci.*, 7(5), 1340–1351, 1987.
109. Smith, E.L., III, Chino, Y.M., Ni, J., Ridder, W.H., III, and Crawford, M.L., Binocular spatial phase tuning characteristics of neurons in the macaque striate cortex, *J. Neurophysiol.,* 78(1), 351–365, 1997.
110. Smith, E.L., III, Chino, Y., Ni, J., and Cheng, H., Binocular combination of contrast signals by striate cortical neurons in the monkey, *J. Neurophysiol.*, 78(1), 366–382, 1997.
111. Wensveen, J.M., Harwerth, R.S., and Smith, E., III, Binocular deficits associated with early anisometropia: I. Behavioral observations, in revision.
112. Zhang, B., Matsuura, K., Mori, T., Wensveen, J.M., Harwerth, R.S., Smith, E., III, and Chino, Y., Binocular deficits associated with early anisometropia. II. Neurophysiological observations, in revision.
113. Blakemore, C. and Hague, B., Evidence for disparity detecting neurones in the human visual system. *J. Physiol.*, 225(2), 437–455, 1972.
114. Felton, T.B., Richards, W., and Smith, R.A., Jr., Disparity processing of spatial frequencies in man, *J. Physiol.*, 225(2), 349–362, 1972.
115. Julesz, B. and Miller, J.E., Independent spatial-frequency-tuned channels in binocular fusion and rivalry, *Perception,* 4, 125–143, 1975.
116. Schor, C.M., Wood, I.C., and Ogawa, J., Spatial tuning of static and dynamic local stereopsis, *Vision Res.*, 24(6), 573–578, 1984.
117. Yang, Y. and Blake, R., Spatial frequency tuning of human stereopsis, *Vision Res.*, 31(7–8), 1177–1189, 1991.
118. Smallman, H.S. and MacLeod, D.I., Size-disparity correlation in stereopsis at contrast threshold, *J. Opt. Soc. Am. A*, 11(8), 2169–2183, 1994.

119. DeAngelis, G.C., Ohzawa, I., and Freeman, R.D., Neuronal mechanisms underlying stereopsis: how do simple cells in the visual cortex encode binocular disparity? *Perception*, 24(1), 3–31, 1995.
120. Cumming, B.G. and Parker, A.J., Responses of primary visual cortical neurons to binocular disparity without depth perception, *Nature*, 389(6648), 280–283, 1997.
121. Cumming, B.G., Shapiro, S.E., and Parker, A.J., Disparity detection in anticorrelated stereograms, *Perception*, 27(11), 1367–1377, 1998.
122. Cumming, B.G. and Parker, A.J., Binocular neurons in V1 of awake monkeys are selective for absolute, not relative, disparity, *J. Neurosci.*, 19(13), 5602–5618, 1999.
123. Cumming, B.G. and Parker, A.J., Local disparity not perceived depth is signaled by binocular neurons in cortical area V1 of the macaque, *J. Neurosci.*, 20(12), 4758–4767, 2000.
124. von Noorden, G.K., *Binocular Vision and Ocular Motility: Theory and Management of Strabismus,* 5th ed., Mosby, St. Louis, 1996.
125. Crawford, M.L.J. and von Noorden, G.K., The effects of short-term experimental strabismus on the visual system in *Macaca mulatta, Invest. Ophthalmol. Vis. Sci.,* 18, 496–505, 1979.
126. Smith, E.L., III, Bennett, M.J., Harwerth, R.S., and Crawford, M.L., Binocularity in kittens reared with optically induced squint, *Science*, 204(4395), 875–877, 1979.
127. Crawford, M.L.J. and von Noorden, G.K., Optically induced commitant strabismus in monkeys, *Invest. Ophthalmol. Vis. Sci.,* 19, 1105–1109, 1980.
128. Kumagami, T., Zhang, B., Smith, E.L., III, and Chino, Y.M., Effect of onset age of strabismus on the binocular responses of neurons in the monkey visual cortex, *Invest. Ophthalmol. Vis. Sci.,* 41(3), 948–954, 2000.
129. Mori, T., Matsuura, K., Zhang, B., Smith, E.L., III, and Chino, Y.M., Effects of the duration of early strabismus on the binocular responses of neurons in the monkey visual cortex (V1), *Invest. Ophthalmol. Vis. Sci.*, 43(4), 1262–1269, 2002.
130. von Noorden, G.K. and Middleditch, P.R., Histology of the monkey lateral geniculate nucleus after unilateral lid closure and experimental strabismus: further observations, *Invest. Ophthalmol.*, 14(9), 674–683, 1975.
131. Crawford, M.L., von Noorden, G.K., Meharg, L.S., Rhodes, J.W., Harwerth, R.S., Smith, E.L., III, and Miller, D.D., Binocular neurons and binocular function in monkeys and children, *Invest. Ophthalmol. Vis. Sci.*, 24(4), 491–495, 1983.
132. Kiorpes, L., Walton, P.J., O'Keefe, L.P., Movshon, J.A., and Lisberger, S.G., Effects of early-onset artificial strabismus on pursuit eye movements and on neuronal responses in area MT of macaque monkeys, *J. Neurosci.,* 16, 6537–6553, 1996.
133. Bi, H., Zhang, B., Mori, T., Sakai, E., Noska, N., Harwerth, R.S., Smith, E.L., III, and Chino, Y., Effects of early strabismus on the development of binocular response properties of neurons in visual area 2 (V2) of rhesus monkeys, *ARVO Abstr.,* 2002.
134. Crawford, M.L., Harwerth, R.S., Smith, E.L., and von Noorden, G.K., Loss of stereopsis in monkeys following prismatic binocular dissociation during infancy, *Behav. Brain Res.*, 79(1–2), 207–218, 1996.
135. Harwerth, R.S., Smith, E.L., III, Crawford, M.L., and von Noorden, G.K., Stereopsis and disparity vergence in monkeys with subnormal binocular vision, *Vision Res.*, 37(4), 483–493, 1997.
136. Ridder, W.H., Psychophysical Investigations of Binocular Summation in Normal and Amblyopic Vision, Ph.D. dissertation, University of Houston, Houston, TX, 1989.
137. Wensveen, J.M., Harwerth, R.S., and Smith, E.L., III, Clinical suppression in monkeys reared with abnormal binocular visual experience, *Vision Res.*, 41(12), 1593–1608, 2001.

138. Wright, K.W., Strabismus management, *Curr. Opin. Ophthalmol.*, 5(5), 25–29, 1994.
139. Birch, E.E., Stager, D.R., and Everett, M.E., Random dot stereoacuity following surgical correction of infantile esotropia, *J. Pediatr. Ophthalmol. Strabismus*, 32(4), 231–235, 1995.
140. Birch, E.E., Fawcett, S., and Stager, D.R., Why does early surgical alignment improve stereoacuity outcomes in infantile esotropia? *J. AAPOS*, 4(1), 10–14, 2000.
141. Levi, M. and Harwerth, R.S., Spatio-temporal interactions in anisometropic and strabismic amblyopia, *Invest. Ophthalmol. Vis. Sci.*, 16(1), 90–95, 1977.
142. Hess, R.F. and Howell, E.R., The threshold contrast sensitivity function in strabismic amblyopia: evidence for two types of classification, *Vision Res.*, 17, 1049–1055, 1977.
143. Levi, D.M. and Klein, S.A., Hyperacuity and amblyopia, *Nature*, 298(5871), 268–270, 1982.
144. Levi, D.M. and Klein, S.A., Vernier acuity, crowding and amblyopia, *Vision Res.*, 25(7), 979–991, 1985.
145. Levi, D.M. and Klein, S.A., Sampling in spatial vision, *Nature*, 320(6060), 360–362, 1986.
146. Sharma, V., Levi, D.M., and Klein, S.A., Undercounting features and missing features: evidence for a high-level deficit in strabismic amblyopia, *Nat. Neurosci.*, 3(5), 496–501, 2000.
147. Sasaki, Y., Cheng, H., Smith, E.L., III, and Chino, Y., Effects of early discordant binocular vision on the postnatal development of parvocellular neurons in the monkey lateral geniculate nucleus, *Exp. Brain Res.*, 118(3), 341–351, 1998.
148. Kiorpes, L., Kiper, D.C., and Movshon, J.A., Contrast sensitivity and vernier acuity in amblyopic monkeys, *Vision Res.* 33(16), 2301–2311, 1993.
149. Chandna, A., Pennefather, P.M., Kovacs, I., and Norcia, A.M., Contour integration deficits in anisometropic amblyopia, *Invest. Ophthalmol. Vis. Sci.*, 42(3), 875–878, 2001.
150. Hess, RF. and Field, D.J., Is the spatial deficit in strabismic amblyopia due to loss of cells or an uncalibrated disarray of cells? *Vision Res.*, 34(24), 3397–3406, 1994.
151. Hess, R.F., McIlhagga, W., and Field, D.J., Contour integration in strabismic amblyopia: the sufficiency of an explanation based on positional uncertainty,*Vision Res.*, 37(22), 3145–3161, 1997.
152. Levi, D.M., Klein, S.A., and Sharma, V., Position jitter and undersampling in pattern perception, *Vision Res.*, 39(3), 445–465, 1999.
153. Crawford, M.L., Smith, E.L., III, Harwerth, R.S., and von Noorden, G.K., Stereoblind monkeys have few binocular neurons, *Invest. Ophthalmol. Vis. Sci.*, 25(7), 779–781, 1984.
154. Qian, J., Kumagami, T., Thornton, M., Smith, E.L., III, Wensveen, J.M., Harwerth, R.S., and Chino, Y., Roles of primary visual cortex (V1) neurons in binocular suppression in strabismic amblyopia, *Invest. Ophthalmol. Vis. Sci.* (Suppl.), 39(4), S327, 1998.
155. Tychsen, L., Motion sensitivity and the origins of infantile strabismus, in *Early Visual Development, Normal and Abnormal,* Simons, K., Ed., Oxford University Press, New York, 1993, 364–390.
156. Sakai, E., Mori, T., Zhang, B., Bi, H., Noska, N., Harwerth, R.S., Smith, E.L., III, and Chino, Y., Directional bias of neurons in visual area 2 (V2) of monkeys reared with early strabismus, *ARVO Abstr.,* 2002.
157. Ohzawa, I. and Freeman, R.D., The binocular organization of simple cells in the cat's visual cortex, *J. Neurophysiol.*, 56(1), 221–242, 1986.

5 Modular Complexity of Area V2 in the Macaque Monkey

Anna W. Roe

CONTENTS

5.1 INTRODUCTION

Area V2 has traditionally been thought of as the second stage of visual cortical processing in the primate. In addition to receiving ascending feedforward inputs

0-8493-1243-4/04/$0.00+$1.50

from primary visual cortex (V1), V2 also receives thalamic input from the pulvinar as well as significant feedback from visual areas in both the ventral and dorsal streams. Thus, as a distribution center for ascending magnocellular, parvocellular, and koniocellular derived inputs from the lateral geniculate nucleus (LGN), V2 is strategically positioned as both an integrator and feedback control point of what and where information.

In both Old World and New World monkeys, when stained for cytochrome oxidase,[1] V2 is characterized by a pattern of alternating dark and light cytochrome oxidase stripes.[2,3] In general, these stripes run perpendicular to the V1/V2 border and alternate in a thin/pale/thick/pale manner from the central (lateral cortex) to peripheral (medial cortex) visual field representations in V2. In the macaque monkey there are approximately 14 sets of thin/pale/thick/pale stripes in dorsal V2, with each set spanning on average 4 mm across. This pattern has also been confirmed by other anatomical staining methods such as Cat-301 staining, which reveals the thick stripes,[4,5] and staining for NOS/NADPH.[6] These stripe organizations are established early in development and, at least from studies thus far, remain in the face of manipulations in visual experience.[7,8] Thus, V2 stripes can be viewed as developmental,[9] evolutionary[10] (cf. Reference 11), and/or functional[12,13] entities.

The long-standing view that the thin, pale, and thick stripes subserve a tripartite division of color, form, and depth information processing in the early visual pathway has been supported by electrophysiological,[12–16] 2-deoxyglucose,[17,18] and optical imaging studies.[13,19–21] This functional segregation in V2 has found compelling parallels in psychophysics of orientation, color, and depth perception.[22,23] However, these broad descriptions by no means suggest uniformity of any single stripe. In fact, it is clear that each stripe contains a mixture of response types,[16,21,24,25] that neuronal responses are multimodal in nature (e.g., Reference 26), and that there is a significant and important degree of form, color, and disparity integration in V2. Unraveling how functional segregation and multimodal integration are incorporated is the key to understanding V2 function.

It is also important to note that, in addition to their uneven and blobby appearance, V2 cytochrome stripes exhibit variations in patterning. At times, dark cytochrome stripes are seen merging with one another or failing to follow the classic thin/thick alternation. Furthermore, given the variability in the widths of dark cytochrome stripes, the relationship of "thick" cytochrome stripes with "disparity" and "thin" cytochrome stripes with "color" is a fickle one (for review, see Reference 27). Despite this, the terms "thin" and "thick" have become strongly associated with the functional terms "color" and "disparity." Following this usage, in this chapter, the terms "thin" and "thick" are based on the conceptual functional view of what thin and thick stripes are (i.e., "color" or "disparity," respectively) and not the thickness of stripe as seen in cytochrome oxidase staining.

5.1.1 Optical Imaging of V2

Perhaps one of the most compelling views of functional division in V2 is visualization of stripe compartments by optical imaging methods. Intrinsic signal optical imaging is a functional mapping method that measures changes in cortical reflectance derived from changes in hemodynamic signals (for review, see References 28 and 29). The primary advantage of this method is its high spatial resolution (tens of microns), which permits the visualization of 50 to 200 μm size functional domains. Because of the limitations of optical penetration and light scatter, detection is thus far limited to the superficial cortical layers (although cf. Reference 30). By definition, optical imaging reveals a representation of the local *population* response; that is, the predominant response summed over many cells. Therefore, (1) imaged response is still consistent with a diversity of single unit responses at a single locale and does not indicate unitary function at any single location, (2) it is useful for revealing local biases in representation that may not be evident from single unit studies, and (3) in combination with single unit studies, it can provide a joint single unit/local population view of the local neuronal response. Despite its limitations (superficial layers only, slow temporal resolution, limited to exposed cortex), intrinsic signal imaging has proved useful for mapping well-established functional organizations, for revealing novel organizations (e.g., References 31 and 32), and promises to be a useful tool for studying cortical activation in the awake-behaving monkey.[33–38]

In the macaque monkey, most of dorsal V2 is buried within the lunate sulcus, leaving a 0.5- to 2-mm-wide strip just posterior to the lunate available on the surface for imaging (Figure 5.1A). Because cells in V2 are predominantly binocular, the V2 border with V1 is clearly demarcated by the lack of ocular dominance columns (Figure 5.1B, upper panel). Functional maps are obtained by imaging the cortex during visual stimulation with stimuli such as achromatic sinusoidal gratings and isoluminant color gratings presented at different orientations. Stimulus-induced activation (usually seen as darkening or decrease in tissue reflectance) of cortical activity reveals functional organizations such as ocular dominance columns, orientation domains, blobs, and interblobs in V1, and stripe structures in V2. As shown in Figure 5.1, the thick/pale stripe location is clearly revealed by mapping for orientation (Figure 5.1B, lower panel). Dark cytochrome oxidase stripes can also be determined by mapping for general activation (Figure 5.1C, middle panel). Together with cytochrome oxidase histology (Figure 5.1C, top panel), which alone often leaves stripe identity uncertain, thick, pale, and thin stripes can be determined with reasonable confidence (Figure 5.1C). Other methods for mapping stripes, such as color or color vs. luminance response (see Figure 5.4 below), monocularity vs. binocularity (see Figure 5.6 below), and disparity response, will be discussed later in this chapter.

Recent studies have furthered our understanding of both the internal organization of V2 and its relationship to V1. These new developments, deriving from studies using anatomical tract tracing, optical imaging, and electrophysiological methods, have led to further modifications and refinement of the long-standing views of V1–V2 connectivity. These new views have developed hand-in-hand with new understandings of V2 architecture and hypotheses regarding its relationship to visual perception. The goal of this chapter is to review and integrate these recent developments in our understanding of V2, and propose a conceptual framework for the role of V2 in visual processing. The studies described center on studies in Old World monkeys (macaque monkey). As there are other excellent reviews describing V2 function and organization (e.g., References 10, 39, and 40), this chapter is not comprehensive,

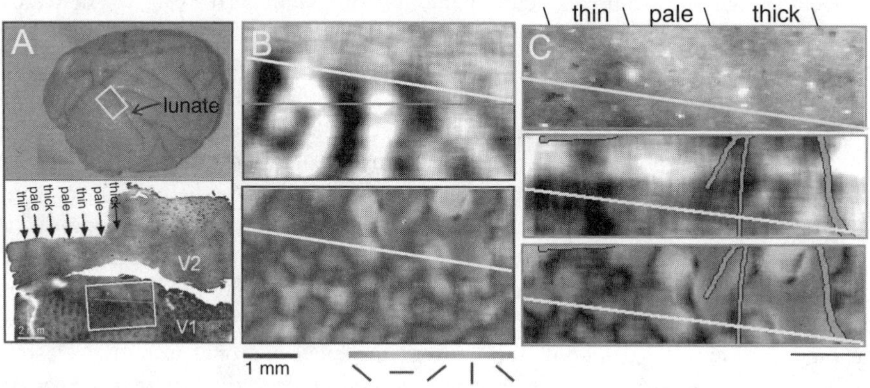

FIGURE 5.1 (Color figure follows p. 142.) Macaque area V2. (A) Top, Macaque brain. Yellow box indicates location of imaged V1/V2 area just posterior to lunate sulcus (arrow); bottom, cytochrome oxidase–stained section through superficial layers of V1 and V2. Blobs are apparent in V1 and thin, pale, and thick stripes in V2 (arrows). A break in this section was caused by unfolding of the tissue at the lunate. Yellow box indicates field of view shown in B. Yellow diagonal line indicates V1/V2 border in all panels. Scale bar: 2 mm. (B) Optical images of V1 and V2. Top, Ocular dominance map. Bottom, Color-coded orientation map. Scale bar: 1 mm. (C) Enlarged views of V2 (indicated by green box in B) show alignment cytochrome oxidase–stained tissue (top) and optical images. Middle, General activation map reveals thin and thick stripe locations. Bottom, Orientation magnitude map (strength of response indicated by saturation level) reveals orientation domains in pale and thick stripes and weak orientation selectivity in thin stripes.

but rather is intended to update by summarizing recent developments on V2 functional organization.

5.2 REDEFINING RELATIONSHIPS WITH V1

5.2.1 Revision of Anatomical Connectivity

Early visual processing comprises the magnocellular, parvocellular, and koniocellular pathways. In broad strokes, the magnocellular pathway carries low-spatial-frequency information, has the fastest conduction velocities, feeds primarily into the dorsal stream, and is concerned with spatial information (where pathway). The parvocellular pathway carries higher-spatial-frequency information and is thought to contribute to fine form vision in the ventral stream (what pathway). The koniocellular pathway, which is least understood, carries color (blue) and low-spatial-frequency information into the ventral stream.[41]

A cornerstone of this framework has been the tripartite division of connectivity between V1 and V2. The landmark studies by Livingstone and Hubel[42,43] demonstrated that the blobs in V1 are connected to the thin stripes in V2, the interblobs in V1 to the pale stripes of V2, and layer 4B in V1 to the thick stripes of V2. However, recent anatomical studies have led to a revision of this model.

Previous studies reported that injections of WGA-HRP into one V2 thick stripe in the macaque monkey and 7 thick stripes in squirrel monkey resulted in layer 4B of V1.[42,43] Thus, the third leg of the tripartite view is primarily based on data from New World monkeys and a single thick stripe injection in the macaque monkey. This has now been reexamined with a greater number of injections and with other retrograde tracers (such as cholera toxin B and fluorescent latex beads), which produce more focal injections and can therefore be more confidently localized to single stripes. The results of such injections have revealed a prominent projection from V1 interblobs to the thick stripes in V2 of the macaque.

The prevalence of this pattern has been convincingly demonstrated by Sincich and Horton[44] who examined label resulting from many well-localized focal stripe injections in Old World monkeys (*Macaca mulatta* and *M. fascicularis*, thin $n = 17$, pale $n = 33$, thick $n = 27$). This preferential interblob–thick stripe connection in the macaque monkey was also reported by Ts'o et al.[21] In this study, green latex beads were injected into a V2 "thick" (disparity) stripe as determined with optical imaging and electrophysiology. This resulted in retrograde transport of label preferentially to interblobs in layers 2/3 of V1 (Figure 5.2A). These data suggest that interblob connectivity is not dominated solely by the pale stripes, but is shared between the thick and the pale stripes.

Given that interblobs project to both thick and pale stripes, the possibility remains that further interblob specialization may exist. The total area occupied by blobs is relatively small (estimated to be 15 to 30%, e.g., Reference 45), leaving open a wide territory to be claimed by competing interests. In fact, uneven labeling of interblob regions from pale stripe injections has been observed in two studies[42,44] and in some instances pale and thick stripe injections exhibit complementary labeling in the interblobs (see Reference 44 and Figure 5.3). Consistent with this idea, Sincich and Horton[44] report that roughly a third of interblob cells project to both pale and thick stripes, while the remainder projects to either pale or thick stripes. The possibility that differential connectivity of interblob regions relates to fluctuation of disparity representation with respect to the ocular dominance map (cf. Reference 46) is an important topic for future study.

A second important revision of V1–V2 connectivity is suggested by Sincich and Horton.[44] In contrast to the layer 2/3-to-thin/pale and layer 4B-to-thick stripe model,[42] they find that each of the three stripe types receives inputs from both layers 2/3 and layer 4B as well as from layer 4A. These inputs from layers 2/3, 4A, and 4B overlie one another in a vertically aligned "columnar" fashion (that is, a column through the superficial and middle layers). Thus, current data from the macaque monkey more strongly support a columnar view of V1–V2 connectivity (cf. Reference 47), one where the "blob column" and "interblob column" project to the thin stripes and thick/pale stripes, respectively.

A third development regards the relative prominence of V1 inputs into the V2 stripes. Sincich and Horton[48] made large injections (spanning multiple blob and interblob regions) of the anterograde tracer proline into V1. In two thirds of their injections (21/29) they found that the heaviest projections from V1 terminate in the pale stripes. Although it is difficult to relate anatomical robustness directly to functional influence, this finding places a slightly different emphasis on V1–V2

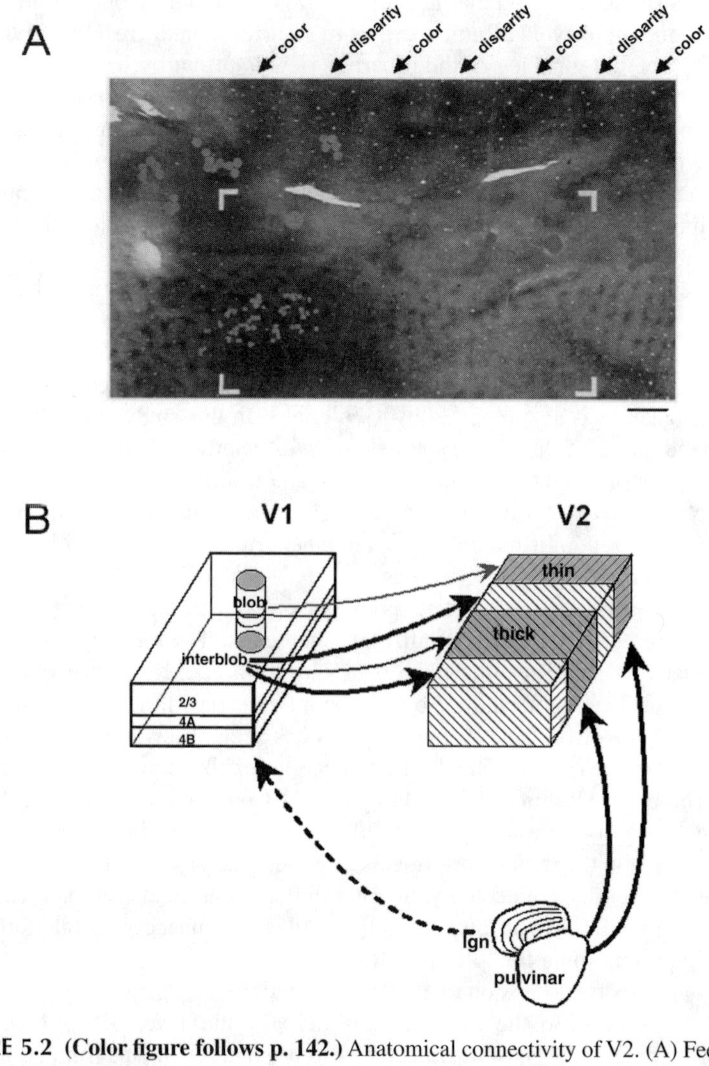

FIGURE 5.2 (Color figure follows p. 142.) Anatomical connectivity of V2. (A) Feedforward V1–V2 connections and intrinsic connections within V2. Cytochrome oxidase histology of a portion of V2, showing several color and disparity stripes and the blobs of V1. Disparity stripe injection (big green dot, green fluorescent latex spheres) results in preferential label in V1 interblobs (small green dots) and extensive label in other V2 stripes (green dots). Color stripe injection (big red dot, red fluorescent latex spheres) also shows extensive V2 labeling but failed to transport to V1. Brackets indicate location of optical imaging. Scale bar: 1 mm. (From Ts'o, D.Y. et al., *Vision Res.,* 41, 1333, 2001. With permission.) (B) Schematic of V2 connectivity. (Based on Reference 44.) Bipartite division of V1–V2 connections: blobs project to thin stripes (red arrow) and interblobs project to both thick and pale stripes (blue arrows). Both projections are columnar in layers 2/3, 4A, and 4B. Pale stripes receive heaviest V1 input (heavy blue arrow). Pulvinar inputs target thin and thick stripes, dividing V2 into pulvinar- (dark stripes) vs. V1-dominated (pale stripes) regions. Lgn, lateral geniculate nucleus.

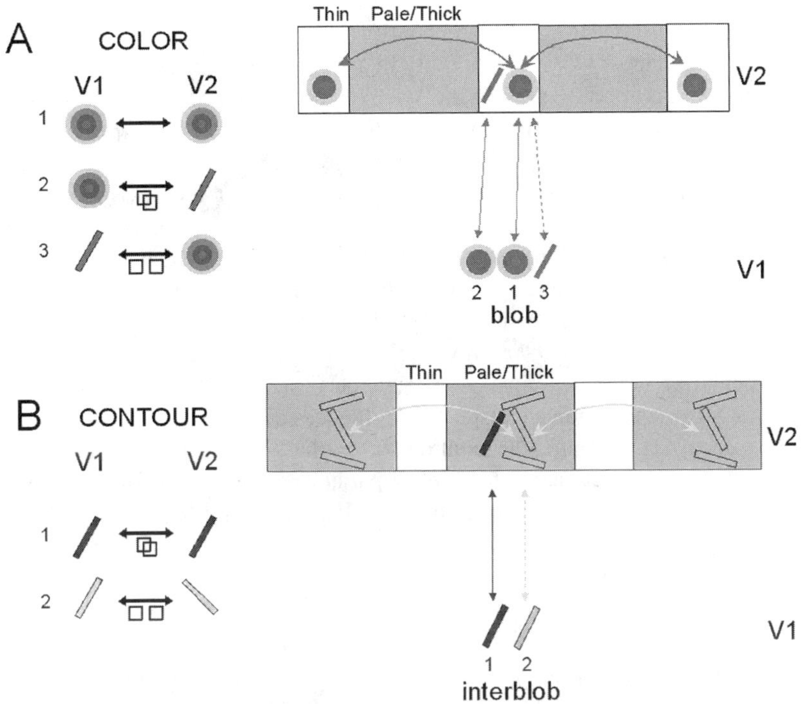

FIGURE 5.3 (Color figure follows p. 142.) Functional interactions between V1 and V2. (A) Three types of color interactions. Left, Three types of functional interactions exhibit different requirements for receptive field overlap. 1, Non-oriented cell pairs do not require receptive field overlap. 2, Non-oriented V1 cells and oriented V2 cells interact only if receptive fields overlap (indicated by overlapped boxes). 3, Oriented V1 cells and non-oriented V2 cells interact only if receptive fields do not overlap (indicated by non-overlapped boxes). Right, Schematic of anatomical connections implied by functional interactions. Gray regions depict thick/pale stripes and white regions thin stripes in V2. Three V1 blob cells (1 to 3) correspond to interactions shown at left (1 to 3). 1, A large surface color network from V1 non-oriented blob cells to nearby and distant V2 thin stripes (via horizontal V2 connections, curved arrows). 2, A small network between non-oriented V1 blob cells gives rise to a small contour building network in nearby V2 thin stripes (red bar). 3, Oriented color cells in V1 blobs contact only distant V2 thin stripes (indicated by dotted arrow). See text. (B) Two types of oriented interactions. (Based on Reference 21.) Left, Two types of functional interactions exhibit different requirements for receptive field overlap. 1, Orientation-matched cell pairs interact only if receptive fields overlap. 2, Orientation-mismatched cell pairs tend to interact when receptive fields do not overlap. Right, Schematic of anatomical connections. 1, Orientation-matched network between V1 interblob cells (dark blue bar) and nearby V2 thick/pale stripes. 2, Orientation-mismatched network between V1 interblob cells (light blue bar) gives rise to orientation-diverse networks reaching distant V2 thick/pale stripes (dotted line). Arrows indicate bidirectional interactions reflecting correlograms centered on zero. (Based on References 79 and 80.)

connectivity, one in which V1-derived input to V2 is dominated by oriented, high-spatial-frequency inputs (e.g., Reference 49).

These data place V1–V2 connectivity in a significantly different light. Two new views emerge (Figure 5.2B). First, the V1 blob vs. interblob dichotomy segregates V2 into *thin* vs. *thick/pale* stripes in V2 (rather than thin vs. pale stripes), a division that is also evident from V2 orientation maps (e.g., Figure 5.1; cf. References 21, 50, and 51). Such thick/pale coupling may be important for disparity boundary computations (see below) and indicates an inherent segregation of contour (thick/pale) vs. surface (thin) feature processing in V2. Although this division is "bipartite" in nature, the suggestion of further distinctions within interblob regions foreshadows a possible return to a "tripartite" view. The second view of V1–V2 connectivity emphasizes a *V1-based* vs. a *thalamically based* division of stripes. The complementarity of the pulvinar-associated dark stripes[52,53] and the heavily V1-driven pale stripes may indicate respective influences of thalamic and tectal pathways (cf. Reference 41). In sum, the confluence of these two sets of complementary connectivities in V2 gives rise to three separate thin, pale, and thick integration zones. Given the connectivities of thin and pale stripes with V4[54] and thick stripes with MT (e.g., References 14, 15, and 55), these new data suggest a more prominent interblob-derived contribution to MT.

5.2.2 Functional V1–V2 Interactions Revealed by Cross-Correlation

Anatomical methods directly reveal the "hard wiring" of a network from which functional processing streams can be inferred. Another view of functional connectivity can be provided by cross-correlation analysis of electrophysiological recordings. Based on the distribution of spike timing coincidences between two neuronal spike trains, it is possible to infer whether or not two neurons are statistically likely to share functional interactions.[56,57] Although functional interactions can result from direct connections, other connectivities, such as polysynaptic relationships and common input, can also result in increased spike firing coincidence. Cell pairs with highly coincident spikes have strong cross-correlation peaks, whereas those exhibiting baseline levels of coincidence produce flat correlograms. In practice, in the cerebral cortex, where networks are highly interconnected, it is not possible to determine specific circuitries between two recorded neurons. However, it is possible to detect presence or absence of interaction and to determine, given a population of recordings, the prevalence of interaction between two cell types compared to other cell types.

These functional interactions can be directly related to known anatomical structures by combining optical imaging and electrophysiology. Functional maps of cortical activity reveal locations of blobs, interblobs, and stripe structures, which are then targeted with microelectrodes. Functional interactions between isolated V1–V2 cell pairs can then be assessed by cross-correlation methods. The prevalence of interaction types can then be discerned by recording many such pairs. This approach achieves a view of functional interactions between single neurons in V1 and V2 as well as their resident structures, thereby providing a bridge between known anatomical circuits and the functional activation of those circuits by specific stimuli.

5.2.3 Blobs-to-Thin Stripes: Three Classes of Color Interactions

Based on anatomical connectivity, blobs in V1 are expected to have prevalent functional interactions with thin stripes in V2. Indeed, cross-correlation of V1–V2 non-oriented color cell pairs reveals that more than 80% of these interactions are localized to blobs and thin stripes.[20] Within blobs and thin stripes, both oriented and non-oriented color cells are recorded, although oriented color response is more commonly seen at thin/pale stripe borders.[13] Not surprisingly, interactions are prevalent between color-matched cell pairs (e.g., red-green to red-green) and not between color-nonmatched (e.g., red-green to blue-yellow) cell pairs.[20] Thus, in parallel with color-specific interactions within V1,[58] V1–V2 interactions also tend to preserve color specificity.

Another advantage of using targeted cross-correlation methods is that interactions can be related to receptive field type and spatial location. Based on color cell type and receptive field separation, three classes of V1–V2 color interactions have been described.[20] Non-oriented color cells have wide-ranging spatial interactions, exhibiting strong peaks even when receptive fields are far apart and even when cells are located several millimeters apart in cortex (Figure 5.3A, left top). Because anatomical V1–V2 connections have fairly localized connection patterns reflected directly across the V1/V2 border,[21,42,44] it is likely that large spatial spread spanning more than a single stripe cycle is mediated by the extensive network of horizontal connections in V2 (Figure 5.3A, right, cell 1, curved arrows in V2; cf. References 21, 50, and 59), as well as by feedback projections from other cortical areas.[60–62] Such spatially extensive interactions are consistent with filling in of surface color properties.

In contrast, interactions between non-oriented V1 color cells and V2-oriented color cells occur only between cells with overlapping receptive fields (Figure 5.3A, left middle and right cell 2), consistent with the building of color contours in a Hubel–Wiesel-type fashion.[63] Finally, oriented V1 cells and non-oriented V2 color cells exhibit interactions only when receptive fields do *not* overlap (Figure 5.3A, left bottom and right cell 3), which has been related to the influence of border contrast on surface color and brightness perception (e.g., References 64 and 65). Thus, three fundamental aspects of color processing: surface color perception, color contour perception, and border-induced color percepts may be initiated in parallel V1–V2 pathways (Figure 5.3A).

5.2.4 Interblobs-to-Thick/Pale Stripes: Two Classes of Oriented Interactions

Orientation networks within V1 tend to involve cells that share similar orientation selectivity.[66–69] However, emergent properties in area V2 require the integration of multiple orientations. These properties include the recognition of inferred contours such as occluded contours, illusory contours, and perceived contours due to disparity capture,[70–73] and of curvature or complex shapes.[74–76] Cells with such emergent properties are reported to be prevalent in the pale and thick stripes of V2.[51,73,77] The complexity of oriented properties in V2 therefore predicts a diversity of orientation interactions between V1 and V2.

Cross-correlation studies of cell pairs recorded from interblobs of V1 and thick/pale stripes of V2 do, in fact, reveal strong V1–V2 interactions between cell pairs of both matching and nonmatching orientation selectivity.[78,79] Similar to color interactions, oriented V1–V2 interactions also exhibit different extents of spatial integration.[79] Orientation-preserving interactions (between cells with matching orientation selectivities) tend to occur between cells with overlapping receptive fields (Figure 5.3B, left top), whereas orientation-diverse interactions (i.e., those with nonmatching orientation selectivity) tend to occur between cells with distant receptive fields (Figure 5.3B, left bottom). Furthermore, correlograms between orientation-mismatched pairs tend to be broader than matched pairs, consistent with greater temporal dispersion of relative spike firings and larger network size. Thus, smaller orientation-specific networks are spatially localized (Figure 5.3B, right, cell 1), whereas larger orientation-diverse networks integrate across visual space (Figure 5.3B, right, cell 2).

The computations required for local vs. global contour computations in V2 are likely to require such diversity of interaction. Computation of local contour orientation requires location specificity, whereas processing of global shape may involve integration of distant and differently oriented points on a curve. Furthermore, as one of the earliest stages of contour generalization, orientation interactions between V1 and V2 must lead to both specific integration (e.g., how to obtain illusory contour percept from orthogonally oriented inducers) and generalized outcome (e.g., how to achieve recognition of horizontal orientation regardless of the nature of cues). Such complex computations are likely to involve integration of diverse orientation networks. Whether there is any preferential relationship to pale or thick stripes remains to be examined.

5.3 MODULAR COMPLEXITY WITHIN V2 STRIPES

The association of the thin, pale, and thick stripes with color, form, and depth/disparity information processing, respectively, has been challenged in a number of studies as overly simplistic. Admittedly, simple characterization is often as much for convenience and brevity as for conceptual organization. For example, color cells are found in disparity stripes and oriented color cells within pale stripes. Quantitatively characterized responses to luminance and isoluminant gratings indicate a clear presence of color-responsive cells in each stripe type.[24,25,80] However, these studies indicate a prevalence of unoriented cells in upper and middle layers of the thin stripes and color-selective cells in the upper layers of the thin and pale stripes. Clearly, color-selective cells are present throughout the V2 stripes but unoriented color cells are preferentially found in superficial layers of the thin stripes.

At the root of the controversy is the fact that V2 is a very complex region. A myriad of receptive field types and single unit responses have been described in V2 (for review, see Reference 27). The real challenge is to answer the question: Why do such functionally diverse cells fall into single stripes? What common computation is being achieved? How do these computations differ from those in V1? One possibility is that V2 is a staging platform for organizing incoming ascending inputs from V1 and from the pulvinar. Subsequent integration and transformations may then be computed before being shuttled to other areas. An alternative characterization may be that all three stripe types perform a common functional transformation, albeit in different modal domains. Finding this commonality may be the key to understanding V2 function.

One organizational feature of V2 that may provide some clues is that single stripes are not uniform structures, but rather are comprised of a collection of functionally distinct subcompartments. V2 stripes have often been described as inhomogeneous or blobby in appearance, following either cytochrome staining[2,3] or 2-deoxyglucose staining.[17,18] Electrophysiological studies have described clustering of cells with similar functional properties, suggestive of some substructure within stripes. Recent studies have strengthened the evidence for the existence of these subcompartments and have further defined their role in visual processing. The following section summarizes studies on these stripe-specific modules. Because some of the strongest evidence demonstrating modular functional organization derives from optical imaging, these imaging studies are the focus.

5.3.1 Thin Stripes: Modular Representation of Surface Color and Brightness

Preferential thin stripe activation detectable by optical imaging has been demonstrated in several ways. Consistent with their higher basal metabolic demands (as indicated by dark cytochrome oxidase staining), both thin and thick stripes exhibit greater general activation during visual stimulation. This permits the dark cytochrome stripes to be mapped by imaging general activation, which is achieved by summing response to all visual conditions (Figure 5.4B). Because of the preferential localization of non-oriented color cells in thin stripes, thin stripes can be mapped with low-spatial-frequency isoluminant red/green or blue/yellow sinusoidal gratings or color + luminance square wave gratings. Their location can also be revealed by lack of orientation-selective response (see Figure 5.1, desaturated areas in orientation map) or by response to monocular stimulation (see Figure 5.6, below). In contrast to V1 blobs, which are best mapped with monocular stimulation, V2 thin stripes can be mapped with either monocular or binocular color stimulation.

Substructure within thin stripes has been shown by imaging for isoluminance vs. luminance gratings.[13,20,21] As shown in Figure 5.4B, general activation reveals a dark 1-mm-wide stripe within V2. Within this stripe are subcompartments preferentially responsive to color (dark) vs. achromatic luminance (light) (Figure 5.4A). These preferences are consistent with neuronal responses recorded from each subcompartment (Figure 5.4D and 5.4E). Cells recorded in the color subcompartment (dark) are modulated by isoluminant color gratings, whereas those in the luminance subcompartment (light) are responsive to achromatic gratings. To examine representation of hue, Felleman and colleagues[81] have used full-field standing isoluminant color/gray gratings to demonstrate a systematic representation of chromatically defined domains across single thin stripes in V2. Hand in hand with this color map, they find a clustering of luminance increment ("ON") and luminance decrement ("OFF") responses in thin stripes[82] (see also Reference 13).

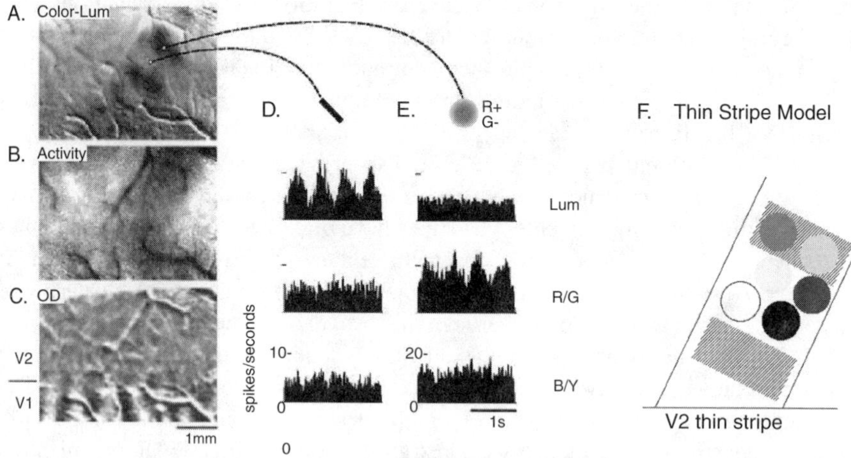

FIGURE 5.4 Modularity of color and brightness representation in thin stripes. (A–C) Optical images of a 4-mm area of V2. Scale bar: 1 mm. (A) Thin stripe substructure. Optical image of V2 thin stripe using sinusoidally modulated isoluminant red–green (dark pixels) vs. achromatic luminance (light pixels) gratings. (B) General activation (sum of all conditions) elicits activation in thin stripe (dark tilted band). (C) Ocular dominance map reveals V1/V2 border. (D) Poststimulus time histograms of single unit recorded from a luminance-preferring subcompartment of thin stripe shown in A. This unit is well modulated by achromatic luminance gratings (Lum, top), but poorly modulated by either red–green gratings (R/G, middle) or blue–yellow gratings (B/Y, bottom). (E) Unit recorded from a color-preferring subcompartment of thin stripe shown in A. This unit is well modulated by isoluminant red–green gratings (R/G, middle), but poorly modulated by either achromatic luminance gratings (Lum, top) or blue–yellow gratings (B/Y, bottom). Scale bar: 1 s. (Parts A through E from Livingstone, M.S. and Hubel, D.H., *J. Neurosci.*, 7, 3416, 1987. With permission.) (F) Schematic depiction of modular organization within thin stripes. Shaded regions represent red–green responsive zone. Thin stripes are composed of a collection of color and brightness modules (each represented by a colored or black/white disk). Hue is systematically represented. Brightness modules (ON or luminance increment, white; OFF or luminance decrement, black) are also represented adjacent to color modules (cf. References 82 and 83).

Taken together, these studies suggest that thin stripes are composed of a collection of surface-processing "modules." As depicted in Figure 5.4F, color-responsive subcompartments (shaded regions) represent surface color with a systematic progression of hue. In achromatic subcompartments of thin stripes (unshaded regions), surface brightness or achromatic contrast is represented in "ON" (light increment) and "OFF" (light decrement) modules. The fact that such ON/OFF segregation is *not* seen in pale stripes[82] suggests a segregated representation of contour and its associated contrast. Thus, the association of contour and surface may occur subsequent to independent encoding of contour and surface properties (cf. Figure 5.10, below). Because evidence suggests that V2 is involved in encoding of figure/ground,[83–85] the association of contour and surface is likely to occur in V2 and may be mediated via interstripe connections within V2. These studies further support the idea that surface features such as color and brightness are represented within thin stripes.

5.3.2 Thick/Pale Stripes: Modular Representation of Complex Contours

The similarity of physiological responses in V2 and V1 is never more evident than response to oriented contours. The same orientation tuning and end-stopping properties can be found in V2 as in V1. Orientation domains in V2, prominent in thick and pale stripes, are larger in size (~500 μm) and appear to be organized in a seemingly continuous manner. As in V1, pinwheels can be found in orientation maps of V2 (see Figure 5.6C, indicated by color wheel). What then differentiates contour representation in V1 from that in V2?

5.3.2.1 Diversity of Higher-Order Contour Cells in V2

Perhaps the most striking hallmark of V2 contour processing is that a proportion, about one third to one half, of neurons in V2 are responsive to the orientation of a contour, whether it is defined by a real line or an inferred contour[70,86] (cf. References 73 and 87; in cat, References 88 through 90). Inferred contours are not defined by luminance contrast but rather by more global features that are perceived only by grouping multiple cues across space. They include occluded contours, abutting line contours, motion discontinuity contours, texture element borders, and have been referred to as higher-order contours, illusory contours, cognitive contours, and anomalous contours.[91] Thus, V2 cells that are responsive to such contours are considered higher order in the sense that they generalize the notion of "orientation" regardless of specific features that cue that orientation. These higher-order contour cells have been hypothesized to underlie figure/ground segregation and to form the basis of contour grouping.

Responses to such illusory contours, of either the abutting grating or occluded contour type, are either absent or extremely sparse in V1.[70,73,84,86] The few reports of illusory contour response in V1 have been found either in the cat where visual cortical organization is significantly different[89] or in studies where offset abutting gratings were incorrectly termed "illusory."[92]

To help provide some structure to the array of contour cell responses in V2, one view of V2 contour response describes a hierarchy of contour representation. As described by Ramsden et al.,[93] when contour cells are characterized using real line and abutting line stimuli (Figure 5.5A), three types of responses can be classified (Figure 5.5B): "real only" (Type 1, oriented cells responsive only to real contours), "inducer-dependent" illusory (Type 2, tuned to illusory contours but only with a specific inducer orientation), and "inducer-independent" illusory (Type 3, those tuned for illusory contours regardless of inducer orientation). Thus, the level of contour generalization increases from Type 1 to Type 3. The circuitry required for building Type 2 or Type 3 responses from Type 1 has been proposed and remains to be investigated (cf. References 94 and 95). Although the existence of these and other response types (e.g., cells with nonmatching real and illusory orientation preference) has been demonstrated, the prevalence of each of these cells types is as yet unknown.

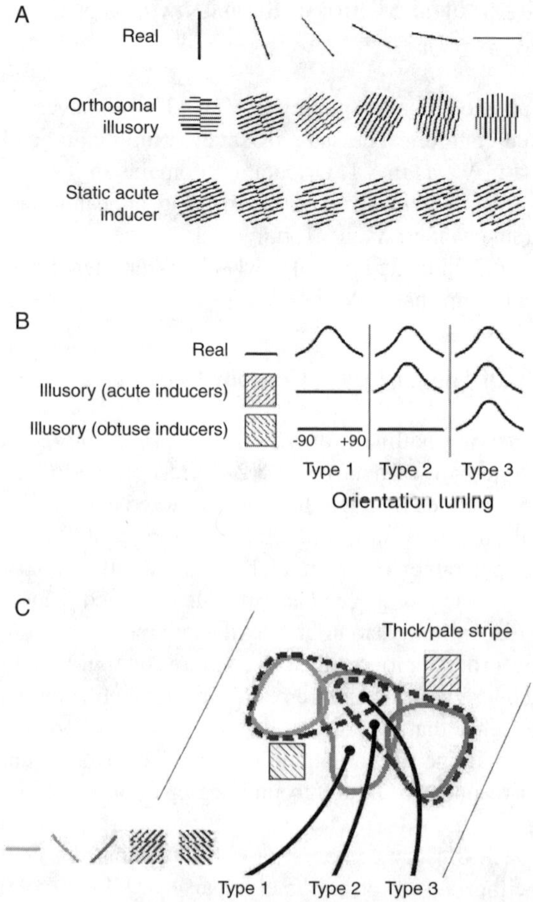

FIGURE 5.5 (Color figure follows p. 142.) Contour representation in thick/pale stripes. (A) Methods of tuning real and illusory contour cells in V2. Top, Real line stimuli. Middle, Illusory contour stimuli with orthogonal inducers. Bottom, Illusory contour stimuli with static inducers. In all illusory contour stimuli, real lines are fixed while illusory contour sweeps back and forth. (B) Hierarchy of illusory contour response. Schematized orientation tuning curves (from –90° to +90° orientation) obtained in response to real and illusory contour stimuli shown at left. Type 1 cells (left) are tuned only to real contour orientation. Type 2 cells (middle) exhibit orientation tuning to real and illusory contours but respond only to contours with either acute (example shown) or obtuse inducers. Type 3 cells (right) are tuned to contour orientation regardless of composition. (C) Proposed organization of orientation domain hierarchy. Each orientation domain (e.g., yellow domain) contains zones of "real only" response (Type 1 response zone, yellow only zone), "inducer dependent" response (Type 2 response zone, region of red and yellow overlap and region of purple and yellow overlap), and "inducer independent" response (Type 3 response zone, region of red, purple, and yellow overlap). Thus, horizontal illusory contours (composed of either acute red or obtuse purple inducers) activate both the domain representing inducer orientation (either acute green or obtuse purple, respectively) and part of the horizontal domain (yellow). Region of red, purple, and yellow overlap delimits most generalized illusory contour response region (Type 3 zone).

5.3.2.2 Localization: Evidence for Complex Orientation Domains in V2

In V2, responses to illusory contours have been reported to be most prominent in V2 pale and thick stripes and rare in the thin stripes.[77,84] End-stopped cells (as defined by responsiveness to line ends and corners), which are proposed to contribute to illusory contour response,[86] are reported to be either evenly distributed throughout the stripes in V2[84] or concentrated in the pale stripes[12,25] or the thin stripes.[16] Optical imaging studies of real and illusory contour response in V2 report the presence of "higher-order" orientation domains, domains that are activated by both real and illusory contour[51] (see upper images in Figure 5.6A). These orientation domains are preferentially located in the thick and pale stripes although occasionally also evident in thin stripes. However, there are also zones activated by real contours without illusory contour activation and zones with nonmatching real and illusory orientation response.[51] How are these complex contour responses organized?

Ramsden et al.[51,96] have proposed a structural layout for real and illusory contour response in V2 consistent with optical imaging and single unit data (Figure 5.5C). In this scheme, each orientation domain in V2 is in essence an "orientation abstraction" region; this region contains all possible combinations of orientations producing a particular oriented percept. Thus, it would contain "real only," "illusory context-dependent," and "illusory context-independent" zones. As schematized in Figure 5.5C, for example, the response to a horizontal illusory contour composed of acute inducers (dotted red line) activates the acute orientation domain (green line) as well as the context-dependent portion of the horizontal domain (yellow line). If inducers are obtuse, then activation (purple dotted line) will include both the obtuse orientation domain (gray line) and a slightly different context-dependent portion of the horizontal domain (yellow). Illusory stimulation would activate orientation domains selective for the orientation of the illusory contour as well as real domains selective for the orientation of the inducers. This results in an apparent extension of the real orientation response domain into the adjacent domain representing the orientation of the illusory contour. Each V2 orientation domain may contain subregions with differing degrees of response to lower (Type 1 response) and higher-order (Type 2 and Type 3 response) contour features, thereby achieving a continuum of alignments of illusory contour response and the real orientation map.

5.3.2.3 Feedback: Changing Balance between Two Orientation Networks

We have seen that the relationship of oriented V1 cells and oriented V2 cells, and by inference their respective orientation domains, falls into two classes: smaller orientation-preserving networks and more extensive orientation-diverse networks (see Figure 5.3). As shown by cross-correlation data, these relationships are equally prevalent in the feedforward as feedback directions between V1 and V2.[78,79,97] This suggests that feedback influences from V2 may also fall into two separate networks. Consistent with these physiological findings, Shmuel et al.[98] and Stettler et al.[99] have elegantly demonstrated, by labeling feedback projections from orientation domains

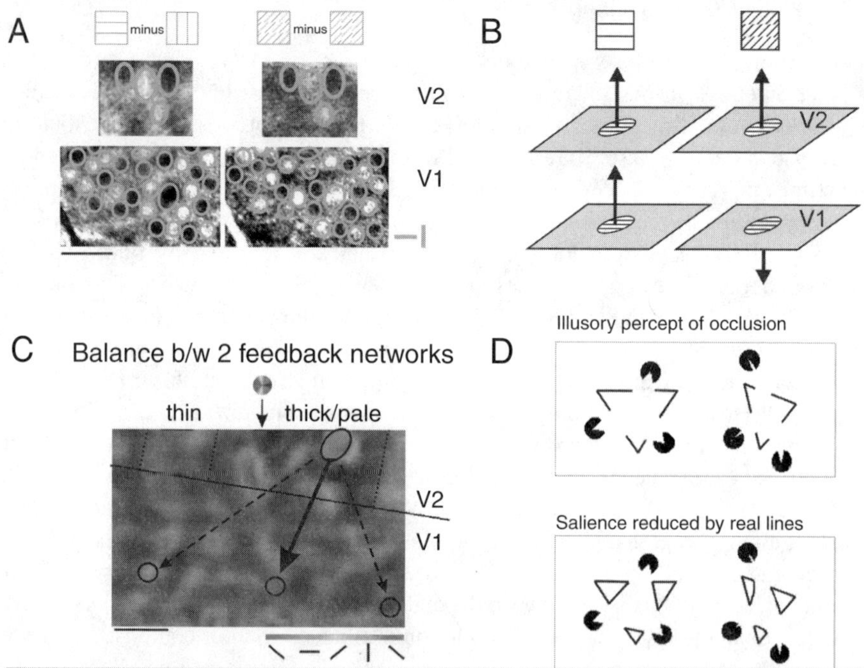

FIGURE 5.6 (Color figure follows p. 142.) Orientation-specific feedback influences from V2 to V1. (A) Optical imaging of real and illusory contour response in V1 and V2. Stimuli are depicted above. Illusory stimuli were designed with identical inducer orientation; horizontal minus vertical illusory contour subtraction removes inducer component of response, leaving illusory specific response. Real and illusory contours have identical spacings. Comparison of real and illusory response reveals very similar maps in V2 (images above), suggesting presence of higher-order contour domains. Horizontal response is circled in yellow and vertical response circled in blue. The same stimulation produces inverted maps in V1 — images below, e.g., yellow (blue) encircles dark (light) pixels with real contour stimulation and light (dark) pixels with illusory contour stimulation. Scale bar: 1 mm. (From Gallant, J.L. et al., *Science,* 259, 100, 1993. With permission.) (B) Summary of V1/V2 response. Real contours produce like activation in V1 and V2. Illusory contours produce activation in V2 and relative suppression in V1. (C) Potential source of V1 map inversion. Color-coded orientation image in V1 and V2 (see color code below). V1/V2 border determined by ocular dominance imaging (solid line). Dotted lines demarcate borders between thin and thick/pale stripes in V2. Pinwheel in V2 lies just below color wheel and arrow. Arrows depict dual feedback projections from V2 to V1, one that is orientation-specific and spatially localized (solid arrow) and one that is orientation-diverse and spatially extensive (dotted arrow). Real contour stimulus preferentially activates orientation-specific feedback. Illusory contour stimulation preferentially activates orientation-diverse feedback. Scale bar: 1 mm. (D) Perceptual effect of real vs. illusory push–pull. Top left, Percept of illusory triangle occluding dark disks around another outlined triangle. Top right, Same except triangle has curved outline. Bottom, Adding real contours diminishes the illusory percept and produces percept of cluttered collection of small triangles and pacmen.

identified by optical imaging, that single V2 orientation domains target a broad range of orientation domains in V1. Thus, feedback projections influence orientation domains of both matching and nonmatching orientation in V1 (schematized in Figure 5.6C, solid and dotted arrows).

What role do these two proposed feedback projections play? Feedback connections have been strongly implicated in context-dependent processing, producing a complex mixture of influences that depend on parameters such as stimulus context, center vs. surround stimulation, and attentional state[100–102] (for review, see References 103 through 106).

The suggestion that feedback projections play a role in orientation-specific modulation of V1 response comes from a comparison of V1 and V2 responses to real vs. illusory contours.[73] The rationale of this study was that, because illusory contour responses are V2-like and not V1-like, any illusory-specific response in V1 likely derives from area V2 or higher. This study reported that, in contrast to real line stimulation, illusory (abutting line) contour stimulation in V1 resulted in a "response inversion." That is, single oriented neurons in V1 are relatively suppressed at the orientation of the illusory contour and relatively activated at the orthogonal orientation, resulting in an illusory contour orientation map that is the exact inversion of the real contour orientation map (Figure 5.6A, lower images). Control studies indicated that this response reversal is not a product of inducer orientation or spatially specific aspects of the stimulus, but rather is due only to the illusory aspects of this stimulus. Thus, real line stimuli produced like-to-like activation of V1 and V2, whereas illusory contour stimuli produced response activation in V2 and relative suppression in V1 (Figure 5.6B). Importantly, this demonstrates a change in the balance of activation between V1 and V2 at the scale of single functional domains.

How does such response inversion in V1 occur? Because the V1 response is specific to the orientation of the illusory contour, it is likely that feedback from V2 (or higher) played a crucial role. Based on response latency analyses, feedback influences have also been demonstrated using flashed Kaniza-like illusory contour stimuli[87] (although cf. Reference 101). I propose that this response inversion results from a change in the balance between "real only" (prominent during contour stimulation) zones in V2. Thus, activating "illusory contour" zones in V2 results in tipping the balance from orientation-preserving to orientation-diverse feedback networks (from solid to dotted arrows in Figure 5.6C), thereby resulting in an inverted orientation map in V1. In sum, V2 can have robust, orientation-specific influences on V1. These influences may be mediated by changing the balance between orientation-preserving and orientation-diverse feedback networks.

A potential role of "changing the balance" on contour perception is illustrated in Figure 5.6D. The illusory contours in the top panel evoke a strong percept of a white triangle (left, straight borders; right, curved borders) occluding three dark disks surrounding another outlined triangle. Adding real lines along the border of the illusory contours (bottom panel) degrades the salience of the occluding white triangle and produces a disjointed percept of multiple objects (three pacmen and three small triangles). This example demonstrates the competitive push–pull between illusory contour and real line percepts, one which may be mediated by the changing balance of "real only" and "illusory contour" domains in V2 and their associated feedback influences on V1. These studies further specify the role of V2 feedback in context-dependent processing and figure/ground segregation.

How feedforward-to-feedback activations work together to produce real or illusory percepts is not fully understood. However, many studies have dispelled the view that "V2 follows V1." Although the shortest latencies in V1 may be shorter than the shortest latencies in V2[107–110] (cf. Reference 111), functional studies show that the predominant interaction, on average, is coactivation (e.g., References 20, 78, 79, 97, 105, and 112). It is thus important to view V1 and V2 as coprocessors rather than serial processors (for review, see References 27, 103, and 113; see Reference 105 for discussion).

5.3.3 Thick Stripes: Modular Representation of Relative Disparity

During natural vision as we move in relation to our surroundings, the spatial location of a stimulus on one retina vs. the other retina (absolute disparity) is constantly changing. However, typically the relative depth relationship between two locations in space (relative disparity) remains fairly constant. It is well known that V1 neurons, which have been classified as "tuned excitatory," "near," "far," and "tuned inhibitory," signal retinal disparity.[114] However, the role of these V1 disparity tuned cells in stereoscopic vision is as yet unclear. That is, these local detectors can encode the offset in spatial location of a stimulus on one retina vs. the other (absolute disparity). But do they encode relative disparity (e.g., near/far distance relative to another object)? Recent evidence suggests that V1 and V2 play different roles in visual depth perception.

Using cleverly designed stimuli (in which the random dot stereogram presented to one eye is reversed in contrast to that presented to the other eye), Cumming and Parker[115–117] suggest that V1 cells respond only to local disparity cues and that global disparity computations are likely to reside outside V1. In contrast, in response to systematic shifts in the relative disparity of center and surround areas of a random dot stimulus (e.g., center patch always appears in front of surround despite changes in absolute perceived depth), some V2 neurons exhibit parallel shifts of their absolute disparity tuning curves, consistent with the encoding of relative disparity.[118,119] Thus, relative disparity encoding appears to be prominent in V2 and nearly absent in V1.

By comparing responses of V1 and V2 cells to contrast-defined vs. random dot stereogram defined square patches, von der Heydt[118] demonstrated that V2 cells (much more so than V1 cells) respond to edges defined by disparity cues (see Figure 5.8B below). As another demonstration of contour generalization in V2, disparity selective V2 cells exhibited similar orientation tuning preferences for contrast-defined vs. disparity-defined borders. (Directionally selective cells do not have a strong presence in V2 but when present tend to be found in the thick stripes of V2.[12,16] A proportion of these are reported to exhibit oriented response to borders produced by differential motion fields, i.e., motion kinetic cells.[120])

Disparity selective cells, especially those termed "obligatory binocular"[121] are preferentially localized in the thick stripes of V2.[12,21,72] Although their functional organization is still poorly understood, optical imaging studies suggest that disparity, like color and contour, is represented in a modular fashion.[21] The orientation of obligatory binocular cells in vertical penetrations tends to be similar, although their disparity tuning shifts slightly with depth (Figure 5.7A). Because disparity representation in V2 is characterized by obligatory binocular cells, thick stripes are less responsive to monocular stimulation than thin stripes, thereby permitting thick stripes to be mapped by monocular vs. binocular interaction (Figure 5.7B). The relationship of the disparity map with the orientation map is not yet clear. However, imaging results indicate that patches of uniform disparity tuning overlie regions of changing orientation preference, resulting in a roughly orthogonal relationship between isodisparity contours and iso-orientation contours.[122] Near disparity patches are observed next to far disparity patches and not adjacent to zero disparity patches. Disparity patches are also observed near color patches, consistent with a close association between surface feature processing and disparity capture (see below). A proposed organization of disparity selectivity is depicted in Figure 5.7C but remains to be explored in detail. Whether the different types of contour cells (disparity defined, motion defined, and illusory contour defined) are located in the same domains in V2 and whether there are contour cells in V2 with true cue invariant responses are topics for future investigation.

5.4 BORDER AND SURFACE CAPTURE: FOUNDATION FOR FIGURE INTEGRATION

Part of the challenge of object identification lies in determining object boundaries and object surfaces regardless of the cues that define them. A second major challenge is to determine how boundaries and surfaces relate to each other, thereby forming the initial stage of figure/ground segregation.

Deciding on "border ownership"[85] is central to determining ambiguous object contour cues. A classic example is that of Rubin's vase, which can be seen as either a Greek vase or the profiles of two faces depending on how the border is interpreted. As illustrated in Figure 5.8A (receptive field indicated by oval on border), both V1 and V2 neurons respond well to oriented contrast borders. However, V2 neurons respond to the border as it relates to its position in figure relative to ground[85,123] (e.g., only when figure is on the left of border). Over half of cells in V2 exhibit significant border ownership preference vs. less than a fifth of cells in V1.[85] Simple line segments placed within receptive field borders failed to elicit responses but when placed in the context of figure/ground produced robust responses, demonstrating that V2 responses were contextually modulated beyond the classical receptive field boundaries. Thus, the "assignment" of the border response to figure is not a local computation but is likely mediated by long-range horizontal connections within V2 — presumably between thin (surface) and thick/pale (border) stripes — or by feedback from higher areas.

V2 is also involved in "surface ownership." In one of the most revealing studies about V2 function, Bakin et al.[72] demonstrated that V2 cells exhibit disparity capture response, a phenomenon in which the border captures the surface. In this study, stimuli consisted of a field of oriented elements presented independently to each

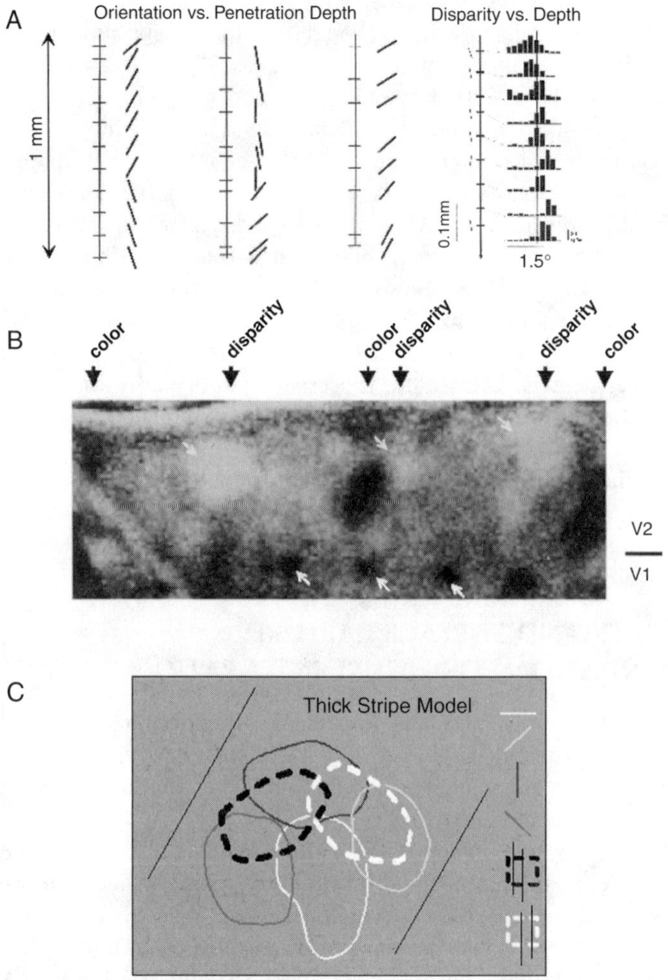

FIGURE 5.7 Modularity of disparity representation in thick stripes. (A) Columnar organization of orientation selectivity and disparity. Disparity representation is similar in superficial layers but shifts with penetration depth. (B) Disparity map in V2. Monocular (dark pixels) minus binocular (light pixels) subtraction. Blue arrows indicate binocular (disparity) zones in V2. Yellow arrows indicate monocular (blobs) in V1. (Parts A and B from Livingstone, M.S. and Hubel, D.H., *J. Neurosci.*, 7, 3416, 1987. With permission.) (C) Model of disparity and orientation response in V2 thick stripe. See text.

eye (Figure 5.8B, below). The ocular disparity of each element could be adjusted to achieve the percept of a square area located in front of or behind the surrounding background. Cells in V2 are known to be tuned for a specific range of such disparities. However, for some cells in V2, even when there was no real disparity for elements in the square region (e.g., small oval receptive field), the square region could appear in front of or behind the background when elements near the border were spatially disparate. That is, the disparity existing at the borders "captured" the elements in

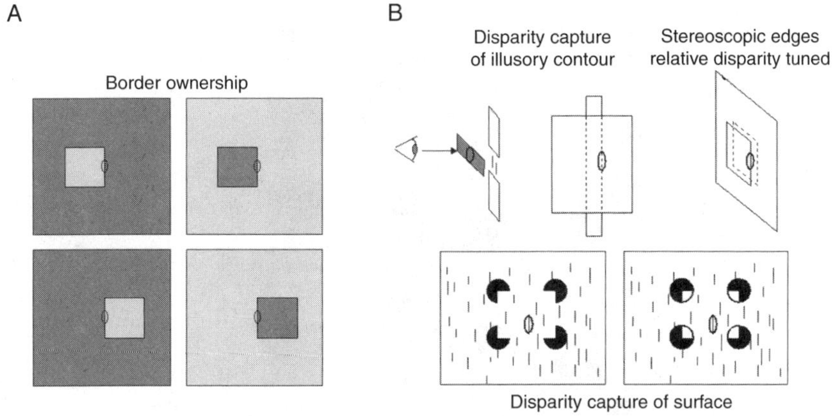

FIGURE 5.8 Association of borders and surfaces. (A) V2 cells exhibit border ownership. (From Peterhans, E. and von der Heydt, R., *J. Neurosci.*, 9, 1749, 1989. With permission.) Single V1 cells respond to presence of contrast border (e.g., both top left and bottom left). Single V2 cells respond to border that belongs to the figure (e.g., both top left and right but not top left and bottom left). (B) V2 cells exhibit disparity capture. (From Ramsden, B.M. et al., *Cereb. Cortex*, 11, 648, 2001. With permission.) Top middle, When bar ends appear in front of surface, V2 cell (receptive field indicated by small oval) exhibits response to illusory disparity-induced border. Top right, V2 cells respond to borders defined by disparity. Bottom, Disparity of oriented elements near borders of illusory rectangle produces strong perception of rectangle in front of four dark circles. V2 cells within rectangle do not have true disparity but respond to illusory disparity induced by surface capture. When rectangle appears behind four circular apertures V2 cell exhibits no disparity response (is not captured).

the center and in essence "assigned" to them the global perceived disparity. This emphasizes the importance of border information in unifying surface information, which in this case resided in oriented line or texture elements.

A third example of surface capture in V2 is illustrated by the Craik–O'Brien–Cornsweet illusion.[64] In this illusion, two equiluminant surfaces appear to differ in brightness because of an intervening border contrast (Figure 5.9A and 5.9B, left column; compare Cornsweet and Real, actual luminance profile below, perceived brightness at far right). Unlike simultaneous contrast stimuli (cf. Reference 124), the Cornsweet is a stimulus that induces a brightness percept purely by virtue of border contrast without accompanying surface luminance contrast. Whereas cells responsive to real brightness contrast are prevalent in both V1 and V2,[125] those responsive to illusory brightness contrast are preferentially located in V2 thin stripes (Figure 5.8 images[125]). These "Cornsweet" cells exhibit modulation of response to perceived brightness modulation induced by border contrast reversal, even though the receptive fields are quite distant from the contrast border (e.g., 3° to 15° away). Thus, the V2 response to "Cornsweet" modulation parallels its response to "Real" brightness modulation (compare Figure 5.9A and 5.9B images), suggesting that V2 thin stripes may encode the percept of brightness contrast regardless of true luminance values (cf. Reference 65). In this case, the border contrast captures the adjacent surfaces by assigning the perceived brightness, thereby unifying the surface percept

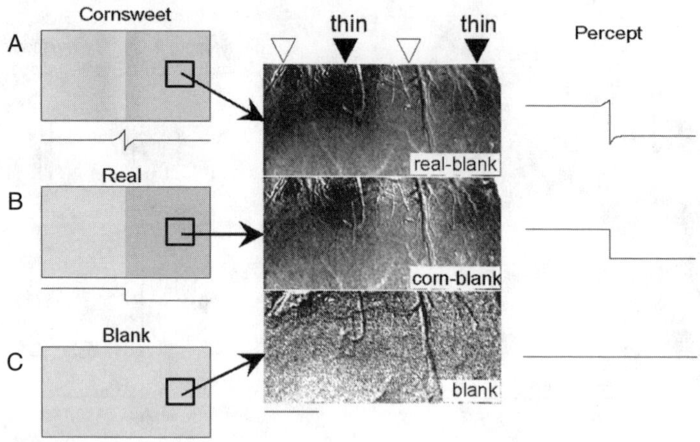

FIGURE 5.9 Similar V2 response to real and illusory brightness (Cornsweet). (A–C) Stimuli are shown in left column with luminance profiles below, optical images shown in middle column, and induced visual percept in right column. Boxed areas in visual stimuli (left column) indicate visual region presented to optically imaged field of view. (A) The Cornsweet stimulus. Sinusoidal (0.5 Hz) alternation of the border contrast produces percept of alternating left-light/right-dark fields. Border contrast selected to evoke same percept as real contrast stimulus. (B) Real luminance contrast stimulus. Sinusoidal (0.5 Hz) alternation of contrast is seen as alternating left-light/right-dark fields. (C) Blank condition. All stimuli have same average luminance. Images show preferential activation of V2 thin stripes by both real and illusory brightness (thin stripes indicated by dark and thick/pale stripes indicated by white triangles at top). Scale bar: 1 mm.

(cf. Reference 126). Long-range horizontal connections within V2, which tend to connect different stripe types, are ideally suited for lateral propagation of border information (cf. Reference 127).

5.5 SUMMARY AND PROPOSAL

This chapter has focused on functional domains in V2 and their interactions with V1. The value of identifying functional compartments within V2 lies in decoding what computations might be important in V2. Understanding the relationship of V2 with V1 can help break down visual processing into specific stages. The studies summarized here suggest two stages of form perception, a "border and surface identification" stage and a "border/surface capture" stage. This two-stage framework attempts to incorporate the many pieces of information about V2 organization. It is presented here as a summary and a proposal.

The first stage — border and surface identification — is proposed to begin in V1 and continue into V2 (mediated via feedforward connections). Elemental features of borders and surfaces (including color, orientation, disparity, and other local cues such as texture, motion) are identified by V1 (Figure 5.10). These elements are subsequently integrated by V1–V2 or intra-V2 convergence (Figure 5.10, large curved arrow) to achieve, for example, generalized contour (e.g., real or illusory contour) or generalized surface (e.g., texture field) identification. Thus, the transfor-

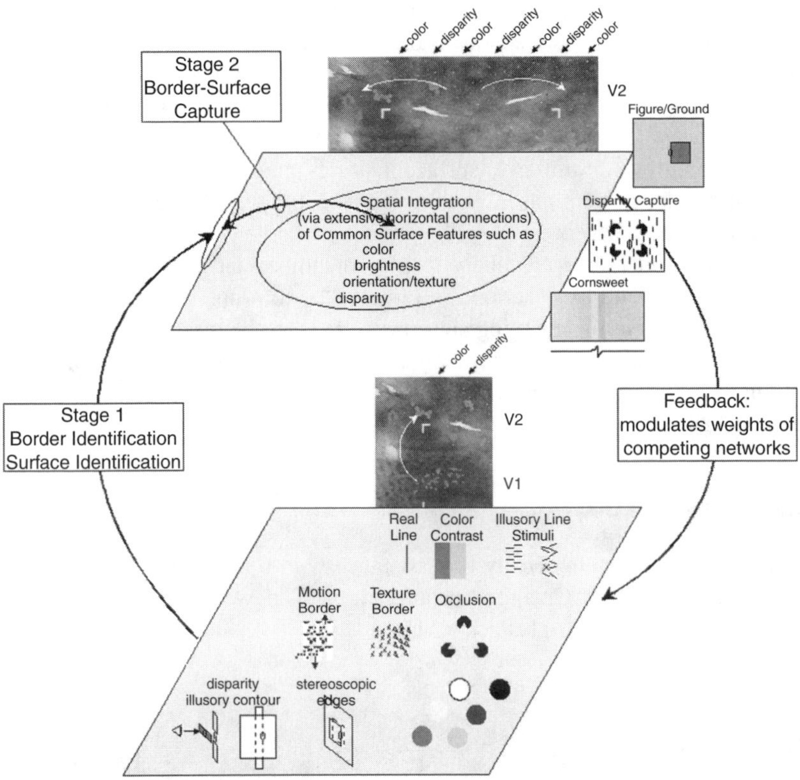

FIGURE 5.10 (Color figure follows p. 142.) Two-stage model of V2 function. Bottom, Stage 1 (border and surface identification): Elemental features of borders and surfaces (including color, orientation, disparity, and other local cues such as texture, motion) are identified by V1. These elements are subsequently integrated by V1–V2 or intra-V2 convergence (big black curved arrow, white arrow on cytochrome oxidase section below) to achieve generalized contour or generalized surface identification. Top, Stage 2 (border and surface capture): Borders and surfaces are associated by horizontal V2 connections (bidirectional arrow, white arrows on cytochrome section of V2) as well as feedback from higher areas. This achieves perceptual unification of the "figure" (e.g., figure/ground, disparity capture, Cornsweet shown at right). Finally, V2–V1 feedback is proposed to modulate weights between competing parallel networks (big curved downward-projecting arrow).[Note that white arrows on cytochrome sections are meant to indicate feedforward (lower panel) and horizontal (upper panel) connections and are not meant to point from or to specific structures.]

mations between V1 and V2 require not only organized modality-specific feedforward transmission of information, but also specific featural and spatial integration.[20,44,79] The V1–V2 connectivities described in this chapter satisfy these demands. They exhibit exquisite anatomical and functional organization and are sufficiently extensive (both spatially and featurally) to underlie higher-order transformations of border and surface identification.

A second stage proposed is that of assigning border and surface ownership (mediated via horizontal connections within V2). This assignment is likely to be carried out by the horizontal connections (typically between different stripe types[21,50,59]) within V2 or via feedback from higher cortical areas. Examples of both border capture and surface capture in V2 have been described.[72,85,125] It is apparent that the global percept captures and, in a sense, overrides local features to unify adjacent surfaces. Global border-defined shape can override the local features signaled by the surface. Similarly, surface features can help unify borders into a conceptual whole. In both cases, accuracy of local detail is often sacrificed at the expense of global coherence.

Feedback connections are likely to play an important role as well. Although this two-stage scheme as described is largely "bottom-up," feedback from V2 is also a key component in achieving appropriate balance between competing parallel circuits. Undoubtedly, perception of figure/ground (e.g., by stimulus specifics, by attentional state) is also effectively modulated by feedback from higher areas (e.g., Reference 128).

ACKNOWLEDGMENTS

The author thanks Francine Healy for assistance with formatting of manuscript and Haidong Lu and Limin Chen for proofreading. The work for this chapter was supported by Grant EY11744, Packard Foundation.

REFERENCES

1. Wong-Riley, M.T.T., Changes in the visual system of monocularly sutured or enucleated cats demonstrable with cytochrome oxidase histochemistry, *Brain Res.*, 171, 11, 1979.
2. Tootell, R.B.H., Silverman, M.S., De Valois, R.L., and Jacobs, G.H., Functional organization of the second cortical visual area in primates, *Science,* 220, 737, 1983.
3. Horton, J.C., Cytochrome oxidase patches: a new cytoarchitectonic feature of monkey visual cortex. *Philos. Trans. R. Soc. Lond.,* 304, 199, 1984.
4. DeYoe, E.A., Hockfield, S., Garren, H., and Van Essen, D.C., Antibody labeling of functional subdivisions in visual cortex: Cat-301 immunoreactivity in striate and extrastriate cortex of the macaque monkey, *Vis. Neurosci.*, 5, 67, 1990.
5. Olavarria, J.F. and Van Essen, D.C., The global pattern of cytochrome oxidase stripes in visual area V2 of the macaque monkey, *Cereb. Cortex*, 7, 395, 1997.
6. Wiencken, A.E. and Casagrande, V.A., The distribution of NADPH diaphorase and nitric oxide synthetase (NOS) in relation to the functional compartments of areas V1 and V2 of primate visual cortex, *Cereb. Cortex*, 10, 499, 2000.
7. Horton, J.C. and Hocking, D.R., An adult-like pattern of ocular dominance columns in striate cortex of newborn monkeys prior to visual experience, *J. Neurosci.*, 16, 1789, 1996.
8. Horton, J.C. and Hocking, D.R., Effect of early monocular enucleation upon ocular dominance columns and cytochrome oxidase activity in monkey and human visual cortex, *Vis. Neurosci.*, 15, 289, 1998.
9. Coogan, T.A. and Van Essen, D.C., Development of connections within and between areas V1 and V2 of macaque monkeys, *J. Comp. Neurol.*, 372, 327, 1996.

10. Rosa, M.G. and Krubitzer, L.A., The evolution of visual cortex: where is V2? *TINS*, 22, 242, 1999.
11. Kaas, J.H., Evolution of multiple areas and modules within neocortex, *Perspect. Dev. Neurobiol.*, 1, 101, 1983.
12. Hubel, D.H. and Livingstone, M.S., Segregation of form, color, and stereopsis in primate area 18, *J. Neurosci.*, 7, 3378, 1987.
13. Roe, A.W. and Ts'o, D.Y., Visual topography in primate V2: multiple representation across functional stripes, *J. Neurosci.*, 15, 689, 1995.
14. DeYoe, E.A. and Van Essen, D.C., Segregation of efferent connections and receptive field properties in visual area V2 of the macaque, *Nature*, 317, 58, 1985.
15. Shipp, S. and Zeki, S., Segregation of pathways leading from area V2 to areas V4 and V5 of macaque monkey, *Nature*, 315, 322, 1985.
16. Shipp, S. and Zeki, S., The functional organization of area V2, I: Specialization across stripes and layers, *Vis. Neurosci.*, 19, 211, 2002.
17. Tootell, R.B. H. and Hamilton, S.L., Functional anatomy of the second visual area (V2) in the macaque, *J. Neurosci.*, 9, 2620, 1989.
18. Vanduffel, W., Tootell, R.B., Schoups, A.A., and Orban, G.A., The organization of orientation selectivity throughout macaque visual cortex, *Cereb. Cortex*, 12, 647, 2002.
19. Ts'o, D.Y., Frostig, R.D., Lieke, E.E., and Grinvald, A., Functional organization of primate visual cortex revealed by high resolution optical imaging, *Science*, 249, 417, 1990.
20. Roe, A.W. and Ts'o, D.Y., Specificity of color connectivity between primate V1 and V2, *J. Neurophysiol.*, 82, 2719, 1999.
21. Ts'o, D.Y., Roe, A.W., and Gilbert, C.D., A hierarchy of the functional organization for color, form and disparity in primate visual area V2, *Vision Res.*, 41, 1333, 2001.
22. Livingstone, M.S. and Hubel, D.H., Psychophysical evidence for separate channels for the perception of form, color, and movement, and depth, *J. Neurosci.*, 7, 3416, 1987.
23. Cavanagh, P., The perception of form and motion, *Curr. Opin. Neurobiol.*, 3, 177, 1993.
24. Levitt, J.B., Kiper, D.C., and Movshon, J.A., Receptive fields and functional architecture of macaque V2, *J. Neurophysiol.*, 71, 2517, 1994.
25. Gegenfurtner, K.R., Kiper, D.C., and Fenstemaker, S.B., Processing of color, form, and motion in macaque area V2, *Vis. Neurosci.*, 13, 161, 1996.
26. Burkhalter, A. and Van Essen, D.C., Processing of color, form, and disparity information in visual areas VP and V2 of ventral extrastriate cortex in the macaque monkey, *J. Neurosci.*, 6, 2327, 1984.
27. Roe, A.W. and Ts'o, D.Y., The functional architecture of Area V2 in the macaque monkey, in *Cerebral Cortex,* Vol. 12: *Extrastriate Cortex in Primates*, Rockland, K., Kaas, J.H., and Peters, A., Eds., Plenum Press, New York, 1997, 295.
28. Grinvald, A., Frostig, R.D., Lieke, E., and Hildesheim, R., Optical imaging of neuronal activity, *Physiol. Rev.*, 68, 1285, 1988.
29. Bonhoeffer, T. and Grinvald, A., Optical imaging based on intrinsic signals. The methodology, in *Brain Mapping. The Methods*, Toga, A.W. and Mazziota, J.C., Eds., Academic Press, San Diego, 1996, 55.
30. Rajagopalan, U., Kamihara, Y., Kadono, H., Homma, R., and Tanifuji, M., Functional imaging of cat primary visual cortex with optical coherence tomography enabling visualization of depth functional structure, *Soc. Neurosci. Abstr.*, 759, 3, 2002.
31. Chen, L.M., Friedman, R.M., Ramsden, B.M., LaMotte, R.H., and Roe, A.W., Fine-scale organization of primary somatosensory cortex (area 3b) in the squirrel monkey revealed with intrinsic optical imaging, *J. Neurophysiol.*, 86, 3011, 2001.

32. Tsunoda, K., Yamane, Y., Nishizaki, M., and Tanifuji, M., Complex objects are represented in macaque inferotemporal cortex by the combination of feature columns, *Nat. Neurosci.*, 4, 832, 2001.
33. Vnek, N., Ramsden, B., Hung, C., Goldman-Rakic, P.S., and Roe, A.W., Optical imaging of functional domains in the cortex of the awake and behaving primate, *Proc. Natl. Acad. Sci. U.S.A.*, 96, 4057, 1999.
34. Shtoyerman, E., Arieli, A., Slovin, H., Vanzetta, I., and Grinvald, A., Long-term optical imaging and spectroscopy reveal mechanisms underlying the intrinsic signal and stability of cortical maps in V1 of behaving monkeys, *J. Neurosci.*, 20, 8111, 2000.
35. Chen, L.M., Heider, B., Healy, F.L., Ramsden, B.R., Williams, G.V., and Roe, A.W., A chamber and artificial dura method for long-term optical imaging in primates, *J. Neurosci. Methods*, 113, 41, 2002.
36. Seidemann, E., Arieli, A., Grinvald, A., and Slovin, H., Dynamics of depolarization and hyperpolarization in the frontal cortex and saccade goal, *Science*, 295,862, 2002.
37. Raffi, M. and Siegel, R.M., A functional architecture of optic flow in the inferior parietal cortex of the behaving monkey investigated with intrinsic optical imaging, *Soc. Neurosci. Abstr.*, 56, 9, 2002.
38. Roe, A.W., Healy, F.L., Friedman, R.M., Heider, B., and Chen, L.M., Differences in SI topography between anesthetized and awake squirrel monkey as revealed by optical imaging, *Soc. Neurosci. Abstr.*, 561.3, 2002.
39. Van Essen, D.C., Functional organization of primate visual cortex, in *Cerebral Cortex,* Vol. 3: *Visual Cortex*, Jones, E.G. and Peters, A.A., Eds., Plenum Press, New York, 1985, 259.
40. DeYoe, E.A. and Van Essen, D.C., Concurrent processing streams in monkey visual cortex, *Trends Neurosci.*, 11, 1988.
41. Hendry, S.H.C. and Reid, R.C., The koniocellular pathway in primate vision, *Annu. Rev. Neurosci.*, 23, 127, 2000.
42. Livingstone, M.S. and Hubel, D.H., Anatomy and physiology of a color system in the primate visual cortex, *J. Neurosci.*, 4, 309, 1984.
43. Livingstone, M.S. and Hubel, D.H., Connections between layer 4B of area 17 and the thick cytochrome oxidase stripes of area 18 in the squirrel monkey, *J. Neurosci.*, 7, 3371, 1987.
44. Sincich, L.C. and Horton, J.C., Divided by cytochrome oxidase: a map of the projections from V1 to V2 in macques, *Science*, 295, 1734, 2002.
45. Purves, D. and LaMantia, A., Development of blobs in the visual cortex of macaques, *J. Comp. Neurol.*, 334, 169, 1993.
46. LeVay, S. and Voigt, T., Ocular dominance and disparity coding in cat visual cortex, *Vis. Neurosci.*, 1, 395, 1988.
47. Wong-Riley, M.T.T., Columnar cortico-cortical interconnections within the visual system of the squirrel and macaque monkeys, *Brain Res.*, 162, 201, 1979.
48. Sincich, L.C. and Horton, J.C., Pale cytochrome oxidase stripes in V2 receive the richest projection from macaque striate cortex, *J. Comp. Neurol.*, 447, 18, 2002.
49. Edwards, D.P., Purpura, K.P., and Kaplan, E., Contrast sensitivity and spatial frequency response of primate cortical neurons in and around the cytochrome blobs, *Vision Res.*, 35, 1501, 1995.
50. Malach, R., Tootell, R.B., and Malonek, D., Relationship between orientation domains, cytochrome oxidase stripes, and intrinsic horizontal connections in squirrel monkey area V2, *Cereb. Cortex*, 4, 151, 1994.
51. Ramsden, B.M., Hung, C.P., and Roe, A.W., Plurality of orientation domains in macaque Area V2, submitted.

52. Levitt, J.B., Yoshioka, T., and Lund, J.S., Connections between the pulvinar complex and cytochrome oxidase-defined compartments in visual area V2 of macaque monkey, *Exp. Brain Res.,* 104, 419, 1995.
53. Adams, M.M., Hof, P.R., Gattass, R., Webster, M.J., and Ungerleider, L.G., Visual cortical projections and chemoarchitecture of macaque monkey pulvinar, *J. Comp. Neurol.*, 419, 377, 2000.
54. Xiao, Y., Zych, A., and Felleman, D.J., Segregation and convergence of functionally defined V2 thin stripe and interstripe compartment projections to area V4 of macques, *Cereb. Cortex*, 9, 792, 1999.
55. Anderson, J.C. and Martin, K.A., Connection from cortical area V2 to MT in macaque monkey, *J. Comp. Neurol.,* 443, 56, 2002.
56. Perkel, D.H., Gerstein, G.L., and Moore, G.P., Neuronal spike trains and stochastic point processes. I. Simultaneous spike train, *Biophys. J.*, 7, 419, 1966.
57. Brody, C.D., Disambiguating different covariation types, *Neural Comput.*, 11, 1527, 1999.
58. Ts'o, D.Y. and Gilbert, C.D., The organization of chromatic and spatial interactions in the primate striate cortex, *J. Neurosci.,* 8, 1712, 1988.
59. Levitt, J.B., Yoshioka, T., and Lund, J.S., Intrinsic cortical connections in macaque visual area V2: evidence for interaction between different functional streams, *J. Comp. Neurol.*, 342, 551, 1994.
60. Rockland, K.S., Saleem, K.S., and Tanaka, K., Divergent feedback connections from areas V4 and TEO in the macaque, *Vis. Neurosci.*, 11, 579, 1994.
61. Rockland, K.S. and Knutson, T., Feedback connections from area MT of the squirrel monkey to areas V1 and V2, *J. Comp. Neurol.*, 425, 345, 2000.
62. Barrone, P., Batardiere, A., Knoblauch, K., and Kennedy, H., Laminar distribution of neurons in extrastriate areas projecting to visual areas V1 and V4 correlates with the hierarchical rank and indicates the operation of a distance rule, *J. Neurosci.*, 20, 3263, 2000.
63. Hubel, D.H. and Wiesel, T.N., Functional architecture of macaque monkey visual cortex, *Proc. R. Soc. Lond. B*, 198, 1, 1977.
64. Cornsweet, T.N., *Visual Perception*, Academic Press, New York, 1970.
65. Hung, C.P., Ramsden, B.M., and Roe, A.W., Building surfaces from borders in areas 17 and 18 of the cat, *Vision Res.*, 41, 1389, 2001.
66. Ts'o, D.Y., Gilbert, C.D., and Wiesel, T.N., Relationships between horizontal interactions and functional architecture in the cat striate cortex as revealed by cross-correlation analysis, *J. Neurosci.*, 6, 1160, 1986.
67. Kisvarday, Z.F., Toth, E., Rausch, M., and Eysel, U.T., Orientation-specific relationship between populations of excitatory and inhibitory lateral connections in the visual cortex of the cat, *Cereb. Cortex*, 7, 605, 1997.
68. Bosking, W.H., Zhang, Y., Schofield, B., and Fitzpatrick, D., Orientation selectivity and the arrangement of horizontal connections in tree shrew striate cortex, *J. Neurosci.*, 17, 2112, 1997.
69. Sincich, L.C. and Blasdel, G.G., Oriented axon projections in primary visual cortex of the monkey, *J. Neurosci.*, 21, 4416, 2001.
70. von der Heydt, R. and Peterhans, E., Mechanisms of contour perception in monkey visual cortex. I. Lines of pattern discontinuity, *J. Neurosci.,* 9, 1731, 1989.
71. Merigan, W.H., Nealey, T.A., and Maunsell, J.H.R.,The effects of lesions of cortical area V2 in macaques, *J. Neurophysiol.,* 13, 3180, 1993.
72. Bakin, J.S., Nakayama, K., and Gilbert, C.D., Visual responses in monkey areas V1 and V2 to three-dimensional surface configurations, *J. Neurosci.*, 20, 8188, 2000.

73. Ramsden, B.M., Hung, C.P., and Roe, A.W., Real and illusory contour processing in area V1 of the primate: a cortical balancing act, *Cereb. Cortex*, 11, 648, 2001.
74. Gallant, J.L., Braun, J., and Van Essen, D.C., Selectivity for polar, hyperbolic, and Cartesian gratings in macaque visual cortex, *Science,* 259, 100, 1993.
75. Hedge, J., and Van Essen, D.C., Selectivity for complex shapes in primate visual area V2, *J. Neurosci.*, 20, RC61, 2000.
76. Mahon, L.E. and De Valois, R.L., Cartesian and non-Cartesian responses in LGN, V1, and V2 cells, *Vis. Neurosci.*, 18, 973, 2001.
77. Peterhans, E. and von der Heydt, R., Functional organization of area V2 in the alert macaque, *Eur J. Neurosci.,* 5, 509, 1993.
78. Roe, A.W. and Ts'o, D.Y., Two networks of curve-building cells? Interactions between orientation-selective cells in primate V1 and V2, *For. Eur. Neurosci. Soc.* (Paris), 2002.
79. Roe, A.W. and Ts'o, D.Y., Specificity of V1-V2 orientation networks in the primate visual cortex, submitted.
80. Kiper, D.C., Fenstemaker, S.B., and Gegenfurtner, K.R., Chromatic properties of neurons in macaque area V2, *Vis. Neurosci.*, 14, 1061, 1997.
81. Xiao, Y., Wang, Y., and Felleman, D.J., Representation of color in V2 thin stripes, *Soc. Neurosci. Abstr.*, 11, 4, 2001.
82. Wang, Y. and Felleman, D.J., Processing of chromatic and luminance surfaces and contours by V2 thin stripes and interstripes, *Soc. Neurosci. Abstr.,* 720, 11, 2002.
83. Baumann, R., van der Zwan, R., and Peterhans, E., Figure-ground segregation at contours: a neural mechanism in the visual cortex of the alert monkey, *Eur. J. Neurosci.*, 9, 1290, 1997.
84. Heider, B., Meskenaite, V., and Peterhans, E., Anatomy and physiology of a neural mechanism defining depth order and contrast polarity at illusory contours, *Eur. J. Neurosci.*, 12, 4117, 2000.
85. Zhou, H., Friedman, H.S., and von der Heydt, R., Coding of border ownership in monkey visual cortex, *J. Neurosci.*, 20, 6594, 2000.
86. Peterhans, E. and von der Heydt, R., Mechanisms of contour perception in monkey visual cortex. II. Contours bridging gaps, *J. Neurosci.*, 9, 1749, 1989.
87. Lee, T.S. and Nguyen, M., Dynamics of subjective contour formation in the early visual cortex, *Proc. Natl. Acad. Sci. U.S.A.*, 98, 1907, 2001.
88. Zhou, Y.X. and Baker, C.L., Jr., Envelope-responsive neurons in areas 17 and 18 of cat, *J. Neurophysiol.*, 72, 2134, 1994.
89. Sheth, B.R., Sharma, J., Rao, S.C., and Sur, R., Orientation maps of subjective contours in visual cortex, *Science*, 274, 2110, 1996.
90. Leventhal, A.G., Wang, Y., Schmolesky, M.T., and Zhou, Y., Neural correlates of boundary perception, *Vis. Neurosci.*, 15, 1107, 1998.
91. Kanisza, G., Contours without gradients or cognitive contours? *Ital. J. Psychol.*, 2, 107, 1974.
92. Grosof, D., Shapley, R.M., and Hawken, M., Macaque V1 neurons can signal "illusory" contours, *Nature*, 365, 550, 1993.
93. Ramsden, B., Hung, C., and Roe, A.W., Organization and circuitry of illusory contour domains in V1 and V2 of the macaque, *Soc. Neurosci. Abstr.*, 24, 1506, 1998.
94. Grossberg, S., Cortical dynamics of three-dimensional figure-ground perception of two-dimensional pictures. *Psychol. Rev.,* 104, 618, 1997.
95. Peterhans, E. and Heitger, F., Simulation of neuronal responses defining depth order and contrast polarity at illusory contours in monkey area V2, *J. Comp. Neurosci.*, 10, 195, 2001.

96. Ramsden, B., Hung, C., Chen, L.M., and Roe, A.W., Sub-domain modulation of intrinsic optical imaging signals associated with illusory contour perception in area V2 of the macaque monkey cortex. *Invest. Ophthamol. Vis. Sci. Suppl.*, 40, S639, 1999.
97. Nowak, L.G., Munk, M.H., James, A.C., Girard, P., and Bullier, J., Cross-correlation study of the temporal interactions between areas V1 and V2 of the macaque monkey, *J. Neurophysiol.*, 81, 1057, 1999.
98. Shmuel, A., Korman, M., Harel, M., Grinvald, A., and Malach, R., Relationship of feedback connections from area V2 to orientation domains in area V1 of the primate, *Soc. Neurosci. Abstr.*, 24, 767, 1998.
99. Stettler, D.D., Das, A., Bennett, J., and Gilbert, C.D., Lateral connectivity and contextual interactions in macaque primary visual cortex, *Neuron*, 36, 739, 2002.
100. Bullier, J., Hupe, J.M., James, A., and Girard, P., Functional interactions between areas V1 and V2 in the monkey, *J. Physiol.*, 90, 217, 1996.
101. Hupe, J.M., James, A.C., Payne, B.R., Lomber, S.G., Girard, P., and Bullier, J., Cortical feedback improves discrimination between figure and background by V1, V2 and V3 neurons, *Nature*, 394, 784, 1998.
102. Hupe, J.M., James, A.C., Girard, P., and Bullier, J., Response modulations by static texture surround in area V1 of the macaque monkey do not depend on feedback connections from V2, *J. Neurophysiol.*, 85, 146, 2001.
103. Bullier, J., Integrated model of visual processing, *Brain Res. Rev.,* 36, 96, 2001.
104. Bullier, J., Hupe, J.M., James, A.C., and Girard, P., The role of feedback connections in shaping the responses of visual cortical neurons, *Prog. Brain Res.*, 134, 193, 2001.
105. Hupe, J.M., James, A.C., Girard, P., Lomber, S.G., Payne, B.R., and Bullier, J., Feedback connections act on the early part of the responses in monkey visual cortex, *J. Neurophysiol.*, 85, 134, 2001.
106. Friston, K.J. and Buchel, C., Attentional modulation of effective connectivity from V2 to V5/MT in humans, *Proc. Natl. Acad. Sci. U.S.A.*, 97, 7591, 2000.
107. Maunsell, J.H.R. and Gibson, J.R., Visual response latencies in striate cortex of the macaque monkey, *J. Neurophysiol.*, 68, 1332, 1992.
108. Nowak, L.G., Munk, M.H., Girard, P., and Bullier, J., Visual latencies in areas V1 and V2 of the macaque monkey, *Vis. Neurosci.*, 12, 371, 1995.
109. Munk, M.H.J., Nowak, L.G., Girard, P., Chounlamountri, N., and Bullier, J., Visual latencies in cytochrome oxidase bands of macaque area V2, *Proc. Natl. Acad. Sci. U.S.A.*, 92, 988, 1995.
110. Schroeder, C.E., Mehta, A.D., and Givre, S.J., A spatiotemporal profile of visual system activation revealed by current source density analysis in the awake monkey, *Cereb. Cortex*, 8, 1047, 1998.
111. Schmolesky, M.T., Wang, Y., Hanes, D.P., Thompson, K.G., Leutgeb, S., Schall, J.D., and Leventhal, A.G., Signal timing across the macaque visual system, *J. Neurophysiol.,* 79, 3272, 1998.
112. Girard, P., Hupe, J.M., and Bullier, J., Feedforward and feedback connections between areas V1 and V2 of the monkey have similar rapid conduction velocities, *J. Neurophys.*, 85, 1328, 2001.
113. Salin, P.A. and Bullier, J., Corticocortical connections in the visual system: structure and function, *Physiol. Rev.*, 75, 107, 1995.
114. Poggio, G.F. and Fischer, B., Binocular interaction and depth sensitivity in striate and prestriate cortex of behaving rhesus monkey, *J. Neurophysiol.*, 40, 1392, 1977.
115. Cumming, B.G. and Parker, A.J., Responses of primary visual cortical neurons to binocular disparity without depth perception, *Nature,* 389, 280, 1997.

116. Cumming, B.G. and Parker, A.J., Binocular neurons in V1 of awake monkeys are selective for absolute, not relative, disparity, *J. Neurosci.*, 19, 5602, 1999.
117. Cumming, B.G. and Parker, A.J., Local disparity not perceived depth is signaled by binocular neurons in cortical area V1 of the macaque, *J. Neurosci.,* 20, 4758, 2000.
118. von der Heydt, R., Zhou, H., and Friedman, H.S., Representation of stereoscopic edges in monkey visual cortex, *Vision Res.*, 40, 1955, 2000.
119. Thomas, O.M., Cumming, B.G., and Parker, A.J., A specialization for relative disparity in V2, *Nat. Neurosci.,* 5, 472, 2002.
120. Marcar, V.L., Raiguel, S.E., Xiao, D., and Orban, G.A., Processing of kinetically defined boundaries in areas V1 and V2 of the macaque monkey, *J. Neurophysiol.*, 84, 2786, 2000.
121. Hubel, D.H. and Wiesel, T.N., Cells sensitive to binocular depth in area 18 of the macaque monkey cortex, *Nature*, 255, 41, 1970.
122. Ts'o, D.Y. and Burkitt, G., Relationships between the hypercolumns of primate visual area V2, *Soc. Neurosci. Abstr.*, 11, 1, 2001.
123. Rossi, A.F., Desimone, R., and Ungerleider, L.G., Contextual modulation in primary visual cortex of macaques, *J. Neurosci.*, 21, 1698, 2001.
124. Rossi, A.F., Rittenhouse, C.D., and Paradiso, M.A., The representation of brightness in primary visual cortex, *Science*, 273, 1104, 1996.
125. Hung, C.P., Ramsden, B.M., and Roe, A.W., Optical imaging and single unit recording of real and illusory brightness response in V1 and V2 of the macaque monkey, *Soc. Neurosci. Abstr.,* 28, 456.5, 2002.
126. Davey, M.P., Maddess, T., and Srinivasan, M.V., The spatiotemporal properties of the Craik-O'Brien-Cornsweet effect are consistent with "filling-in," *Vision Res.*, 38, 2037, 1998.
127. Hung, C.P., Ramsden, B.M., and Roe, A.W., Interactions between contrast borders and surfaces in cat areas 17 and 18 during processing of real and Cornsweet brightness stimuli, *Soc. Neurosci. Abstr.*, 286.1, 2001.
128. Lee, T.S., Yang, C.F., Romero, R.D., and Mumford, D., Neural activity in early visual cortex reflects behavioral experience and higher-order perceptual saliency, *Nat. Neurosci.,* 5, 589, 2002.

6 Early Visual Areas: V1, V2, V3, DM, DL, and MT

Jon H. Kaas

CONTENTS

6.1 INTRODUCTION

Arguably, the most basic step toward the scientific goal of understanding how the visual system of humans mediates vision is to determine the parts of the system (the nuclei and cortical areas) and how they interconnect. Modern neuroscientists widely accept the well-supported premise of Brodmann (1909)[1] that the neocortex is divided into a number of areas or "organs of the brain" with more or less sharp and definable boundaries and distinct functions. The task of neuroscientists, then, is to determine the distinct functions of each definable part, and determine how they work together to mediate the various aspects of vision. Clearly, there has been a good start, as proposals of how the primate visual system is organized have become increasingly complex (e.g., Reference 2), the functions of subdivisions of the system have been elaborated (e.g., Reference 3), and subsystems or streams of processing for aspects of visions have been proposed (e.g., References 4 through 6).

The progress has been so substantial that many researchers have the impression that the organization of the visual system is well understood, and the emphasis should now be on understanding the functions of the system, and the roles of its parts. A greatly improved understanding of the human visual cortex has gradually emerged from the early theories of Brodmann.[1] Subsequent early investigators postulated a central primary area (area 17 or V1) surrounded by successive ringlike

0-8493-1243-4/04/$0.00+$1.50

secondary fields (area 18 or V2 and area 19 or V3, bordered by various bimodal or multimodal association fields (e.g., References 7 through 9 for review). As modern methods of recording from the brain and revealing brain anatomy started to provide further insights some 30 to 40 years ago, investigators revised and elaborated these early theories to produce an array of modern proposals (e.g., References 2 and 9 through 13). These proposals are largely focused on the results of experimental studies on monkeys, especially macaque monkeys, but interpretations of how the visual cortex of humans is organized have been increasingly influenced and guided by the results of noninvasive imaging of brain activity patterns in humans (e.g., References 14 through 17).

At least superficially, the modern theories all seem quite similar to each other. Many resemblances occur because current proposals have been greatly influenced by earlier theories. This reliance on previous pronouncements sometimes serves as a brake on progress or creates confusion. We now know, for example, that the proposed subdivisions of neocortex of Brodmann,[1] such as areas 18 and 19, do not correspond to any of the subdivisions of modern proposals of visual cortex organizations, but the terms area 18 and area 19 have been redefined and retained in modern use as the architectonic equivalents of V2 and V3. Even more recently proposed visual areas have been redefined. The proposed area V4, for example, has been extended from its original dorsolateral position well into the temporal lobe (for example, compare V4 of Reference 18 with V4 of Reference 2), and V4 has become disassociated from its original position relative to other reference areas in current proposals of visual cortex organization in humans (see References 19 and 20). As another example, the central nucleus of the inferior pulvinar (IPc) has been redefined by some investigators as only a part of the original (for review, see Reference 21 and Stepniewska, Chapter 3, this volume). More recently, we have adjusted the caudal border of the dorsomedial visual area (DM) to allow for a narrow V3[11] (see below). Thus, proposals may resemble each other in terminologies, while differing in how areas of the same name are defined and located.

Apparent differences in proposals may be misleading as well. A confusing, but less troublesome practice has been to use different names for the same visual area. The most notable example is the widespread use of two names, the middle temporal visual area (MT) and the visual area 5 (V5), for the same well-defined visual area in the upper temporal lobe. Other examples exist, such as the use of the terms dorsolateral visual area (DL) and V4 for roughly the same region of visual cortex, although no subdivision of visual cortex is well defined in this region. Thus, the different names may usefully distinguish somewhat different concepts. There are a number of other examples of the practice of renaming or redefining cortical regions. The parieto-occipital area (PO) of Cowey et al.[22] has been renamed area V6A (e.g., Reference 23), and the medial intraparietal area (MIP) of Colby et al.[24] has been incorporated into a larger parietal reach region (PRR) (e.g., Reference 25). To add further complications, Cowey et al.[22] and Colby et al.[24] considered PO to be the same as the previously defined medial visual area, M,[26] although others have argued that PO (V6) corresponds to the dorsomedial visual area, DM, of Allman and Kaas[27] (e.g., Reference 26; see Rosa and Tweedale, Chapter 11, this volume).

Part of the reason that investigators have applied multiple names to the same cortical region, and have moved the locations of named areas, is that few cortical areas are so well defined that there is widespread agreement over the existence of areas and the locations of borders. In perhaps the most widely referenced review of visual cortex organization in monkeys, Felleman and Van Essen[2] described the locations and connections of 32 visual areas, while rating each area on a confidence level of high (1) to low (3) over validity and known borders. While most readers remember from the review that there are a lot of visual areas and that the connection diagram is exceedingly complex, few seem to recognize that Felleman and Van Essen rated only five visual areas (V1, V2, V3d, VP, and MT) as well defined. A more appropriate conclusion might be that only V1, V2, and MT are well defined, as they are the universal components of all proposals. As a number of investigators consider V3d and VP to be parts of a single visual area, a redefined V3[11] (see below), VP and V3d cannot be considered well-defined areas.

This chapter evaluates the evidence for six visual areas that are early in the processing hierarchy of visual areas of primates. Although there is widespread agreement that visual areas of primates can be assigned to different levels of processing from early to late, and that it is useful to do so, current proposals of hierarchical levels and arrangements are based on incomplete information about the identities of areas, the nature of their connections, and the functional roles of these connections (see Bullier, Chapter 8, this volume). Even when based on the same conclusions about areas and connections, hierarchies may differ in detail according to the premises held (see Reference 28 for review). Yet, all proposals place V1 at the first cortical level, area V2 at a second level, and areas MT, DM, and DL or their equivalents are placed at a higher level. Areas early (lower) in the hierarchy are reviewed here because these areas are better defined, and more is known about them. There is evidence that they exist in all primates and that they evolved early in mammalian or primate history, as basic components of all primate visual systems. These areas are likely to be the most extensively studied in the near future. Here we examine the evidence for the validity of these proposed areas, and describe their major organizational features.

6.2 PRIMARY VISUAL CORTEX, V1

All current neuroscientists accept V1 as a valid visual area and agree on its location and borders. Because of the distinctive histological appearance of V1 (area 17 of Brodmann[1]), it was the first cortical area clearly identified in the human brain (by Gennari in 1782; see Reference 29 for review). V1 is unique among visual areas in that it has been identified and delimited in all studied mammals. Even the "blind" subterranean mole-rat, with a rudimentary visual system has a small V1 (see Reference 30). V1 can be identified with assurance by its architectonic distinctiveness, relative position in the caudal pole of the cerebral hemisphere (Figure 6.1 through Figure 6.3), and a pattern of topographic inputs from the dorsal lateral geniculate nucleus (LGN) of the visual thalamus (providing a pattern of retinotopic activation). Of course, the LGN can be identified with certainty by direct inputs from the two eyes, as well as by its relative position in the thalamus and its distinctive architecture.

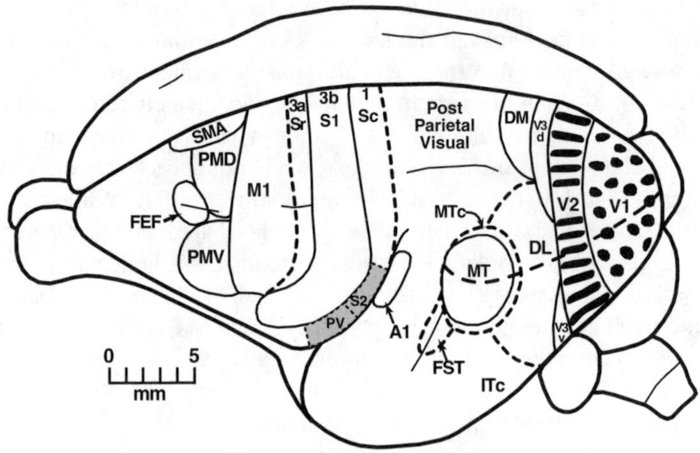

FIGURE 6.1 A dorsolateral view of the brain of a prosimian primate. This galago (bushbaby) brain shows the locations of proposed visual areas, including primary visual cortex (V1 or area 17), the second visual area (V2), the third visual area (V3), and the middle temporal visual area (MT). V1 is characterized by patterns of CO-dense patches (the CO blobs) whereas V2 has an array of CO-dense bands that cross the area. Less well established visual areas include the dorsolateral visual area (DL; cortex in the region is also called V4), the dorsomedial visual area (DM), the middle temporal crescent (MTc), the fundal area of the superior temporal sulcus (FST), and the caudal division of inferotemporal cortex (ITc). Much of posterior parietal cortex is also visual, and additional visual areas may be identified here and in the temporal lobe. For reference, primary auditory cortex (A1) is identified, other somatosensory areas include rostral and caudal somatosensory belts (Sr or area 3a and Sc or area 1), the second somatosensory area (S2), and the parietal ventral somatosensory area (PV). Premotor areas include the supplementary motor area (SMA), the dorsal (PMD) and ventral premotor areas (PMV), and the frontal eye field (FEF).

While V1 has pronounced "sensory" features in primates, such as a well-differentiated layer 4 of small neurons, such features are much less marked in mammals with less emphasis on vision such as rats and hedgehogs. Yet, V1 is architecturally distinct enough to be clearly identified in these mammals. The nearly universal existence of V1 among mammals means that this visual area emerged with the first mammals, or the reptilian ancestors of mammals, and was retained in nearly all subsequent lines of evolution.

The histological distinctiveness of V1 also varies among primates. Remarkably, primary visual cortex is not the most differentiated in humans. Early primates diverged into three groups leading to present day prosimians (galagos and lemurs), tarsiers, and anthropoids (monkeys, apes, and humans). Tarsiers have the most distinctive layers and sublayers in V1 of any primate,[31] while prosimian galagos have rather poor lamination. Humans and apes lack sublaminar features of layer 3 of monkeys[32] (see Preuss, Chapter 10, this volume). Unfortunately, the deeper part of layer 3 of anthropoids is usually assigned to sublayers of layer 4, 4A and 4B, as a result of early misidentifications (see Reference 33 for review). Other sublaminar features distinguish V1 of apes from V1 in humans.[32] Primates also vary in how

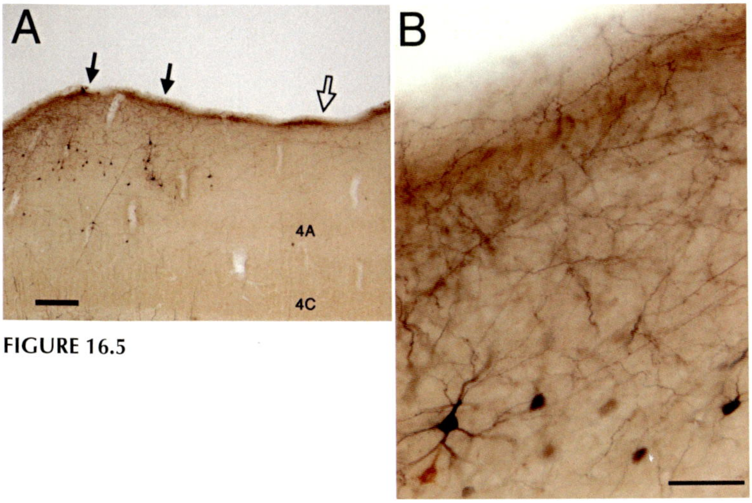

FIGURE 16.5

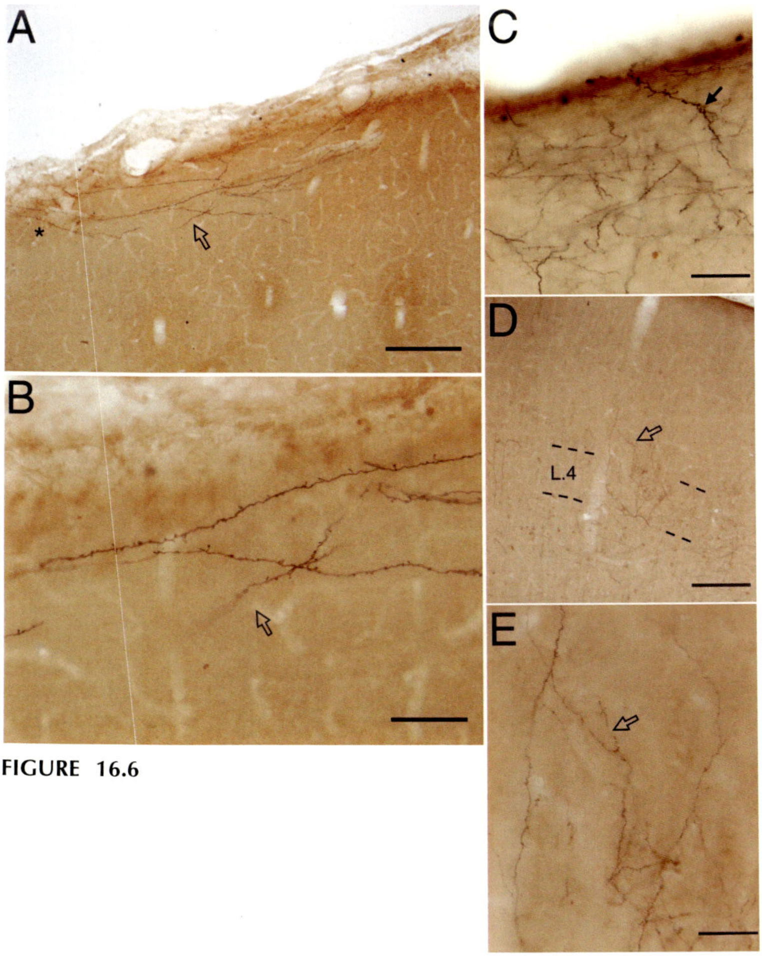

FIGURE 16.6

October 3, 2003

Dear Customer:

Thank you for your purchase of ***The Primate Visual System*** (#1243) edited by Jon H. Kaas and Christine E. Collins.

The color versions of Figures 16.5 and 16.6 that appear on the reverse side of this page replace Figures 16.5 and 16.6 of the color insert that follows p. 142. The figure captions are unchanged.

We sincerely regret any inconvenience this may have caused you. Please let us know if we can be of any assistance regarding this title or any other titles that CRC Press publishes.

Best regards,

CRC Press LLC

#1243/0-8493-1243-4

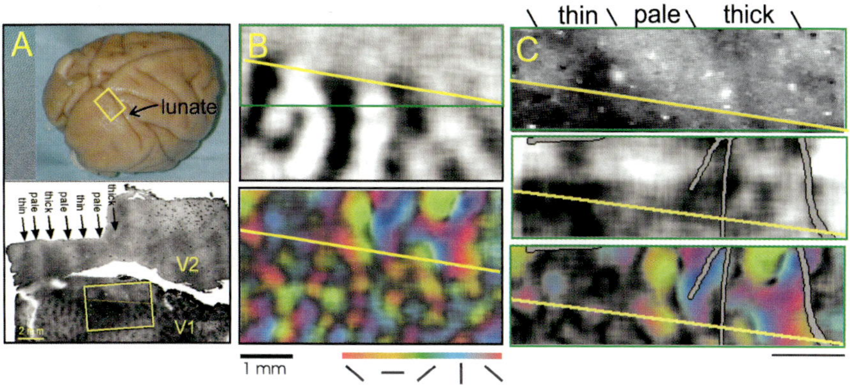

FIGURE 5.1 Macaque area V2. (A) Top, Macaque brain. Yellow box indicates location of imaged V1/V2 area just posterior to lunate sulcus (arrow); bottom, cytochrome oxidase–stained section through superficial layers of V1 and V2. Blobs are apparent in V1 and thin, pale, and thick stripes in V2 (arrows). A break in this section was caused by unfolding of the tissue at the lunate. Yellow box indicates field of view shown in B. Yellow diagonal line indicates V1/V2 border in all panels. Scale bar: 2 mm. (B) Optical images of V1 and V2. Top, Ocular dominance map. Bottom, Color-coded orientation map. Scale bar: 1 mm. (C) Enlarged views of V2 (indicated by green box in B) show alignment cytochrome oxidase–stained tissue (top) and optical images. Middle, General activation map reveals thin and thick stripe locations. Bottom, Orientation magnitude map (strength of response indicated by saturation level) reveals orientation domains in pale and thick stripes and weak orientation selectivity in thin stripes.

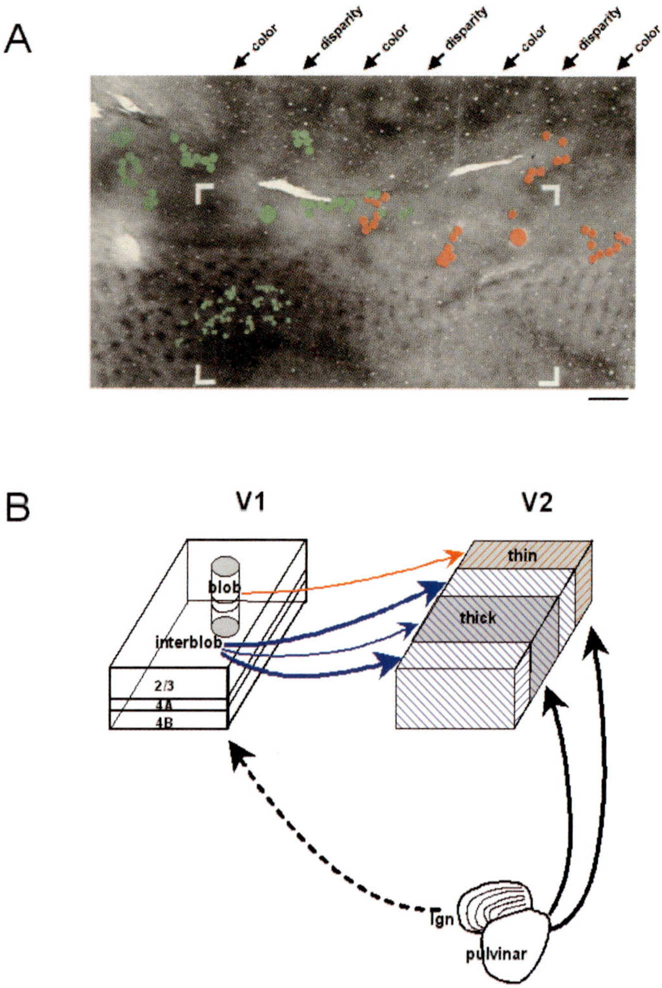

FIGURE 5.2 Anatomical connectivity of V2. (A) Feedforward V1–V2 connections and intrinsic connections within V2. Cytochrome oxidase histology of a portion of V2, showing several color and disparity stripes and the blobs of V1. Disparity stripe injection (big green dot, green fluorescent latex spheres) results in preferential label in V1 interblobs (small green dots) and extensive label in other V2 stripes (green dots). Color stripe injection (big red dot, red fluorescent latex spheres) also shows extensive V2 labeling but failed to transport to V1. Brackets indicate location of optical imaging. Scale bar: 1 mm. (From Ts'o, D.Y. et al., *Vision Res.*, 41, 1333, 2001. With permission.) (B) Schematic of V2 connectivity. (Based on Reference 44.) Bipartite division of V1–V2 connections: blobs project to thin stripes (red arrow) and interblobs project to both thick and pale stripes (blue arrows). Both projections are columnar in layers 2/3, 4A, and 4B. Pale stripes receive heaviest V1 input (heavy blue arrow). Pulvinar inputs target thin and thick stripes, dividing V2 into pulvinar- (dark stripes) vs. V1-dominated (pale stripes) regions. Lgn, lateral geniculate nucleus.

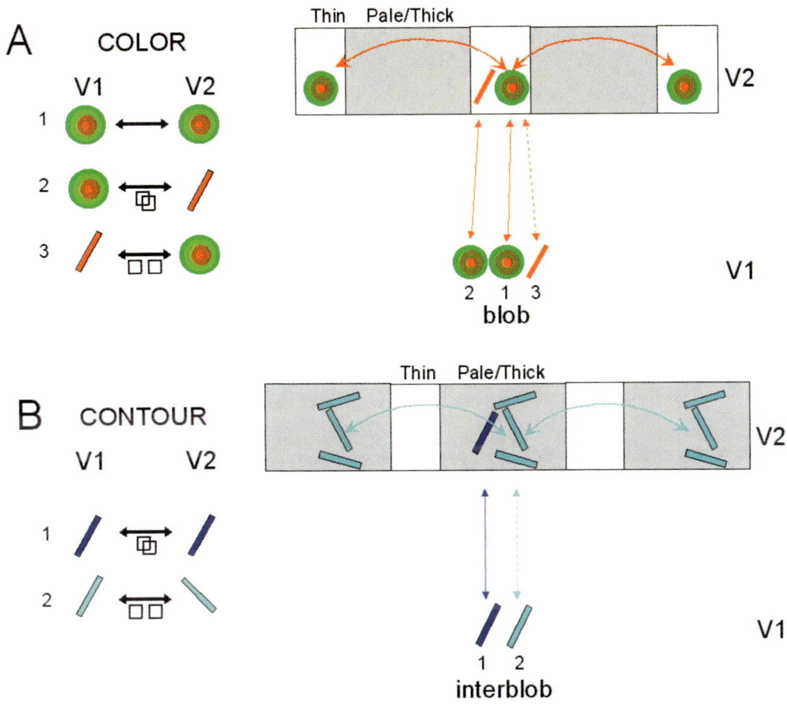

FIGURE 5.3 Functional interactions between V1 and V2. (A) Three types of color interactions. Left, Three types of functional interactions exhibit different requirements for receptive field overlap. 1, Non-oriented cell pairs do not require receptive field overlap. 2, Non-oriented V1 cells and oriented V2 cells interact only if receptive fields overlap (indicated by overlapped boxes). 3, Oriented V1 cells and non-oriented V2 cells interact only if receptive fields do not overlap (indicated by non-overlapped boxes). Right, Schematic of anatomical connections implied by functional interactions. Gray regions depict thick/pale stripes and white regions thin stripes in V2. Three V1 blob cells (1 to 3) correspond to interactions shown at left (1 to 3). 1, A large surface color network from V1 non-oriented blob cells to nearby and distant V2 thin stripes (via horizontal V2 connections, curved arrows). 2, A small network between non-oriented V1 blob cells gives rise to a small contour building network in nearby V2 thin stripes (red bar). 3, Oriented color cells in V1 blobs contact only distant V2 thin stripes (indicated by dotted arrow). See text. (B) Two types of oriented interactions. (Based on Reference 21.) Left, Two types of functional interactions exhibit different requirements for receptive field overlap. 1, Orientation-matched cell pairs interact only if receptive fields overlap. 2, Orientation-mismatched cell pairs tend to interact when receptive fields do not overlap. Right, Schematic of anatomical connections. 1, Orientation-matched network between V1 interblob cells (dark blue bar) and nearby V2 thick/pale stripes. 2, Orientation-mismatched network between V1 interblob cells (light blue bar) gives rise to orientation-diverse networks reaching distant V2 thick/pale stripes (dotted line). Arrows indicate bidirectional interactions reflecting correlograms centered on zero. (Based on References 79 and 80.)

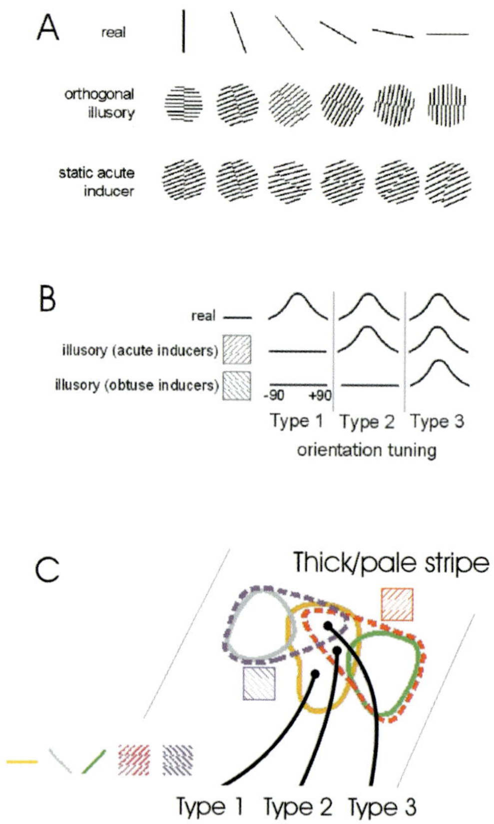

FIGURE 5.5 Contour representation in thick/pale stripes. (A) Methods of tuning real and illusory contour cells in V2. Top, Real line stimuli. Middle, Illusory contour stimuli with orthogonal inducers. Bottom, Illusory contour stimuli with static inducers. In all illusory contour stimuli, real lines are fixed while illusory contour sweeps back and forth. (B) Hierarchy of illusory contour response. Schematized orientation tuning curves (from –90° to +90° orientation) obtained in response to real and illusory contour stimuli shown at left. Type 1 cells (left) are tuned only to real contour orientation. Type 2 cells (middle) exhibit orientation tuning to real and illusory contours but respond only to contours with either acute (example shown) or obtuse inducers. Type 3 cells (right) are tuned to contour orientation regardless of composition. (C) Proposed organization of orientation domain hierarchy. Each orientation domain (e.g., yellow domain) contains zones of "real only" response (Type 1 response zone, yellow only zone), "inducer dependent" response (Type 2 response zone, region of red and yellow overlap and region of purple and yellow overlap), and "inducer independent" response (Type 3 response zone, region of red, purple, and yellow overlap). Thus, horizontal illusory contours (composed of either acute red or obtuse purple inducers) activate both the domain representing inducer orientation (either acute green or obtuse purple, respectively) and part of the horizontal domain (yellow). Region of red, purple, and yellow overlap delimits most generalized illusory contour response region (Type 3 zone).

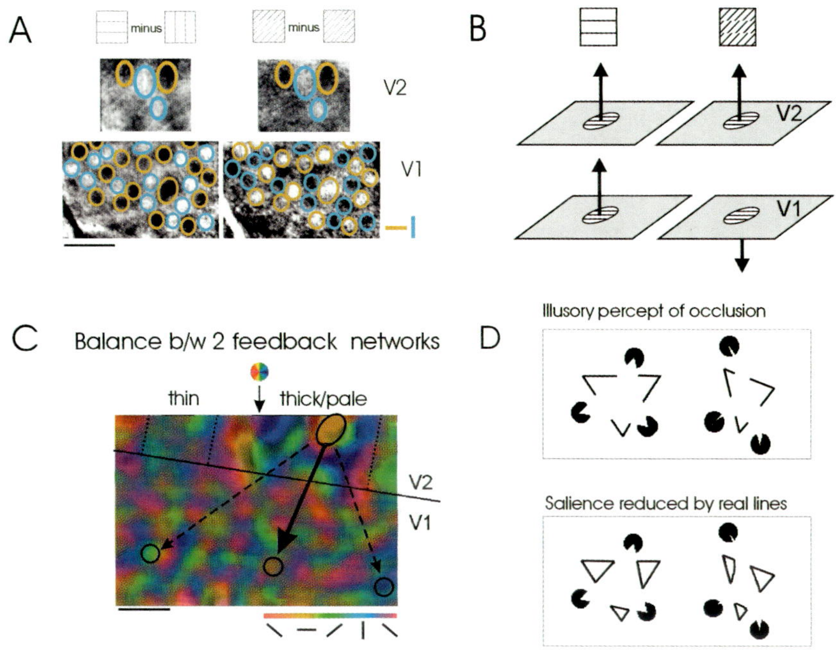

FIGURE 5.6 Orientation-specific feedback influences from V2 to V1. (A) Optical imaging of real and illusory contour response in V1 and V2. Stimuli are depicted above. Illusory stimuli were designed with identical inducer orientation; horizontal minus vertical illusory contour subtraction removes inducer component of response, leaving illusory specific response. Real and illusory contours have identical spacings. Comparison of real and illusory response reveals very similar maps in V2 (images above), suggesting presence of higher-order contour domains. Horizontal response is circled in yellow and vertical response circled in blue. The same stimulation produces inverted maps in V1—images below, e.g., yellow (blue) encircles dark (light) pixels with real contour stimulation and light (dark) pixels with illusory contour stimulation. Scale bar: 1 mm. (From Gallant, J.L. et al., *Science*, 259, 100, 1993. With permission.) (B) Summary of V1/V2 response. Real contours produce like activation in V1 and V2. Illusory contours produce activation in V2 and relative suppression in V1. (C) Potential source of V1 map inversion. Color-coded orientation image in V1 and V2 (see color code below). V1/V2 border determined by ocular dominance imaging (solid line). Dotted lines demarcate borders between thin and thick/pale stripes in V2. Pinwheel in V2 lies just below color wheel and arrow. Arrows depict dual feedback projections from V2 to V1, one that is orientation-specific and spatially localized (solid arrow) and one that is orientation-diverse and spatially extensive (dotted arrow). Real contour stimulus preferentially activates orientation-specific feedback. Illusory contour stimulation preferentially activates orientation-diverse feedback. Scale bar: 1 mm. (D) Perceptual effect of real vs. illusory push–pull. Top left, Percept of illusory triangle occluding dark disks around another outlined triangle. Top right, Same except triangle has curved outline. Bottom, Adding real contours diminishes the illusory percept and produces percept of cluttered collection of small triangles and pacmen.

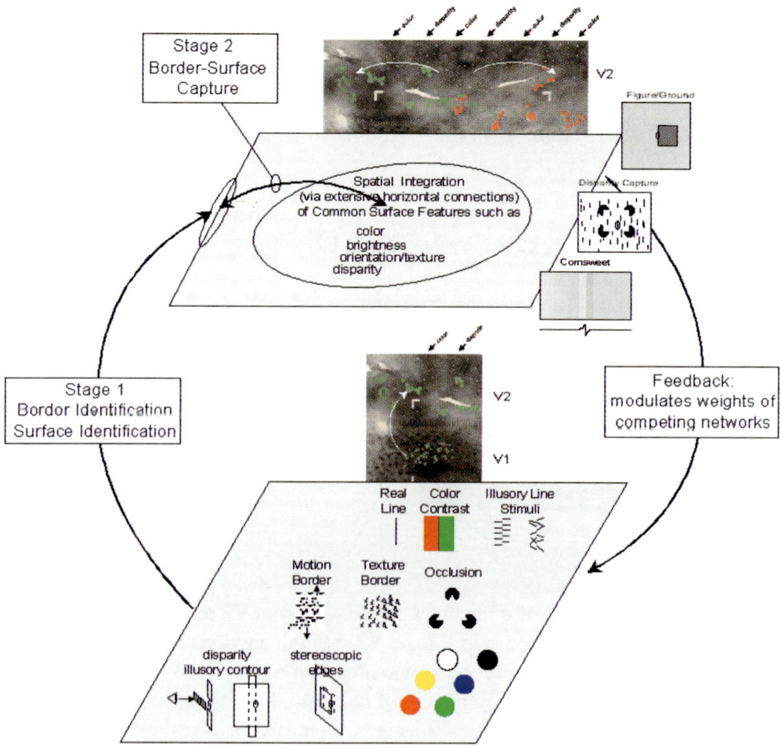

FIGURE 5.10 Two-stage model of V2 function. Bottom, Stage 1 (border and surface identification): Elemental features of borders and surfaces (including color, orientation, disparity, and other local cues such as texture, motion) are identified by V1. These elements are subsequently integrated by V1–V2 or intra-V2 convergence (big black curved arrow, white arrow on cytochrome oxidase section below) to achieve generalized contour or generalized surface identification. Top, Stage 2 (border and surface capture): Borders and surfaces are associated by horizontal V2 connections (bidirectional arrow, white arrows on cytochrome section of V2) as well as feedback from higher areas. This achieves perceptual unification of the "figure" (e.g., figure/ground, disparity capture, Cornsweet shown at right). Finally, V2–V1 feedback is proposed to modulate weights between competing parallel networks (big curved downward-projecting arrow). [Note that white arrows on cytochrome sections are meant to indicate feedforward (lower panel) and horizontal (upper panel) connections and are not meant to point from or to specific structures.]

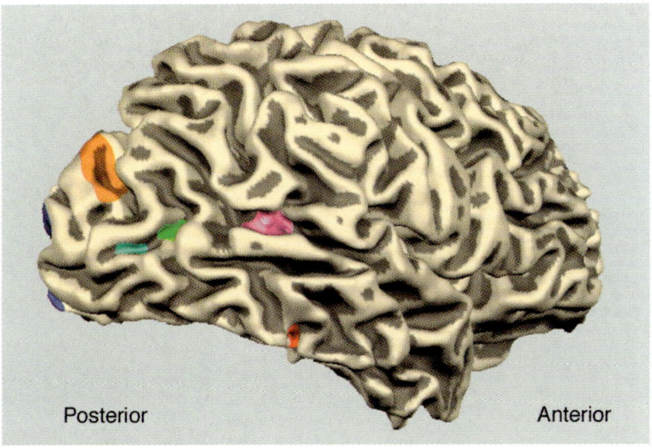

FIGURE 13.2 Human brain with colored regions denoting approximate locations of several well-studied motion-responsive areas of the human brain, areas that are discussed in this chapter. These regions are shown superimposed on a high-resolution surface map of the right hemisphere, reconstructed at the level of the white matter/gray matter junction. These motion-responsive visual areas are color-coded as follows: purple = V1 (shown here is just the small part of V1 visible in lateral view; the bulk of V1 is located on the medial surface of the hemisphere and, therefore, is largely hidden in this lateral view); orange = V3a; cyan = kinetic occipital region (KO); green = MT+ complex (sometimes referred to as V5); pink = posterior part of the STS (STSp); red = fusiform face area (typically localized by contrasting two viewing conditions: faces vs. objects). There are additional motion-responsive areas that have been identified using fMRI, located in more dorsal and more anterior regions of the brain and, as well, in the cerebellum (not pictured). For a comprehensive map of these multiple motion areas, see Sunaert et al.[56]

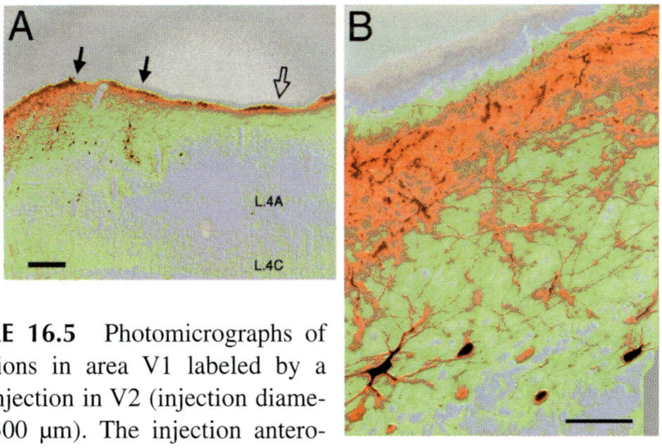

FIGURE 16.5 Photomicrographs of projections in area V1 labeled by a BDA injection in V2 (injection diameter ≤ 300 μm). The injection anterogradely labels feedback terminations, mainly in layer 1; and a few neurons in V1 have been retrogradely labeled by the same injection. Note the widespread extent of the feedback axons, which extend beyond (hollow arrow) and between patches (solid arrows) of retrogradely filled neurons. Scale bars: 200 μm in A; 5 μm in B.

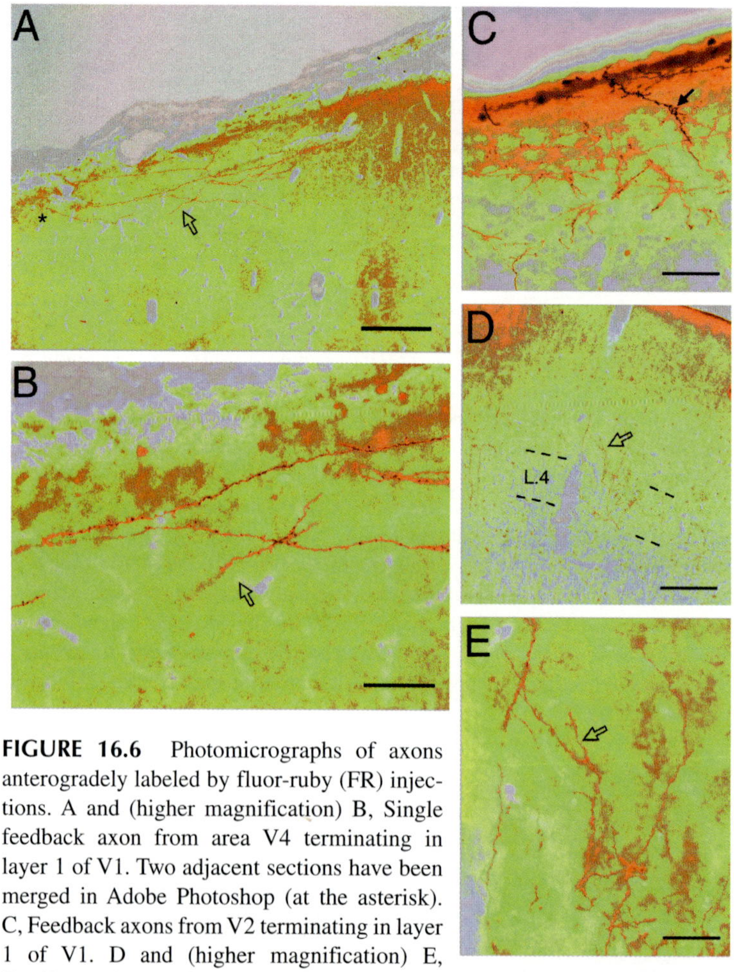

FIGURE 16.6 Photomicrographs of axons anterogradely labeled by fluor-ruby (FR) injections. A and (higher magnification) B, Single feedback axon from area V4 terminating in layer 1 of V1. Two adjacent sections have been merged in Adobe Photoshop (at the asterisk). C, Feedback axons from V2 terminating in layer 1 of V1. D and (higher magnification) E, Feedforward axons from V2 terminating in layers 3 and 4 of area V4 (from the same brain as the image in C). Note the generally similar mix of stalked and beaded terminations. One segment, indicated by the solid arrow in C, is distinguishable by its higher density of stalked boutons. Hollow arrows mark corresponding points in A and B, D and E. Scale bars: 200 μm in A, D; 50 μm in B, C, E.

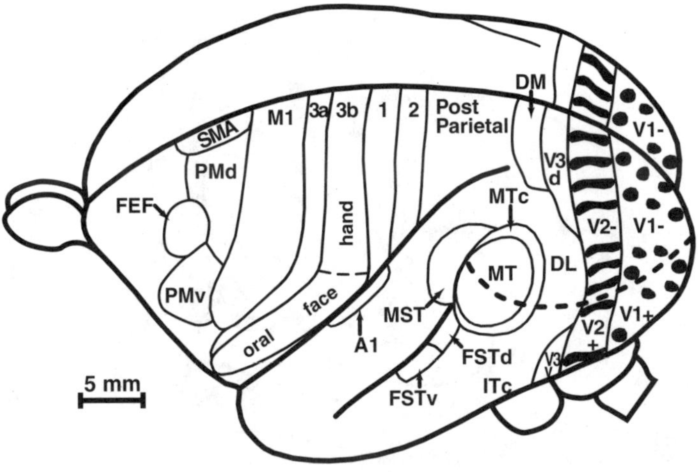

FIGURE 6.2 Visual and other areas shown on a dorsolateral view of an owl monkey (a small New World monkey) brain. The representations of the upper, and (+) and lower (–) visual quadrants in V1 and V2 are indicated. The dashed line corresponds to the representation of the horizontal meridian, which also separates representations of the upper and lower quadrants in DL, MTc, and MT. FST has ventral (FSTv) and dorsal (FSTd) subdivisions, and somatosensory cortex contains a distinct area 2. A middle superior temporal (MST) visual area has been identified. Representations of the oral cavity, face, and hand in area 3b (S1) are indicated. See Figure 6.1 for abbreviations.

inputs from the LGN that are activated by the contralateral or ipsilateral eye are segregated into the "ocular dominance columns" or stripes, with humans and most anthropoids having well-defined stripes, but some monkeys and prosimians having variably defined, poorly defined, or no obvious stripes. (see Reference 34 for review). All or nearly all primates (see Reference 35), have another distinctive histological feature of V1, a periodic distribution of ovals of tissue in the upper layer of cortex that are most commonly revealed as zones or "blobs" expressing high levels of the metabolic enzyme cytochrome oxidase (CO). These blobs (Figure 6.1 through Figure 6.3) mark specialized processing zones or columns in V1 (see Reference 33; Casagrande and Royal, Chapter 1, this volume), and they are not found in the nearest relatives of primates (tree shrews, rodents), although bloblike structures also occur in V1 of some other mammals, such as cats. We conclude that blobs emerged as a feature of V1 in early primates or their immediate ancestors and have been retained in subsequent lines of descent.

The overall size, proportional representation of central and peripheral parts of the retina, and connection patterns with other areas of cortex are also variable features of V1. Thus, V1 is a subdivision of visual cortex that is homologous in all mammals (that is, inherited as an area from a common ancestor), but one that has also differentiated in various ways in the branches of mammalian evolution to have different features and, by implication, modified functions. If the variable of size is considered, the smallest prosimian primates have a V1 of approximately 50 mm^2 in surface area, while human V1 is around 3000 mm^2 in surface area, or about 60 times

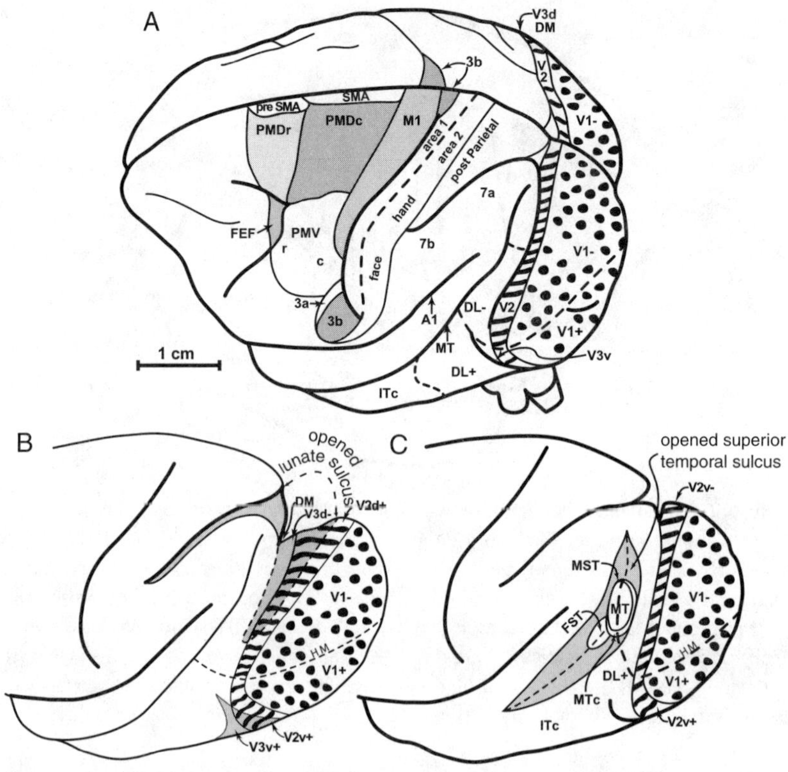

FIGURE 6.3 Visual and other areas shown on a dorsolateral view of the brain of an Old World macaque monkey. (A) The intact brain. Because this brain has deep fissures, many of the visual areas are not apparent on the surface of the brain. (B) The caudal half of the brain with the lunate sulcus opened to show the location of V3d, bordered by DM. Ventrally, the inferior occipital sulcus has been opened to show the location of V3v. (C) The caudal half of the brain with the superior temporal sulcus opened to show the locations of MT, MTc, and FST. MST borders MT in the location indicated. In A, the locations of the "area" 7a and "area" 7b are included, and PMD and PMV have rostral (R) and caudal (c) subdivisions. A pre-SMA area has been identified. See Figures 6.1 and 6.2 for abbreviations and conventions.

larger. Because neurons do not increase in size proportionally as areas increase in size, larger areas have many more neurons. This means that individual neurons have connections with a smaller proportion of the total number of neurons in the area and across less of the extent of the area. Such alterations in intrinsic connections must have functional consequences. Even a small change in the size of an area implies a significant change in functional role.[36,37] Because their intrinsic connections involve large extents of the representations, small visual areas are more suited for global comparisons of visual stimuli, whereas large areas are restricted to local comparisons. The small V1 of a mouse can function as a general-purpose visual area, but the huge V1 of a human is extremely specialized for detecting the details of visual images. Without additional, smaller visual areas to perform global comparisons, human V1 would be nearly useless.

Most visual functions in primates depend on the distribution of information from V1 to nearby visual areas and subsequent relays of visual information (see Reference 38). Humans without primary visual cortex consider themselves blind, although some visual functions remain (e.g., Reference 39). The full extent of the functions that remain after damage to V1, sometimes called blindsight, is somewhat uncertain as some of the reported abilities may depend on preserved parts of V1[38,40] (see also Collins and Kaas, Chapter 7, this volume). In monkeys, V1 projects to about ten visual areas, although most of the projections are to V2 (~50%), V3 (~20%), and MT (~10%).[41] Other connections of V1 are to the LGN, pulvinar, superior colliculus, and pons (see Reference 33).

In summary, V1 is a visual area that is a basic component of the mammalian brain, and yet it is quite variable in size, architecture, connections, representational emphasis, and presumably function. In primates, outputs from V1 directly or indirectly activate most other visual areas (see Reference 38), and thus most aspects of primate visual perception depend on V1. Yet V1, by itself, would be of little use in humans and most primates, because it is specialized for local feature detection rather than global vision.

6.3 SECOND VISUAL AREA, V2

The second visual area, V2, is another visual area that seems to have been present in the common ancestor of all mammals, and retained in nearly all.[42] As a likely exception, some species of shrews appear to have lost V2 and other extrastriate visual areas as a consequence of evolving an extremely small body and brain size (see Reference 43). The premise that nearly all mammals have V2 is not universally accepted, as V2 is not always histologically distinct. Thus, evidence for V2 often depends on experimental studies of brain connections and retinotopy, and the results from such studies can be ambiguous. Some investigators have argued that the V2 region of rats and other rodents actually consists of a number of much smaller visual areas (see Reference 42 for review). Fortunately, in anthropoid primates an unambiguous histological identification of V2 is possible (see below).

V2 has several characteristics that help identify the area. V2 forms a long belt along most of the border of V1 (Figure 6.1 through Figure 6.3). V2 constitutes a second systematic representation of the contralateral visual hemifield (after V1) that has been described as a mirror reversal of the retinotopic map in V1. However, V2 is actually a highly distorted, split mirror image of V1, as V1 is an oval while V2 is a long, narrow strip of cortex. To conform to a strip, the zero vertical meridian of the contralateral hemifield is represented along the common V1/V2 border, whereas the representation of the rest of the hemifield is split in one of several different ways (see Reference 42). In primates, the split is along all but the central portion of the horizontal midline (the zero horizontal meridian) of the retina through the fovea or center of the eye. Thus, the outer border of V2 represents the horizontal meridian along the lower visual quadrant in dorsal V2 and along the upper visual quadrant in ventral V2, with more central vision represented near the middle of the

V2 belt and peripheral vision toward the ends. This "split" representation has been called a second-order transformation of the hemifield.[44]

Clinical evidence had long suggested the existence of secondary visual "psychic" or "association" cortex next to V2.[9] The first experimental evidence for a second systematic representation of the visual hemifield next to V1 came from recordings of visually evoked potentials in cortex of cats.[45] Somewhat later, similar evidence for V2 in primates came from the recordings from the portion of V2 of squirrel monkeys on the dorsolateral surface of the occipital lobe.[46] The first complete map of the retinotopic organization of V2 was later obtained from owl monkeys.[44] Subsequent maps have been derived from microelectrode recordings in macaques,[47] cebus monkeys,[48] marmosets,[12] and galagos.[49] The non-invasive imaging of visually evoked activity patterns in human visual cortex (functional magnetic response imaging, fMRI) has provided a crude portrayal of the retinotopy of V2 in humans (e.g., References 15, 16, and 50).

Many mammals appear to have some type of repeating modular organization in V2. Such modules are often reflected by a patchy or bandlike projection pattern from V1 to V2 (e.g., References 51 and 52) or a periodic disruption of V2 by narrow bands of interhemispheric connections.[53] In V2 of monkeys, periodic bandlike arrays of cells projecting to DL[54] or V4[55] alternate with those projecting to MT.[50,55] Fortunately, these functional subdivisions of V2 can be visualized by processing brain sections for CO or myelin. In brain sections cut parallel to the cortical surface, an array of CO dark bands separated by CO light interbands cross V2 in a periodic pattern (see References 56 and 57 for early descriptions of the CO pattern and Reference 58 for the myelin pattern). The functional significance of these bands and their differing inputs and outputs are discussed elsewhere in this volume (see Roe, Chapter 5). The major practical value of this easily seen pattern of bands is that it can be used to unequivocally identify the extent of V2 (e.g., Reference 59). In traditional architectonic preparations, the architectonic distinctions between V2 and rostrally bordering areas, such as V3, are slight (see Reference 44), but the CO or myelin dense bands clearly mark this border in all anthropoid primates that have been studied (*Saimiri, Callithrix, Cebus, Aotus, Macaca, Homo*). Unfortunately, the bands are faint or absent in prosimian primates (see Reference 35). However, for anthropoid primates, V2 joins V1 as one of the areas with precisely defined borders.

Although V2 has been assumed to be co-extensive with area 18 of Brodmann,[1] this is not the case for Old World monkeys where V2 occupies only the inner third or so of the region designated as area 18. However, Brodmann's area 18 in a New World monkey was close to V2 in width. What this means is that the region depicted as area 18 occupied various amounts of visual cortex in primates (and other mammals), and it did not constantly reflect any visual area or set of areas. Some investigators now use the term "area 18" for the region of cortex that is co-extensive with V2.

As V2 receives about half of the output of V1, it is only the major but not the sole target of V1 cortical projections. Thus, V2 is not equal to V1 as a second step in cortical processing, but is only one of the several "second" steps. However, these other V1 targets are assigned to higher processing than V2 levels as they also receive feedforward projections from V2 and/or other cortical areas (see Bullier, Chapter 8, this volume). As V2 is much larger in the large human brain[60] than in the small brains of some prosimians, V2 (as for V1) can be expected to be more specialized for processing local features of the visual scene in humans than in small primates. Somewhat unexpectedly, there is some evidence that V2 is proportionately larger in large brains than in small brains. V2 is about 20% of the size of V1 in galagos[49] and about the size of V1 in macaques.[59,61]

In summary, V2 is a visual area that is present in most mammals. In primates, V2 forms a second-order "split" representation of the visual hemifield in a band of tissue along the outer border of most of V1. By receiving about half of the cortical projections of V1, V2 is the most significant of the ten or so cortical targets of V1. In anthropoid primates, the extent of V2 can be unequivocally determined from its characteristic pattern CO dense bands. V2 varies from about the size of V1 to one fifth the size of V1, and V2 is quite variable in size with overall brain size. This implies that V2 functions have been modified as hominoid primates with large brains evolved.

6.4 MIDDLE TEMPORAL VISUAL AREA, MT

The middle temporal visual area was the third systematic representation of the visual hemifield to be discovered in visual cortex of primates.[62] Cortex in the region of MT in the upper temporal lobe was known to be visually responsive from the early recording studies of Woolsey et al.[63] in 1955 in marmosets. In 1965, Kuypers et al.[64] and Myers[65] first described V1 projections to the region, but these and other early descriptions of the V1 projections[66–68] failed to reveal any topographic (retinotopic) pattern of input. However, microelectrode recordings revealed a systematic representation of the visual hemifield in the upper temporal lobe of owl monkeys that was named the middle temporal visual area, MT.[62] Although no previous architectonic study had identified a field corresponding in location and extent to MT, the representation was precisely aligned with a densely myelinated oval of cortex about one tenth the size of V1 (Figure 6.1 through Figure 6.3). This myelin dense oval is also CO rich[60] and it is strongly labeled by the Cat-301 antibody.[69] Soon thereafter, MT was identified as a retinotopic representation within a densely myelinated oval of cortex in a number of primates: galagos,[70] macaques,[71,72] marmosets,[73] and cebus monkeys.[74] In humans, MT has been located in CO, myelin, and Cat-301 preparations[60] and in fMRI data suggestive of a retinotopic map.[75] In addition, a number of studies revealed that connections of V1 with MT are topographic and retinotopically congruent (e.g., References 41, 61, and 76 through 80).

Another defining characteristic of MT is that its population of neurons is more selective for the direction of movement of visual stimuli than are neurons of other visual areas. While Allman and Kaas[62] noted that moving bars were more effective in stimulating MT neurons than the onset or offset of stationary stimuli, Dubner and Zeki[81] stressed that neurons in the region are especially responsive to direction of motion. Subsequent comparisons revealed that MT has a higher proportion of direction-selective neurons than any of the other studied visual areas.[82,83] Across studies of MT, 70 to 85% of the recorded neurons responded at least twice as well to movement in the preferred as compared to the opposite direction, and most of the remaining neurons responded to the preferred axis of movement (see Reference 84). Less than a third of neurons in other areas were strongly direction selective. Such an emphasis on neurons with selectivity for direction of movement in MT obviously has functional implications, and there is considerable evidence that MT is involved in the perception of motion (e.g., References 3 and 85). In addition, this feature provides another way of identifying MT, one that has been used to locate the MT region in fMRI studies in humans (e.g., Reference 17).

In summary, MT is a well-defined visual area that can be identified by very distinctive architectonic characteristics (dense myelination and CO), a retinotopic map, topographic inputs from V1 (and V2), and a uniquely high proportion of neurons selective for direction of movement of visual stimuli. The area has been identified in all studied primate species, but a clear homologue of MT is not apparent in the closest relatives of primates (tree shrews and rodents). Thus, it appears that MT emerged as a distinctive visual area with the advent of primates. MT is thought to play a major role in the processing of visual motion, and it is considered to be part of the dorsal stream of visual processing. As a relatively small visual area (about one tenth the size of V1), MT is well suited to comparing local motion with global (background) motion.[86] Although the functions of MT may vary somewhat across its range of sizes from about 12.88 mm^2 in galagos[87] and 13.3 mm^2 in marmosets[73] to 180 mm^2 in humans,[60] human MT is only 15 times larger than galago MT, whereas the surface area of human neocortex is 300 times larger than galago neocortex. Thus, MT has remained a relatively small area, even in the largest brains.

6.5 AREAS BORDERING MT: FST, MTc, AND MST

Several areas bordering MT are only briefly considered here, as they are at higher levels in the processing hierarchy. Nevertheless, each of these areas is likely to be a basic component of the primate visual system. One of these areas, the fundal area of the superior temporal sulcus (FST), was first distinguished by Ungerleider and Desimone[88] as the ventral component of the projection pattern of MT in macaque monkeys (the name is somewhat unfortunate because it is inconveniently long, and it is inappropriate and confusing when applied to primates without a superior temporal sulcus). Kaas and Morel[89] later divided the region into a dorsal FSTd with dense MT connections and a ventral FSTv without MT connections, but connections instead with a crescentlike belt of cortex surrounding most of MT, the MT crescent (MTc). The MTc consists of a series of periodic patches of cortex that are dark in myelin and CO, as described originally by Tootell et al.[90] in owl monkeys and more recently by Tootell and Taylor[60] in macaque and green monkeys. MTc includes the territory of area V4t of Ungerleider and Desimone,[88] as V4t was described as a narrow area along the part of MT representing the lower visual quadrant. However, MTc is a longer area that includes representations of both the upper and lower visual quadrants.[91] Finally, injections in MT labeled an adjacent projection zone that was called the medial superior temporal region (or area), MST, by Maunsell and Van Essen.[92] Later, Ungerleider and Desimone[88] divided the bordering projection zone into a dorsal zone, still termed MST, and a ventral zone, FST. Both the MST and the FST projection zones were included in the superior temporal (ST) region of MT projections in owl monkeys.[93] However, Weller et al.[93] described the dorsal ST (MST) and ventral ST (FSTd) projection zones as separate. In addition, some of the targets of the two zones were found to differ. The MST targets included posterior parietal cortex (the "intraparietal" region) and the frontal eye field, whereas the FSTd projections included MST, posterior parietal cortex, the frontal eye fields, and dorsal occipital cortex, including the dorsomedial visual area DM. There is clear evidence from MT projection patterns that MST and FSTd exist in galagos,[87,94] and we have recently obtained connectional evidence for MTc in galagos. The evidence for MST, FSTd, and MTc in prosimians, New World monkeys, and Old World monkeys suggests that these three areas are present in all primates.

6.6 THIRD VISUAL AREA, V3

The third visual area, V3, has been a proposed part of the visual system of primates for a long time, but the evidence has only recently become convincing (for review, see Reference 11). After V3 was described in cortex identified as area 19 along the lateral border of V2 in cats,[95] some investigators generally assumed that the area 19 that had been defined in monkeys and other primates would also correspond to V3 (e.g., Reference 8). As in cats, the presumptive V3 in primates was conceptualized as an area along the outer border of V2 that formed a mirror reversal map of the retinotopy of V2. Thus, the "split" horizontal meridian formed the V2/V3 border, whereas the outer border of the ringlike V3 was formed by the representation of the zero vertical meridian. Initial studies of V1 projection patterns from dorsolateral V1 (central 5° or so of the lower visual quadrant) seemed to support this contention in that two separate loci of terminations from V1 were described in dorsal occipital cortex, one central to the other, that could correspond to projections to a V2 and a V3, respectively (see References 66 and 67). The notable problem, however, was that both loci were well within area 18 of Brodmann,[1] so the expected association of evidence for V3 and V2 with the classical beltlike areas of 19 and 18 did not occur. A second problem is that the projections only provided evidence for the dorsal half of the proposed V3, and subsequent studies failed to provide conclusive evidence for connections of ventral V1 to the ventral V3 region. Although microelectrode recordings in the proposed dorsal and ventral V3 regions in both Old World[96,97] and New World[27,98] monkeys revealed representations of the lower and upper visual quadrants, respectively, this evidence was open to other interpretations.

One interpretation was that the dorsal V3 region is a valid visual area by itself, even though it represented only the lower visual quadrant. This area was called V3d. The corresponding, ventral V3 region was considered to be another visual area, the ventral posterior area (VP), even though VP represented only the upper visual quadrant (e.g., Reference 96). This led to the common practice of calling these regions of the human brain V3d and VP (e.g., Reference 14), even though the fMRI data on retinotopy was equally consistent with the concept of a single V3 with dorsal and ventral halves. In monkeys, further differences in VP and V3 were reported in other connections, neuron response properties, and cortical architecture in support of the premise that these regions constitute separate visual areas rather than parts of V3 (e.g., References 99 through 101). However, such described differences in VP and V3d could result if recordings or anatomical results were contaminated and reflected cortex adjoining as well as in VP and V3d. Such contamination seems possible in that quite different interpretations of the width and extent of V3, as well as differing descriptions of architectonic features, emerged in various reports.

Another interpretation was based on evidence from microelectrode recordings in owl monkeys that an upper visual quadrant existed next to the lower visual quadrant representation in the dorsal V3 region.[27] Similar recording results have been obtained from other species of monkeys.[12] The two quadrant representations were considered parts of a single systematic hemifield representation, the dorsomedial area, DM. DM appeared densely to moderately myelinated throughout, and DM was found to receive inputs from both dorsal and ventral parts of V1 in both New World and Old World monkeys.[102] Thus, there was good reason to conclude that the evidence from V1 projections for dorsal V3 was simply a misinterpretation of part of the projection pattern to DM. Furthermore, it seems likely that ventral V3 or VP was also part of a larger, complete visual area, one without V1 inputs. This interpretation, by invalidating VP and V3d as visual areas, avoided the troublesome concept that some areas represent only the upper or only the lower visual quadrant. These partial representations were called "Improbable Areas,"[103] although Zeki[20] noted that "it is not the improbability of such a view that is surprising; rather, it is its ready acceptance on the basis of questionable evidence."

Three interpretations of the V3 regions were possible because the connection patterns of V1 had not been investigated in enough detail. In a series of studies including prosimian galagos, macaque monkeys, marmosets, owl monkeys, titi monkeys, and squirrel monkeys,[41,79,80,104] injections of tracers were placed in several locations in V1 of the same lobe, including the less accessible ventral V1 as well as dorsal V1, and the results were studied after cortex had been separated from the rest of the brain, flattened, and cut parallel to the surface, so that the details of surface-view patterns of connections could be appreciated. The results provided compelling evidence for V3 in all these primates. These results redefined V3 as an area smaller than the V3 of earlier proposals. This redefined V3 was about half the width of V2 and less long, but it contained both ventral and dorsal halves in the positions of VP and V3d, respectively. This V3 was characterized by a CO banding pattern that joined the bordering banding pattern in V2, but the bands were thicker and less obvious. Such a banding pattern in human visual cortex had been previously recognized and attributed to dorsal V3.[105]

In summary, there is good evidence for a V3 in all studied primates, including humans. This V3 is smaller by at least half than originally proposed, and thus V3 cannot be considered to be a third step equal to V2 in serial processing. As V3 also receives inputs from V2, V3 can be considered to be at a third level of processing, but one comparable to MT and DM (see below), both also with direct V1 and V2 inputs. V3 occupies much of the territories of VP and V3d, or VP and the caudal third of DM, of previous reports. Although additional connections and neuron response properties of V3 have been described (see Reference 11 for review), uncertainties about V3 borders in these studies make conclusions uncertain.

6.7 DORSOMEDIAL AREA, DM

The dorsomedial area, DM, was defined in owl monkeys as a systematic representation of the contralateral hemifield within a field of moderately dense myelination,[27] with major connections with MT and adjoining posterior parietal (intraparietal) cortex.[106] Inputs from V1 were soon demonstrated.[78] The original caudal border of DM was thought to be retinotopically congruent with a portion of dorsal V2. As noted above, we now place V3d between dorsal V2 and DM (Figure. 6.1 through Figure 6.3). Most likely, some of the recordings sites previously placed along the estimated V2/DM border with lower quadrant receptive fields were in V3d, and likely the widths of both V2 and DM were overestimated to eliminate the narrow V3d. However, after using V1 connection patterns to delimit V3, DM remains as a slightly smaller area with its own topographic connection pattern with V1.[41,79,80,104] The connection pattern suggests that the representation of the upper visual quadrant largely lies rostral to that part of the lower visual quadrant in DM, but it approaches the caudal border laterally, where more central vision is represented. The use of microelectrode recordings alone to determine the retinotopy of DM is complicated by the widespread existence of retinotopically matched borders between areas, making the exact locations of borders difficult to detect, and by the evidence for as many as four representations of the upper visual quadrant in the DM region (e.g., Reference 12). Thus, the connections of DM with V1 distinguish the retinotopy of V3d from the retinotopy in DM, and serve as a strong guide as to the extent of DM. Optical imaging of visually evoked activity patterns has also been used to distinguish the retinotopy of V3d from DM.[107]

Other investigators have usually placed DM within other more recently proposed fields. By location and retinotopy, DM appears to correspond to most of the visual area of macaques named V3a.[108,109] Cortex in a similar location has been termed V3a in humans (e.g., Reference 60). A difficulty in assuming that DM and V3a are basically the same area is that, in all studied primates, DM is connected with V1, although less densely than V3. According to Zeki,[108] V3 receives a direct input from V1, whereas “V3a does not.” However, it seems likely from the proposed locations of V3a and DM that they are largely the same area, and that the projections V1 to V3a have been missed. Alternatively, DM may correspond to some other proposed visual area, such as V6 as suggested by Rosa and Tweedale.[6] V6 is medial to V3a, and it receives inputs from V2 and V3,[16] as does DM.[102] Thus, uncertainties remain.

In summary, there is clear evidence for DM as a roughly retinotopic representation of the visual hemifield with topographic connections with V1 and V2 in cortex just rostral to part of dorsal V3. Other connections are with MT and rostrally adjoining posterior parietal cortex. DM appears to exist in all primates. Cortex in the region of DM may correspond to much of V3a or V6 of other proposals.

6.8 DORSOLATERAL AREA, DL

The dorsolateral area, DL, was originally defined in owl monkeys as a representation of the contralateral visual hemifield in cortex located between central V2 and MT.[110] Subsequent studies of connections and architecture provided evidence that the part of DL immediately adjacent and around MT was part of another visual area, MTc.[13,89] Studies of connections also provided evidence that rostral and caudal halves of the remaining bulk of DL formed two parallel maps, each with the upper quadrant represented in cortex lateral to that for the upper quadrant. The caudal representation, DLc, received the most dense projections from V2, whereas the rostral representation, DLr, was more densely connected with MT and DM.[21,111–113] Sereno and Allman[13] and Rosa and Tweedale[6] have similarly divided the DL region of monkeys into parallel rostral and caudal representations with somewhat different names.

Earlier, Zeki[114] in 1971 suggested on the basis of projection patterns from V2 and V3 that two areas exist in the DL region, V4 andV4a. The term V4 has come into common use, but V4a has not. In addition, microelectrode mapping studies have suggested that the upper field representation in V4 (e.g., Reference 97) extends much further ventrally and onto the ventral side of the hemifield than previously portrayed, although the precise boundaries of this V4 have not been determined. This would seem to raise some questions, as it makes the upper field representation much larger than the lower field representation. In locating a V4 in human visual cortex, some investigators have been more extreme and have transposed V4 from between central (foveal) V2 and MT to a more rostroventral location on the lingual gyrus (e.g., References 20 and 114 for review), a puzzling transformation given that other visual areas hold their relative locations across primate taxa. Others have argued against this interpretation, suggesting instead that the human "V4" is another visual area, V8.[83,115] We have proposed elsewhere[9,21] that differences in interpretation, in monkeys at least, can be resolved if DLc/V4 is defined as the major feedforward projection zone of V2. Much of this projection zone is known, but not the full extent. In macaque monkeys, Stepniewska and Kaas[21] estimated that the DLc zone with V2 inputs extends about 30 mm along the V2 border, and is 7.8 mm in width. The retinotopic mapping results alone can be misinterpreted so that retinotopically similar regions may be misassigned to V4.

In summary, the DL/V4 region appears to contain two parallel representations with different connections, and thereby different functional roles. The caudal representation, DLc, receives the bulk of the V2 outputs and serves as a relay of this information into the temporal lobe for object identification and visual perception.[4,116] The rostral representation, DLr, receives far less of the output of V2, and has connections with MT and DM that suggest a greater role in the dorsal stream of visual processing for action. The full extents of these areas have not been determined, but further studies of connection patterns could help resolve differences in current proposals.

6.9 CONCLUSIONS

To understand fully how the functions of the visual system are mediated, it is essential to describe the system in terms of its parts and connections. We have seen progress toward this understanding over the last 30 years that is both impressive and disappointing. We have moved from thinking about visual cortex as containing a visual sensory area, a visual perceptual area, and multimodal association areas to the realization that anthropoid primates have a large number of visual areas (including visuomotor areas and areas that are multimodal but predominantly visual) and that each of these areas is interconnected with a number of other visual areas, especially adjoining and nearby visual areas. This progress in understanding is impressive, and it is solidly based on a great many experimental studies from a number of laboratories. But this understanding is of only the broad outline of the system. What is disappointing is that we do not know exactly how visual cortex is divided into functionally unique areas or processing stations, and thus we cannot know the connections, the properties of neurons within areas, or the histochemical characteristics of the areas. Without knowing the extents of areas for certain, the results from all such studies are likely contaminated by mixtures of results from more than one area, and descriptions of projections from one to another area are confounded by uncertainties about boundaries of areas and the validity of proposed areas. For now, we can be fairly certain about the existence and extents of the visual areas, V1, V2, and MT. At least eight or more visual areas (V3, DM, MST, FSTd, FSTv, DLr, DLc, and MTc) appear ready to join this list, but a continuation of studies focused on identifying visual areas and resolving discrepancies and differences in proposals is greatly needed.

REFERENCES

1. Brodmann, K., *Vergleichende Lokalisationslehre der Grosshirmrinde,* Barth, Leipzig, 1909.
2. Felleman, D.J. and Van Essen, D.C., Distributed hierarchical processing in the primate cerebral cortex, *Cereb. Cortex,* 1, 1–47, 1991.
3. Bisley, J.W. and Pasternak, T., The multiple roles of visual cortical areas MT/MST in remembering the direction of visual motion, *Cereb. Cortex,* 10, 1053–1065, 2000.
4. Ungerleider, L.G. and Mishkin, M., Two cortical visual systems, in *Analysis of Visual Behavior,* Ingle, D.J., Goodale, M.A., and Masfield, R.J.W., Eds., MIT Press, Cambridge, MA, 1982.
5. Ungerleider, L.G. and Haxby, J., "What" and "where" in the human brain, *Curr. Opin. Neurobiol.,* 4, 157–165, 1994.
6. Rosa, M.G.P. and Tweedale, R., The dorsomedial visual areas in New World and Old World monkeys: homology and function, *Eur. J. Neurosci.,* 13, 421–427, 2001.
7. McCulloch, W.S., Functional organization of cerebral cortex, *Physiol. Rev.,* 24, 390–407, 1944.
8. Konorski, J., *Integrative Activity of the Brain,* University of Chicago Press, Chicago, 1967.
9. Kaas, J.H., Theories of visual cortex organization in primates, in *Extrastriate Cortex in Primates,* Rockland, K.S., Kaas, J. H., and Peters, A., Eds., Plenum Press, New York, 1997, 91–125.
10. Desimone, R. and Ungerleider, L.G., Multiple visual areas in the caudal superior temporal sulcus of the macaque, *J. Comp. Neurol.,* 246, 164–189, 1986.
11. Kaas, J.H. and Lyon, D.C., Visual cortex organization in primates: theories of V3 and adjoining visual areas, *Prog. Brain Res.,* 134, 285–295, 2001.
12. Rosa, M.G.P. and Schmid, L.M., Visual areas in the dorsal and medial extrastriate cortices of the marmoset, *J. Comp. Neurol.,* 359, 272–299, 1995.

13. Sereno, M.I. and Allman, J.M., Cortical visual areas in mammals, in *Vision and Visual Dysfunction: The Neural Basis of Visual Function*, Leventhal, A.G., Ed., Macmillan, London, 1991.
14. DeYoe, E., Carman, G.J., Bandettini, P., Glickman, S., Wieser, J., and Cox, R., Mapping striate and extrastriate visual areas in human cerebral cortex, *Proc. Natl. Acad. Sci. U.S.A.,* 93, 2382–2386, 1996.
15. Engel, S.A., Glover, G.H., and Wandell, B.A., Retinotopic organization in human visual cortex and the spatial precision of functional MRI, *Cereb. Cortex,* 7, 181–192, 1997.
16. Shipp, S., Watson, J.D.G., Frackowiak, R.S.V., and Zeki, S.M., Retinotopic maps in human prestriate visual cortex: the demarcation of areas V2 and V3, *Neuroimage,* 2, 125–132, 1995.
17. Tootell, R.B.H., Reppas, J.B., Kwong, K.K., Malach, R., Born, R., Brady, T.V., Rosen, B.R., and Belliveau, J.W., Functional analysis of human MT and related visual cortical areas using magnetic resonance imaging, *J. Neurosci.,* 15, 3215–3230, 1995.
18. Van Essen, D.C. and Maunsell, J.H.R., Hierarchical organization and the functional streams in visual cortex, *Trends Neurosci.,* 6, 370–375, 1983.
19. Tootell, R.B.H. and Hadjikhani, N., Has a new color area been discovered? (Reply), *Nat. Neurosci.,* 1, 335–336, 1998.
20. Zeki, S.M., Improbable areas in the visual brain, *Trends Neurosci.,* 26, 23–26, 2003.
21. Stepniewska, I. and Kaas, J.H., Topographic patterns of V2 cortical connections in macaque monkeys, *J. Comp. Neurol.,* 371, 129–152, 1996.
22. Cowey, E., Gattass, R., and Gross, C.G., A new visual area in the parieto-occipital sulcus of the macaque, *Soc. Neurosci. Abstr.,* 8, 681, 1982.
23. Galletti, C., Fattori, P., Gamberini, M., and Kutz, D.F., The cortical visual area V6: brain location and visual topography, *Eur. J. Neurosci.,* 11, 3922–3936, 1999.
24. Colby, C.L., Gattass, R., Olson, C.R., and Gross, C.G., Topographical organization of cortical afferents to extrastriate visual area PO in the macaque: a dual tracer study, *J. Comp. Neurol.,* 269, 392–419, 1988.
25. Snyder, L.H., Batista, A.P., and Andersen, R.A., Intention related activity in the posterior parietal cortex: a review, *Vision Res.,* 40, 1433–1441, 2000.
26. Allman, J.M. and Kaas, J.H., Representation of the visual field on the medial wall of occipital-parietal cortex in the owl monkey, *Science,* 191, 572–575, 1976.
27. Allman, J.M. and Kaas, J.H., The dorsomedial cortical visual area: a third tier area in the occipital lobe of the owl monkey (*Aotus trivirgatus*), *Brain Res.,* 100, 473–487, 1975.
28. Hilgetag, C.C., O'Neill, M.A., and Young, M.P., Hierarchical organization of macaque and cat cortical sensory systems explored with a novel network processor, *Philos. Trans. R. Soc. Lond. B,* 355, 71–89, 2000.
29. Gross, C.G., From Imhotep to Hubel and Wiesel: the story of visual cortex, in *Extrastriate Cortex in Primates*, Rockland, K.S., Kaas, J.H., and Peters, A., Eds. Plenum Press, New York, 1997, 1–58.
30. Cooper, H.M., Herbin, M., and Nevd, E., Visual system of a naturally micropthalmic mammal: the blind mole rat, *Spalax ehrenbergi*, *J. Comp. Neurol.,* 328, 313–350, 1993.
31. Collins, C.E., Hendrickson, A.E., and Kaas, J.H., Distribution of calbindin, parvalbumin and wisteria in primary visual cortex of *Tarsius*, *Soc. Neurosci. Abstr.,* 27, 2001.
32. Preuss, T.M., Qi, H.-X., and Kaas, J.H., Distinctive compartmental organization of human primary visual cortex, *Proc. Natl. Acad. Sci. U.S.A.,* 96, 11601–11606, 1999.

33. Casagrande, V.A. and Kaas, J.H., The afferent, intrinsic, and efferent connections of primary visual cortex in primates, in *Cerebral Cortex*, Peters, A. and Rockland, K., Eds., Plenum Press, New York, 1994, 201–259.
34. Horton, J.C. and Hocking, D.R., Anatomical demonstration of ocular dominance columns in striate cortex of the squirrel monkey, *J. Neurosci.,* 16, 5510–5522, 1996.
35. Preuss, T.M. and Kaas, J.H., Cytochrome oxidase "blobs" and other characteristics of primary visual cortex in a lemuriform primate, *Cheirogaleus medius*, *Brain Behav. Evol.,* 47, 103–112, 1996.
36. Kaas, J.H. and Preuss, T.M., Human brain evolution, in *Fundamental Neuroscience*, 2nd ed., Squire, L.R., Ed., Academic Press, San Diego, 2002, 1147–1166.
37. Kaas, J.H., Why is brain size so important: design problems and solutions as neocortex gets bigger or smaller, *Brain Mind,* 1, 7–23, 2000.
38. Collins, C.E., Lyon, D.C., and Kaas, J.H., Responses of neurons in MT after long-standing lesions of V1 in adult New World monkeys, *J. Neurosci*, 231, 2251–2264, 2003.
39. Weiskrantz, L., *Blindsight,* Clarendon Press, Oxford, U.K.,1986.
40. Fendrich, R., Wessinger, C.M., and Gazzaniga, M.S., Speculations on the neural basis of islands of blindsight, *Prog. Brain Res.,* 134, 353–366, 2001.
41. Lyon, D.C. and Kaas, J.H., Evidence from V1 connections for both dorsal and ventral subdivisions of V3 in three species of New World monkeys, *J. Comp. Neurol.,* 449, 281–297, 2002.
42. Rosa, M.G.P. and Krubitzer, L.A., The evolution of visual cortex: where is V2? *Trends Neurosci.,* 22, 242–248, 1999.
43. Catania, K.C., Lyon, D.C., Mock, O.B., and Kaas, J.H., Cortical organization in shrews: evidence from five species, *J. Comp. Neurol.,* 410, 55–72, 1999.
44. Allman, J.M. and Kaas, J.H., The organization of the second visual area (V-II) in the owl monkey: a second-order transformation of the visual hemifield, *Brain Res.,* 76, 247–265, 1974.
45. Talbot, S.A., A lateral localization in cat's visual cortex, *Fed. Proc.,* 1, 84, 1941.
46. Cowey, E., Projection of the retina on to striate and pre striate cortex in the squirrel monkey (*Saimiri sciureus*), *J. Neurophysiol.,* 27, 366–393, 1964.
47. Gattass, R., Gross, C.G., and Sandell, J., Visual topography of V2 in the macaque, *J. Comp. Neurol.,* 201, 519–539, 1981.
48. Rosa, M.G.P., Sousa, A.P.B., and Gattass, R., Representation of the visual field in the second visual area in the *Cebus* monkey, *J. Comp. Neurol.,* 275, 326–345, 1988.
49. Rosa, M.G.P., Casagrande, V.A., Preuss, T.M., and Kaas, J.H., Visual field representation in striate and prestriate cortices of a prosimian primate (*Galago gainetti*), *J. Neurophysiol.,* 77, 3193–3217, 1997.
50. DeYoe, E. and Van Essen, D.C., Segregation of efferent connections and receptive field properties in visual area V2 of the macaque, *Nature,* 317, 58–61, 1985.
51. Sesma, M.A., Casagrande, V.A., and Kaas, J.H., Cortical connections of area 17 in tree shrews, *J. Comp. Neurol.,* 230, 337–351, 1984.
52. Lyon, D.C., Jain, N., and Kaas, J.H., Cortical connections of striate and extrastriate visual areas in the tree shrew, *J. Comp. Neurol.,* 401, 109–128, 1998.
53. Cusick, C.G. and Kaas, J.H., Interhemispheric connections of cortical sensory and motor maps in primates, in *Two Hemispheres One Brain*, Lepore, F., Ptito, M., and Jasper, H.H., Eds., Alan R. Liss, New York, 1986, 83–102.
54. Weller, R.E. and Kaas, J.H., Cortical projections of the dorsolateral visual area in owl monkeys: the prestriate relay to inferior temporal cortex, *J. Comp. Neurol.,* 234, 35–59, 1985.

55. Shipp, S. and Zeki, S.M., Segregation of pathways leading from area V2 to areas V4 and V5 of macaque monkey visual cortex, *Nature,* 315, 322–325, 1985.
56. Livingstone, M.S. and Hubel, D.H., Thalamic inputs to cytochrome oxidase-rich regions in monkey visual cortex, *Proc. Natl. Acad. Sci. U.S.A.,* 79, 6098–6101, 1982.
57. Tootell, R.B.H., Silverman, M.S., De Valois, R.L., and Jacobs, G.H., Functional organization of the second visual area in primates, *Science,* 220, 737–739, 1983.
58. Krubitzer, L.A. and Kaas, J.H., Cortical integration of parallel pathways in the visual system of primates, *Brain Res.,* 478, 161–165, 1989.
59. Olavarria, J.F. and Van Essen, D.C., The global pattern of cytochrome oxidase stripes in visual area of V2 of the macaque monkey, *Cereb. Cortex,* 7, 395–404, 1997.
60. Tootell, R.B.H. and Taylor, J.B., Anatomical evidence for MT and additional cortical visual areas in humans, *Cereb. Cortex,* 1, 39–55, 1995.
61. Weller, R.E. and Kaas, J.H., Retinotopic patterns of connections of area 17 with visual areas VII and MT in macaque monkeys, *J. Comp. Neurol.,* 220, 253–279, 1983.
62. Allman, J.M. and Kaas, J.H., A representation of the visual field in the caudal third of the middle temporal gyrus of the owl monkey (*Aotus trivirgatus*), *Brain Res.,* 31, 85–105, 1971.
63. Woolsey, C.N., Akert, K., Benjamin, R.M., Leibowitz, H., and Welker, W.I., Visual cortex in the marmoset, *Fed. Proc.,* 14, 166, 1955.
64. Kuypers, H.G.J.M., Szwarcbart, M.K., Mishkin, M., and Rosvold, H.E., Occipitotemporal corticocortical connections in the rhesus monkey, *Exp. Neurol.,* 11, 245–262, 1965.
65. Myers, R.E., Organization of visual pathways, in *Functions of the Corpus Callosum,* Ettlinger, E.G., Ed., Churchill Press, London, 1965, 133.
66. Zeki, S.M., Representations of central visual fields in prestriate cortex of monkey, *Brain Res.,* 14, 271–291, 1969.
67. Cragg, B.G., The topography of the afferent projections in the circumstriate visual cortex of the monkey studied by the Nauta method, *Vision Res.,* 9, 722–747, 1969.
68. Spatz, W.B., Tigges, J., and Tigges, M., Subcortical projections, cortical associations and some intrinsic interlaminar connections of the striate cortex in the squirrel monkey (*Saimiri*), *J. Comp. Neurol.,* 140, 155–174, 1970.
69. DeYoe, E., Hockfield, S., and Van Essen, D.C., Antibody labeling of functional subdivisions in visual cortex: cat-301 immunoreactivity in striate and extrastriate cortex of the macaque monkey, *Vis. Neurosci.,* 5, 67–81, 1990.
70. Allman, J.M., Kaas, J.H., and Lane, R.H., The middle temporal visual area (MT) in the bushbaby, *Galago senegalensis, Brain Res.,* 57, 197–202, 1973.
71. Gattass, R. and Gross, C.G., Visual topography of striate projection zone (MT) in posterior superior temporal sulcus of the macaque, *J. Neurophysiol.,* 46, 621–638, 1981.
72. Van Essen, D.C., Maunsell, J.H.R., and Bixby, J.L., The middle temporal visual area in the macaque: myeloarchitecture, connections, functional properties and topographic organization, *J. Comp. Neurol.,* 199, 293–326, 1981.
73. Rosa, M.G.P. and Elston, G.N., Visuotopic organisation and neuronal response selectivity for direction of motion in visual areas of the caudal temporal lobe of the marmoset monkey (*Callithrix jacchus*): middle temporal area, middle temporal crescent, and surrounding cortex, *J. Comp. Neurol.,* 393, 505–527, 1998.
74. Fiorani, J.M., Gattass, R., Rosa, M.G.P., and Sousa, A.P.B., Visual area MT in the *Cebus* monkey: location, visuotopic, organization and variability, *J. Comp. Neurol.,* 287, 98–118, 1989.
75. Huk, A.C., Dougherty, R.F., and Heeger, D.J., Retinotopy and functional subdivision of human areas MT and MST, *J. Neurophysiol.,* 22, 7195–7205, 2002.

76. Spatz, W.B., Topographically organized reciprocal connections between areas 17 and MT (visual area of superior temporal sulcus) in the marmoset *Callithrix jacchus*, *Exp. Brain Res.*, 27, 559–572, 1977.
77. Symonds, L.L. and Kaas, J.H., Connections of striate cortex in the prosimian, *Galago senegalensis*, *J. Comp. Neurol.*, 181, 477–512, 1978.
78. Lin, C.S., Weller, R.E., and Kaas, J.H., Cortical connections of striate cortex in the owl monkey, *J. Comp. Neurol.*, 211, 165–176, 1982.
79. Lyon, D.C. and Kaas, J.H., Connectional and architectonic evidence for dorsal and ventral V3 and area DM in marmoset monkeys, *J. Neurosci.*, 21, 249–261, 2001.
80. Lyon, D.C. and Kaas, J.H., Connectional evidence for dorsal and ventral V3, and other extrastriate areas in the prosimian primate, *Galago garnetti*, *Brain Behav. Evol.*, 59, 114–129, 2002.
81. Dubner, R. and Zeki, S.M., Response properties and receptive fields of cells in an anatomically defined region of the superior temporal sulcus in the monkey, *Brain Res.*, 35, 528–532, 1971.
82. Baker, J.F., Peterson, S.E., Newsome, W.T., and Allman, J.M., Visual response properties of neurons in four extrastriate visual areas of the owl monkey (*Aotus trivergatus*): a quantitative comparison of medial, dorsomedial, dorsolateral, and middle temporal areas, *J. Neurophysiol.*, 45, 397–416, 1981.
83. Cowey, E. and Heywood, C.A., There's more to colour than meets the eye, *Brain Res.*, 71, 89–100, 1995.
84. Felleman, D.J. and Kaas, J.H., Receptive-field properties of neurons in middle temporal visual area (MT) of owl monkeys, *J. Neurophysiol.*, 52, 488–513, 1984.
85. Salzman, C.D., Britten, K.C., and Newsome, W.T., Cortical microstimulation influences perceptual judgements of motion direction, *Nature*, 346, 174–177, 1990.
86. Born, R.T. and Tootell, R.B.H., Segregation of global and local motion processing in primate middle temporal visual area, *Nature*, 357, 497–499, 1992.
87. Krubitzer, L.A. and Kaas, J.H., Cortical connections of MT in four species of primates: areal, modular and retinotopic patterns, *Vis. Neurosci.*, 5(2), 165–204, 1990.
88. Ungerleider, L.G. and Desimone, R., Projections to the superior temporal sulcus from the central and peripheral field representations of V1 and V2, *J. Comp. Neurol.*, 248, 147–163, 1986.
89. Kaas, J.H. and Morel, A., Connections of visual areas of the upper temporal lobe of owl monkeys: the MT crescent and dorsal and ventral subdivisions of FST, *J. Neurosci.*, 13, 534–546, 1993.
90. Tootell, R.B.H., Hamilton, S.L., and Silverman, M.S., Topography of cytochrome activity in owl monkey cortex, *J. Neurosci.*, 5, 2786–2800, 1985.
91. Rosa, M.G.P. and Tweedale, R., Visual areas in lateral and ventral extrastriate cortices of the marmoset monkey, *J. Comp. Neurol.*, 422, 621–651, 2000.
92. Maunsell, J.H.R. and Van Essen, D.C., The connections of the middle temporal visual area (MT) and their relationship to a cortical hierarchy in the macaque monkey, *J. Neurosci.*, 3, 2563–2586, 1983.
93. Weller, R.E., Wall, J.T., and Kaas, J.H., Cortical connections of the middle temporal visual area (MT) and the superior temporal cortex in owl monkeys, *J. Comp. Neurol.*, 228, 81–104, 1984.
94. Wall, J.T., Symonds, L.L., and Kaas, J.H., Cortical and subcortical projections of the middle temporal area (MT) and adjacent cortex in galagos, *J. Comp. Neurol.*, 211, 193–214, 1982.
95. Hubel, D.H. and Wiesel, T.N., Receptive fields and functional architecture in two nonstriate visual areas (18 and 19) of the cat, *J. Neurophysiol.*, 28, 229–289, 1965.

96. Newsome, W.T., Maunsell, J.H.R., and Van Essen, D.C., Ventral posterior visual area of the macaque: visual topography and areal boundaries, *J. Comp. Neurol.,* 252, 139–153, 1986.
97. Gattass, R., Sousa, A.P.B., and Gross, C.G., Visuotopic organization and extent of V3 and V4 of the macaque, *J. Neurosci.,* 8, 1831–1845, 1988.
98. Newsome, W.T. and Allman, J.M., Interhemispheric connections of the visual cortex in the owl monkey, *Aotus trivirgatus*, and the bush baby, *Galago senegalensis*, *J. Comp. Neurol.,* 194, 209–233, 1980.
99. Burkhalter, A. and Van Essen, D.C., Processing of color, form, and disparity information in visual areas VP and V2 of ventral extrastriate cortex in the macaque monkey, *J. Neurosci.,* 6, 2327–2351, 1986.
100. Burkhalter, A., Felleman, D.J., and Van Essen, D.C., Anatomical and physiological asymmetries related to visual areas V3 and VP in macaque extrastriate cortex, *Vision Res.,* 26, 63–80, 1986.
101. Felleman, D.J., Burkhalter, A., and Van Essen, D.C., Cortical connections of areas V3 and VP of macaque monkey extrastriate visual cortex, *J. Comp. Neurol.,* 379, 21–47, 1997.
102. Krubitzer, L.A. and Kaas, J.H., The dorsomedial visual area of owl monkeys: connections, myeloarchitecture, and homologies in other primates, *J. Cogn. Neurosci.,* 334, 497–528, 1993.
103. Kaas, J.H., The organization of visual cortex in primates: problems, conclusions, and the use of comparative studies in understanding the human brain, in *The Functional Organization of the Human Visual Cortex*, Gulyas, B., Ottoson, D., and Roland, P.E., Eds., Pergamon Press, Oxford, 1993, 1–11.
104. Lyon, D.C. and Kaas, J.H., Evidence for a modified V3 with dorsal and ventral halves in macaque monkeys, *Neuron,* 33, 453–461, 2002.
105. Tootell, R.B.H., Dale, A.M., Sereno, M.I., and Malach, R., New images from human visual cortex, *Trends Neurosci.,* 19, 481–489, 1996.
106. Wagor, E., Lin, C.S., and Kaas, J.H., Some cortical projections of the dorsomedial visual area (DM) of association cortex in the owl monkey, *Aotus trivirgatus*, *J. Comp. Neurol.,* 163, 227–250, 1975.
107. Lyon, D.C., Xu, X., Casagrande, V.A., Stefansic, J.D., Shima, D., and Kaas, J.H., Optical imaging reveals retinoptic organization of dorsal V3 in New World owl monkeys, *Proc. Natl. Acad. Sci. U.S.A.,* 99, 15735–15742, 2002.
108. Zeki, S.M., The third visual complex of rhesus monkey prestriate cortex, *J. Physiol.,* 277, 245–272, 1978.
109. Van Essen, D.C. and Zeki, S.M., The topographic organization of rhesus monkey prestriate cortex, *J. Physiol.,* 277, 193–226, 1978.
110. Allman, J.M. and Kaas, J.H., A crescent-shaped cortical visual area surrounding the middle temporal (MT) in the owl monkey (*Aotus trivirgatus*), *Brain Res.,* 81, 199–213, 1974.
111. Steele, G.E., Weller, R.E., and Cusick, C.G., Cortical connections of the caudal subdivision of the dorsolateral area (V4) in monkeys, *J. Comp. Neurol.,* 306, 495–520, 1991.
112. Weller, R.E., Steele, G.E., and Cusick, C.G., Cortical connections of dorsal cortex rostral to V II in squirrel monkeys, *J. Comp. Neurol.,* 306, 521–537, 1991.
113. Cusick, C.G. and Kaas, J.H., Cortical connections of area 18 and dorsolateral visual cortex in squirrel monkeys, *Vis. Neurosci.,* 1, 211–237, 1988.
114. Zeki, S.M., Convergent input from the striate cortex (area 17) to the cortex of the superior temporal sulcus in the rhesus monkey, *Brain Res.,* 28, 338–340, 1971.

115. Hadjikhani, N., Liu, A.K., Dale, A.M., Cavanagh, P., and Tootell, R.B.H., Retinotopy and color sensitivity in human visual cortical area V8, *Nat. Neurosci.,* 1, 235–241, 1998.

116. Goodale, M.A. and Milner, A.D., Separate visual pathways for perception and action, *Trends Neurosci.,* 15, 20–25, 1992.

7 Plasticity of Visual Cortex in Adult Primates

Christine E. Collins and Jon H. Kaas

CONTENTS

7.1 INTRODUCTION

This chapter considers the evidence that the mature visual system of primates and other mammals is capable of extensive reorganization as the roles of inputs and pathways are altered by visual experience, sensory loss, or cortical lesions. These reorganizations seem to depend on cellular mechanisms that alter synaptic strengths, sometimes quite rapidly, and quite likely the growth of new connections in many situations. The perceptual consequences are sometimes useful, as in perceptual learning and the recovery of lost abilities, and sometimes irrelevant, such as in the perceptual filling-in of missing parts of the visual field. This chapter concentrates on changes in primary visual cortex and other areas that are early in the hierarchy of structures involved in visual processing because the internal organizations of these structures are best understood, and changes in organization can be demonstrated. Of course, this means that we neglect most of the more than 30 visual areas that are thought to exist in the brains of humans and other anthropoid primates, but principles derived from the better-studied visual areas may apply broadly.

There have been relatively few studies of the plasticity of the mature visual system in mammals, and most of those studies have been in cats, monkeys, or humans.

0-8493-1243-4/04/$0.00+$1.50

Thus, this chapter is rather restricted by the sparseness of relevant information. Interest in studying plasticity in the mature visual system probably has been limited because of the wide acceptance of the early conclusion of Sperry[1] and others that the mature visual system is quite stable in organization (as behavior after a rotated eye in amphibians suggested), and the subsequent evidence that many of the alterations in the anatomy and physiology of the visual system that can be produced by abnormal visual experience occurred largely or only during an early critical or sensitive period of development (e.g., References 2 and 3). Of course, the ability of individuals to improve on visual and other tasks with practice has long been recognized, and here improvements were assumed to be mediated by the plasticity of the central nervous system (e.g., Reference 4). Similarly, clinicians and investigators have long known that patients and subjects perceptually adapt to distorted images produced by glasses or retinal detachments so that visual images become normal. Such perceptual adjustments to distorted images also were attributed to the plasticity of the visual system (e.g., Reference 5). Finally, patients and monkeys often recover, in part, from behavioral impairments induced by lesions of visual cortex. For example, the scotoma that follows a restricted lesion of primary visual cortex (V1 or area 17) in monkeys gradually shrinks in size,[6] and behavioral impairments in visual tracking after partial lesions of the middle temporal visual area, MT, disappear or are greatly reduced over a period of days to weeks, according to the size of the lesion.[7] Such recoveries are compatible with the presumption that neurons around the lesion become responsive to the inputs that formerly activated the missing neurons, allowing surviving neurons to compensate for the visual field deficit. Thus, there have been long-standing assumptions that the mature visual system is plastic and capable of functional reorganization and modification, as well as assumptions that the organization is fixed and inflexible. What was needed to merge these different perspectives was anatomical and physiological evidence on how the system does or does not change in adults, along with evidence that any observed anatomical and physiological alterations relate to behavioral changes, improvements, and recoveries. Some evidence has been obtained and it is considered here.

We start by describing rapid changes in the functional organization of the visual system that reflect the built-in ability of a dynamic system to adjust to altered inputs. Such rapid adjustments reflect the disinhibition or unmasking of "hidden" synapses in the connectional framework.

7.2 PSEUDOPLASTICITY: RAPID ADJUSTMENTS IN RECEPTIVE FIELD PARAMETERS

Many of the changes in receptive field locations and properties that have been reported after sensory loss or manipulation occur immediately, and thus they reflect increases and decreases in the effectiveness of previously existing synapses. Such changes in the central somatosensory system after damage to sensory afferents have been called the "unmasking" of previous "silent" (undetected) synapses,[8] or "disinhibition,"[9] as this is a potential and likely mechanism. In the auditory system it has been called the "iceberg effect," the "residue" response,[10] or pseudoplasticity.[11] The immediate changes reflect a rebalancing of excitatory and inhibitory influences on the recorded neurons. These changes in balance can be mediated in various ways. A restricted lesion of the retina, for example, both eliminates excitatory drive on a central core of deprived neurons in primary visual cortex, so that these neurons fail to respond to visual stimuli, but also eliminates the drive on laterally projecting neurons from the core that activate inhibitory neurons in a fringe around the core. Thus, neurons in the fringe with formerly subthreshold excitatory inputs become activated as the result of reduced inhibition (see Reference 9). We do not consider such immediate changes in responsiveness and receptive field extents as true plasticity, but only as a rebalancing of excitatory and inhibitory inputs to reveal other aspects of a dynamically maintained receptive field.

Such rebalancing may be produced by any of a number of changes in the delivery of stimuli to the retina. Investigators now commonly refer to the "classical" receptive field of cortical neurons as the region where stimuli consistently activate the neuron (also called the minimum receptive field), and a larger region, the "nonclassical" receptive field, which includes a surround where stimuli modulate the response.[12] The boundary between the classical and nonclassical receptive field is not fixed, but depends on stimulus features that affect the balance between inhibition and facilitation of the recorded neurons.[13] In visual cortex, optical imaging has been used to reveal the considerable extent of subthreshold (for spikes) activity for stimuli outside the classical receptive field.[14,15] The extent of this fringe of subthreshold activation around the suprathreshold activation of the core is comparable to the local range of the lateral intrinsic connections in cortex.[14] The progression from subthreshold to suprathreshold activation is likely to be responsible for much of the rapid perceptual filling-in of small, natural cortical "scotomas" produced by retinal lesions[16] and the so-called artificial scotomas of inactivity in cortex resulting from viewing a hole in a texture field.[17–19] Perceptual filling-in also occurs in the natural blind spot, produced by the optic nerve head (disk) during monocular viewing (see Reference 20). During binocular viewing, neurons in the cortical representation of the optic disk are activated from the retina of the other eye. During monocular viewing, these neurons are activated by stimuli outside the cortical representation of the optic disk, and this activity in turn activates neurons within the representation,[21,22] presumably via horizontal connections in V1 (see Reference 21 for this and other possibilities). Again, we do not regard this as plasticity, but as an aspect of the normal functioning of cortical circuits. Of course, these circuits in the region of the representation of the optic disk may have developed differently from those in cortical regions with binocularly matched inputs, as a result of developmental "plasticity."

7.3 PERCEPTUAL LEARNING AND HEBBIAN CHANGES IN SYNAPTIC STRENGTHS

In humans, training on a wide range of perceptual tasks results in improved performance that has been called perceptual learning.[23] Because this is the case with many types of perceptual tasks, a number of different brain structures are likely to be involved. However, some types of perceptual learning depend on stimulus features that closely match those used in the training. For example, the ability to discriminate between narrow bars of slightly different orientations improves greatly with practice, but the gains are restricted to the location in the visual field where the training stimuli were presented, and to stimulus orientations corresponding to the training stimuli (e.g., References 24 through 26). In addition, monocular training sometimes transfers poorly to the other eye[25,27,28] (however, see Reference 29 for a critical review). Such observations suggest that the relevant changes in the visual system take place in primary visual cortex where there is a detailed retinotopic map, most neurons are orientation selective, and large populations of neurons are monocular. This logical assumption has been supported by observed changes in V1 responsiveness as a result of perceptual learning in both monkeys and humans.

In monkeys, perceptual learning occurred in animals trained to determine whether a central line of three short lines was closer to one or the other of the outside lines.[30] The stimuli were presented while the monkeys were fixated on a nearby point so that the lines were always in the same locations in the visual field. The monkeys improved in performance with training. After training, neurons with receptive fields near the trained location were recorded during training and during a simple fixation task. During fixation, a blinking bar outside the receptive field reduced the response to a bar in the receptive field, but during training the outside bar facilitated the response. Thus, the training produced changes in the receptive field properties of neurons in V1, and these changes were dependent on the nature of the visual task. This suggests that training modified the local effectiveness of horizontal connections in a specific region of V1, but in a way that could be switched on or off by an awareness of the nature of the visual task. This switch would seem to depend on other modulatory influences, possibly from higher visual areas (see Reference 14).

In humans, the effects of perceptual learning on visual cortex were visualized in a functional magnetic resonance imaging (fMRI) study by monocularly presenting a texture-discrimination task to a small part of the visual field for either the trained or untrained eye so that responses to training could be compared to responses to untrained stimuli.[31] The same stimuli in the trained eye produced more activity than in the untrained eye in a retinotopically corresponding region of V1. Thus, training altered cortical circuits so that the trained eye became more effective in evoking fMRI activity in V1.

In brief, some types of perceptual learning appear to depend on changes in connection strengths in localized portions of V1, and even within specific cell classes in V1 (those selective for a given orientation or eye). Part of the argument for this conclusion is based on an understanding of the functional organization of V1, and the evidence that V1 is the only place with organizational features that are consistent with such selective learning. However, there is additional direct evidence from recording from neurons in V1 of monkeys and fMRI of humans during perceptual learning that V1 is altered by such learning. The general assumption (e.g., Reference 25) is that the changes in V1 involve use-dependent synaptic enhancements that are induced by neural activity patterns (e.g., Hebbian plasticity). However, the recordings in monkeys[30] suggest that the changes in synaptic strengths are context dependent, and this would presumably require context-dependent differences in the activation of local inputs to V1 (feedback) from higher-order visual areas during testing. Other types of focused activity in V1 that depend on higher-order visual areas may be important in training. A brief deactivation of parietal cortex in humans interferes with perceptual learning but not performance after learning.[32]

7.4 EFFECTS OF RETINAL LESIONS

Small retinal lesions sometimes occur for various reasons in humans. In addition, such lesions are deliberately placed in the retinas of some individuals as a treatment to reduce the possibility of more serious retinal damage. Obviously, these lesions result in retinal scotomas, regions of blindness where retinal receptors are missing. What is not so clear is what happens at subsequent levels of visual processing. However, as for the natural blind spot corresponding to the optic disk of the retina, there is perceptual filling-in of the retinal scotoma so that the observer is unaware of the blind spot, even with monocular viewing when the missing information cannot be supplied by the other eye. Fortunately, research on cats and monkeys with retinal lesions has provided evidence that the perceptual filling-in is a consequence of cortical plasticity.

Retinal lesions can be placed in only one eye, or in visuotopically matched locations in both eyes (Figure 7.1). A restricted monocular lesion deprives portions

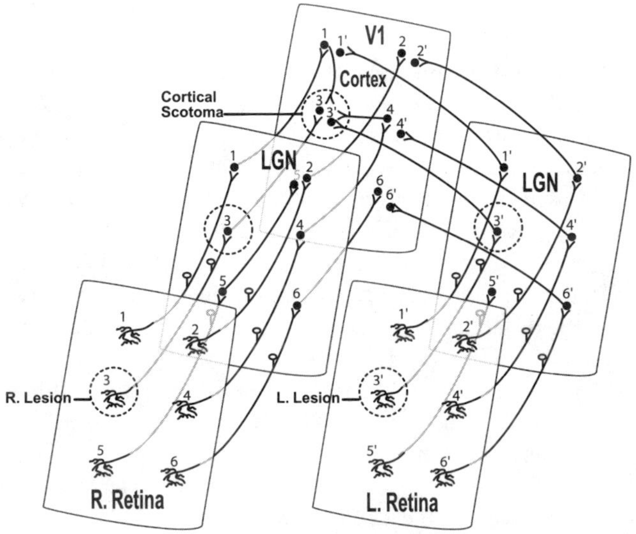

FIGURE 7.1 A schematic of the visual system from the retina to cortex and the consequences of restricted lesions of the retina. Lesions of a single eye would totally deprive zones in layers of the LGN related to that eye, but cortex with binocular inputs would have a zone of neurons deprived of activation only from the lesioned eye. Retinotopically matched lesions in both eyes would deprive zones of neurons in both sets of LGN layers, and totally deprive neurons in a zone of cortex (dashed circles). Although layers for the visuotopically congruent hemiretinas for the right (R) and left (L) eyes project to stacked, retintotopically aligned layers in the same LGN, the layers for the two eyes are side by side in the figure for visualization. Numbers correspond to different retinal or retinotopic locations.

of each of the lateral geniculate nucleus layers that receive input from that eye. In cats, this would involve one of the two "A" layers, and one or more of the "C" layers, while in monkeys a monocular lesion would involve one of the two complete parvocellular layers (and its sublayers), one of the two magnocellular layers, and zones of koniocellular inputs. For simplicity, these layers for each eye are depicted in Figure 7.1 as a single layer, and the single layers for each eye are shown as adjacent rather than stacked, as they are in the lateral geniculate nucleus (LGN). The two layers (e.g., sets of layers) then project in matching patterns to primary visual cortex, V1. A monocular lesion obviously deprives each set of monocular LGN layers of all visual activation in a portion of the retinotopic map corresponding to the lesion, while the equivalent part of V1 is deprived of activation from only one eye. Obviously, the converging inputs from undeprived geniculate layers can compensate for the missing inputs. Matching lesions of both retinas, however, totally deprive both portions of LGN layers and a portion of V1. The basic experimental question addressed in the studies in cats and monkeys is what happens to the deprived neurons in the LGN and V1 after retinal lesions?

7.4.1 Plasticity in the LGN

Neurons in affected portions of the LGN are totally deprived of visual activation by monocular lesions, and one adds nothing by placing lesions in both eyes. In early landmark experiments in cats, well before the effects of retinal lesions in cortex were known, Eysel and co-workers[33,34] found that small retinal lesions left a retinotopically matched zone of deprived neurons in the LGN unresponsive to visual stimuli, and this lack of responsiveness appeared to be permanent. The core of unresponsive neurons was bordered by a narrow fringe of neurons that acquired new receptive fields based on inputs from ganglion cells just outside the lesion of the retina. This limited reactivation of a fringe of deprived neurons does not seem to depend on any notable sprouting of preserved retinal afferents into the denervated zone,[35–37] but rather on an increased synaptic effectiveness of preserved retinal terminations on the distal dendrites of deprived neurons that extend into innervated parts of the LGN. The mechanisms responsible for this increased effectiveness are unknown, but a reduction of GABAergic inhibition is possible. Although Somogyi et al.[38] did not detect a reduction of GABA-positive neurons or synaptic density after retinal lesions, later studies did provide evidence for a reduction of GABAergic inhibition.[39] In addition, increases in growth-associated protein (GAP-43) and synapsin have been observed in the deprived zones of the LGN.[40] As synapsin is a marker of synaptic density and GAP-43 is expressed in growing axons, these molecular changes suggest that new synapses are being formed, and that synapses are being modified on LGN neurons.

As in cats, lesions of the retina in monkeys render zones of LGN neurons unresponsive to visual stimulation; apparently without recovery.[41] Thus, only limited plasticity occurs at the level of the LGN in cats and monkeys, and this plasticity appears to depend on synaptic rearrangements rather than on axon growth. The plasticity that does occur in the LGN is not extensive enough to account fully for the cortical reactivation that follows such lesions.[41]

7.4.2 Plasticity in V1 Following Binocular Lesions

A small zone of primary visual cortex can be deprived of its normal sources of visual activation by carefully placing restricted lesions in matching locations in the retinas of the two eyes, as shown in Figure 7.1, or a small lesion in one eye can be paired with a larger or total lesion of the other eye. In either case, a zone of the retinotopic map in V1 is totally deprived of its normal sources of visual activation. The surprising result of this type of experiment is that over time the deprived neurons become responsive to visual stimuli. Their former receptive fields in the lesioned parts of the retinas are permanently lost, but they acquire new receptive fields based on inputs from the intact fringe of retina around the lesion. The new receptive fields are roughly normal in size; response properties appear to be normal; and responses to visual stimuli are vigorous. Some neurons have bipartite receptive fields with domains on each side of the lesion. Similar results have been reported after binocular lesions in cats[41–45] and monkeys.[41,43]

A larger zone of binocular deprivation of V1 in monkeys, produced by a lesion of the foveal retina of each eye, resulted in incomplete recovery.[46] Nevertheless, many neurons acquired new receptive fields in the parafoveal retina. These reactivated neurons often had receptive fields with diffuse borders, and responses were less vigorous than those of normal neurons. Thus, large zones of total deactivation may not fully recover responsiveness to visual stimulation.

Although the time course of the recovery of visually evoked activity after total deprivation has not been adequately studied, maximal reactivation does appear to take at least weeks and perhaps months to emerge. Immediately after such lesions, the larger core of the deprived zone of cortex appears to be unresponsive to visual stimuli, although neurons along the margin of the deprived zone have receptive fields that are shifted in location to represent parts of the visual field just outside of the retinal lesions. These new receptive fields are often larger than normal receptive fields.[43,46,47] The more extensive recoveries, including neurons throughout the deprived core, have been obtained several months after the placement of lesions.

The response properties of deprived neurons also appear to take some time to recover. Immediately after lesions, the responses of the few visually activated neurons have been described as weak, inconsistent, and habituating to repeated stimuli.[47] After weeks to months of recovery, the responses of such neurons have been characterized as roughly normal.[42,43,48] In more systematic and quantitative studies of response properties after about 3 months of recovery, reactivated neurons had normal orientation, spatial frequency, and direction selectivity tuning, but somewhat elevated stimulus contrast thresholds.[44] Longer recoveries of more than 3 years led to nearly normal stimulus contrast thresholds.[49]

The immediate or rapid recoveries of responsiveness after retinal lesions reflect synaptic connections that provided subthreshold activation in the intact animal. In intact animals, larger-than-normal receptive fields can be revealed by procedures that reduce cortical inhibition, such as by producing an artificial scotoma by masking a portion of the visual field during stimulation.[19,22,50] Retinal lesions would similarly remove the drive of inhibitory neurons in the deprived zone of cortex, and allow subthreshold sources of activation to be expressed.

Other mechanisms are likely to mediate the gradually emerging recoveries. Periods of visual stimulation could potentiate weak connections via associative conditioning (Hebbian) mechanisms.[49,51,52] During the period of reduced neural activity in the deprived cortex, the inhibitory (GABA) neurotransmitters and receptors would be downregulated,[39,53] allowing subthreshold inputs to be expressed and strengthened by visual experience. The deprived cortex also expresses neurotrophins that may promote the formation of new connections, and molecular markers of axon growth and synapse formation emerge.[54] In addition, there is direct evidence for the growth of horizontal axons within and near the deprived cortex.[45] Thus, a reduction in the neural activity in the deprived cortex appears to trigger gene expressions that result in molecular expressions promoting recovery.

7.4.3 Plasticity after Restricted Monocular Lesions

Restricted lesions of only one retina will totally deprive zones of the LGN layers related to that eye, in the same manner as matching lesions of both eyes, but the altered cortical zone will only be deprived of visual activation from the lesioned eye. As the majority of cortical neurons are binocular and are effectively activated by either eye, powerful visual activation would continue to exist. Furthermore, any weak, subthreshold activation via horizontal connections from nearby, normally active neurons outside the deprived zone would be inhibited as a result of the disparity in the receptive fields for each eye. The monocularly deprived cortical neurons would have a normally located receptive field for the intact eye, whereas any activation from the lesioned eye would reflect a displaced receptive field. Spatially mismatched inputs from the two eyes should lead to suppression of the weaker of the two inputs, that from the displaced receptive field for the lesioned eye (see Reference 55 for review). Thus, neurons in the zone of cortex deprived by a monocular lesion would be expected to be activated only by the intact eye and show no receptive field plasticity, at least during binocular viewing.

The expected result does occur after monocular lesions.[42,43,48,56] The monocular lesions appeared to reveal only the effects of removing one of the two sources of activation, rather than any cortical reorganization. Deprived neurons responded normally to stimulations of the intact eye, and did not respond to stimulation in the lesioned eye. Yet, a framework for activation via displaced receptive fields could be revealed under some experimental conditions. Most notably, removal of the intact eye immediately allowed displaced receptive fields to be expressed.[48] Clearly the inhibition evoked by the intact eye suppressed weak inputs from the lesion surround in the lesioned eye. The lateral connections in visual cortex that mediated these responses likely exist, at least in part, in normal cortex, but they likely have been strengthened in the recovery period after the monocular lesion (see below). Other investigators have shown that displaced receptive fields can be revealed, even with the intact eye in place,[47,57] but these displaced receptive fields appear to emerge when the inhibitory influence of the intact eye is reduced during short periods of visual deprivation (lack of stimulation in a dim room).

Immediately after a monocular retinal lesion, displaced receptive fields have not been apparent,[57] suggesting that the lateral framework of excitatory connections is insufficient to activate deprived neurons from the lesioned eye. However, some responsiveness to the lesioned eye emerges after hours of stimulation of the lesioned eye. Thus, stimulation appears to potentiate weak horizontal connections over periods of time as short as several hours. Weeks or months of normal visual experience during recovery may strengthen these connections even further. When monocular lesions were placed in immature cats, and long recovery times were allowed,[58] deprived neurons were quite responsive to stimuli within displaced receptive fields. The neuron response properties for stimuli in the displaced receptive fields of the lesioned eye were similar to those evoked from normally located receptive fields from the intact eye. Dichoptic stimulations revealed both excitatory and inhibitory effects. Possibly inhibition dominates when the two receptive fields for individual neurons are most disparate.

7.5 EFFECTS OF AREA 17 LESIONS

Primary visual cortex of each cerebral hemisphere contains a complete retinotopic map of the contralateral visual hemifield. Human patients with varying degrees of V1 damage have blind areas (scotoma) in their contralateral visual hemifield corresponding to the visuotopic location of the cortical lesion, suggesting that an intact primary visual cortex is necessary for visual perception. Yet over recovery times, patients with focal visual cortex lesions can exhibit a decrease in the size of cortical scotoma, or in some cases a complete filling-in of the blind area can occur. In addition, some patients, even those with extensive cortical damage, appear to retain some residual visual function in parts of their visual field made blind by V1 damage (blindsight).[59,60] Moreover, restoration of vision in a scotoma can be promoted by repetitive visual stimulation or visual training in areas of the visual field with residual function. These areas of residual function are probably only partially deprived of inputs, and are typically the regions bordering the blind area (e.g., Reference 61). Thus, it appears that considerable plasticity exists in the mature visual system of humans. The mechanisms underlying such plasticity have been studied in monkeys and cats.

7.5.1 Local Effects of Cortical Damage

In adult cats, a reorganization of the damaged retinotopic map after focal chemical or heat lesions of V1 is first revealed by physiological changes in neurons along the border of the lesioned area of cortex. Two days after a heat-induced V1 lesion in adult cats, cells in preserved cortex in the immediate surround of the lesion show lower-than-normal levels of spontaneous activity. This region of reduced activity is surrounded by a zone of reduced inhibition, where neurons have increased levels of spontaneous activity and are hyperexcitable.[62] These hyperexcitable neurons have abnormal response properties and often exhibit irregular, bursty spontaneous activity (reviewed in Reference 63). Although the dimensions of receptive fields for these neurons appear unchanged at this time, Schweigert and Eysel[64] have recently demonstrated that 1 h of visual training of neurons in the hyperexcitable, perilesional zone is sufficient to promote an enlargement of receptive fields in about 50%

of neurons 2 days after cortical lesions. The enlarged receptive fields fill in some of the blind portion of the visual field, thus reducing the size of the scotoma. However, receptive field enlargements after training were modest (see below) compared with those occurring months after the lesion, suggesting that different mechanisms were responsible.[64] Similar results were obtained in adult rats: 1 to 5 days after lesions of primary somatosensory cortex in adult rats, a zone of hyperactivity was described in the near-surround of the lesion. Recordings from neurons 2 to 3 mm from the lesion core revealed a breakdown of the intracortical GABAergic system, and an increase in NMDA-receptor-mediated activity.[65] Thus, cortex lesions appear to create an imbalance in the local excitatory and inhibitory systems. GABAergic systems are suppressed in the near-surround of the lesion, and the activity of glutamatergic systems is enhanced. These changes lead to hyperactivation near the border of the lesion. In the short term, the hyperactivation may disrupt local function, but it may promote synaptic plasticity and reorganization by facilitating mechanisms like long-term potentiation (LTP).[65,66]

In adult cats with cortex lesions, recordings from cells in the hyperexcitable, perilesional zone 2 months after V1 lesions revealed major increases in receptive field size and a cortical scotoma that was greatly reduced in size or absent; i.e., the area of the visual field formerly represented by the lesioned portion of cortex was now represented by cells along the border of the lesion.[67] This is similar to the long-term reorganization reported following lesions of the palm representation in primary somatosensory cortex of owl monkeys, where cells in the immediate surround of the lesion enlarged their receptive fields to represent the missing portion of the somatotopic map after more than 4 months of recovery.[68]

7.5.2 Effects of V1 Lesions on Extrastriate Cortex

Deactivation or removal of V1 (area 17 or striate cortex) in adult primates results in neurons in visuotopically matched portions of many extrastriate cortical areas being deprived of their main source of activating input. Reports of spared visual capabilities in humans with V1 lesions (blindsight; see Reference 60) and recent fMRI studies showing activation in cortical areas thought to be silenced by V1 lesions (e.g., Reference 69) have reinvigorated the question of whether all parts of the extrastriate cortex are dependent on V1 for activation. In the absence of activating input fromV1, are there V1-independent pathways that are sufficient to maintain normal activation of extrastriate cortical areas or can remaining activation be accounted for by surviving parts of the main visual pathway through V1? Several experiments have addressed this question in adult monkeys and in adult human patients with damage to V1, but contradictory results have prevented a clear resolution of the issue.

One factor that appears to account for much of the variation in experimental results is the size of the lesion or reversibly inactivated zone in V1. The fact that lesion size is a factor at all implies that surviving parts of V1 may have a role in any recovery of function. In addition, each V1 cell projecting to extrastriate areas, such as MT, has terminations spread over a large region of these areas (e.g., Reference 70), making the effects of a small V1 ablation potentially negligible in terms

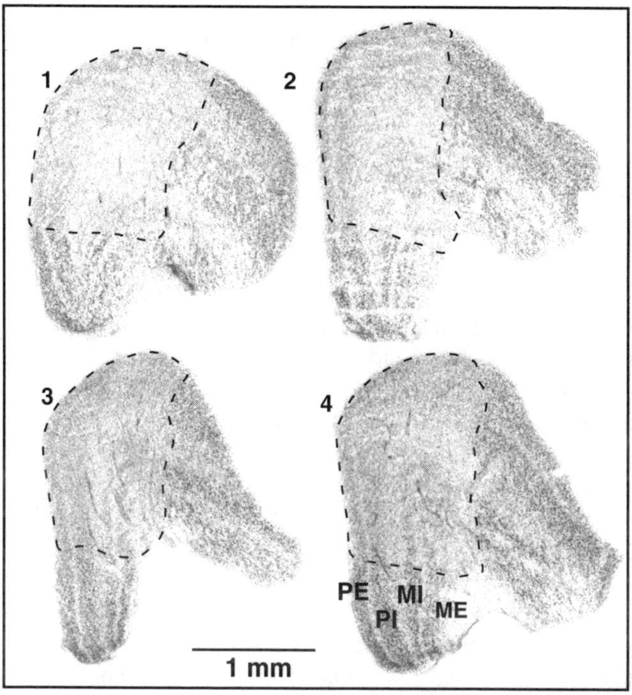

FIGURE 7.2 Series of coronal sections from the LGN of a marmoset monkey 96 days after an ipsilateral V1 lesion. Dashed lines outline the borders of the retrograde degeneration resulting from the large V1 ablation. Sections are numbered sequentially from rostral to caudal. (From Collins, C.E. et al., *J. Neurosci.*, 23, 2251–2264, 2003. With permission.)

of the responses of neurons in the extrastriate areas. The impact of these widely ramifying projections is magnified in areas that receive indirect inputs from V1. Thus, much of the recovery of visual activity in extrastriate areas of visual cortex may depend on intact portions of V1.

Another possibility is that direct LGN inputs to the extrastriate cortex substitute for missing V1 projections. The LGN has significant projections to extrastriate cortex in cats,[71,72] for example, and in several other mammals, so that considerable vision remains when V1 is removed or damaged. In primates, however, the LGN projects almost exclusively to V1 and undergoes extensive retrograde degeneration after V1 damage (Figure 7.2), appearing almost completely devoid of neurons in areas corresponding to the retinotopic position of the V1 lesion.[73] The 80% of retinal ganglion cells that project to the degenerated parvocellular layers of the LGN ultimately are lost as well. Nevertheless, a small number of LGN neurons project to extrastriate cortex (e.g., Reference 74), including MT,[75] and these may survive V1 lesions and activate extrastriate neurons.

There are several other routes by which a representation of the visual scene might reach extrastriate cortex. Targets of direct retinal input in primates include the superior colliculus and parts of the pulvinar complex (e.g., Reference 76). These pathways provide possible sources of cortical activation in the absence of V1, but there is some evidence that the activity of pulvinar neurons is reduced or eliminated as a result of V1 lesions (e.g., Reference 77). The pathway that has been repeatedly proposed as a substrate for remaining activation in extrastriate cortex is the relay of visual information from the retina to the superior colliculus, to the pulvinar complex, and then to cortex (e.g., References 78 and 79).

Despite the existence of additional visual pathways to cortex, there is widespread agreement that some areas of extrastriate cortex are totally dependent on V1 for activation. Rocha-Miranda et al.[80] demonstrated that V1 lesions totally deactivate neurons in inferotemporal cortex of macaque monkeys when the corpus callosum and anterior commissure were severed. These experiments demonstrated the involvement of callosal connections in responsiveness remaining after unilateral V1 lesions, and the lack of a sufficient, V1-independent pathway to IT cortex. Additionally, Schiller and Malpeli[81] demonstrated a complete deactivation of neurons in retinotopically corresponding parts of V2 following reversible inactivation of V1 by cooling in macaque monkeys. This finding was later confirmed by Girard and Bullier.[82] Further experiments by Girard et al.[83,84] using macaque monkeys reported a lack of responsiveness to visual stimuli in the third visual area, V3,[83] and in the fourth visual area, V4 (also known as the dorsolateral area, DL).[84]

Two areas of extrastriate cortex, V3a (also known as the dorsomedial area, DM)[83] and especially the middle temporal area, MT (also known as V5), remain controversial in the extent to which they rely on V1 for activation (see Figure 7.2). In 1989, Rodman et al.[85] lesioned V1 of Old World macaque monkeys, then recorded from MT neurons after at least 5 weeks of recovery. They reported weakly responsive neurons in the region of MT affected by the lesion, but no completely unresponsive zone in MT as a result of the lesion. A study by Girard and colleagues,[86] examining the effects of V1 deactivation by cooling, again reported responsive neurons in portions of MT retinotopically matched to deactivated parts of V1. The authors concluded that an alternative source of activation via the superior colliculus and pulvinar is likely to account for the remaining responses in MT.[86] A further study by Rodman et al.[79] recorded from MT in monkeys with superior colliculus lesions combined with V1 lesions and found no remaining responsiveness in MT of macaque monkeys when both areas were damaged. When the superior colliculus alone was damaged, there were minimal effects on MT neuron responses, suggesting that V1 is the normal source of activation.

In contrast, Kaas and Krubitzer[87] found that V1 lesions deactivated portions of MT in New World owl monkeys. In this study, the visuotopic map of MT was determined with electrophysiological recordings, a portion of the visuotopic map in V1 was removed by aspiration, then recorded sites in MT were immediately revisited to assess the effect of the lesion. Kaas and Krubitzer[87] reported a complete lack of responsiveness of neurons in the V1-deprived portion of MT, and some neurons with receptive fields slightly shifted from pre-lesion locations in the surround of the scotoma. Neurons in parts of MT unaffected by the lesion, however, were normally responsive to visual stimulation. From these data, Kaas and Krubitzer concluded that MT neurons in owl monkeys are completely dependent on V1 for activation.[87]

Several possibilities have been suggested as causes of the differences in results in owl and macaque monkeys.[78] First, any cortical lesions may cause a short-term general depression of activity in cortical neurons, and MT was explored in owl monkeys within hours of the V1 lesion. The possibility of general depression is eliminated, however, as the parts of MT subserved by intact portions of V1 remained normally responsive in these experiments. Another suggestion was that differences in anesthetic protocols caused MT neuron activity to be more depressed in owl monkeys than in macaque monkeys (see below). A third suggestion was that MT neurons in owl monkeys, but not in macaque monkeys, are dependent on V1 for activation. Maunsell and colleagues[88] reported a complete lack of responsiveness of MT neurons to visual stimuli following deactivating lidocaine injections into parvocellular (P) and magnocellular (M) layers of the LGN in macaque monkeys. When M layers alone were deactivated, MT neuron responsiveness was reduced to approximately 10% of normal, strongly suggesting that MT neurons in macaque monkeys are reliant on activating inputs from the LGN via V1.[88] As the LGN projects mainly to V1, the deactivation of the LGN can be seen as equivalent to deactivating V1.

In contrast to Kaas and Krubitzer,[87] Rosa and colleagues[78] reported that a small fraction of MT neurons remain responsive to visual stimuli in the absence of inputs from V1 in New World marmoset monkeys. They also reported encountering an abnormal population of neurons that rapidly habituated to repeated visual stimuli, and neurons that were unresponsive to visual stimulation. Rosa et al.[78] proposed that the receptive fields of at least a third of neurons in deprived parts of MT were shifted from predicted locations to represent the part of the visual field surrounding the scotoma, providing evidence for a significant reorganization of the representation of the visual field in MT within weeks of V1 lesions. However, only 13 neurons were described as "unequivocally" responsive to visual stimulation restricted to the blind part of the visual field (stimulation inside the scotoma). The authors concluded that an alternative input pathway to MT, possibly via the superior colliculus and pulvinar complex, is likely to be responsible for the remaining responsiveness of MT neurons.

A number of issues remain to be addressed to reach a conclusion about the reliance of MT on V1. Given that the responsiveness of MT neurons to visual stimuli was found to be greatly reduced, or eliminated, by LGN deactivation in macaques,[88] it may be that MT does depend on V1 in macaque monkeys. In macaque monkeys, the exposed portion of V1 represents less than 5° of the central visual field,[89] and all of area MT is buried in the superior temporal sulcus (STS). This makes it difficult to lesion (or deactivate by cooling) a significant portion of the retinotopic map in V1, and the convoluted cortex of macaques makes it much more difficult to accurately reconstruct and confirm that recording sites are in MT. In addition, the visuotopic map in MT of macaque monkeys is complex,[90] creating a further obstacle to correct interpretations of physiological recordings. The study of Rosa et al.[78] on marmosets presents uncertainties about the extents of V1 lesions, and thus V1 may have had a more extensive role in remaining responses than proposed. There are also problems in estimating the displacements of receptive fields from normal by relating the locations of recorded neurons to locations in a standard visuotopic map.

A recent study by Collins et al.[91] attempted to resolve some of the remaining uncertainty over the effects of V1 lesions on MT neurons, by testing MT neuron responsiveness in owl monkeys with long-standing V1 ablations. Long recovery times would maximize the possibility of reactivations via pathways that are independent of V1. In addition, New World owl monkeys have much smaller brains compared with macaque monkeys, with fewer fissures, and the visuotopic maps of V1 and MT are well understood. The central 10° of the lower visual field is represented on the exposed dorsolateral surface of V1, in addition to the representation of the paracentral upper visual field. Even more importantly, the majority of MT is exposed on the surface of the cortex at the tip of the STS, making it easily accessible for microelectrode recordings. In owl monkeys with significant lesions involving as much as half of V1 and recoveries of 2 months or more, most of the deprived portions of MT were totally unresponsive to visual stimuli (Figure 7.3). Neurons in a fringe of cortex around the unresponsive zone had reduced responsiveness, and possibly new receptive fields corresponding to intact portions of V1, suggesting a limited reorganization of MT that depended on V1. Smaller lesions failed to produce unresponsive zones in MT, but no receptive fields corresponded to the missing parts of V1. Thus, some plasticity was evident, but there was no evidence for any effective input other than that directly or indirectly from V1. Given the mixed results from macaques, and those from marmosets, it seems reasonable to conclude that MT is highly dependent on V1 in all monkeys (and this is also likely in humans), that intact portions of V1 can activate larger than normal portions of MT, and that alternate sources of activation may exist, but they are not very effective and are not always apparent.

7.6 MT AS A SOURCE OF BLINDSIGHT IN HUMANS

Some human patients with visual scotoma as a result of damage to V1 exhibit residual visual abilities within the blind portion of their visual field (blindsight), suggesting preserved function in the absence of V1. Preserved visual abilities in one human patient, G.Y., include sensitivity to moving stimuli (e.g., References 92 and 93). The retinotectal pathway through the pulvinar to extrastriate cortex (especially visual area MT, which is involved in the perception of motion) and direct projections from the LGN to extrastriate areas have been proposed as underlying the remaining visual activation. However, a recent fMRI investigation in the same human patient failed to demonstrate a significant activation of MT+/V5 to moving stimuli.[94] The authors reported that remaining visual activation was restricted to a part of the visual field corresponding to a spared part of V1, but that the residual vision in this patient was likely due to callosal projections from the intact contralateral V1. There is some evidence from other patients to suggest that blindsight can result from small "islands" of spared V1 that are locally isolated within V1, but that maintain connectivity with distant targets, such as MT. Fendrich and co-workers[95] suggest that residual vision is reliant on spared parts of V1 and their remaining connections with parts of extrastriate cortex. These investigators have demonstrated the existence of such islands in a number of human patients that exhibit blindsight. Thus, the visual abilities that remain after lesions of V1 in humans may largely depend on intact portions of V1, and may not involve alternative pathways to extrastriate cortex.

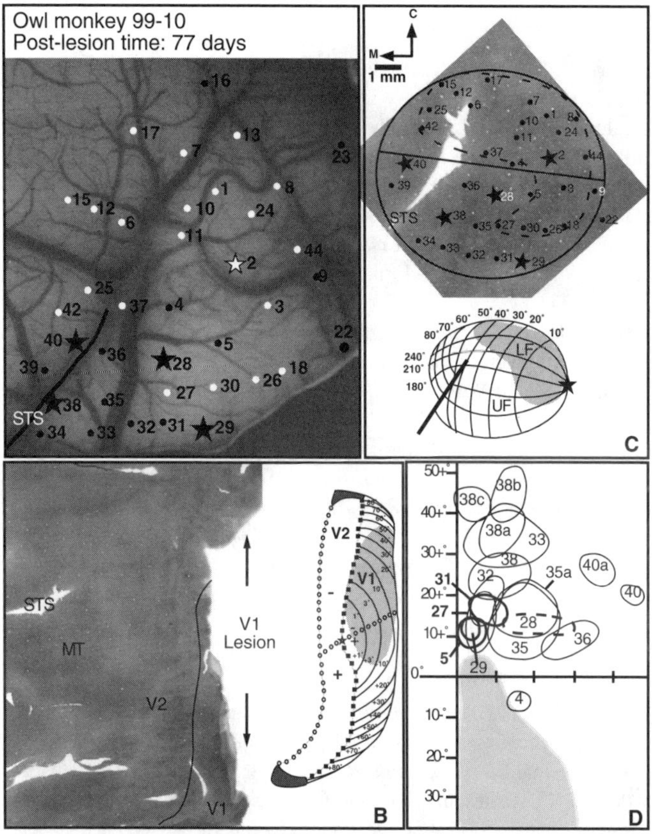

FIGURE 7.3 V1 lesion and physiological mapping data from an owl monkey with a large lesion of V1. (A) Photograph of the cortical surface used for recording electrode penetration sites during mapping. Penetration sites are marked with numbered dots, and locations of lesions made at five recording sites are marked with stars. White dots (or white stars) indicate sites where neurons were unresponsive to visual stimuli. Black dots or stars mark sites where neurons were responsive. (B) A section of flattened cortex stained for cytochrome oxidase (CO) illustrates the location of the V1 lesion. Spared parts of the upper field (UF) representation in V1 are evident. Additional spared V1 located medially is not pictured. The visuotopic location of the lesion is represented by gray shading on a schematic of unfolded V1 and V2. V2 is not damaged. (C) A gray-shaded zone on a schematic of the visuotopic map of MT illustrates the part of MT affected by the V1 ablation. An enlarged CO-stained section of MT shows electrode penetration sites with numbered black dots and recording sites with electrolytic lesions marked with stars. The 20 electrode penetration sites where there were no responses to visual stimuli are encircled by a dashed line. Numbers shown in white indicate penetrations where neurons were only transiently responsive to visual stimuli. (D) Receptive field locations corresponding to recording sites in MT. RF 28 is outlined with a dashed line to indicate that neurons at that location were only transiently responsive and the RF has less precise borders. (From Collins, C.E. et al., *J. Neurosci.*, 23, 2251–2264, 2003. With permission.)

REFERENCES

1. Sperry, R.W., Effect of 180 rotation of the retinal field on visuomotor coordination, *J. Exp. Zool.,* 92, 263–279, 1943.
2. Hubel, D.H. and Wiesel, T.N., The period of susceptibility to the physiological effects of unilateral eye closure in kittens, *J. Physiol.* (London), 206, 419–436, 1970.
3. Levitt, F.B. and Van Sluyters, R.C., The sensitive period for strabismus in the kitten, *Dev. Brain Res.,* 3, 323–327, 1982.
4. James, W., *Psychology: Briefer Course,* Harvard University Press, Cambridge, MA, 1890.
5. Kohler, I., The formation and transformation of the perceptual world, *Psych. Issues,* 3, 1–173, 1964.
6. Weiskrantz, L. and Cowey, A., Comparison of the effects of striate cortex and retinal lesions on visual acuity in the monkey, *Science,* 155, 104–106, 1967.
7. Yamasaki, D.S. and Wurtz, R.H., Recovery of function after lesions in the superior temporal sulcus in the monkey, *J. Neurophysiol.,* 66, 651–673, 1991.
8. Wall, P.D., The presence of ineffective synapses and the circumstances which unmask them, *Philos. Trans. R. Soc. Lond. B Biol. Sci.,* 278, 361–372, 1977.
9. Calford, M.B. and Tweedale, R., C-fibres provide a source of masking inhibition to primary somatosensory cortex, *Proc. R. Soc. Lond. B Biol. Sci.,* 243, 269–275, 1991.
10. Rajan, R. and Irvine, D.R., Neuronal responses across cortical field A1 in plasticity induced by peripheral auditory organ damage, *Audiol. Neurootol.,* 3, 123–144, 1998.
11. Kaltenbach, J.A., Czaja, J.M., and Kaplan, C.R., Changes in the tonotopic map of the dorsal cochlear nucleus following induction of cochlear lesions by exposure to intense sound, *Hear. Res.* 59, 213–223, 1992.
12. Allman, J., Miezen, F., and McGuinness, E., Stimulus specific responses from beyond the classical receptive field: Neurophysiological mechanisms for local-global comparisons in visual neurons, *Annu. Rev. Neurosci.,* 8, 407–430, 1985.
13. Kapadia, M.K., Westheimer, G., and Gilbert, C.D., Dynamics of spatial summation in primary visual cortex of alert monkeys, *Proc. Natl. Acad. Sci. U.S.A.,* 96, 12073–12078, 1999.
14. Stettler, D.D., Das, A., Bennett, J., and Gilbert, C.D., Lateral connectivity and contextual interactions in macaque primary visual cortex, *Neuron,* 36, 739–750, 2002.
15. Toth, L.J., Rao, C., Kim, D.-S., Somers, D., and Sur, M., Subthreshold facilitation and suppression in primary visual cortex revealed by intrinsic signal imaging, *Proc. Natl. Acad. Sci. U.S.A.,* 93, 9869–9874, 1996.
16. Tripathy, S.P. and Levi, D.M., Looking behind a pathological blind spot in human retina, *Vision Res.,* 39, 1917–1925, 1999.
17. Ramachandran, V.S. and Gregory, R.L., Perceptual filling in of artificially induced scotomas in human vision, *Nature,* 350, 699–702, 1991.
18. De Weerd, P., Gattass, R., Desimone, R., and Ungerleider, L.G., Responses of cells in monkey visual cortex during perceptual filling-in of an artificial scotoma, *Nature,* 377, 731, 1995.
19. Pettet, M.W. and Gilbert, C.D., Dynamic changes in receptive-field size in cat primary visual cortex, *Proc. Natl. Acad. Sci. U.S.A.,* 89, 8366–8370, 1992.
20. Pessoa, L., Thompson, E., and Noe, A., Finding out about filling-in: A guide to perceptual completion for visual science and the philosophy of perception, *Behav. Brain Sci.,* 21, 723–748, 1998.
21. Komatsu, H., Konoshita, M., and Murakami, I., Neural responses in the retinotopic representation of the blind spot in the macaque V1 to stimuli for perceptual filling-in, *J. Neurosci.,* 20, 9310–9319, 2000.

22. Fiorani, M., Rosa, M.G.P., Gattass, R., and Rocha-Miranda, C.E., Dynamic surrounds of receptive fields in primate striate cortex: a physiological basis for perceptual completion? *Proc. Natl. Acad. Sci. U.S.A.,* 89, 8547–8551, 1992.
23. Gilbert, C.D., Early perceptual learning, *Proc. Natl. Acad. Sci. U.S.A.,* 91, 1195–1197, 1994.
24. Fiorentini, A. and Berardi, N., Perceptual learning specific for orientation and spatial frequency, *Nature,* 287, 43–44, 1980.
25. Karni, A. and Sagi, D., Where practice makes perfect in texture discrimination: cortex plasticity, *Proc. Natl. Acad. Sci. U.S.A.,* 88, 4966–4970, 1991.
26. Ramachandran, V.S. and Braddick, O., Orientation-specific learning in stereopsis, *Perception,* 2, 371–376, 1973.
27. Poggio, T., Fahle, M., and Edelman, S., Fast perceptual learning in visual hyperacuity, *Science,* 256, 1018–1021, 1992.
28. Gais, S., Plihal, W., Wagner, U., and Born, J., Early sleep triggers memory for early visual discrimination skills, *Nat. Neurosci.,* 3, 1335–1339, 2000.
29. Gilbert, C.D., Sigman, M., and Crist, R.E., The neural basis of perceptual learning, *Neuron,* 31, 681–697, 2001.
30. Crist, R.E., Li, W., and Gilbert, C.D., Learning to see: experience and attention in primary visual cortex, *Nat. Neurosci.,* 4, 519–525, 2001.
31. Schwartz, S., Maquet, P., and Frith, C., Neural correlates of perceptual learning: a functional MRI study of visual texture discrimination, *Proc. Natl. Acad. Sci. U.S.A.,* 99, 17137–17142, 2002.
32. Walsh, V., Ellison, A., Ashbridge, E., and Cowey, A., The role of the parietal cortex in visual attention — hemispheric asymmetries and the effects of learning: a magnetic stimulation study, *Neuropsychologia,* 37, 245–251, 1999.
33. Eysel, U.T., Gonzalez-Aguilar, F., and Mayer, U., Time-dependent decrease in the extent of visual deafferentation in the lateral geniculate nucleus of adult cats with small retinal lesions, *Exp. Brain Res.,* 41, 256–263, 1981.
34. Eysel, U.T., Gonzalez-Aguilar, F., and Mayer, U., A functional sign of reorganization in the visual system of adult cats: lateral geniculate neurons with displaced receptive fields after lesions of the nasal retina, *Brain Res.,* 181, 285–300, 1980.
35. Stelzner, D.J. and Keating, E.G., Lack of intralaminar sprouting of retinal axons in monkey LGN, *Brain Res.,* 126, 201–210, 1977.
36. Eysel, U.T., Functional reconnections without new axonal growth in a partially denervated visual relay nucleus, *Nature,* 299, 442–444, 1982.
37. Baisden, R.H., Polley, E.H., Goodman, D.C., and Wolf, E.D., Absence of sprouting by retinogeniculate axons after chronic focal lesions in the adult cat retina, *Neurosci. Lett.,* 17, 33–38, 1980.
38. Somogyi, P., Eysel, U.T., and Hamori, J., A quantitative study of morphological reorganization following chronic optic deafferentation in the adult cat dorsal lateral geniculate nucleus, *J. Comp. Neurol.,* 255, 341–350, 1987.
39. Arckens, L., Eysel, U.T., Vanderhaeghen, J.J., Orban, G.A., and Vandesande, F., Effect of sensory deafferentation on the GABAergic circuitry of the adult cat visual system, *Neuroscience,* 83, 381–391, 1998.
40. Baekelandt, V., Arckens, L., Annaert, W., Eysel, U.T., Orban, G.A., and Vandesande, F., Alterations in GAP-43 and synapsin immunoreactivity provide evidence for synaptic reorganization in adult cat dorsal lateral geniculate nucleus following retinal lesions, *Eur. J. Neurosci.,* 6, 754–765, 1994.
41. Darian-Smith, C. and Gilbert, C.D., Topographic reorganization in the striate cortex of the adult cat and monkey is cortically mediated, *J. Neurosci.,* 15, 1631–1647, 1995.

42. Kaas, J.H., Krubitzer, L.A., Chino, Y.M., Langston, A.L., Polley, E.H., and Blair, N., Reorganization of retinotopic cortical maps in adult mammals after lesions of the retina, *Science,* 248, 229–231, 1990.
43. Gilbert, C.D. and Wiesel, T.N., Receptive field dynamics in adult primary visual cortex, *Nature,* 356, 150–152, 1992.
44. Chino, Y.M., Smith, E.L., III, Kaas, J.H., Sasaki, Y., and Cheng, H., Receptive-field properties of deafferentated visual cortical neurons after topographic map reorganization in adult cats, *J. Neurosci.,* 15, 2417–2433, 1995.
45. Darian-Smith, C. and Gilbert, C.D., Axonal sprouting accompanies functional reorganization in adult cat striate cortex, *Nature,* 368, 737–740, 1994.
46. Heinen, S.J. and Skavenski, A.A., Recovery of visual responses in foveal V1 neurons following bilateral foveal lesions in adult monkey, *Exp. Brain Res.,* 83, 670–674, 1991.
47. Schmid, L.M., Rosa, M.G., Calford, M.B., and Ambler, J.S., Visuotopic reorganization in the primary visual cortex of adult cats following monocular and binocular retinal lesions, *Cereb. Cortex,* 6, 388–405, 1996.
48. Chino, Y.M., Kaas, J.H., Smith, E.L., III, Langston, A.L., and Cheng, H., Rapid reorganization of cortical maps in adult cats following restricted deafferentation in retina, *Vision Res.,* 32, 789–796, 1992.
49. Matsuura, K., Zhang, B., Mori, T., Smith, E.L., III, Kaas, J.H., and Chino, Y., Topographic map reorganization in cat area 17 after early monocular retinal lesions, *Vis. Neurosci.,* 19, 85–96, 2002.
50. Volchan, E. and Gilbert, C.D., Interocular transfer of receptive field expansion in cat visual cortex, *Vision Res.,* 35, 1–6, 1995.
51. McLean, J. and Palmer, L.A., Plasticity of neuronal response properties in adult cat striate cortex, *Vis. Neurosci.,* 15, 177–196, 1998.
52. Eysel, U.T., Eyding, D., and Schweigart, G., Repetitive optical stimulation elicits fast receptive field changes in mature visual cortex, *NeuroReport,* 9, 949–954, 1998.
53. Rosier, A.M., Arckens, L., Demeulemeester, H., Orban, G.A., Eysel, U.T., Wu, Y.J., and Vandesande, F., Effect of sensory deafferentation on immunoreactivity of GABAergic cells and on GABA receptors in the adult cat visual cortex, *J. Comp. Neurol.,* 359, 476–489, 1995.
54. Obata, S., Obata, J., Das, A., and Gilbert, C.D., Molecular correlates of topographic reorganization in primary visual cortex following retinal lesions, *Cereb. Cortex,* 9, 238–48, 1999.
55. Thiele, A., Bremmer, F., Ilg, U.J., and Hoffmann, K.-P., Visual responses of neurons from areas V1 and MT in a monkey with late onset strabismus: a case study, *Vision Res.,* 37, 853–863, 1997.
56. Komatsu, H. and Murakami, I., Behavioral evidence of filling-in at the blind spot of the monkey, *Vis. Neurosci.,* 11, 1103–1113, 1994.
57. Calford, M.B., Schmid, L.M., and Rosa, M.G., Monocular focal retinal lesions induce short-term topographic plasticity in adult cat visual cortex, *Proc. R. Soc. Lond. B Biol. Sci.,* 266, 499–507, 1999.
58. Mori, T., Matsuura, K., Zhang, B., Smith, E.L., III, and Chino, Y.M., Effects of the duration of early strabismus on the binocular responses of neurons in the monkey visual cortex (V1), *Invest. Ophthalmol. Vis. Sci.,* 43, 1262–1269, 2002.
59. Weiskrantz, L., Warrington, E.K., Sanders, M.D., and Marshall, J., Visual capacity in the hemianopic field following a restricted occipital ablation, *Brain,* 97, 709–728, 1974.
60. Weiskrantz, L., *Blindsight: A Case Study and Implications,* Clarendon Press, Oxford, U.K., 1986.

61. Kasten, E., Poggel, D.A., and Sabel, B.A., Computer-based training of stimulus detection improves color and simple pattern recognition in the defective field of hemianopic subjects, *J. Cogn. Neurosci.,* 12, 1001–1012, 2000.
62. Eysel, U.T. and Schmidt-Kastner, R., Neuronal dysfunction at the border of focal lesions in cat visual cortex, *Neurosci. Lett.,* 131, 45–48, 1991.
63. Eysel, U.T., Perilesional cortical dysfunction and reorganization, *Adv. Neurol.,* 73, 195–206, 1997.
64. Schweigert, G. and Eysel, U.T., Activity-dependent receptive field changes in the surround of adult cat visual cortex lesions, *Eur. J. Neurosci.,* 15, 1585–1596, 2002.
65. Mittmann, T., Luhmann, H.J., Schmidt-Kastner, R., Eysel, U.T., Weigel, H., and Heinemann, U., Lesion-induced transient suppression of inhibitory function in rat neocortex *in vitro*, *Neuroscience,* 60, 891–906, 1994.
66. Mittmann, T. and Eysel, U.T., Increased synaptic plasticity in the surround of visual cortex lesions in rats, *NeuroReport,* 12, 3341–3347, 2001.
67. Eysel, U.T. and Schweigart, G., Increased receptive field size in the surround of chronic lesions in the adult cat visual cortex, *Cereb. Cortex,* 9, 101–109, 1999.
68. Jenkins, W.M. and Merzenich, M.M., Reorganization of neocortical representations after brain injury: a neurophysiological model of the bases of recovery from stroke, *Prog. Brain Res.,* 71, 249–266, 1987.
69. Goebel, R., Muckli, L., Friedhelm, E.Z., Singer, W., and Stoerig, P., Sustained extrastriate cortical activation without visual awareness revealed by fMRI studies of hemianopic patients, *Vision Res.,* 41, 1459–1474, 2001.
70. Rockland, K.S., Bistratified distribution of terminal arbors of individual axons projecting from area V1 to middle temporal area (MT) in the macaque monkey, *Vis. Neurosci.,* 3, 155–170, 1989.
71. Doty, R.W., Survival of pattern vision after removal of striate cortex in the adult cat, *J. Comp. Neurol.,* 143, 341–369, 1971.
72. Winans, S.S., Visual form discrimination after removal of the visual cortex in cats, *Science,* 158, 944–946, 1967.
73. Weller, R.E. and Kaas, J.H., Parameters affecting the loss of ganglion cells of the retina following ablations of striate cortex in primates, *Vis. Neurosci.,* 3, 327–349, 1989.
74. Kisvárday, Z.F., Cowey, A., Stoerig, P., and Somogyi, P., Direct and indirect retinal input into degenerated dorsal lateral geniculate nucleus after striate cortex removal in monkey: implications for residual vision, *Exp. Brain Res.,* 86, 271–292, 1991.
75. Stepniewska, I., Qi, H.X., and Kaas, J.H., Do superior colliculus projection zones in the inferior pulvinar project to MT in primates? *Eur. J. Neurosci.,* 11, 469–480, 1999.
76. O'Brien, B.J., Abel, P.L., and Olavarria, J.F., The retinal input to calbindin-D28k-defined subdivisions in macaque inferior pulvinar, *Neurosci. Lett.,* 312, 145–148, 2001.
77. Bender, D.B., Visual activation of neurons in the primate pulvinar depends on cortex but not colliculus, *Brain Res.,* 279, 258–261, 1983.
78. Rosa, M.G.P., Tweedale, R., and Elston, G.N., Visual responses of neurons in the middle temporal area of New World monkeys after lesions of striate cortex, *J. Neurosci.,* 20, 5552–5563, 2000.
79. Rodman, H.R., Gross, C.G., and Albright, T.D., Afferent basis of visual response properties in area MT of the macaque. II. Effects of superior colliculus removal, *J. Neurosci.,* 10, 1154–1164, 1990.
80. Rocha-Miranda, C.E., Bender, D.B., Gross, C.G., and Mishkin, M., Visual activation of neurons in inferotemporal cortex depends on striate cortex and forebrain commissures, *J. Neurophysiol.,* 38, 475–491, 1975.

81. Schiller, P.H. and Malpeli, J.G., The effect of striate cortex cooling on area 18 cells in the monkey, *Brain Res.,* 126, 366–369, 1977.
82. Girard, P. and Bullier, J., Visual activity in area V2 during reversible inactivation of area 17 in the macaque monkey, *J. Neurophysiol.,* 62, 1287–1302, 1989.
83. Girard, P., Salin, P.A., and Bullier, J., Visual activity in areas V3a and V3 during reversible inactivation of area V1 in the macaque monkey, *J. Neurophysiol.,* 66, 1493–503, 1991.
84. Girard, P., Salin, P.A., and Bullier, J., Visual activity in macaque area V4 depends on area 17 input, *NeuroReport* 2, 81–84, 1991.
85. Rodman, H.R., Gross, C.G., and Albright, T.D., Afferent basis of visual response properties in area MT of the macaque. I. Effects of striate cortex removal, *J. Neurosci.,* 9, 2033–2050, 1989.
86. Girard, P., Salin, P.A., and Bullier, J., Response selectivity of neurons in area MT of the macaque monkey during reversible inactivation of area V1, *J. Neurophysiol.,* 67, 1437–1446, 1992.
87. Kaas, J.H. and Krubitzer, L.A., Area 17 lesions deactivate area MT in owl monkeys, *Vis. Neurosci.,* 9, 399–407, 1992.
88. Maunsell, J.H., Nealey, T.A., and DePriest, D.D., Magnocellular and parvocellular contributions to responses in the middle temporal visual area (MT) of the macaque monkey, *J. Neurosci.,* 10, 3323–3334, 1990.
89. Adams, D.L. and Horton, J.C., A precise retinotopic map of the primate striate cortex generated from angioscotomas, *Soc. Neurosci. Abstr.,* 27, 2001.
90. Van Essen, D.C., Maunsell, J.H., and Bixby, J.L., The middle temporal visual area in the macaque: myeloarchitecture, connections, functional properties and topographic organization, *J. Comp. Neurol.,* 199, 293–326, 1981.
91. Collins, C.E., Lyon, D.C., and Kaas, J.H., Responses of neurons in MT after long-standing lesions of V1 in adult New World monkeys, *J. Neurosci.*, 23, 2251–2264, 2003.
92. ffytche, D.H., Guy, C.N., and Zeki, S., Motion specific responses from a blind hemifield, *Brain,* 119(6), 1971–1982, 1996.
93. Barbur, J.L., Watson, J.D., Frackowiak, R.S., and Zeki, S., Conscious visual perception without V1, *Brain,* 116(6), 1293–1302, 1993.
94. Baseler, H.A., Morland, A.B., and Wandell, B.A., Topographic organization of human visual areas in the absence of input from primary cortex, *J. Neurosci.*, 19, 2619–2627, 1999.
95. Fendrich, R., Wessinger, C.M., and Gazzaniga, M.S., Speculations on the neural basis of islands of blindsight, *Prog. Brain Res.,* 134, 353–366, 2001.

8 Hierarchies of Cortical Areas

Jean Bullier

CONTENTS

8.1 HISTORICAL ASPECTS OF THE HIERARCHICAL MODEL

8.1.1 Processing in a Hierarchy or in Parallel

The idea that the cortical surface in mammals is made up of separate regions organized in a hierarchy dates from the distinction by Flechsig[1] in the late 19th century between the primary sensory areas and the association cortex. The cortex was then viewed as the central element in an elaborate reflex system interposed between the sensory input and the motor output. The sensory inputs were thought to reach the cortex through the primary areas of different modalities and to be combined in the association areas before sending orders to the motor cortex. With the advent of tracing studies using anterograde degeneration of nerve fibers, it became clear that, within the association cortex, there are large cortical territories mostly devoted specifically to vision, audition, and the somatic senses. The organizations of these sensory regions were investigated by Jones and Powell[2] who

0-8493-1243-4/04/$0.00+$1.50

followed the connections from the primary areas into the association cortex. With the exception of projections to the frontal cortex, they concluded that connections in each sensory domain follow a neighbor-to -neighbor rule and that, consequently, the information progresses step by step from the primary areas to high-order areas such as the temporal pole and the frontal cortex.

In these early models of the sensory cortex, only the primary areas were thought to be organized according to the topography of the surface of the recipient structure (retina, cochlea, or body surface). Thus, in the visual system, it was thought that retinotopic organization was lost beyond area 17. In the 1940s the discovery by Talbot and Marshall[3] of a retinotopic organization in area 18 in the cat visual cortex introduced the idea that there can be more than one functional area with retinotopic organization in the mammalian visual cortex. Before the complete mapping of visual areas was achieved in the 1980s with the work of Tusa, Palmer, and Rosenquist[4] in the cat and Allman and Kaas[5] in the owl monkey, Hubel and Wiesel[6] made the early discovery of a third visual area in the cat visual cortex. They introduced the first hierarchical model to explain the step-by-step transformation of the thalamic inputs from the lateral geniculate nucleus (LGN) to the simple cell in area V1 and from the simple cell to the complex cell. Simple cells were supposed to be limited to area V1 and complex cells were found to be more numerous in area V2 (area 18 in the cat). Combination of complex cell inputs was supposed to lead to a higher processing stage with the hypercomplex cells that were thought to be limited to area V3 (area 19 in the cat). Thus the simple to complex to hypercomplex hierarchy corresponded to a progression in the cortical areas from V1 to V2 to V3. This first version of the hierarchical model was refuted by the discoveries that hypercomplex cells are also found in area 17,[7] that complex cells are found in areas 17, 18, and 19, that LGN fibers contact directly simple as well as complex cells,[8,9] and that destruction or inactivation of area 17 does not lead to a silence of area 18 neurons.[10,11] Later investigations confirmed that complex cells as well as simple cells receive direct input from the LGN[12] and showed that some visual stimuli can drive complex cells without driving simple cells[13] and that simple cells can be transformed into complex cells by manipulating the strength of inhibitory inputs.[14] In 1980, all that remained of the hierarchical model of the cat visual areas was that hypercomplex cells tended to be more numerous in area 19 than in areas 17 and 18.

In the cat, the hierarchical model was replaced in the 1980s by a parallel organization. This followed the discovery by the Australian school that the retina of mammals is not composed of a homogeneous population of ganglion cells but of three major types, the X, Y, and W ganglion cells.[15,16] The parallel model stated that each of these populations drives one of the three cortical areas 17, 18, and 19 in parallel: X cells constitute the main drive to area V1, Y cells are mostly connected to area V2 (area 18), and W to area V3 (area 19).[17] Ironically, the concept of a hierarchical organization of the cortical areas, which was first proposed for the cat, has not been retained for this species and the question remains whether the properties of neurons in area V2 are principally due to their Y thalamic input or to the drive from V1 neurons. Because the organization of the mammalian visual cortex generally resembles that of the cat,[18] it may be that the hierarchicoparallel model of organization of the cat visual system is more the rule than the exception and that the much-studied primate visual system is actually an exception among the numerous species with well-developed vision.

After the parallel model had replaced the hierarchical model for the cat visual cortex, the idea of a hierarchical organization of cortical areas reemerged for the primate. The main reason was the discovery by Kuypers and collaborators[19] that information does not circulate only in one direction, from primary areas to higher-order areas, but that there are connections in the other direction. Rockland and Pandya[20] called these connections "caudally directed" and they showed that the laminar distribution of the neuron of origin and the level of termination of their axons belong to different cortical laminae from what they called the "rostrally directed connections." Rostrally directed connections are issued by neurons in the supragranular layers (layers 2 and 3) and terminate in and around layer 4, whereas caudally directed connections arise from infragranular layer neurons (layers 5 and 6) and terminate outside of layer 4. This distinction between rostrally directed and caudally directed connections was confirmed by Maunsell and Van Essen[21] who traced the connections from and into area MT in the macaque monkey. They called them feedforward and feedback connections and found that most connections that they observed from and into area MT could be classified in one of these two categories. Using this distinction, they organized the primate cortical areas in a hierarchy, beginning with area V1 that provides feedforward connections to the areas it connects to (areas V2, V3d, and MT) and receives feedback connections from them. At the next level is V2 that receives feedforward connections from V1 and feedback connections from the other areas. With this bootstrap method, they arrived at a classification of cortical areas known as the hierarchy of cortical areas (Figure 8.1). Areas located at the same level in the hierarchy are interconnected by a third type of connections called "lateral," which are characterized by widely distributed levels of axon termination in the cortical depth.

8.1.2 Modern Versions of the Hierarchical Model

This method of classification was later applied to a larger number of cortical areas to arrive at the hierarchical model known to all students of vision.[22] Connections between a large number of cortical areas were examined. The reciprocity rule was confirmed in most cases: if area A sends feedforward connections to area B, area B sends feedback connections to area A. The transitivity rule was also examined: if area A sends feedforward (feedback) connections to area B that sends feedforward (feedback) connections to area C, then area A sends feedforward (feedback) connections to area C. A similar diagram, oriented laterally instead of vertically and giving more emphasis to the separation between the ventral and the dorsal streams, was provided by Ungerleider and Desimone.[23]

Such diagrams are useful for organizing the large number of cortical areas of the monkey visual system in a small number of distinct levels but they have two disadvantages. (1) They constitute anatomical classifications that say little about the functional interactions between the areas. We will come back to this question in Section 8.4. (2) Because they are based on discrete information (area A is connected or not to area B, with no indication of the strength of the projection; only three

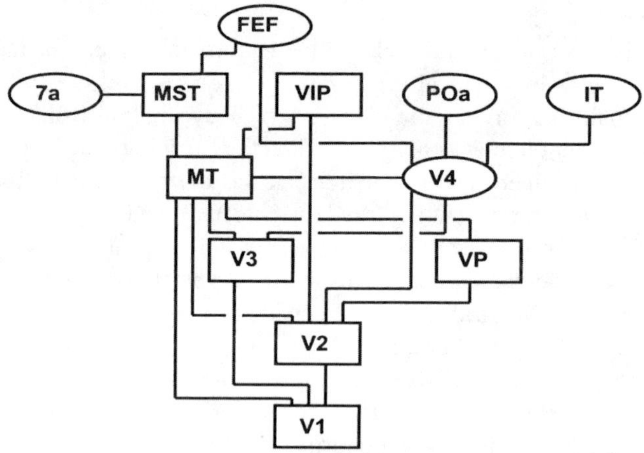

FIGURE 8.1 The early version of the hierarchical organization of cortical areas of the macaque monkey visual system. (Modified from Reference 21; for more recent versions, see Reference 81.)

classes of connections are recognized, feedforward, feedback, lateral), they are largely underdetermined as pointed out by Malcolm Young and associates.[24] This means that the set of rules used to construct the hierarchy does not produce a unique solution for a given set of areas. Optimal solutions are those that lead to the smallest numbers of rule violations, but it is usually impossible to identify the optimal configurations, because of the lack of knowledge concerning many connections. The consequence of the underdetermined character of the hierarchical scheme is that a given cortical area such as area MT can occupy any level between 5 and 10 in the hierarchy. For most pairs of cortical areas beyond V1, V2, and V3,* it is impossible to be certain that one area is at a lower or at a higher level than the other, or that they belong to the same level in the hierarchy.

An attempt has been made recently to reduce this underdetermined character by using a continuous variable to characterize each set of connections. This was done by quantifying the proportion of source neurons in supragranular vs. infragranular layers.[25] When this is done, a reasonably good match to the original hierarchy of Maunsell and Van Essen[21] is obtained, and, provided a few adjustments are made, the proportion of labeled neurons in supragranular layers is a good predictor of the level in the hierarchy. However, when the hierarchy obtained by Young and colleagues by modeling[24] and that obtained by quantifying the proportion of supragranular layer neurons[25] are compared with those published by Van Essen and colleagues, there are serious discrepancies between the levels at which a given area belongs. For example, the frontal eye field (FEF) belongs to the highest level in the original map of Maunsell and Van Essen[21] (Figure 8.1), to the eighth level in the 1991 version of the hierarchy[22] and is brought down to the fourth level when

* For these areas the numbers of possible levels are small.

the hierarchy is built using the proportion of supragranular layer neurons.[25] A similar discrepancy concerning the position of the FEF area with respect to other areas was also found when connections between cortical areas were modeled using graph theory.[26] The underdetermined character and the variability in the levels occupied by areas in different classification schemes suggest therefore that caution should be applied when attempts are made at deriving functional interpretations from the hierarchical classification.

Another interpretation of the laminar organization of neurons and axon terminals of different corticocortical connections has been proposed by Barbas[27,28] for the connections between different areas in the monkey frontal cortex. According to this interpretation, the laminar organization of a given connection can be predicted by the laminar organizations of the source and target areas. Most sensory areas have a classic organization with six layers and a well-identified layer 4 (or granular layer). In the frontal cortex, there are also agranular areas, lacking the characteristic layer 4, and several areas belonging to the limbic type of cortex with a dense population of neurons concentrated in the deep layers. It has been argued that evolutionary steps can be traced by studying the laminar organization of different cortical regions, and that areas can be classed in different types, from ancient limbic-type cortex, to agranular and to more recent granular cortex.[29] Barbas[27] showed that the laminar organizations of corticocortical connections are related to the types of the source and target areas. Thus, areas of the granular type project from the supragranular layers and terminate mostly in the middle and deep layers to less-laminated areas, whereas the return projections originate from the deep layers and preferentially target the upper layers.[28] Thus, the differences between feedforward and feedback connections in occipital, parietal, and temporal cortex may simply reflect differences in organization between areas of the occipital cortex with well-differentiated structures (such as areas V1 and V2) and areas with less-differentiated structures in the parietal and temporal cortices. One advantage of this model is that it predicts the observed progressive shift toward projections from infragranular layers terminating at higher levels in the target area with increasing distance between the areas (distance in terms of levels of organization or, equivalently, distance on the cortical surface).

8.2 WHAT IS VALID IN THE HIERARCHICAL MODEL?

According to the functional interpretation of the hierarchical model, the response selectivity of neurons in one area is mainly due to the ordered convergence of neurons located in a lower-order area that sends feedforward connections to this area. To give an example, the direction selectivity of a neuron in layer 4 of area MT is observed over a relatively large region corresponding to the receptive field center. This direction selectivity is assumed to result from the converging inputs of properly chosen direction-selective neurons in areas V1 and V2 with smaller receptive fields and same optimal direction.[30] This explanation is similar to that advanced for the orientation selectivity of simple cells in V1 by the converging inputs of LGN cells with receptive fields aligned with the optimal orientation.[6] There are several lines of evidence to support this explanation: The progressive increase in the sizes of the receptive fields of neurons situated at successively higher levels of the hierarchy is in keeping with the principle of ordered convergence of neurons of lower-order areas onto neurons of higher-order areas. Area V1 with its very small receptive fields is at the lowest level in the hierarchy in all species and the highest levels are occupied by regions of the parietal and inferotemporal cortex where neurons have very large receptive fields. Although a direct correlation between RF size and level in the hierarchy has not been made, there are numerous examples for which the sizes fit in the ordered sequence of areas from low to high levels in the hierarchy. This is particularly evident in the ventral stream where RF sizes progressively increase as one examines V1, V2, V4, TEO, IT in a sequence corresponding to the successively higher levels in the hierarchy.

Some functional predictions of the ordered convergence rules have also been confirmed. It has been reported that the V1 neurons projecting to area MT belong to a specific type with homogeneous properties.[30] This result is comparable to an earlier work by Henry and collaborators[31] that also demonstrated the homogeneity of the functional properties of the neurons projecting from area 17 to PMLS in the cat. Also, the well-known patchy distribution of feedforward terminals[32] presumably corresponds to the convergence of axon terminals of neurons with some common properties. This is suggested by the results of Sherk[33] in the LS area of the cat, an area that contains numerous direction-selective neurons. Using a neurotoxin, Sherk killed the neurons in a small region of that area and recorded from what presumably correspond to the terminals of afferent axons. She found that axon terminals group together according to their direction selectivity, thus suggesting that afferents with the same optimal direction terminate in common patches. The neurons contacted by this axon group presumably inherit the property of direction selectivity transmitted by the converging feedforward axons.

Thus, the progressive increase in RF size and functional specificity observed at progressively higher levels of the hierarchy appear to result from the rule of ordered convergence that is central to the hierarchical model.

8.3 FEEDFORWARD/FEEDBACK VS. DRIVERS/MODULATORS

The functional interpretation of the hierarchical model assumes a fundamental difference between feedforward and feedback axons: feedforward axons are supposed to act as "drivers," whereas feedback axons would play the role of "modulators." This distinction between drivers and modulators has recently been reintroduced in the study of cortical connections.[34,35] Driver connections are supposed to be small in number and have strong, focused, and rapid effects on their target neurons. Modulators have symmetrical properties: they are numerous and have weak, slow, and diffuse actions. It is likely that such a dichotomy is artificial and that there is a continuum of connection types from the strongest and most rapid ones to the very weak and slow. However, as argued by Crick and Koch,[35] it is probably important to distinguish between strong and weak connections for understanding the function of a network and addressing the question of its stability. Typical examples of drivers and modulators are provided by comparing the thalamocortical and corticothalamic connections for relay neurons of the thalamus. For example, LGN inputs into layer 4 concern a small number of synaptic contacts compared to the cortical connections,[36,37] but they appear to have a strong driving force, as evidenced by the effects of reversible inactivation of the LGN[38,39] and the large sizes of the synaptic potentials.[40] In contrast, corticothalamic axons are extremely numerous,[41] but have a weaker effect, as evidenced by the variable and weak results obtained by reversible inactivation of cortex on thalamic firing (see review in Reference 42). The effects also appear to be slow since most thalamocortical axons are approximately ten times slower than corticothalamic axons (review in Reference 43) and activate metabotropic receptors that produce responses delayed over several tens or hundreds of milliseconds.[44–46]

8.3.1 Are Feedforward Connections Drivers?

The distinction between drivers and modulators that appears so clear between thalamocortical and corticothalamic connections has been extended to the feedforward and feedback corticocortical connections by Crick and Koch.[35] Feedforward connections, like thalamocortical connections, terminate in layer 4, whereas feedback connections terminate outside layer 4, in layers that may be less essential for driving a cortical area. There is evidence to support the idea that feedforward connections are of the driver type. Inactivation of area V1 leads to a complete cessation of activity in area V2, which receives a strong feedforward connection from V1.[47,48] Because V2 is an important feedforward input to area V4, the loss of activity in V4 during inactivation of V1[49] also suggests that the feedforward connections from V2 to V4 are of a driver type. The lack of activity in inferotemporal cortex following lesion of area V1 is also in keeping with the idea that feedforward connections are drivers (see review in Reference 18).

It may be, however, that this equation between feedforward and driver connections is an oversimplification because of the small number of feedforward connections that have been tested. In particular, the fact that data from primates come only from inactivation of area V1 may bias the conclusions because this action stops or strongly decreases firing in many areas. Thus, what is interpreted as the result of inactivating the feedforward connection from V1 to area A may in fact reflect the influence of inactivating V1 on the whole network to which area A belongs. For example, it would be interesting to test whether area V4 stops responding when V2 instead of V1 is inactivated. Furthermore, the conclusion that feedforward connections carry the major drive to a cortical area is questioned by the residual activity observed in MT after inactivation or lesion of V1 in macaque monkeys (see review in Reference 18). In that case, is the input from V1 a driver or a modulator input? Consideration of results in other species also calls for caution since inactivation of feedforward connections often does not lead to much change in the target area (as in the case of area 17 to area 18 connections in the cat; see review in Reference 18).

In conclusion, the idea that feedforward connections have a strong driver-type effect on the target area may be true only in such special cases as the thalamic inputs from relay nuclei into layer 4 of primary cortical areas, or in the case of the projection from V1 to V2, which is the main target of V1. Results from the V1 to MT connections in macaque monkeys and from the feedforward connections in the cat suggest that for other types of corticocortical connections, the situation may be less clear-cut.

8.3.2 Are Feedback Connections Modulators?

Less is known concerning the action of feedback connections. As mentioned above, the published accounts of effects of inactivating the corticothalamic pathway suggest that the role of this input is not to drive the thalamic neurons, but to modulate the activity driven by the retinal inputs.[45] It is not certain, however, that the cortical feedback connections necessarily have the same functional role as the corticothalamic connections. For example, there is a clear difference in terms of timing between corticothalamic and corticocortical connections. Most thalamocortical connections are much faster than the corticothalamic connections that conduct information in the reverse direction.[42,43] In contrast, among corticocortical connections, the conduction speed of feedback connections is as fast as that of feedforward connections.[50,51] This suggests that the role of feedback connections is not limited to that of a slow and diffuse modulating action on the activity of neurons in lower-order areas. Our finding of a small number of cells in area V2 that completely stopped responding to visual stimulation during inactivation of area MT[52] also suggests that feedback connections can have a very strong effect on their target neurons. The results of Miyashita and colleagues,[53] demonstrating neural drive by feedback connections from the frontal cortex to IT neurons during memory recall, also illustrate the possibility for feedback connections to act as drivers under certain conditions. Finally, the recent demonstration that most feedback connections are as patchy as feedforward connections[54] runs counter to the generally accepted idea that feedback connections are diffuse and act nonselectively on large populations of neurons. Such an action is likely to be the case for the feedback connections terminating in layer 1, but is unlikely to be the general rule.

Some results of inactivation of cortical areas are consistent with a driver/modulator difference between feedforward and feedback connections. Consider what is observed in area V2 when area V1 is inactivated: despite the presence of a strong residual activity in area MT and numerous feedback connections from MT to V2, there is no response to visual stimulation in V2 neurons.[48] On the other hand, we found that inactivating area MT changes the responses of a number of neurons in V2[52] when V1 is active. This suggests that feedback connections from MT cannot drive V2 neurons in the absence of a feedforward input from V1 but can modulate the responses of neurons in V2 when they are driven by their feedforward inputs from V1. Consideration of the responses of V1/V2/V3 neurons to visual stimuli during MT inactivation also points to the same conclusion: when MT is inactivated, the response changes are observed only on the part of the response corresponding to stimulation of the RF center mechanism, which presumably reflects the feedforward input to the neuron.[55] This suggests that feedback connections act as a gain control of the response to the feedforward input. It remains to be seen whether this role is specific to the feedback connections or whether feedforward or horizontal connections can also act as gain controls to activation of other inputs. This will need to be tested on other sets of connections than those arising from V1, as V1 input is essential to the activation of the entire network.

Anatomical differences between feedforward and feedback connections suggest that they use different types of receptors and/or that their axon terminals contact different parts of the neuron. Evidence has been presented that feedforward and feedback connections have different synaptic mechanisms.[50,56] However, the validity of these results is limited by the use of electrical stimulation that does not distinguish between a direct orthodromic drive and an antidromic activation of recurrent collaterals.[50] It will be necessary to target specific feedforward and feedback circuits with activation techniques other than electrical stimulation before we understand the differences between feedforward and feedback connections at the synaptic level.

It has been known for a long time that axons of different functional groups of neurons conduct at different speeds. Although this has been a major theme in the study of the retinogeniculate connections,[57] it is only recently that it has been considered as an important aspect of corticocortical connectivity. It is interesting that the thickest axons that are known to conduct faster usually have very large terminal boutons, suggesting that their action on the target neuron is very potent in terms of synaptic activation. This has been observed in thalamocortical connections[42,58] and in corticocortical connections.[59,60] Thus, the driver-type connections may correspond to the thick axons with heavy terminal boutons that can "drive" a target neuron even when its membrane potential is far from the spike threshold. Thick axons and heavy terminal boutons are indeed found in the terminals of thalamocortical axons in layer 4 of V1,[61,62] in keeping with their classification as drivers. In contrast, most corticothalamic axons have thin axons, small boutons "en passant,"[42,58] in keeping with their being classified as modulators.

If thick axons always correspond to driver connections, the presence of a few corticopulvinar axons originating in layer 5 with large axons and heavy terminal boutons[42,58] suggests that not all corticothalamic connections should be regarded as modulators. Indeed, as argued by Sherman and Guillery,[63] such corticothalamic drivers with thick axons may be an important component of the feedforward transfer of information between visual cortical areas. A similar case may be made for the fastest feedback connections that may correspond to drivers. For example, it is known that area FEF is activated very early (see below), much earlier than some of the inferotemporal cortex regions to which it is connected.[64] It would be interesting to determine whether some of the FEF input into IT cortex involves rapid axons of the driver type.

In conclusion, although feedback connections in general appear to produce a nonlinear effect on their targets by acting on the gain of the response to the feedforward inputs, and thus act as modulators, it remains to be seen whether this is true also for the thickest axons. Feedback connections are very diverse in their morphology, their topographical organization on the cortical surface, and the laminar distribution of their terminals.[32] It is likely that one cannot speak of the functional role of feedback connections in general, but that it will be necessary to distinguish between different sets of feedback connections, which play different roles in the cortical processing of visual information. This will require isolating specific circuits and activating and inactivating different inputs to a recorded neuron.

8.4 LIMITATIONS OF THE HIERARCHICAL MODEL

There are three major limitations in using the hierarchical model to explain the responses of neurons in the visual cortex:

1. It has been shown, at least in area V1, that local connections play a major role in setting the functional selectivity of cortical neurons.
2. Recent results show that the function of feedback connections is much more important than the simple modulatory role generally attributed to them.
3. The timing of activation in different cortical areas does not follow the order suggested by the hierarchical model.

These three limitations are examined in the sections below.

8.4.1 Role of Horizontal Connections

There is evidence that local horizontal connections play an important role in shaping the selectivity of cortical neurons. For example, several reports by the group of Eysel[65] clearly demonstrate a major role for the network of local horizontal connections in shaping the orientation selectivity of neurons in area V1. An example of their results (Figure 8.2) shows the orientation tuning curves of two neurons recorded at the site RS in area V1 of the cat and the effects on these tuning curves of reversible inactivation of the sites IS1 and IS2 by GABA iontophoresis. The first neuron (Figure 8.2A to 8.2D) has an optimal orientation of 30° with respect to the horizontal and is strongly direction selective (Figure 8.2A). The cortical site IS1 with a preferred orientation orthogonal to that of the recorded neuron is inactivated with GABA. This leads to a considerable broadening of the orientation selectivity of the recorded neuron (Figure 8.2B) that totally recovers after the effects of GABA have vanished (Figure 8.2C). When the site IS2 with optimal orientation parallel to that of the recorded neuron is inactivated, practically no effect is observed (Figure 8.2D). A similar specific effect is observed for another neuron also recorded at the site RS (Figure 8.2E to 8.2H). Again, the inactivation of the site with orthogonal optimal orientation leads to an almost complete loss of orientation selectivity (Figure 8.2F), whereas very little effect is observed for inactivation of the iso-oriented site IS2 (Figure 8.2H). Because GABA reversibly silences cortical neurons within a few hundred microns from the iontophoresis micropipette and does not affect axons, these results demonstrate that a proper combination of afferent thalamic axons is not sufficient to achieve orientation selectivity in a cortical neuron but that a major role is devoted to local excitatory and inhibitory connections. Given that inactivation of the IS1 and IS2 sites broadens the tuning curves but does not change the optimal orientation, it is likely that the optimal orientation of a cortical neuron is determined by the selection of thalamocortical axons that drive it. However, it is equally obvious that the sharp orientation selectivities displayed by the cortical neurons illustrated in Figure 8.2 are mainly the result of processing by the local network of horizontal connections. Note that the effects of inactivation of local connections are specific to regions with optimal orientation orthogonal to that of the recorded neuron (Figure

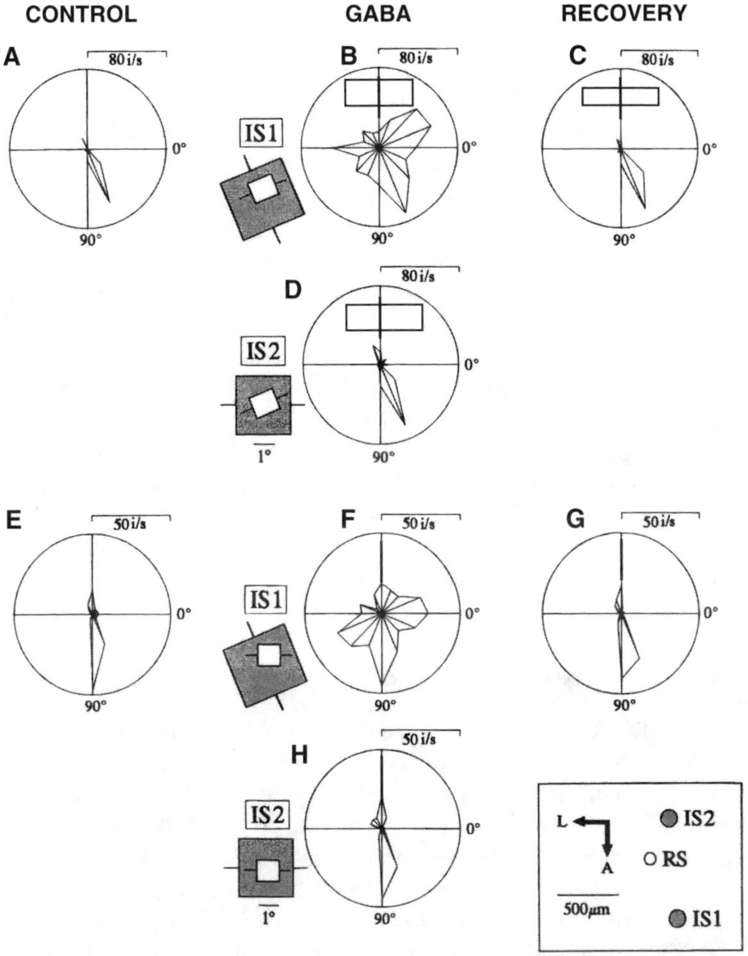

FIGURE 8.2 Effects of inactivation by GABA iontophoresis of the local network of horizontal connections on orientation tuning in area V1. (Modified from Reference 65.) (A) Polar plot of the orientation tuning of a V1 neuron recorded at the site RS (sites illustrated in the lower right insert). (B) Orientation tuning of the neuron during GABA inactivation at the site IS1 with optimal orientation orthogonal to that of the recorded neuron. There is a substantial broadening of the tuning curve. (C) Recovery run. (D) Effect of inactivating the site IS2 with optimal orientation close to that of the recorded neuron: no effect is observed on the response strength or the orientation tuning curve of the recorded neuron. (E) Orientation tuning curve of another neuron recorded at the same site RS. (F) Effect of inactivating the site IS1 with optimal orientation close to orthogonal to that of the recorded neuron: there is an almost complete loss of orientation selectivity (but note that for both neurons in RS, direction selectivity is more robust). (G) Recovery run. (H) Effect of inactivating the site IS2 with same optimal orientation as the recorded neuron: again, no effect is observed on the orientation tuning curve of the recorded neuron.

8.2B, 8.2D, 8.2F, and 8.2H) and that the broadening of the tuning curve is achieved by response increases at non optimal orientations. This strongly suggests that the orientation tuning curve is mainly shaped by inhibitory mechanisms for non-optimal orientations, in keeping with the early conclusions of Sillito[66] and the results of Borg-Graham and colleagues.[67]

The importance of the local connectivity network in shaping the responses of neurons in the visual cortex is further demonstrated by a recent report from the group of Grinvald[68] that shows that the emission of spikes by a cortical neuron is controlled by the activity of its neighbor neurons. The distribution of activity on the cortical surface surrounding the recorded neuron that is associated with the appearance of a spike is the same whether the spike is emitted in response to a visual stimulus at the optimal orientation or during spontaneous firing. This suggests that the spatiotemporal pattern of local cortical activity acts as gate, in such a way that a strong cortical response to a given thalamic input will be produced only when the local cortical network is in a permissive state. Thus, the local network of intrinsic connections most likely plays a major role in modulating the excitability of a cortical neuron and the feedforward input may be more properly compared to a trigger than to an input determining the receptive field selectivity. Such a trigger role is consistent with the small number of feedforward thalamic inputs to a given neuron of layer 4 in V1 compared to the very rich array of connections from local neighbor neurons.[36,37,69]

8.4.2 Role of Feedback Connections

The role of feedback connections has to a large extent been ignored in the hierarchical model, in which they are supposed to play a slow modulatory role, possibly connected to retrieval of memories or attention shifting. Despite the fact that they are as numerous as the feedforward connections, it is only recently that the role of feedback connections has been tested by inactivating high-order areas and recording neuronal responses in lower-order areas receiving feedback connections from the inactivated area. These experiments have been performed for two sets of feedback connections: from V2 to V1, and from MT to areas V1, V2, and V3. In the first set of experiments,[55] neurons in V2 were inactivated by pressure ejection of the inhibitory neurotransmitter GABA through several micropipettes; recordings in V1 were made from neurons with receptive fields overlapping those of neurons at the inactivated V2 region. The visual stimulus used resembled that of Knierim and Van Essen[70] and Nothdurft et al.,[71] and consisted of a high contrast bar of optimal orientation flashed in the receptive field center and a set of bars of varying orientations in the surround. The main effect of inactivating area V2 was observed in a small number of neurons in V1 and consisted of a decrease in the responses to the central bar, suggesting that feedback connections from area V2 normally act to enhance the response of V1 neurons. These results confirmed previous conclusions that feedback connections act mainly to enhance the responses of neurons in low-order areas.[72,73]

The role of feedback connections was also tested by inactivating the superior temporal sulcus by cooling and recording in areas V1, V2, and V3. The entire extent of MT (and surrounding areas) was inactivated by a cooling probe positioned in the superior temporal sulcus at the level of MT. When MT was inactivated by cooling, a clear decrease of the response to a moving bar stimulating the RF center was observed for approximately 40% of the neurons,[52,74] thus confirming the conclusion drawn from the V2 experiments on the facilitatory effect of feedback connections onto the center mechanism.

It is known that the response of cells in V1 (and extrastriate cortex) can be modified by contextual stimuli lying far outside the neurons' receptive field.[75] These effects are attributed to stimulation of a region located outside the receptive field center and called surround, although it often does not entirely surround the center as is the case in the retina.[76,77] Simultaneous activation of the center and surround most often produces a decrease in the neuron response to the stimulus. These surround effects are generally thought to be mediated by the horizontal connections.[78] However, Angelucci and collaborators[54] recently compared the total extent of the surround in visual space to that corresponding to the horizontal connections in V1 of the macaque monkey. They concluded that the horizontal connections cover a region corresponding to the RF center and the proximal region of the surround but cannot constitute the basis of all center–surround interactions. In contrast, feedback connections can transfer information from long distances in the visual field[54] and are therefore likely to play an important role in the generation of center–surround interactions.

The role of feedback connections in center–surround interactions was tested for the V2→V1 and MT→V1-V2-V3 connections. For the MT experiments, center–surround interactions were studied by comparing the responses to the moving bar in the center, with the responses to the same bar moving together with a background stimulus consisting of a checkerboard of low contrast light and dark gray bars of similar length and width as the center stimulus (Figure 8.3). Because the background stimulus did not generate any response when moved alone across the RF but strongly suppressed the response to the center bar when it moved together with it, it can be assumed that it activated mostly the surround mechanism. By comparing the responses to these two stimuli it is possible to estimate the strength of the inhibitory effect of the surround upon the center mechanism. This effect was measured as the percentage decrease of the response to the center stimulus when the background stimulus was added. Figure 8.3 presents a scattergram of the values of background suppression for the control condition, and for the MT inactivation condition. It is clear that in most cases background suppression was diminished by MT inactivation and sometimes suppression was changed into facilitation. This suggests that feedback connections from MT exert a potent effect on the center–surround suppression of neurons in lower-order areas. One important parameter in the effect of MT inactivation is the salience of the center stimulus. This was measured as the ratio of the contrasts of the center and surround stimuli.[52] When the salience was high, the center bar was clearly visible above the background; when it was low, the center stimulus was only visible when moving against the stationary background. Note in Figure 8.3 that the strongest effect of inactivating area MT occurred for low-salience stimuli, and that only mild effects were observed for high-salience stimuli.

The effect of V2 feedback on center–surround interactions was also tested for V1 neurons by comparing the response of neurons before and during V2 inactivation

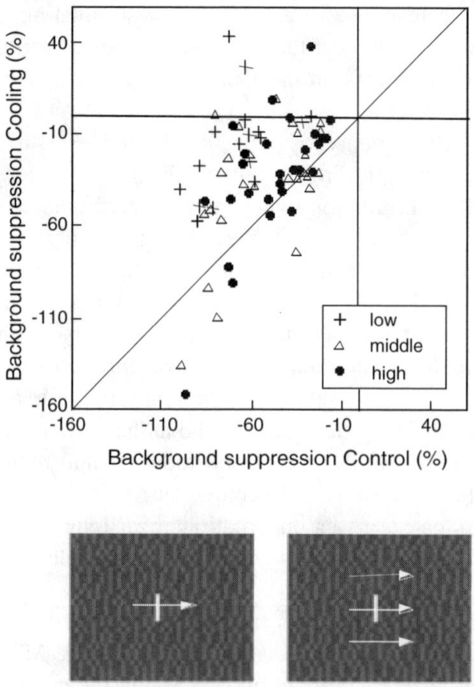

FIGURE 8.3 Effect of inactivating MT on the suppression of the response to a moving bar activating the center mechanism by a moving low-contrast background pattern activating the surround. The stimuli used in this comparison are illustrated below the scattergram. Background suppression represents the percent difference between the responses to the moving bar with the stationary or with the moving background. The scattergram presents the results of MT inactivation on surround suppression: apart from cases for which no effect is observed (along the first diagonal) or the few cases for which there is an increase in suppression (below the diagonal), most neurons show a decrease in background suppression during MT inactivation. Note that the effects on surround suppression are stronger for low-salience stimuli (see Reference 92 for details).

to stimuli that activated only the center (an optimally oriented bar flashed in the center) and center + surround (a set of optimally oriented bars covering center and surround). In that case we did not observe any effect of inactivating V2 on surround suppression in V1 neurons.[55] There are several possible reasons for the discrepancy between the results obtained in the MT and the V2 experiments:

1. Stimuli were suprathreshold and high contrast in the V2 experiments and were therefore likely to activate all the cortical areas providing feedback connections to V1 neurons. On the contrary, stimuli in the MT experiments were better designed to activate primarily MT (moving low contrast background). The failure to observe effects in the V2 experiments may simply result from the fact that other feedback systems were not inactivated.

2. It may be that feedback and horizontal connections work in synergy and that the horizontal connections may be dominant in mediating surround effects demonstrated with the stimuli used by Knierim and Van Essen.[70] More experiments need to be done to tease out the role of horizontal and feedback connections in mediating center–surround interactions.

Thus, feedback connections enhance the center response and increase the suppressive effects of surround stimuli in neurons of low-order areas. This double effect of feedback connections makes it possible to modify the filtering properties of neurons in low-order areas according to activities of neurons in higher-order areas. The predominant role given to feedforward connections in the hierarchical model is therefore an oversimplification that leaves aside a number of important properties of neurons, such as the shape of the tuning curve, which appears to be under the influence of the horizontal connections, and the center–surround interactions, which appear to be under the influence of the horizontal and feedback connections.

8.4.3 Spatial and Temporal Properties of Feedback Connections

There is another reason feedforward connections are thought to be the main determinant of the functional properties of cortical neurons. As mentioned in Section 8.3, it is assumed that feedforward connections are fast and precise, whereas feedback connections are slow and diffuse and therefore unlikely to be involved in processing visual information. The view of the feedback connections as diffuse stems from the early studies of feedback connections using anterograde tracers with relatively large injections. Such injections led to a dense mesh of axon terminals in upper layers, which extended over long distances and did not have a patchy organization.[21,23] However, when more limited injections are made with iontophoresis, a different picture emerges: besides the diffuse system of feedback connections terminating in layer 1, there are feedback connections to layer 2–3 neurons that are patchy and precisely reciprocal with the patches of neurons providing feedforward connections.[54] Such an organization suggests the presence of precise interactions between cortical columns, a role that is likely to differ from that of the feedback connections terminating in layer 1.

The late action of feedback connections is thought to result from the supposedly late activation of higher cortical areas and the slow nature of feedback connections. This is in keeping with the idea that feedback connections mediate the late modulation of V1 neuron responses for stimuli consisting of a figure against a background[79] (but see Reference 80), and the general assumption that higher-order areas have delayed responses in comparison with lower-order areas. However, a literature survey reveals that neuronal responses in some higher-order areas of the macaque visual system are delayed by only a few milliseconds relative to the responses of neurons in areas V1 and V2[43,81] (Figure 8.4A). In addition, feedback connections have been shown to be as rapid as feedforward connections.[50,51] Figure 8.4B shows the conduction velocities of feedforward and feedback axons between macaque areas V1 and V2, as deduced from the orthodromic latencies (antidromic latencies gave similar values[51]). Both types of connections are made of fast-conducting axons, usually between 2 and 6 m/s. In contrast, horizontal connections have slow conduction velocities, most of them being between 0.1 and 0.4 m/s, i.e., about ten times slower than the feedforward and feedback connections (Figure 8.4B).

The rapid conduction velocity of feedback axons is consistent with the lack of delay observed in the effect of inactivating area MT on the responses of neurons in areas V1, V2, and V3, or of inactivating area V2 on the responses of V1 neurons.[55,74] If the effects of feedback connections were systematically delayed, as suggested by the results of Lamme et al.,[82] one would expect to observe the effects of inactivation of a higher-order area to show in the late part of the response. The latency of the effect of feedback connections onto the center responses was directly examined for both systems of feedback connections; in both cases, these effects were found not to be delayed by more than 10 ms.[55,74] A similar observation was made for the timing of center–surround interactions.

These results therefore disprove common assumptions on the slow effects of feedback connections, and instead point to the slow nature of horizontal connections. This is in keeping with the large caliber of feedback axons[83] and the thin, unmyelinated character of most horizontal connections. It is clear, therefore, that feedback connections are much better suited than horizontal connections for rapid transfer of signals between neurons with distant receptive fields.

8.5 CONCLUSION: HIERARCHY OF VISUAL CUES

The hierarchical model assumes that processing starts at the local level in the low-order areas and then progressively builds up to more global aspects in higher-order areas of the parietal and temporal lobes. This assumption accords with the small sizes of the receptive fields in areas V1 and V2 and the progressive increase in RF sizes as one ascends to higher levels in the hierarchy of cortical areas. However, we saw in the preceding sections that the activities of neurons occur almost simultaneously in many areas (for example, in areas MT and V2; see Figure 8.4A), and it is therefore possible for neurons located at different levels in the hierarchy to exchange information in both directions. This possibility is facilitated by the similar conduction speeds of feedforward and feedback axons. In other words it takes the same time to ascend or descend the hierarchy of cortical areas.

Although the local-to-global progression has been advocated in the study of machine vision[84] and is usually thought to occur in the primate visual system, it is not necessarily the optimal way to process visual information. In a purely feedforward model, the small receptive fields of V1 and V2 neurons function independently of what occurs in distant regions of the visual field. This may be sufficient for unambiguous stimuli used in the laboratory but is inefficient for real-life stimuli, which are ambiguous because of light and shadow artifacts and occlusion of one object by others located closer to the eye. To properly segment a visual scene in a number of different elements, it is important to be able to combine global and local views.[85] Such global-to-local interactions can be transferred by the feedback connections, and it is likely that the potentiation of center and surround mechanisms illustrated above (see Figure 8.3) corresponds to such a role. Global influences on

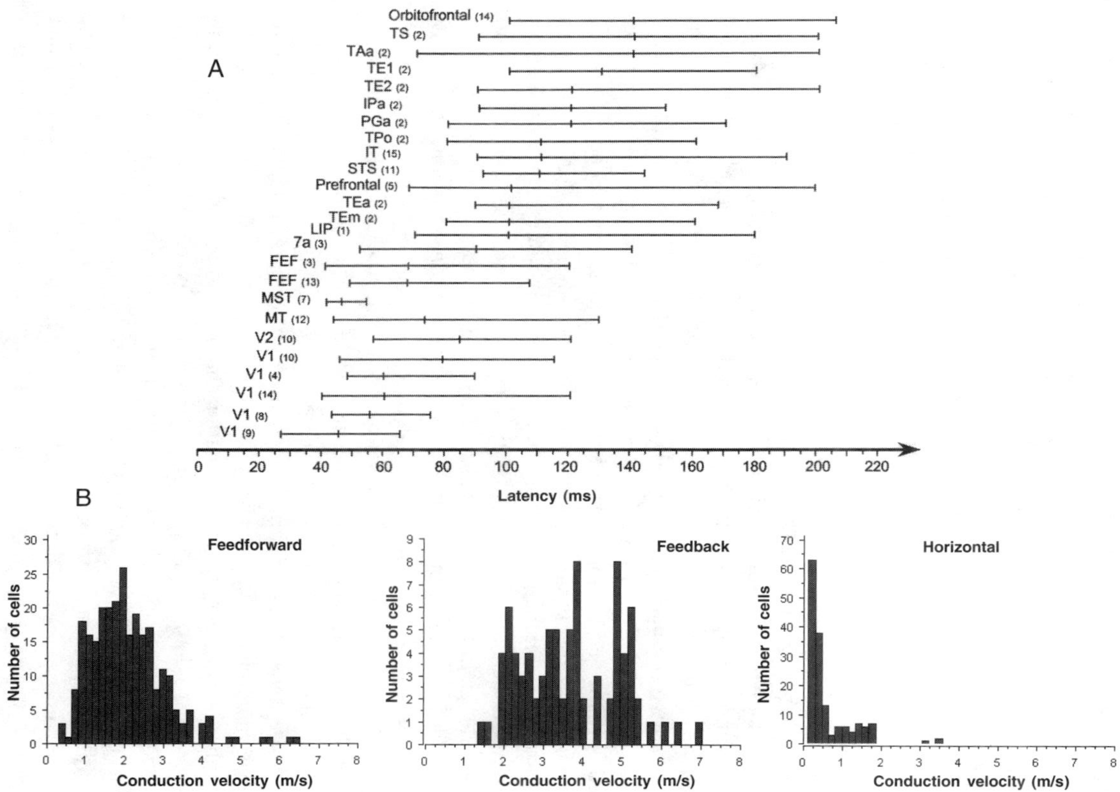

FIGURE 8.4. Temporal aspects of interactions between cortical areas. (A) Latencies of neurons in different cortical areas of the macaque monkey. Numbers refer to published papers; for details and references of papers see Reference 86. Note that the FEF latencies are similar to those in V1 despite the large distance between these two areas in the cortical hierarchy (see Figure 8.1); same remark for V2 and MT. Thus the cortical hierarchy is a poor predictor of the timing of activation of neurons in different cortical areas. (B) Conduction velocities of feedforward and feedback connections between areas V1 and V2 of the monkey and horizontal connections in monkey V1. Conduction velocities are based on orthodromic latencies; antidromic latencies return similar values. Note that horizontal connections are ten times slower than feedforward and feedback connections that both conduct at similar speeds. For details see Reference 51.

local processing needs fast access to long distances in the visual field. As shown above, this is unlikely to be done solely by horizontal connections and I suggest that this is one of the major roles of feedback connections.

Another common assumption concerning the primate visual system is that all information reaches area V1 simultaneously, similar to what occurs in machine vision for which all the information is transferred simultaneously from the sensors to the computer for processing in parallel. This appears not to be the case in biological vision: It has been shown that information from the magnocellular layers of the LGN reaches area V1 20 ms earlier than that transferred by the axons of the parvocellular neurons and that there is a further delay for the koniocellular input.[43] This asynchrony of the different inputs into V1 has led to the suggestion that the magnocellular input is rapidly sent to areas of the dorsal stream for a first-pass processing and that the results of this first computation are sent back through feedback connections to adapt the processing of information carried by the parvocellular neurons.[86] It may be that the koniocellular input benefits from a similar preprocessing by the magnocellular and parvocellular inputs.

It appears that, in addition to a hierarchy of cortical areas, there may be a hierarchy of parameters used for the processing of visual information. The idea is that not all parameters (for example, depth, movement, color) are processed at the same time. This is suggested by the different streams of processing in the visual cortex (the M, P, and K streams) and by the fact that processing some parameters first may greatly reduce the computation load. For example, a red sphere placed in front of a green background can be segmented either by using color-selective neurons or neurons sensitive to differences in speed because small head movements generate different speeds of the red sphere vs. the distant background. Several questions arise: Are the two segmentation methods (color border and speed border) used simultaneously and combined together? Is one parameter processed first and the second used only for verification? These are important questions that need to be addressed.

The early processing of magnocellular input in areas of the dorsal stream suggests that differential movement that represents a major cue to separation of objects in three dimensions may be the first cue to be processed. The later arrival of parvocellular input suggests that fine detail is processed later and uses results of computations done with the differential movement information. Finally, the recent finding that at least some of the blue-yellow channel is transferred through the koniocellular pathway[87–89] suggests that processing of color information is further delayed. This is in keeping with the late responses of neurons in the thin cytochrome oxidase bands in V2 that are, at least partially, selective for color.[90]

More generally, one should think more in terms of a hierarchy of visual cues than a hierarchy of visual areas. The visual system constantly produces working hypotheses from the retinal information that is used to update its internal model of the world. To do this, many different cues have to be combined in the computations needed to segment the visual scene. In other words, one should no longer think in terms of local processing before global, which is linked to the way machines process images, but more in terms of what needs to be processed first. This depends on two parameters: processing time and priority. Processing time differs from one cue to another because some computations require more time than others and it is important to rapidly compute some essential variables that can be combined with other cues. Apart for some recent findings,[91] we know very little about computation time in the visual cortex.

For segmentation, motion cues appear to have a high priority because "what belongs together moves together." Because we are constantly moving our eyes and our bodies, differential movement is probably the most efficient cue for grouping together elements belonging to the same three-dimensional object. Because of the large receptive fields in areas MT and MST, differential movement cues can be computed over large distances in the visual field and can rapidly produce first a three-dimensional segmentation sketch. This cue is also the first to be processed because the dorsal stream is activated much earlier than the rest of the visual system (Figure 8.4A). Beyond this first-pass analysis, we have no understanding of the dynamics of segmentation by other cues and how these segmentation operations are governed by the first sketch drawn from differential motion cues.

In conclusion, consideration of the temporal aspects of cortical processing and of the role of feedback connections leads to the suggestion that cortical processing of visual information does not proceed sequentially following a local-to-global gradient along the different areas of the hierarchy. Instead, it is suggested that different cues are processed with different priorities and interact in a complex fashion and that such processing involves many areas of the hierarchy at the same time, with information flowing in the feedforward as well as in the feedback direction.

ACKNOWLEDGMENTS

The author thanks Frédéric Sarrato for his help with the iconography.

REFERENCES

1. Flechsig, P., *Die Lokalisation der geistigen Vorgänge insbesondere der Sinnesempfindungen des Menschen,* Veit, Leipzig, 1896.
2. Jones, E.G. and Powell, T.P., An anatomical study of converging sensory pathways within the cerebral cortex of the monkey, *Brain,* 93(4), 793–820, 1970.
3. Talbot, S.A. and Marshall, W.H., Physiological studies of neural mechanisms of visual localization and discrimination, *Am. J. Ophthalmol.*, 24, 1255–1264, 1941.
4. Tusa, R.J., Palmer, L.A., and Rosenquist, A.C., *Multiple Cortical Visual Areas: Visual Field Topography in the Cat,* Humana Press, Totawa, NJ, 1981.
5. Kaas, J.H., The organization of visual cortex in primates, in *Sensory Systems of Primates,* Noback, C.R., Ed., Plenum Press, New York, 1978, 151–179.
6. Hubel, D.H. and Wiesel, T.N., Receptive fields, binocular interaction and functional architecture in the cat's visual cortex, *J. Physiol.,* 160, 106–154, 1962.
7. Dreher, B., Hypercomplex cells in the cat's striate cortex, *Invest. Ophthalmol.,* 11(5), 355–356, 1972.
8. Singer, W., Tretter, F., and Cynader, M., Organization of cat striate cortex: a correlation of receptive-field properties with afferent and efferent connections, *J. Neurophysiol.,* 38(5), 1080–1098, 1975.
9. Bullier, J. and Henry, G.H., Ordinal position of neurons in cat striate cortex, *J. Neurophysiol.,* 42(5), 1251–1263, 1979.

10. Dreher, B. and Cottee, L.J., Visual receptive-field properties of cells in area 18 of cat's cerebral cortex before and after acute lesions in area 17, *J. Neurophysiol.,* 38(4), 735–750, 1975.
11. Sherk, H., Area 18 cell responses in cat during reversible inactivation of area 17, *J. Neurophysiol.,* 41(1), 204–215, 1978.
12. Tanaka, K., Cross-correlation analysis of geniculostriate neuronal relationships in cats, *J. Neurophysiol.,* 49(6), 1303–1318, 1983.
13. Hammond, P. and McKay, D.M., Differential responses of cat visual cortical cells to textured stimuli, *Exp. Brain Res.,* 22, 427–430, 1975.
14. Sillito, A.M., The contribution of inhibitory mechanisms to the receptive field properties of neurones in the striate cortex of the cat, *J. Physiol.,* 250(2), 305–329, 1975.
15. Cleland, B.G., Dubin, M.W., and Levick, W.R., Sustained and transient neurones in the cat's retina and lateral geniculate nucleus, *J. Physiol.,* 217(2), 473–496, 1971.
16. Wilson, P.D., Rowe, M.H., and Stone, J., Properties of relay cells in cat's lateral geniculate nucleus: a comparison of W-cells with X- and Y-cells, *J. Neurophysiol.,* 39(6), 1193–1209, 1976.
17. Stone, J. and Dreher, B., Parallel processing of information in the visual pathways, *TINS,* Dec., 441–446, 1982.
18. Bullier, J., Girard, P., and Salin, P.A., The role of area 17 in the transfer of information to extrastriate visual cortex, in *Primary Visual Cortex in Primates,* Peters, A. and Rockland, K.S., Eds., Plenum Press, New York, 1994, 301–330.
19. Kuypers, H.G., Szwarcbart, M.K., Mishkin, M., and Rosvold, H.E., Occipitotemporal corticocortical connections in the rhesus monkey, *Exp. Neurol.,* 11, 245–262, 1965.
20. Rockland, K.S. and Pandya, D.N., Laminar origins and terminations of cortical connections of the occipital lobe in the rhesus monkey, *Brain Res.,* 179(1), 3–20, 1979.
21. Maunsell, J.H. and Van Essen, D.C., The connections of the middle temporal visual area (MT) and their relationship to a cortical hierarchy in the macaque monkey, *J. Neurosci.,* 3(12), 2563–2586, 1983.
22. Felleman, D.J. and Van Essen, D.C., Distributed hierarchical processing in the primate cerebral cortex, *Cereb. Cortex,* 1(1), 1–47, 1991.
23. Ungerleider, L.G. and Desimone, R., Cortical connections of visual area MT in the macaque, *J. Comp. Neurol.,* 248(2), 190–222, 1986.
24. Hilgetag, C.C., O'Neill, M.A., and Young, M.P., Indeterminate organization of the visual system, *Science,* 271(5250), 776–777, 1996.
25. Barone, P., Batardiere, A., Knoblauch, K., and Kennedy, H., Laminar distribution of neurons in extrastriate areas projecting to visual areas V1 and V4 correlates with the hierarchical rank and indicates the operation of a distance rule, *J. Neurosci.,* 20(9), 3263–3281, 2000.
26. Jouve, B., Rosenstiehl, P., and Imbert, M., A mathematical approach to the connectivity between the cortical visual areas of the macaque monkey, *Cereb. Cortex,* 8(1), 28–39, 1998.
27. Barbas, H., Pattern in the laminar origin of corticocortical connections, *J. Comp. Neurol.,* 252(3), 415–422, 1986.
28. Barbas, H. and Rempel-Clower, N., Cortical structure predicts the pattern of corticocortical connections, *Cereb. Cortex,* 7(7), 635–646, 1997.
29. Sanides, F., Functional architecture of motor and sensory cortices in primates in the light of a new concept of neocortex evolution, in *The Primate Brain: Advances in Primatology,* Noback, C.R. and Montagna, W., Eds., Appleton, New York, 1970, 137–208.

30. Movshon, J.A. and Newsome, W.T., Visual response properties of striate cortical neurons projecting to area MT in macaque monkeys, *J. Neurosci.,* 16(23), 7733–7741, 1996.
31. Henry, G.H., Salin, P.A., and Bullier, J., Projections from area 18 and 19 to cat striate cortex: divergence and laminar specificity, *Eur. J. Neurosci.,* 3, 186–200, 1991.
32. Salin, P.A. and Bullier, J., Corticocortical connections in the visual system structure and function, *Physiol. Rev.,* 75, 107–154, 1995.
33. Sherk, H., Functional organization of input from areas 17 and 18 to an extrastriate area in the cat, *J. Neurosci.,* 10(8), 2780–2790, 1990.
34. Sherman, S.M. and Guillery, R.W., On the actions that one nerve cell can have on another: distinguishing "drivers" from "modulators," *Proc. Natl. Acad. Sci. U.S.A.,* 95(12), 7121–7126, 1998.
35. Crick, F. and Koch, C., Constraints on cortical and thalamic projections: the no-strong-loops hypothesis, *Nature,* 391(6664), 245–250, 1998.
36. Peters, A., Payne, B.R., and Budd, J., A numerical analysis of the geniculocortical input to striate cortex in the monkey, *Cereb. Cortex,* 4(3), 215–229, 1994.
37. Latawiec, D., Martin, K.A., and Meskenaite, V., Termination of the geniculocortical projection in the striate cortex of macaque monkey: a quantitative immunoelectron microscopic study, *J. Comp. Neurol.,* 419(3), 306–319, 2000.
38. Malpeli, J.G., Activity of cells in area 17 of the cat in absence of input from layer a of lateral geniculate nucleus, *J. Neurophysiol.,* 49(3), 595–610, 1983.
39. Nealey, T.A. and Maunsell, J.H., Magnocellular and parvocellular contributions to the responses of neurons in macaque striate cortex, *J. Neurosci.,* 14(4), 2069–2079, 1994.
40. Stratford, K.J., Tarczy-Hornoch, K., Martin, K.A., Bannister, N.J., and Jack, J.J., Excitatory synaptic inputs to spiny stellate cells in cat visual cortex, *Nature,* 382(6588), 258–261, 1996.
41. Sherman, S.M. and Koch, C., The control of retinogeniculate transmission in the mammalian lateral geniculate nucleus, *Exp. Brain Res.,* 63(1), 1–20, 1986.
42. Steriade, M., Jones, E.G., and McCormick, D.A., *Thalamus,* Vol. 1: *Organisation and Function,* Elsevier, Amsterdam, 1997.
43. Nowak, L.G. and Bullier, J., The timing of information transfer in the visual system, in *Extrastriate Visual Cortex in Primates*, Rockland, K.S., Kaas, J.H., and Peters, A., Eds., Plenum Press, New York, 1997, 205–241.
44. McCormick, D.A. and von Krosigk, M., Corticothalamic activation modulates thalamic firing through glutamate "metabotropic" receptors, *Proc. Natl. Acad. Sci. U.S.A.,* 89(7), 2774–2778, 1992.
45. Rivadulla, C., Martinez, L.M., Varela, C., and Cudeiro, J., Completing the corticofugal loop: a visual role for the corticogeniculate type 1 metabotropic glutamate receptor, *J. Neurosci.,* 22(7), 2956–2962, 2002.
46. Godwin, D.W., Vaughan, J.W., and Sherman, S.M., Metabotropic glutamate receptors switch visual response mode of lateral geniculate nucleus cells from burst to tonic, *J. Neurophysiol.,* 76(3), 1800–1816, 1996.
47. Schiller, P.H. and Malpeli, J.G., The effect of striate cortex cooling on area 18 cells in the monkey, *Brain Res.,* 126(2), 366–369, 1977.
48. Girard, P. and Bullier, J., Visual activity in area V2 during reversible inactivation of area 17 in the macaque monkey, *J. Neurophysiol.,* 62(6), 1287–1302, 1989.
49. Girard, P., Salin, P.A., and Bullier, J., Visual activity in macaque area V4 depends on area 17 input, *NeuroReport,* 2(2), 81–84, 1991.
50. Nowak, L.G., James, A.C., and Bullier, J., Corticocortical connections between visual areas 17 and 18a of the rat studied *in vitro*: spatial and temporal organisation of functional synaptic responses, *Exp. Brain Res.,* 117(2), 219–241, 1997.

51. Girard, P., Hupe, J.M., and Bullier, J., Feedforward and feedback connections between areas v1 and v2 of the monkey have similar rapid conduction velocities, *J. Neurophysiol.,* 85(3), 1328–1331, 2001.
52. Hupe, J.M., James, A.C., Payne, B.R., Lomber, S.G., Girard, P., and Bullier, J., Cortical feedback improves discrimination between figure and background by V1, V2 and V3 neurons, *Nature,* 394(6695), 784–787, 1998.
53. Tomita, H., Ohbayashi, M., Nakahara, K., Hasegawa, I., and Miyashita, Y., Top-down signal from prefrontal cortex in executive control of memory retrieval [see comments], *Nature,* 401(6754), 699–703, 1999.
54. Angelucci, A., Levitt, J.B., Walton, E.J., Hupe, J.M., Bullier, J., and Lund, J.S., Circuits for local and global signal integration in primary visual cortex, *J. Neurosci.,* 22(19), 8633–8646, 2002.
55. Hupe, J.M., James, A.C., Girard, P., and Bullier, J., Response modulations by static texture surround in area V1 of the macaque monkey do not depend on feedback connections from V2, *J. Neurophysiol.,* 85(1), 146–163, 2001.
56. Shao, Z. and Burkhalter, A., Different balance of excitation and inhibition in forward and feedback circuits of rat visual cortex, *J. Neurosci.,* 16(22), 7353–7365, 1996.
57. Stone, J., *Parallel Processing in the Visual System. The Classification of Retinal Ganglion Cells and Its Impact on the Neurobiology of Vision*, Plenum Press, New York, 1983.
58. Rockland, K.S., Convergence and branching patterns of round, type 2 corticopulvinar axons, *J. Comp. Neurol.,* 390(4), 515–536, 1998.
59. Rockland, K.S., Bistratified distribution of terminal arbors of individual axons projecting from area V1 to middle temporal area (MT) in the macaque monkey, *Vis. Neurosci.,* 3, 155–170, 1989.
60. Rockland, K.S., Morphology of individual axons projecting from area V2 to MT in the macaque, *J. Comp. Neurol.,* 355(1), 15–26, 1995.
61. Humphrey, A.L., Sur, M., Uhlrich, D.J., and Sherman, S.M., Projection patterns of individual X- and Y-cell axons from the lateral geniculate nucleus to cortical area 17 in the cat, *J. Comp. Neurol.,* 233(2), 159–189, 1985.
62. Freund, T.F., Martin, K.A., Soltesz, I., Somogyi, P., and Whitteridge, D., Arborisation pattern and postsynaptic targets of physiologically identified thalamocortical afferents in striate cortex of the macaque monkey, *J. Comp. Neurol.,* 289(2), 315–336, 1989.
63. Sherman, S.M. and Guillery, R.W., *Exploring the Thalamus,* Academic Press, San Diego, 2001.
64. Bullier, J., Schall, J.D., and Morel, A., Functional streams in occipito-frontal connections in the monkey, *Behav. Brain Res.,* 76(1–2), 89–97, 1996.
65. Crook, J.M., Kisvarday, Z.F., and Eysel, U.T., GABA-induced inactivation of functionally characterized sites in cat striate cortex: effects on orientation tuning and direction selectivity, *Vis. Neurosci.,* 14(1), 141–158, 1997.
66. Sillito, A.M., Kemp, J.A., Milson, J.A., and Berardi, N., A re-evaluation of the mechanisms underlying simple cell orientation selectivity, *Brain Res.,* 194(2), 517–520, 1980.
67. Borg-Graham, L.J., Monier, C., and Fregnac, Y., Visual input evokes transient and strong shunting inhibition in visual cortical neurons, *Nature,* 393(6683), 369–373, 1998.
68. Tsodyks, M., Kenet, T., Grinvald, A., and Arieli, A., Linking spontaneous activity of single cortical neurons and the underlying functional architecture, *Science,* 286(5446), 1943–1946, 1999.
69. Douglas, R.J., Koch, C., Mahowald, M., Martin, K.A.C., and Suarez, H.H., Recurrent excitation in neocortical circuits, *Science,* 269, 981–985, 1995.

70. Knierim, J.J. and Van Essen, D.C., Neuronal responses to static texture patterns in area V1 of the alert macaque monkey, *J. Neurophysiol.,* 67(4), 961–980, 1992.
71. Nothdurft, H.C., Gallant, J.L., and Van Essen, D.C., Response modulation by texture surround in primate area V1: correlates of "popout" under anesthesia, *Vis. Neurosci.,* 16(1), 15–34, 1999.
72. Sandell, J.H. and Schiller, P.H., Effect of cooling area 18 on striate cortex cells in the squirrel monkey, *J. Neurophysiol.,* 48(1), 38–48, 1982.
73. Mignard, M. and Malpeli, J.G., Paths of information flow through visual cortex, *Science,* 251(4998), 1249–1251, 1991.
74. Hupe, J.M., James, A.C., Girard, P., Lomber, S.G., Payne, B.R., and Bullier, J., Feedback connections act on the early part of the responses in monkey visual cortex, *J. Neurophysiol.,* 85(1), 134–145, 2001.
75. Allman, J., Miezin, F., and McGuinness, E., Stimulus specific responses from beyond the classical receptive field: neurophysiological mechanisms for local-global comparisons in visual neurons, *Annu. Rev. Neurosci.,* 8, 407–430, 1985.
76. Xiao, D.K., Raiguel, S., Marcar, V., and Orban, G.A., The spatial distribution of the antagonistic surround of MT/V5 neurons, *Cereb. Cortex,* 7(7), 662–677, 1997.
77. Walker, G.A., Ohzawa, I., and Freeman, R.D., Asymmetric suppression outside the classical receptive field of the visual cortex, *J. Neurosci.,* 19(23), 10536–10553, 1999.
78. Stettler, D.D., Das, A., Bennett, J., and Gilbert, C.D., Lateral connectivity and contextual interactions in macaque primary visual cortex, *Neuron,* 36(4), 739–750, 2002.
79. Zipser, K., Lamme, V.A., and Schiller, P.H., Contextual modulation in primary visual cortex, *J. Neurosci.,* 16(22), 7376–7389, 1996.
80. Rossi, A.F., Desimone, R., and Ungerleider, L.G., Contextual modulation in primary visual cortex of macaques, *J. Neurosci.,* 21(5), 1698–1709, 2001.
81. Bullier, J., Communications between cortical areas of the visual system, in *The Visual Neurosciences*, Chalupa, L. and Werner, J., Eds., MIT Press, Cambridge, MA, 2002.
82. Lamme, V.A., The neurophysiology of figure-ground segregation in primary visual cortex, *J. Neurosci.,* 15(2), 1605–1615, 1995.
83. Rockland, K.S. and Virga, A., Terminal arbors of individual "feedback" axons projecting from area V2 to V1 in the macaque monkey: a study using immunohistochemistry of anterogradely transported *Phaseolus vulgaris*-leucoagglutinin, *J. Comp. Neurol.,* 285(1), 54–72, 1989.
84. Marr, D., *Vision,* W.H. Freeman, San Francisco, 1982.
85. Mumford, D., Neuronal architectures for pattern-theoretic problems, in *Large Scale Neuronal Theories of the Brain*, Koch, C. and Davis, J.L., Eds., MIT Press, Cambridge, MA, 1993, 521–554.
86. Bullier, J., Integrated model of visual processing, *Brain Res. Brain Res. Rev.,* 36(2–3), 96–107, 2001.
87. Martin, P.R., White, A.J., Goodchild, A.K., Wilder, H.D., and Sefton, A.E., Evidence that blue-on cells are part of the third geniculocortical pathway in primates, *Eur J. Neurosci.,* 9(7), 1536–1541, 1997.
88. Calkins, D.J. and Sterling, P., Evidence that circuits for spatial and color vision segregate at the first retinal synapse, *Neuron,* 24(2), 313–321, 1999.
89. Dacey, D.M., Parallel pathways for spectral coding in primate retina, *Annu. Rev. Neurosci.,* 23, 743–775, 2000.
90. Munk, M.H., Nowak, L.G., Girard, P., Chounlamountri, N., and Bullier, J., Visual latencies in cytochrome oxidase bands of macaque area V2, *Proc. Natl. Acad. Sci. U.S.A.,* 92(4), 988–992, 1995.

91. Pack, C.C. and Born, R.T., Temporal dynamics of a neural solution to the aperture problem in visual area MT of macaque brain, *Nature,* 409(6823), 1040–1042, 2001.
92. Bullier, J., Hupe, J.M., James, A.C., and Girard, P., The role of feedback connections in shaping the responses of visual cortical neurons, *Prog. Brain Res.,* 134, 193–204, 2001.

9 Visual Processing in the Macaque Frontal Eye Field

Jeffrey D. Schall, Kirk G. Thompson, Narcisse P. Bichot, Aditya Murthy, and Takashi R. Sato

CONTENTS

9.1 INTRODUCTION

Following Ferrier's[1] original demonstration that electrical stimulation of certain parts of frontal cortex elicits eye movements, the frontal eye field (FEF) has been regarded most commonly as part of the oculomotor system. The evidence for this is beyond dispute. Low-intensity electrical stimulation of FEF elicits saccadic eye movements.[2] Saccades are the rapid shifts of gaze that redirect the focus of vision to different locations in an image. Reversible inactivation of FEF prevents saccade production,[3,4] complementing earlier observations that ablation of FEF causes an initially severe impairment in saccade production that recovers in some but not all respects over time.[5–8] The direct influence of FEF on saccade production seems to be mediated by neurons in FEF that are activated specifically before and during saccades.[9–13] Two kinds of neurons that control gaze have been distinguished. In general, *movement* neurons contribute to gaze shifting, and *fixation* neurons contribute to gaze holding. Neurons in FEF that generate movement-related or fixation-related activity are located in layer 5 and innervate the superior colliculus[14–16] and parts of the neural

0-8493-1243-4/04/$0.00+$1.50

circuit in the brainstem that generate saccades.[17] Physiological recordings indicate that these neurons, in concert with a network including the superior colliculus[18–22] produce signals necessary to produce saccadic eye movements. In fact, a recent experiment has demonstrated that inactivation of the superior colliculus disrupts or prevents microstimulation of FEF from eliciting saccades.[23] The function of FEF in gaze control has been reviewed previously.[24–27]

Experiments probing the control of saccades in monkeys provide compelling evidence for the sufficiency of the activity of movement and fixation neurons in FEF to specify whether and when saccades will be produced.[11,13] In monkeys performing a countermanding task that requires control over whether a saccade is initiated, particular neurons in the FEF are modulated in a manner sufficient to control gaze. Specifically, when a partially prepared saccade is canceled because of a stop signal, neurons with movement and fixation activity exhibit a marked deviation of the modulation that occurs early enough to cancel the saccade (Figure 9.1).

However, certain other neurons in FEF are not modulated in a manner sufficient to be said to control gaze. As shown in Figure 9.1, *visual* neurons in FEF are modulated not at all or too late when planned saccades are canceled. These neurons respond to the visual target, but nothing about their modulation can control whether or not the eyes move. Therefore, not every neuron in the FEF is linked directly to the oculomotor system.

In this chapter we review the evidence that we have obtained over the last 10 years that FEF should be regarded as part of the visual pathway. We focus on the role of the FEF in the selection of targets for covert and overt orienting. We would note that other laboratories have made important observations about the role of the FEF in selecting the target for saccadic and pursuit eye movements.[28–31] It is also crucial to note that the kinds of neural signals reviewed for FEF have been or most likely will be observed in related structures such as the superior colliculus or posterior parietal cortex. Thus, the processes that are described must be regarded as occurring concurrently in a network of interconnected structures.

9.2 VISUAL INPUTS TO THE FRONTAL EYE FIELDS

The evidence that the FEF is involved in visual processing is compelling. One source of visual signals to FEF is the central thalamus. The FEF is innervated mainly by the lateral segment of the mediodorsal nucleus as well as parts of neighboring thalamic nuclei.[32] Neurons in these nuclei can convey a diversity of signals to FEF including visual afferents.[33–35] These thalamic nuclei receive visual signals most directly from the intermediate layers of the superior colliculus.

FEF also receives abundant inputs from a multitude of visual cortical areas in both the dorsal and ventral streams[36–39] (Figure 9.2). In fact, FEF is unique in the extent of its connectivity with extrastriate visual cortex.[40] However, it should not be overlooked that FEF provides reciprocal connections to equally many extrastriate visual areas. In fact, according to a recent analysis of intracortical connectivity, FEF may be in a feedforward anatomical relation to prestriate areas like V4.[41] Thus, FEF can influence the activation of neurons in extrastriate visual cortex.

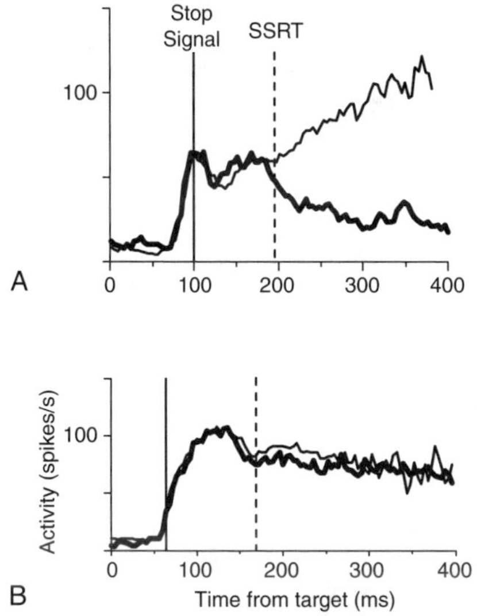

FIGURE 9.1 Relationship between FEF neural activity and canceling a partially prepared saccade. (A) Activity of a FEF neuron with movement-related activity in trials in which the movement was produced but would have been canceled if the stop signal had been presented (thin line) is compared with activity on trials when the planned saccade was canceled because the stop signal appeared (thick line). The time of the stop signal is indicated by the solid vertical line. The time needed to cancel the planned movement — stop signal reaction time (SSRT) — is indicated by the dashed vertical line. When the movement was canceled, neural activation decayed rapidly. This modulation occurred within an interval known as the stop signal reaction time, which is an estimate obtained from behavioral data of the time needed to cancel the movement. Observing the strong modulation of the discharge rate within the stop signal reaction time demonstrates that this neuron conveys a signal sufficient to control whether the eyes move. (B) Activity of a neuron in FEF with a visual response but no saccade-related modulation when saccades were initiated or canceled. The discharge rate of this neuron did not differ between trials when the saccade was produced and trials when the saccade was withheld. Therefore, neurons like this do not produce a signal sufficient to control gaze. (Modified from Reference 13.)

The connectivity of FEF with visual areas caudal to the central sulcus is topographically organized.[38] The more ventrolateral portion of FEF, which is responsible for generating shorter saccades, is interconnected with the perifoveal representation in retinotopically organized areas, from areas that represent central vision in inferotemporal cortex and from other areas having no retinotopic order. In contrast, mediodorsal FEF, which is responsible for generating longer saccades, is interconnected with the peripheral visual field representation of retinotopically organized areas, from areas that emphasize peripheral vision or are multimodal and from other areas that have no retinotopic order.

The concept of hierarchy as an organizing principle has been applied to the extensive network of connectivity among visual cortical areas,[42] but it has been shown that the precise organization based on connectivity is indeterminate.[43] The

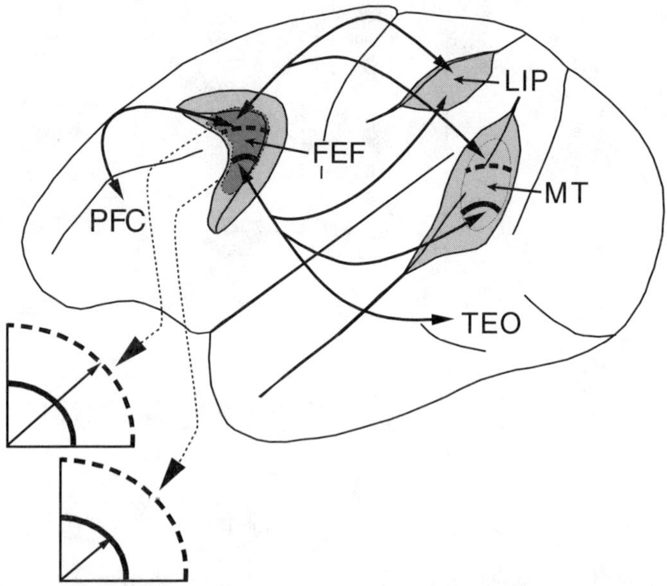

FIGURE 9.2 Cortical connectivity of the FEF. The FEF contributes to the preparation and initiation of saccades through projections to the superior colliculus and brain stem saccade generator. There is a rough map of saccade amplitude in FEF; shorter saccades are generated ventrally, and longer saccades, medially. FEF is reciprocally connected with a multitude of extrastriate visual areas in both the dorsal and ventral streams. The projections are topographically organized; the foveal representation of retinotopic areas projects to the ventrolateral part of FEF (which produces shorter saccades), and the peripheral representation projects to the dorsomedial part of FEF (which produces longer saccades). These diverse visual inputs convey an elaborate representation of the image that contributes to specifying which saccade to produce. The FEF is also interconnected with areas in prefrontal cortex. These connections convey the influence of context, which can supplement or override the outcome of visual processing.

indeterminacy may be resolved to some extent, however, with converging evidence. A recent study measured the visual response latency of neurons in several parts of the visual pathway[44] (Figure 9.3). The results show clearly that activation is earliest in the magnocellular followed by the parvocellular layers of the dorsal lateral geniculate nucleus (LGNd). In the cortex, visual responses are earliest in V1, particularly in layer 4 of V1.[45–47] The next wave of activation occurs effectively concurrently in several areas including MT and FEF as well as V3 and MST (not shown). This activation is rather early. For example, at the time when 50% of FEF neurons have responded — ~70 ms — 25% of V1 neurons have yet to respond to the visual stimulus. Thus, areas associated with the dorsal stream of processing are activated by visual stimuli early and concurrently. In contrast, neurons in areas V2 and V4 are activated later and sequentially in these data.

The hierarchical schemes of visual area organization commonly include multiple levels. For example, FEF is commonly regarded as situated rather high in the hierarchy of visual areas (level 8 in Reference 42). The appearance of the hierarchy invites the conception that visual processing occurs in a sequential manner,

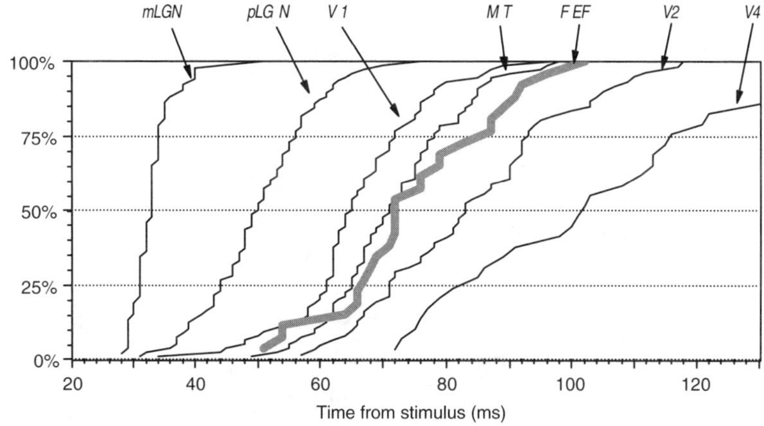

FIGURE 9.3 Times of first response to an optimal visual stimulus are plotted for the indicated stages of the visual pathway. The earliest visual responses are observed in the magnocellular layers (mLGN) and the parvocellular layers of the dorsal lateral geniculate nucleus (pLGN) followed by activation in V1. Area MT and the FEF respond over approximately the same time period followed in order by V2 and V4. (Modified from Reference 44.)

percolating up the network. However, the visual response latency data are not entirely consistent with this. These data were incorporated into a new analysis of the relationship of the visual areas, and the hierarchy collapses if based on a rule of the shortest path to accommodate this independent, converging evidence.[48]

Because of the extensive convergence of afferents from the thalamus and multiple extrastriate visual areas, the FEF receives signals representing the color, form, direction of motion, and stereoscopic depth of objects in the image. Unlike neurons in occipital and temporal visual areas, neurons in FEF typically do not exhibit feature selectivity.[49] However, under particular conditions reviewed below, neurons in FEF can exhibit a form of selectivity based on color, form, and motion. At least some neurons in FEF are responsive to acoustic stimuli.[10,50] In addition, a recent study has shown that neurons in FEF are sensitive to disparity.[51] This may be related to new evidence that FEF or neighboring cortex contributes to vergence eye movements.[52]

Responses of neurons to single stimuli are informative, but they cannot reveal all that is involved in visual processing because rarely is an organism presented with a single stimulus. The rest of this chapter reviews the evidence that FEF contributes to selecting a target for a saccade among distracting stimuli. It is shown that the extensive anatomical convergence contributes to a system that can select targets for gaze shifts regardless of the visual properties of the stimuli.

9.3 SELECTION OF A TARGET AMONG UNIFORM DISTRACTORS

The visual search paradigm has been used extensively to investigate visual selection and attention.[53] The results of many experiments distinguish two general modes of

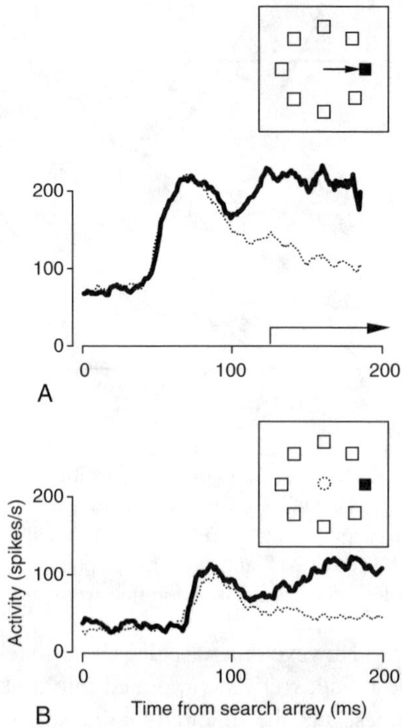

FIGURE 9.4 Activity of a FEF visual neuron following presentation of a popout search array when the monkey produced (A) or withheld (B) a saccade to the singleton. Each plot shows the average activation when the singleton stimulus appeared in the receptive field (solid line) and when only distractors appeared in the receptive field (dotted line). When a saccade was produced to foveate the target, the initial response to the search array did not discriminate whether the target or distractors were in the receptive field. However, after ~100 ms the activation evolved such that the neural representation of the distractors was suppressed, and the activation representing the location of the target was sustained or elevated. When no saccade was produced (B), the overall level of activation was attenuated, but the same selection process was observed. Thus, the neural selection process was not contingent on production of the saccade. (Modified from References 68 and 72.)

visual search. One mode is the efficient search for, say, a black spot among several gray spots (Figure 9.4). The second mode is the less efficient, more effortful search for, say, a randomly oriented T among randomly oriented Ls.

To investigate how the brain selects targets for visually guided saccades, we have recorded the activity of neurons in the FEF of monkeys trained to shift gaze to the oddball target in either of two complementary popout visual search arrays.[54,55] Most visually responsive cells in FEF responded initially indiscriminately to the target or the distractor of the search array in their receptive field (Figure 9.4A). The absence of a feature-selective response in FEF during visual search is consistent with the original observation that neurons in FEF are not feature selective.[49] However, before gaze shifted, a selection process transpired by which most visually responsive cells in FEF ultimately signaled the location of the oddball target stimulus.

A visual target selection process has been observed in FEF during natural scanning eye movements as well.[56] Similar results have also been obtained under somewhat different conditions in the superior colliculus[57–60] and posterior parietal cortex.[61,62] The selection of the target expressed by visuomotor structures such as FEF must be based on the selection process observed in extrastriate visual cortex areas selective for color or form.[63–67]

The evolution of activity over time from nonselective to selective for the location of the target does not distinguish whether this selection process corresponds to explicit visual selection or instead to saccade preparation. A series of experiments has been conducted to evaluate these alternative hypotheses. In one study, FEF activity was recorded while monkeys maintained fixation during presentation of a search array with a single conspicuous oddball.[68] Although no saccade was made to the oddball, FEF neurons still discriminated the oddball from distractors at the same time and to the same degree as when a gaze shift was produced (Figure 9.4B). Thus, the visual selection observed in FEF does not require saccade execution. This study also concluded that saccade preparation was not happening because the saccade made after the trial was rarely directed to the location where the oddball had been.

Another experiment created a condition in which monkeys frequently shifted gaze to a location different from that occupied by a target.[69] A search-step task combines a standard visual search task with the classic double-step saccade task.[70,71] On most trials (referred to as no-step trials) monkeys were rewarded for making a saccade to a color oddball target among distractors. On the remaining trials (step-trials) the target and one distractor unexpectedly swapped positions after presentation of the array. When the target stepped from its original position to a new position, monkeys were rewarded for directing gaze to the new target location (compensated trials). However, monkeys often failed to compensate for the target step and made a saccade to the original target location (noncompensated trials). This behavior was not rewarded. Noncompensated saccade trials provided data to test the dissociation of visual target selection from saccade preparation. Even when gaze shifted away from the popout oddball of a search array, visual neurons in FEF represented the current location of the target (Figure 9.5). Given the evidence that attention is allocated automatically to the conspicuous oddball in a search array, these findings are consistent with the hypothesis that the activation of visually responsive neurons in FEF corresponds to or guides the covert orienting of visual attention.[72]

9.4 RELIABILITY OF TARGET SELECTION BY FEF NEURONS

The conclusions just reviewed were drawn from a comparison of the average activity of neurons contrasted between trials with the target or the distractors in the response field. However, this kind of analysis does not reveal how reliably neurons signal the different stimulus conditions given the variability in discharges of cortical neurons under identical conditions.

A recent study measured the reliability of individual neurons in signaling the target location in correct trials using a neuron–antineuron approach within a winner-take-all architecture.[73] Specifically, the number of trials or neurons that needed to

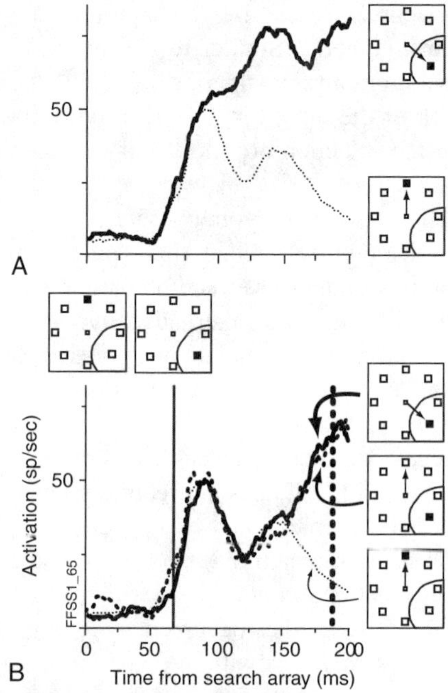

FIGURE 9.5 Response of FEF neuron during search-step task. (A) Activation when the target (black line) or distractors (dotted line) fell in the receptive field (indicated by the boundary in the stimulus arrays). Following the initial 100 ms of activation that did not discriminate target from distractor, the activity was modulated strongly before the saccade to the target. The response to the distractor was suppressed, and the response to the target grew. (B) Averaged activity in compensated (solid black) and noncompensated (thick dotted black) target-step trials when the distractor in the receptive field unexpectedly became the target compared with activity on no-step trials when distractors remained in the receptive field (thin dotted line). In both compensated and noncompensated trials the neuron responded equally strongly to the unexpected appearance of the target in the receptive field. The fact that the activity before the noncompensated saccades was indistinguishable for the search-step trials with opposing saccade directions means that the activity of this neuron could not be involved directly in saccade production. (Modified from Reference 69.)

be combined to match performance was determined through a random sampling procedure. Before the activity representing the target and various distractors became different, the target could not be found in the activity across pools of any size. As the selection process began, the activity of large pools of neurons could locate the target with near perfect reliability (Figure 9.6). As the selection process ran to completion, the activity of pools of around ten neurons represented the location of the target among distractors with high reliability. At the asymptote, on average, the activity from a pool of seven neurons was sufficient to support nearly perfect performance in the easiest search, and pools of about 14 trials generated signals sufficient for nearly perfect localization of the target in the most difficult search. However, monkeys did not achieve nearly perfect performance. In fact, the actual

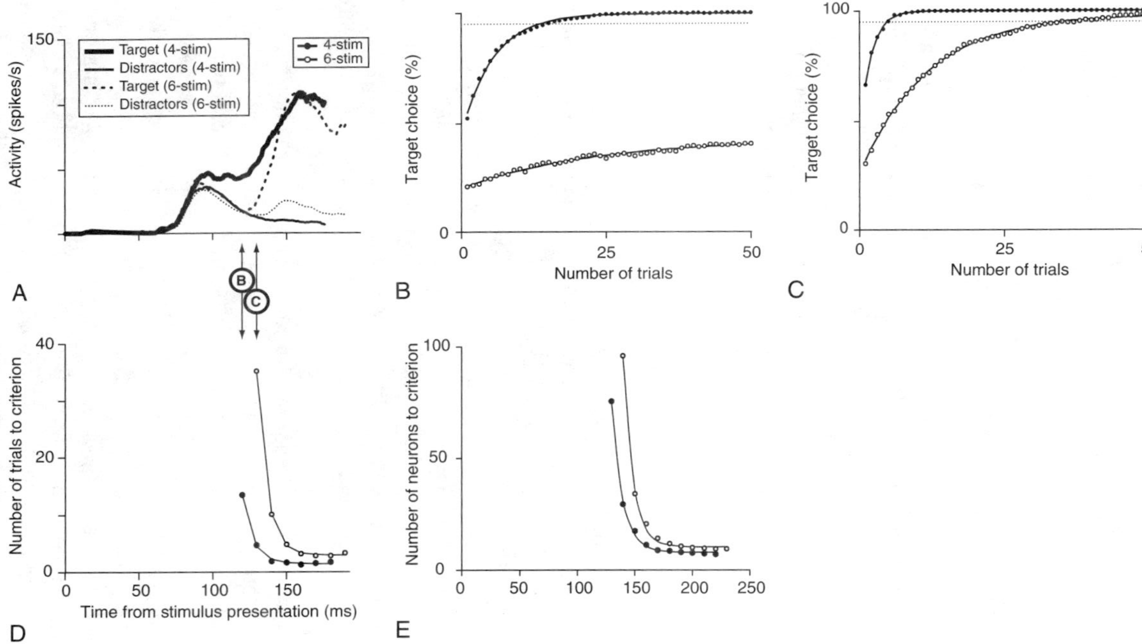

FIGURE 9.6 Reliability of FEF target selection. (A) Average activation of a FEF neuron during search for a target (thick) among distractors (thin) in four-element (solid) and six-element (dotted) arrays. Spike density functions were aligned on stimulus presentation at time 0 and are plotted up to the mean saccade latency during each search condition. Only spikes that occurred before saccade initiation were used in the calculations. (B, C) Probability of target choice as a function of number of trials drawn randomly from the set of trials contributing to panel A at the two time points shown between A and D. Filled circles represent simulations for search with four stimuli, and open circles represent simulations for search with six stimuli. The number of trials needed to reach a fixed criterion of 95% (indicated by dotted line) was calculated. (D) The number of trials required to reach the criterion level is plotted as a function of time from stimulus presentation for search with four (solid) and six (open) element arrays. The best-fit exponential curves are shown overlaid on the data points. (E) The number of neurons randomly sampled from the entire data set required to reach the criterion level is plotted as a function of time from stimulus presentation for search with four (solid) and six (open) element arrays. In the asymptote fewer than ten trials or neurons suffice to locate the target with 95% accuracy. (Modified from Reference 73.)

behavioral accuracy across the range of feature and conjunction search tasks employed was approximated when only six trials or neurons were combined.

The pool sizes obtained in this study were consistent with findings of several studies of neural coding in various visual and other cortical areas reporting pool sizes ranging from 5 to 40 neurons.[74–78] In contrast, the pool sizes determined in this study of FEF were one or two orders of magnitude smaller than those determined by Shadlen and co-workers[79] in their analysis of the relationship between neural and behavioral responses to visual motion. Several factors may contribute to this difference. In addition to basic differences in task requirements and area of the cortex, perhaps the most important difference is that the analysis of area MT activity included neurons with optimal directions of motion different from the one being discriminated, whereas all of the FEF neurons contributing to the sample provided clear information about the location of the target and distractors. Concern about distinguishing between signals in pools of 10 or 100 neurons pales in significance when viewed from the perspective of the total number of neurons comprising the causal chain between stimulus and response. The apparent potency or relevance of so few among so many neurons reveals a profound principle of brain function. The importance of small signals in guiding visually guided behavior is emphasized in the next topic reviewed.

9.5 SELECTION OF INVISIBLE TARGETS

The search for neural correlates of visual awareness has received considerable interest. Bistable stimuli have been used to dissociate the presentation of a stimulus from perception of that stimulus in neurophysiological,[80,81] neuroimaging,[82–84] and event-related potential studies.[85–87] Implicit in such studies of awareness is the subject's voluntary response to an environmental event that allows an inference about the subject's perceptual state. We investigated the neurophysiological link between sensation and action by training macaque monkeys to shift gaze to a visual target that was rendered intermittently perceptible by backward masking (Figure 9.7).[88,89] The experiment was designed to discourage guessing by requiring monkeys to report either the perceived presence or absence of a target.

When monkeys fixated a central spot, a dim target appeared at one of eight possible target locations followed by a bright white mask stimulus at all possible locations. On a high fraction of trials no target was presented. The interval between the appearance of the target and the appearance of the mask was varied according to monkeys' performance. Monkeys were rewarded for correctly reporting with a shift of gaze whether or not the target was present. Monkeys indicated "yes" (target present) by shifting gaze to the target location. Monkeys indicated "no" (target not present) by maintaining fixation on the central spot. Single trials were scored according to the conventions of signal detection theory as either hits (correctly responding "yes"), misses (incorrectly responding "no"), correct rejections (correctly responding "no"), or false alarms (incorrectly responding "yes").

It has been suggested that visual responses in frontal cortex occur specifically for stimuli engaged by action and awareness.[90] Hence, we were surprised to find that visual neurons in FEF responded to undetected masked stimuli,[88,89] This

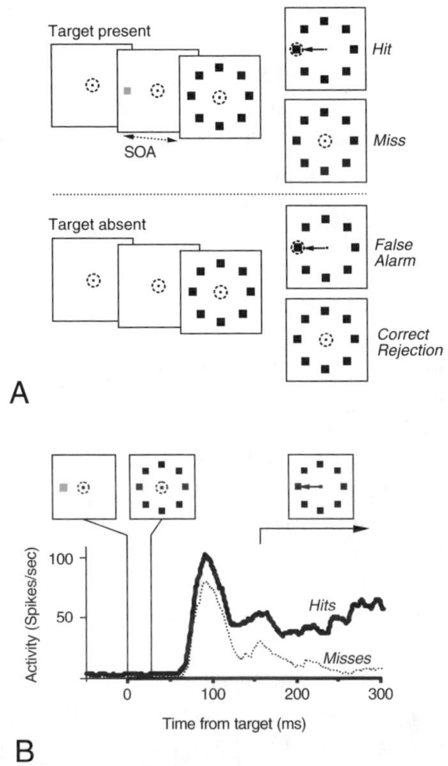

FIGURE 9.7 Response of FEF to a masked target. (A) Visual backward masking task. The target appeared followed after an adjustable stimulus onset asynchrony (SOA) by a mask in target-present trials. A saccade directly to the location where the target had been was scored a *Hit*. Continued fixation was scored a *Miss*. Only the mask appeared on target-absent trials. A saccade to any location was scored a *False Alarm*. Continued fixation was scored a *Correct Rejection*. Monkeys were rewarded for hits and correct rejections. (B) Activity of a FEF visual neuron. The panel compares the activity during target-present trials leading to hits (thick) and misses (thin). Trials are aligned on the time of target presentation in the left panels and on time of mask presentation in the right panels. The differences between alternative trials are shown in the bottom panels. (Modified from Reference 89.)

unexpected result challenges the view that neuronal responses to undetected stimuli that are masked by light do not get out of the retina[91] and indicates that a psychophysical response criterion must be reached for target detection. In fact, we observed that monkeys shifted gaze to the masked stimulus when the initial visual response was only slightly stronger than otherwise (Figure 9.7). It is most plausible that the response of FEF neurons to the target independent of the overt report is conveyed by the afferents from areas MT and MST in which neurons are sensitive to dim, low-contrast stimuli.[92,93] The observation that a difference in activity amounting to just one or two spikes per neuron predicts the ultimate overt response is consistent with observations in area MT.[94] In fact, a small difference in the activation of MT neurons can predict subsequent reports even if no stimulus is present. Likewise, we found that the same small difference in the initial response of neurons in FEF was present even in false-alarm trials.

When saccades were produced, we found that the saccade was preceded by prolonged selective activation of many FEF neurons. The selective pattern of activation of visual neurons preceding the saccade to a detected masked stimulus resembled the selective activation preceding the saccade to a visual search target. However, the prolonged elevated activation was observed in false-alarm trials as well. What is the function of this prolonged elevated activity? Perhaps the late, enhanced activation of visual neurons when the target was detected is related to preparation of the eye movement? In a weak sense this must be the case because the late activation of the visual neurons occurs specifically before the saccade is made. However, several lines of evidence suggest that the relationship between the late visual activation and saccade production is more distal. First, as described above, visual neurons do not play a direct role in controlling gaze; when a planned saccade is canceled, visual neurons are modulated not at all or too late to participate in the act of control.[13] Second, many (but not all) neurons participating in visual selection during search are located in the supragranular layers of FEF.[72] Neurons in the superficial layers do not innervate subcortical oculomotor structures. Third, the remoteness of the visual cells in FEF relative to the motor system is confirmed by the fact that to elicit saccades with microstimulation, higher currents are needed at the sites of visual neurons (>50 μA).[2]

Another interpretation of the late visual activation in FEF is guided by the fact that FEF provides a strong feedback projection to extrastriate visual cortex.[38] A number of studies have suggested that prolonged activation of certain neurons in extrastriate visual cortical areas is a correlate of visual awareness.[80,81] Accordingly, under the conditions of the masking task, it is possible that the selective postmask activation in FEF is correlated not only with "yes" responses, but also with a perceptual experience of the target. The prolonged postmask activity when the monkeys responded "yes" satisfies the condition that activity is of necessary magnitude and duration to be related to awareness.[95,96]

Of course, we would not suggest that FEF is uniquely responsible for visual awareness. Nevertheless, evidence from several studies indicates that prefrontal cortex plays some role in awareness. First, functional imaging studies have shown that areas of prefrontal cortex, possibly including FEF, exhibit activation associated with binocular rivalry[83] even when subjects make no overt motor report.[84] Second, in another functional imaging study, activation of prefrontal cortex including FEF was observed specifically when a blindsight patient reported being aware of presented stimuli.[97] Thus, even if FEF is not directly responsible for generating visual awareness, a plausible hypothesis suggests that activation similar to the selective modulation observed in FEF visual neurons occurs in whichever brain regions are related to the production of visual awareness.

9.6 CHRONOMETRY OF TARGET SELECTION

Explaining the duration and variability of response times is a central problem in psychology.[98–100] A general hypothesis guiding cognitive psychology holds that behavioral response times are occupied by more or less distinct stages of processing.[101,102] If it were possible to identify physiological markers for the end of one stage and the beginning of another, then the alternative hypotheses about the existence of stages and the time course of activation could be distinguished. A series of studies has investigated how the time of visual target selection relates to the total time taken to initiate the saccade. During search for a single, conspicuous target in a search array the large majority of visually responsive neurons in FEF discriminated the target from distractors at a constant interval after search array presentation[72,103] (Figure 9.8). This finding indicates that at least under the conditions of efficient,

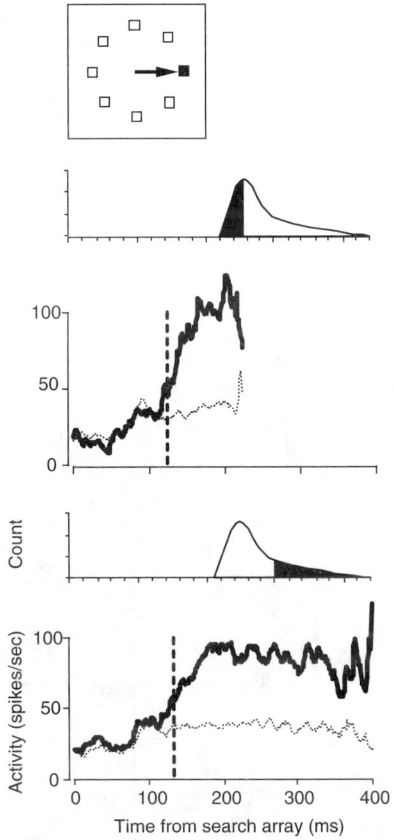

FIGURE 9.8 Relation of time of neural target selection to time of saccade initiation during efficient search for a green target among red distractors. The activity of a FEF neuron representing the target (thick) or distractors (thin) is shown during trials with saccades of the shortest (top) or longest (bottom) latencies. The upper plots in each panel indicate the distribution of saccade latencies with the range selected for the analysis of activity shaded. The time at which the activity distinguished whether a target or distractor was in the receptive field is marked by the dashed vertical line. The neuron discriminated the target from distractors following a relatively constant interval after presentation of the search array. (Modified from Reference 103.)

popout search, a relatively constant period of time is needed to locate potential targets, and the additional variability in saccade latency is introduced by the time needed to prepare and execute the eye movement. When the discrimination of the target is more difficult because the target more closely resembles distractors and search is less efficient,[104] the time taken by FEF neurons to locate the target increases.[103] This increase in time taken to locate targets among more similar distractors is accompanied by an increase in the variability of the selection time across trials. Consequently, the variability in the visual selection time accounts for a larger fraction of the variability in saccade latency. This occurs because the production of an accurate saccade cannot proceed until the target is located (Figure 9.9).

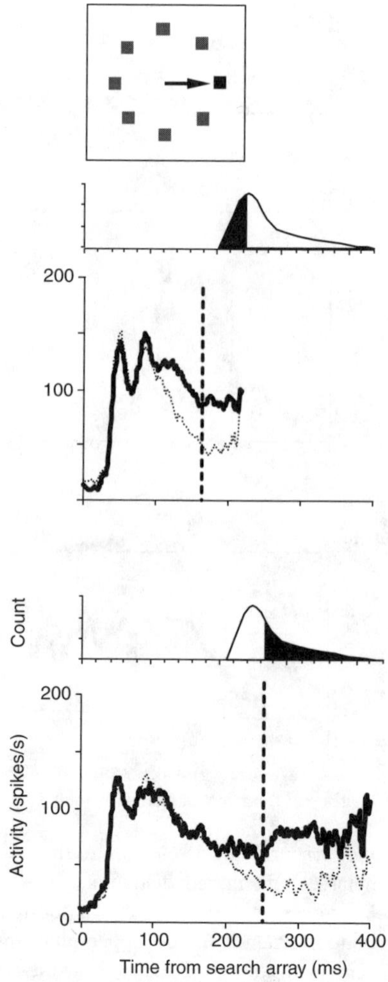

FIGURE 9.9 Activity of an FEF neuron during trials searching for a target that was difficult to distinguish from the distractors. The neural selection of the target is delayed in proportion to the increase of reaction time. Conventions as in Figure 9.8. (Modified from Reference 103.)

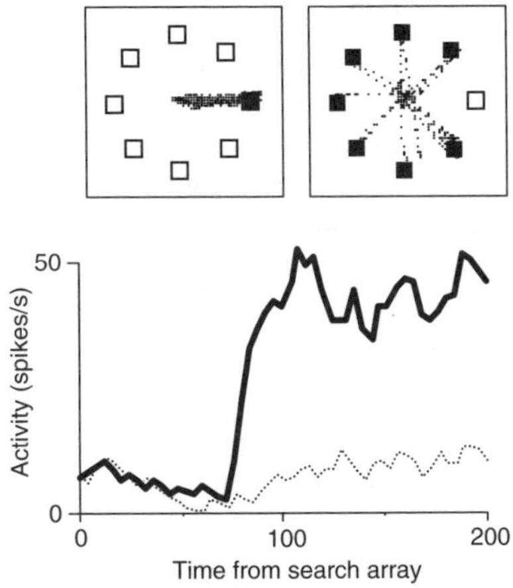

FIGURE 9.10 Effect of cognitive strategy on target selection in FEF. Top panels illustrate performance of a monkey that had been trained on just one visual search array. When presented an array in which the target and distractor colors were switched, instead of looking at the conspicuous singleton (left panel), this monkey looked at a distractor that was the same color as the target in the complementary array (right panel). The time course of activation of a single FEF neuron during the visual search task when the target was in the receptive field (thick) and when distractors were in the receptive field (thin) is shown in the lower panel. Unlike what was observed in monkeys trained on both complements of the search array, in these particularly trained monkeys half of the neurons in FEF exhibited a selective response from the earliest spikes that could be measured. (Modified from Reference 106.)

9.7 SELECTION OF A TARGET REQUIRING KNOWLEDGE

Several lines of evidence demonstrate that gaze can be guided as much by knowledge as by the visual features of stimuli. First, cognitive strategies can override both covert[105] and overt[106] selection of a single oddball in a search array. Second, experts are more likely than novices to ignore conspicuous but irrelevant parts of a visual image from their field of expertise.[107–109] Finally, the pattern of visual fixation can be influenced by verbal instruction.[110]

To study the effects of training experience on gaze behavior and associated neural activity in FEF, monkeys were trained exclusively with search arrays that contained a single item of a constant color among distractor items of another constant color (for example, always a red target among green distractors or always a green target among red distractors).[106] Control monkeys were trained to shift gaze to the oddball of both configurations of the search array (that is, alternating between red among green and green among red). The control monkeys shifted gaze to the oddball stimulus, regardless of the feature that defined it. In contrast, experimental monkeys persistently directed gaze to stimuli with the color of the target even when the configuration of the array was switched for a few trials. In other words, when the experimental monkeys were presented with the search array complementary to that with which they had been trained, they shifted gaze to one of the distractors (that was the color of the overlearned target) and not to the target (even though it was the oddball). As described above, FEF neurons in control monkeys did not exhibit feature selectivity, but their activity evolved to signal the location of the oddball stimulus. In monkeys trained exclusively with a search array with constant target and distractor colors, however, about half of FEF neurons exhibited selectivity for the color of the stimuli in the initial response (Figure 9.10). That is, if the overlearned target fell in the receptive field, the neurons responded strongly, but if the overlearned distractors fell in the receptive field, the neurons responded significantly more weakly or not at all. Subsequently, the appearance of selectivity for stimulus features that consistently guide eye movements has been reported for neurons in the parietal cortex[111,112] and superior colliculus.[113] It appears that the visuomotor system can commit itself to particular interpretations of the image to guide saccadic eye movements. The mechanisms underlying this form of plasticity deserve investigation.

In addition to these long-term changes, target selection during visual search is influenced by shorter-term implicit memory representations arising through repetition of location or stimulus features affecting covert[114,115] and overt[116,117] orienting. Evidence for this is provided by particular changes in performance following sequential changes in stimulus features and target location. Until recently, the neuronal mechanisms underlying such sequential effects had not been examined. Single neurons were recorded in the FEF of monkeys performing a popout search during which stimulus features and target position changed unpredictably across trials.[118] In keeping with previous studies, repetition of stimulus features improved performance (Figure 9.11). This feature-based facilitation of return was manifested in the target discrimination process in FEF. Neurons discriminated the target from distractors earlier and better following repetition of stimulus features, corresponding to improvements in saccade latency and accuracy, respectively. The change of neuronal target selection occurred through both target enhancement and distractor suppression. This result shows adjustments of the target selection process in FEF corresponding to and therefore possibly contributing to the changes in performance across trials due to sequential regularities in display properties.

Knowledge gained through experience is necessary when objects of interest cannot be located based solely on their visual features. Such cases are exemplified by a search for a conjunction of features such as color and shape in which an explicit memory representation is needed to identify the target.[119] A recent study investigated how the brain combines knowledge with visual processing to locate targets for eye movements by training monkeys to perform a visual search for a target defined by a unique combination of color and shape (e.g., red cross). The color–shape combinations that defined the target were varied pseudo-randomly between sessions. Two separate, contextual influences were exerted on gaze behavior and the neural selection process: visual similarity to the target and the history of target properties.[116,120] The evidence for the influence of visual similarity was that monkeys made occasional errant saccades during this conjunction search, which tended to direct gaze to distractors that resembled the current target. Similar observations have been made with human observers during covert[121] and overt orienting.[122,123] When monkeys correctly shifted gaze to the target, FEF neurons not only discriminated the target from distractors but also discriminated among the nonselected distractors resulting

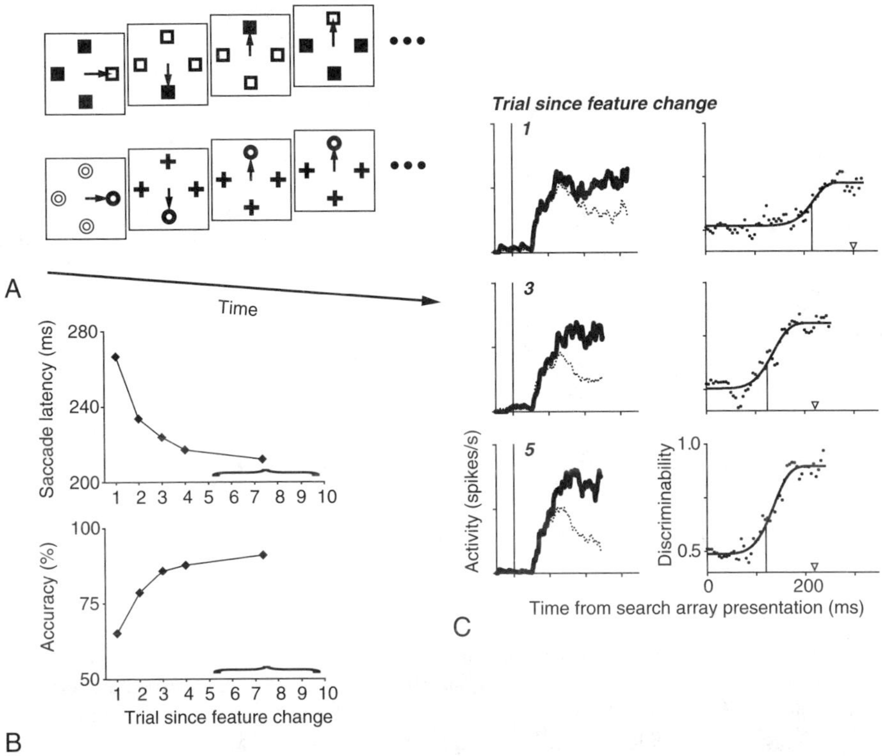

FIGURE 9.11 Priming in search and FEF. (A) Popout search task. The monkeys' task was to shift gaze to the single target defined by color (top) or shape (bottom). The arrow indicates the saccade to the target. Top panel illustrates a change from a search for green among red to a search for red among green. Bottom panel illustrates a change from a search for a red circle among green circles to a search for a red circle among red crosses. Stimuli are not drawn to scale. (B) Variation of average saccade latency (top) and accuracy (bottom) as a function of the number of trials following the change of features in the search array. Immediately following a change, performance is slower and more error prone. After trials, performance improves to asymptote. (C) Effect of feature change during popout search on the activity of one FEF neuron. Left panels illustrate average activity when the target (thick) or distractors (thin) appeared in the receptive field. Right panels illustrate the time course and magnitude of target selection through a measurement of discriminability (0.5 indicates no difference of activity; 1.0 indicates perfect discrimination of target from distractor) derived from the activity when the target or distractors appeared in the receptive field. In each plot of discriminability the vertical line marks the time of target discrimination. The open arrowhead above the abscissa marks the median saccade latency. The three rows show activity for increasing numbers of trials after the feature change with the top panels illustrating activity during the first trial after the change, and the bottom row showing the activity during the fifth trial after the change. The speed and quality of target selection improves with increasing number of trials from the feature change. (Modified from Reference 118.)

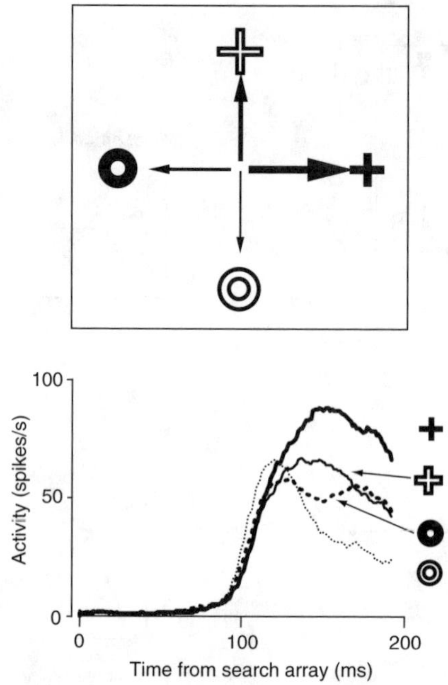

FIGURE 9.12 Visual selection of a remembered target during conjunction search. The assignment of the patterns of neural activation and the incidence of saccades to the alternative stimuli are indicated in the diagram of the search array. The width of the arrows in the search array represents the incidence of saccades to the different stimuli. Most saccades were made to the target (black cross). Occasional errant saccades were directed to distractors that were the same shape or color as the target (black circle, white cross) more often than to the distractor that shared neither feature with the target (white circle). In addition, errant saccades exhibited an additional tendency to shift gaze to the distractor that had been the target in the previous experimental session (white cross). The evolution of activity of an FEF neuron is shown during conjunction search when the target stimulus (thick solid), same-color distractors (intermediate dashed), same-shape distractors (intermediate solid), and the opposite distractors (thin dotted) fell in the receptive field. The initial response did not distinguish the target from the various kinds of distractors, but shortly thereafter the activation for the target became greater, while the activation for the distractors was reduced. The degree of suppression of the distractor activation varied according to whether the distractors resembled the target or had been the target in the previous session. (Modified from Reference 120.)

in more activation for distractors that shared a target feature than for distractors that shared none (Figure 9.12). Thus, the pattern of neural discrimination among non-selected distractors corresponded to the pattern of errors that reveals the allocation of attention. Evidently, a template of the target held in memory influenced performance and activity.

During conjunction search, the history of stimulus presentation across sessions also affected the selection process.[116] If an error was made, monkeys showed a significant tendency (in addition to the visual similarity tendency just described) to shift gaze to the distractors that had been the target in the previous session.

Recordings from FEF neurons during trials with correct saccades to the conjunction target revealed a corresponding discrimination among distractors with more activation for distractors that had been the target during the previous session. This effect was evident across sessions that were more than a day apart and persisted throughout experimental sessions. The longer duration of this influence distinguishes it from the short-term priming during popout search that lasts for about ten trials or 30 s in humans,[114,115] as well as monkeys.[116,117]

In the aforementioned studies, the target was present in the search array on every trial. This means that one possible explanation of the modulation involves a direct comparison of the distractor features with the target. To determine more directly whether a memory representation spanning across trials can influence the selection of FEF neurons, an experiment with singleton search was carried out in which no target and only distractors were presented on a fraction of trials and monkeys were rewarded for maintaining fixation on the central spot.[124] Even in trials with no target present, the activation of FEF neurons in response to distractors was proportional to the visual similarity of the distractors to the target. This observation suggests that a template of the target held in memory can influence the target selection process in FEF. Although the precise relationship between memory and search is not entirely clear,[125–127] the necessity of some kind of target template seems clear.[128–130]

The source of these various expressions of contextual modulation observed in FEF is not known. Current thought would hold that the target template is represented in areas 12 and 46 of prefrontal cortex,[131] which are connected with FEF.[36,132] Recent findings have prompted the suggestion that dorsolateral prefrontal cortex encodes rules for guiding behavior.[133,134] Such contingent activation seems a necessary basis for the modulation observed in FEF. The activity of neurons in dorsolateral prefrontal areas rostral to FEF has been described during visual search or selection tasks,[135–139] but the selection was more all or none because the responses began typically after the selection process was completed. Thus, under the general conditions of search used in this work, nontarget stimuli did not activate cells in prefrontal areas 12 and 46. Much more research is needed to understand how arbitrary rules influence saccade target selection.

ACKNOWLEDGMENTS

The work for this chapter was supported by Grants R01-EY08890, R01-MH55806, P30-EY08126, the McDonnell-Pew Program in Cognitive Neuroscience, and by the McKnight Endowment Fund for Neuroscience. We are grateful to Gale Fox, Elizabeth Gotcher, and Jennifer Jewett for assistance in preparing the manuscript.

REFERENCES

1. Ferrier, D., The localization of function in brain, *Proc. R.. Soc. Lond.*, 22, 229–232, 1874.
2. Bruce, C.J., Goldberg, M.E., Bushnell, C., and Stanton, G.B., Primate frontal eye fields II: Physiological and anatomical correlates of electrically evoked eye movements, *J. Neurophysiol.*, 54, 714–734, 1985.

3. Dias, E.C., Kiesau, M., and Segraves, M.A., Acute activation and inactivation of macaque frontal eye field with GABA-related drugs, *J. Neurophysiol.,* 74, 2744–2748, 1995.
4. Sommer, M.A. and Tehovnik, E.J., Reversible inactivation of macaque frontal eye field, *Exp. Brain Res.,* 116, 229–249, 1997.
5. Schiller, P.H., Sandell, J.H., and Maunsell, J.H.R., The effect of frontal eye field and superior colliculus lesions on saccadic latencies in the rhesus monkey, *J. Neurophysiol.*, 57, 1033–1049, 1987.
6. Schiller, P.H. and Chou, I.H., The effects of frontal eye field and dorsomedial frontal cortex lesions on visually guided eye movements, *Nat. Neurosci.*, 1, 248–253, 1998.
7. Schiller, P.H. and Chou, I., The effects of anterior arcuate and dorsomedial frontal cortex lesions on visually guided eye movements in the rhesus monkey: 1. Single and sequential targets, *Vision Res.*, 40, 1609–26, 2000.
8. Schiller, P.H. and Chou, I., The effects of anterior arcuate and dorsomedial frontal cortex lesions on visually guided eye movements: 2. Paired and multiple targets, *Vision Res.*, 40, 1627–38, 2000.
9. Bruce, C.J. and Goldberg, M.E., Primate frontal eye fields. I: Single neurons discharging before saccades, *J. Neurophysiol.*, 53, 603–635, 1985.
10. Schall, J.D., Neuronal activity related to visually guided saccades in the frontal eye fields of rhesus monkeys: comparison with supplementary eye fields, *J. Neurophysiol.*, 66, 559–579, 1991.
11. Hanes, D.P. and Schall, J.D., Neural control of voluntary movement initiation, *Science,* 274, 427–430, 1996.
12. Hanes, D.P., Thompson, K.G., and Schall, J.D., Relationship of presaccadic activity in frontal and supplementary eye field to saccade initiation in macaque: Poisson spike train analysis, *Exp. Brain Res.*, 103, 85–96, 1995.
13. Hanes, D.P., Patterson, W.F., and Schall, J.D., The role of frontal eye field in countermanding saccades: visual, movement and fixation activity, *J. Neurophysiol.*, 79, 817–834, 1998.
14. Segraves, M.A. and Goldberg, M.E., Functional properties of corticotectal neurons in the monkey's frontal eye fields, *J. Neurophysiol.*, 58, 1387–1419, 1987.
15. Sommer, M.A. and Wurtz, R.H., Composition and topographic organization of signals sent from the frontal eye field to the superior colliculus, *J. Neurophysiol.*, 83, 1979–2001, 1998.
16. Sommer, M.A. and Wurtz, R.H., Frontal eye field sends delay activity related to movement, memory, and vision to the superior colliculus, *J. Neurophysiol.*, 85, 1673–1685, 2001.
17. Segraves, M.A., Activity of monkey frontal eye field neurons projecting to oculomotor regions of the pons, *J. Neurophysiol.*, 68, 1967–1985, 1992.
18. Sparks, D.L., Functional properties of neurons in the monkey superior colliculus: coupling of neuronal activity and saccade onset, *Brain Res.*, 156, 1–16, 1978.
19. Munoz, D.P. and Wurtz, R.H., Fixation cells in monkey superior colliculus. I. Characteristics of cell discharge, *J. Neurophysiol.*, 70, 559–575, 1993.
20. Munoz, D.P. and Wurtz, R.H., Saccade-related activity in monkey superior colliculus. I. Characteristics of burst and buildup cells, *J. Neurophysiol.*, 73, 2313–2333, 1995.
21. Dorris, M.C., Pare, M., and Munoz, D.P., Neuronal activity in monkey superior colliculus related to the initiation of saccadic eye movements, *J. Neurosci.*, 17, 8566–8579, 1997.

22. Dorris, M.C. and Munoz, D.P., Saccadic probability influences motor preparation signals and time to saccadic initiation, *J. Neurosci.*, 18, 7015–7026, 1998.
23. Hanes, D.P. and Wurtz, R.H., Interaction of the frontal eye field and superior colliculus for saccade generation, *J. Neurophysiol.*, 85, 804–815, 2001.
24 Schall, J.D., Visuomotor areas of the frontal lobe, in *Extrastriate Cortex of Primates*, Vol. 12, *Cerebral Cortex*, Rockland, K., Peters, A., and Kaas, J.H., Eds., Plenum Press, New York, 1997, 527–638.
25. Schall, J.D. and Thompson, K.G., Neural selection and control of visually guided eye movements, *Annu. Rev. Neurosci.*, 22, 241–259, 1999.
26. Munoz, D.P. and Schall, J.D., Concurrent distributed control of saccades, in *The Oculomotor System: New Approaches for Studying Sensorimotor Integration*, Hall, W.C. and Moschovakis, A.K., Eds., CRC Press, Boca Raton, FL, 2003.
27. Tehovnik, E.J., Sommer, M.A., Chou, I.H., Slocum, W.M., and Schiller, P.H., Eye fields in the frontal lobes of primates, *Brain Res. Rev.,* 32, 413–448, 2000.
28. Kim, J.N. and Shadlen, M.N., Neural correlates of a decision in the dorsolateral prefrontal cortex of the macaque, *Nat. Neurosci.,* 2, 176–185, 1999.
29. Gold, J.I. and Shadlen, M.N., Representation of a perceptual decision in developing oculomotor commands, *Nature*, 404, 390–394, 2000.
30. Coe, B., Tomihara, K., Matsuzawa, M., and Hikosaka, O., Visual and anticipatory bias in three cortical eye fields of the monkey during an adaptive decision-making task, *J. Neurosci.*, 22, 5081–5090, 2002.
31. Gardner, J.L. and, Lisberger, S.G., Serial linkage of target selection for orienting and tracking eye movements, *Nat. Neurosci.,* 5, 892–899, 2002.
32. Huerta, M.F., Krubitzer, L.A., and Kaas, J.H., Frontal eye field as defined by intracortical microstimulation in squirrel monkeys, owl monkeys and macaque monkeys. I. Subcortical connections, *J. Comp. Neurol.*, 253, 415–439, 1986.
33. Schlag, J. and Schlag-Rey, M., Visuomotor functions of central thalamus in monkey. II. Unit activity related to visual events, targeting, and fixation, *J. Neurophysiol.*, 51, 1175–1195, 1984.
34. Sommer, M.A. and Wurtz, R.H., Frontal eye field neurons orthodromically activated from the superior colliculus, *J. Neurophysiol.*, 80, 3331–3335, 1998.
35. Sommer, M.A. and Wurtz, R.H., Frontal eye field sends delay activity related to movement, memory, and vision to the superior colliculus, *J. Neurophysiol.*, 85, 1673–1685, 2001.
36. Huerta, M.F., Krubitzer, L.A., and Kaas, J.H., Frontal eye field as defined by intracortical microstimulation in squirrel monkeys, owl monkeys and macaque monkeys. II. Cortical connections, *J. Comp. Neurol.*, 265, 332–361, 1987.
37. Baizer, J.S., Ungerleider, L.G., and Desimone, R., Organization of visual inputs to the inferior temporal and posterior parietal cortex in macaques, *J. Neurosci.,* 11,168–190, 1991.
38. Schall, J.D., Morel, A., King, D.J., and Bullier, J., Topography of visual cortical afferents to frontal eye field in macaque: functional convergence and segregation of processing streams, *J. Neurosci.,* 15, 4464–4487, 1995.
39. Stanton, G.B., Bruce, C.J., and Goldberg, M.E., Topography of projections to posterior cortical areas from the macaque frontal eye fields, *J. Comp. Neurol.,* 353, 291–305, 1995.
40. Jouve, B., Rosenstiehl, P., and Imbert, M., A mathematical approach to the connectivity between the cortical visual areas of the macaque monkey, *Cereb. Cortex*, 8, 28–39, 1998.

41. Barone, P., Batardiere, A., Knoblauch, K., and Kennedy, H., Laminar distribution of neurons in extrastriate areas projecting to visual areas V1 and V4 correlates with the hierarchical rank and indicates the operation of a distance rule, *J. Neurosci.*, 20, 3263–3281, 2000.
42. Felleman, D.J. and Van Essen, D.C., Distributed hierarchical processing in the primate cerebral cortex, *Cereb. Cortex,* 1, 1–47, 1991.
43. Hilgetag, C.C., O'Neill, M.A., and Young, M.P., Indeterminate organization of the visual system, *Science*, 271, 776–777, 1996.
44. Schmolesky, M.T., Wang, Y., Hanes, D P., Thompson, K.G., Leutgeb, S., Schall, J.D., and Leventhal A.G., Signal timing across the macaque visual system, *J. Neurophysiol.*, 79, 3272–3278, 1998.
45. Maunsell, J.H. and Gibson, J.R., Visual response latencies in striate cortex of the macaque monkey, *J. Neurophysiol.,* 68, 1332–1344, 1992.
46. Nowak, L.G., Munk, M.H., Girard, P., and Bullier, J., Visual latencies in areas V1 and V2 of the macaque monkey, *Vis. Neurosci.*, 12, 371–384, 1995.
47. Bullier, J., Integrated model of visual processing, *Brain Res. Rev.*, 36, 96–107, 2001.
48. Petroni, F., Panzeri, S., Hilgetag, C.C., Kotter, R., and Young, M.P., Simultaneity of responses in a hierarchical visual network, *NeuroReport*, 12, 2753–2759, 2001.
49. Mohler, C.W., Goldberg, M.E., and Wurtz, R.H., Visual receptive fields of frontal eye field neurons, *Brain Res.*, 61, 385–389, 1973.
50. Vaadia, E., Benson, D.A., Hienz, R.D., and Goldstein, M.H., Unit study of monkey frontal cortex: active localization of auditory and of visual stimuli, *J. Neurophysiol.*, 56, 934–952, 1986.
51. Ferraina, S., Pare, M., and Wurtz, R.H., Disparity sensitivity of frontal eye field neurons, *J. Neurophysiol.*, 83, 625–629, 2000.
52. Gamlin, P.D. and Yoon, K., An area for vergence eye movement in primate frontal cortex, *Nature*, 407, 1003–1007, 2000.
53. Wolfe, J.M., Visual search, in *Attention*, Pashler, H., Ed., Psychological Press, Hove, U.K., 1998, 13–74.
54. Schall, J.D. and Hanes, D.P., Neural basis of saccade target selection in frontal eye field during visual search, *Nature,* 366, 467–469, 1993.
55. Schall, J.D., Hanes, D.P., Thompson, K.G., and King, D.J., Saccade target selection in frontal eye field of macaque. I: Visual and premovement activation, *J. Neurosci.*, 15, 6905–6918, 1995.
56. Burman, D.D. and Segraves, M.A., Primate frontal eye field activity during natural scanning eye movements, *J. Neurophysiol.*, 71, 1266–1271, 1994.
57. Ottes, F.P., Van Gisbergen, J.A.M., and Eggermont, J.J., Collicular involvement in a saccadic colour discrimination task, *Exp. Brain Res.,* 66, 465–478, 1987.
58. Basso, M.A. and Wurtz, R.H., Modulation of neuronal activity in superior colliculus by changes in target probability, *J. Neurosci.,* 18, 7519–7534, 1998.
59. Krauzlis, R. and Dill, N., Neural correlates of target choice for pursuit and saccades in the primate superior colliculus, *Neuron,* 35, 355–363, 2002.
60. McPeek, R.M. and Keller, E.L., Saccade target selection in the superior colliculus during a visual search task, *J. Neurophysiol.*, 88, 2019–2034, 2002.
61. Gottlieb, J.P., Kusunoki, M., and Goldberg, M.E., The representation of visual salience in monkey parietal cortex, *Nature*, 391, 481–484, 1998.
62. Constantinidis, C. and Steinmetz, M.A. Neuronal responses in area 7a to multiple-stimulus displays: I. Neurons encode the location of the salient stimulus, *Cereb. Cortex*, 11, 581–591, 2001.

63. Motter, B.C., Neural correlates of attentive selection for color or luminance in extrastriate area V4, *J. Neurosci.,* 14, 2178–2189, 1994.
64. Motter, B.C., Neural correlates of feature selective memory and pop-out in extrastriate area V4, *J. Neurosci.,* 14, 2190–2199, 1994.
65. Chelazzi, L., Miller, E.K., Duncan, J., and Desimone, R., A neural basis for visual search in inferior temporal cortex, *Nature,* 363, 345–347, 1993.
66. Chelazzi, L., Duncan, J., Miller, E.K., and Desimone, R., Responses of neurons in inferior temporal cortex during memory-guided visual search, *J. Neurophysiol.,* 80, 2918–2940, 1998.
67. Luck, S.J., Chelazzi, L., Hillyard, S.A., and Desimone, R., Neural mechanisms of spatial selective attention in areas V1, V2, and V4 of macaque visual cortex, *J. Neurophysiol.,* 77, 24–42, 1997.
68. Thompson, K.G., Bichot, N.P., and Schall, J.D., Dissociation of target selection from saccade planning in macaque frontal eye field, *J. Neurophysiol.,* 77, 1046–1050, 1997.
69. Murthy, A., Thompson, K.G., and Schall, J.D., Dynamic dissociation of visual selection from saccade programming in frontal eye field, *J. Neurophysiol.,* 86, 2634–2637, 2001.
70. Lisberger, S.G., Fuchs, A.F., King, W.M., and Evinger, L.C., Effect of mean reaction time on saccadic responses to two-step stimuli with horizontal and vertical components, *Vision Res.*, 15, 1021–1025, 1975.
71. Becker, W. and Jurgens, R., An analysis of the saccadic system by means of double step stimuli, *Vision Res.*, 19, 967–983, 1979.
72. Thompson, K.G., Hanes, D.P., Bichot, N.P., and Schall, J.D., Perceptual and motor processing stages identified in the activity of macaque frontal eye field neurons during visual search, *J. Neurophysiol.,* 76, 4040–4055, 1996.
73. Bichot, N.P., Thompson, K.G., Chenchal Rao, S., and Schall, J.D., Reliability of macaque frontal eye field neurons signaling saccade targets during visual search, *J. Neurosci.,* 21, 713–725, 2001.
74. Tolhurst, D.J., Movshon, J.A., and Dean, A.F., The statistical reliability of signals in single neurons in cat and monkey visual cortex, *Vision Res.*, 23, 775–785, 1983.
75. Optican, L.M. and Richmond, B.J., Temporal encoding of two-dimensional patterns by single units in primate inferior temporal cortex, III. Information theoretic analysis, *J. Neurophysiol.*, 57, 162–178, 1987.
76. Gawne, T.J. and Richmond, B.J., How independent are the messages carried by adjacent inferior temporal cortical neurons, *J. Neurosci.*, 13, 2758–2771, 1993.
77. Rolls, E.T., Treves, A., and Tovee, M.J., The representational capacity of the distributed encoding of information provided by populations of neurons in primate temporal visual cortex, *Exp. Brain Res.*, 11, 4149–4162, 1997.
78. Lee, D., Port, D.L., Kruse, W., and Georgopoulos, A.P., Neuronal population coding multielectrode recordings in primate cerebral cortex, in *Neuronal Ensembles Strategies for Recording and Decoding*, Eichenbaum, H. and Davis, J., Eds., Wiley, New York, 1998, 117–136.
79. Shadlen, M.N., Britten, K.H., Newsome, W.T., and Movshon, J.A., A computational analysis of the relationship between neuronal and behavioral responses to visual motion, *J. Neurosci.*, 16, 1486–1510, 1996.
80. Logothetis, N.K. and Schall, J.D., Neuronal correlates of subjective visual perception, *Science,* 245, 761–763, 1989.
81. Sheinberg, D.L. and Logothetis, N.K., The role of temporal cortical areas in perceptual organization, *Proc. Natl. Acad. Sci. U.S.A.,* 94, 3408–3413, 1997.

82. Tong, F., Nakayama, K., Vaughan, J.T., and Kanwisher, N., Binocular rivalry and visual awareness in human extrastriate cortex, *Neuron*, 21, 753–759, 1998.
83. Lumer, E.D., Friston, K J., and Rees, G., Neural correlates of perceptual rivalry in the human brain, *Science,* 280, 1930–1934, 1998.
84. Lumer, E.D. and Rees, G., Covariation of activity in visual and prefrontal cortex associated with subjective visual perception, *Proc. Natl. Acad. Sci. U.S.A.*, 96, 1669–1673, 1999.
85. Brown, R.J. and Norcia, A.M., A method for investigating binocular rivalry in real-time with the steady-state VEP, *Vision Res.*, 37, 2401–2408, 1997.
86. Tononi, G., Srinivasan, R., Russell, D.P., and Edelman, G.M., Investigating neural correlates of conscious perception by frequency-tagged neuromagnetic responses, *Proc. Natl. Acad. Sci. U.S.A.*, 95, 3198–3203, 1998.
87. Kaernbach, C., Schroger, E., Jacobsen, T., and Roeber, U., Effects of consciousness on human brain waves following binocular rivalry, *NeuroReport*, 10, 713–716, 1999.
88. Thompson, K.G. and Schall, J.D., The detection of visual signals by macaque frontal eye field during masking, *Nat. Neurosci.*, 2, 283–288, 1999.
89. Thompson, K.G. and Schall, J.D., Antecedents and correlates of visual detection and awareness in macaque prefrontal cortex, *Vision Res.*, 40, 1523–1538, 2000.
90. Crick, F. and Koch, C., Are we aware of neural activity in primary visual cortex, *Nature*, 375, 121–123, 1995.
91. Breitmeyer, B.G., *Visual Masking: An Integrative Approach*, Oxford University Press, New York, 1984.
92. Sclar, G., Maunsell, J.H.R., and Lennie, P., Coding of image contrast in central visual pathways of the macaque monkey, *Vision Res.*, 30, 1–10, 1990.
93. Cheng, K., Hasegawa, T., Saleem, K.S., and Tanaka, K., Comparison of neuronal selectivity for stimulus speed, length, and contrast in the prestriate visual cortical areas V4 and MT of the macaque monkey, *J. Neurophysiol.*, 71, 2269–2280, 1994.
94. Britten, K.H., Newsome, W.T., Shadlen, M.N., Celebrini, S., and Movshon, J.A., A relationship between behavioral choice and the visual responses of neurons in macaque MT, *Vis. Neurosci.*, 13, 87–100, 1996.
95. Libet, B., Pearl, D.K., Morledge, D.E., Gleason, C.A., Hosobuchi, Y., and Barbaro, N.M., Control of the transition from sensory detection to sensory awareness in man by the duration of a thalamic stimulus. The cerebral "time-on" factor, *Brain,* 114, 1731–1757, 1991.
96. Koch, C. and Crick, F., Some further ideas regarding the neuronal basis of awareness, in *Large-Scale Neuronal Theories of the Brain*, Koch, C. and Davis, J.L., Eds., MIT Press, Cambridge, MA, 1994, 93–109.
97. Sahraie, A., Weiskrantz, L., Barbur, J.L., Simmons, A., Williams, S.C., and Brammer, M.J., Pattern of neuronal activity associated with conscious and unconscious processing of visual signals, *Proc. Natl. Acad. Sci. U.S.A.*, 94, 9406–9411, 1997.
98. Posner, M.I., *Chronometric Explorations of Mind*, L. Erlbaum Associates, Hillsdale, NJ, 1978.
99. Luce, R.D., *Response Times: Their Role in Inferring Elementary Mental Organization*, Oxford University Press, Oxford, U.K., 1986.
100. Meyer, D.E., Osman, A.M., Irwin, D.E., and Yantis, S., Modern mental chronometry, *Biol. Psychol.*, 26, 3–67, 1988.
101. Donders, F.C., On the speed of mental processes, in *Attention and Performance II*, translated by W.G. Koster, 1969, North-Holland, Amsterdam, 1868, 412–431.

102. Sternberg, S., Separate modifiability, mental modules, and the use of pure and composite measures to reveal them, *Acta Psychol.*, 106, 147–246, 2001.
103. Sato, T., Murthy, A., Thompson, K.G., and Schall, J.D., Search efficiency but not response interference affects visual selection in frontal eye field, *Neuron*, 30, 583–591, 2001.
104. Duncan, J. and Humphreys, G.W., Visual search and stimulus similarity, *Psychol. Rev.*, 96, 433–458, 1989.
105. Bacon, W.F. and Egeth, H.E., Overriding stimulus-driven attentional capture, *Percept. Psychophys.*, 55, 485–496, 1994.
106. Bichot, N.P., Schall, J.D., and Thompson, K.G., Visual feature selectivity in frontal eye fields induced by experience in mature macaques, *Nature,* 381, 697–699, 1996.
107. Nodine, C.F., Kundel, H.L., Lauver, S.C., and Toto, L.C., Nature of expertise in searching mammograms for breast masses, *Acad. Radiol.*, 3, 1000–1006, 1996.
108. Chapman, P. and Underwood, G., Visual search of driving situations: danger and experience, *Perception,* 27, 951–964, 1998.
109. Nodine, C.F. and Krupinski, E.A., Perceptual skill, radiology expertise, and visual test performance with NINA and WALDO, *Acad. Radiol.,* 5, 603–612, 1998.
110. Yarbus, A.L., *Eye Movements and Vision*, Plenum Press, New York, 1967.
111. Sereno, A.B. and Maunsell, J.H., Shape selectivity in primate lateral intraparietal cortex, *Nature*, 395, 500–503, 1998.
112. Toth, L.J. and Assad, J.A., Dynamic coding of behaviourally relevant stimuli in parietal cortex, *Nature,* 415, 165–168, 2002.
113. Horwitz, G.D. and Newsome, W.T., Target selection for saccadic eye movements: direction-selective visual responses in the superior colliculus, *J. Neurophysiol.,* 86, 2527–2542, 2001.
114. Maljkovic, V. and Nakayama, K., Priming of pop-out: I. Role of features, *Mem. Cognit.,* 22, 657–672, 1994.
115. Maljkovic, V. and Nakayama, K., Priming of pop-put: II. The role of position, *Percept. Psychophys.,* 58, 977–991, 1996.
116. Bichot, N.P. and Schall, J.D., Saccade target selection in macaque during feature and conjunction visual search, *Vis. Neurosci.,* 16, 81–89, 1999.
117. McPeek, R.M., Maljkovic, V., and Nakayama, D., Saccades require focal attention and are facilitated by a short-term memory system, *Vision Res.,* 39, 1555–1566, 1999.
118. Bichot, N.P. and Schall, J.D., Priming in macaque frontal cortex during popout visual search: feature-based facilitation and location-based inhibition of return, *J. Neurosci.*, 22, 4675–8546, 2002.
119. Treisman, A. and Sato, S., Conjunction search revisited, *J. Exp. Psychol. Hum. Percept. Perform.,* 16, 459–478, 1990.
120. Bichot, N.P. and Schall, J.D., Effects of similarity and history on neural mechanisms of visual selection, *Nat. Neurosci.,* 2, 549–554, 1999.
121. Kim, M.S. and Cave, K.R., Spatial attention in search for features and feature conjunctions, *Psych. Sci.*, 6, 376–380, 1999.
122. Findlay, J.M., Saccade target selection during visual search, *Vision Res.,* 37, 617–631, 1995.
123. Motter, B.C. and Belky, E.J., The guidance of eye movements during active visual search, *Vision Res.*, 38, 1805–1815, 1998.
124. Sato, T., Watanabe, K., Thompson, K.G., and Schall, J.D., Effect of singleton similarity on stimulus representation in frontal eye field, *Exp. Brain Res.*, in press.
125. Chun, M.M. and Jiang, Y., Contextual cuing implicit learning and memory of visual context guides spatial attention, *Cogn. Psychol.*, 36, 28–71, 1998.

126. Horowitz, T.S. and Wolfe, J.M., Visual search has no memory, *Nature*, 39, 4575–4577, 1998.
127. Woodman, G.F., Vogel, E.K., and Luck, S.J., Visual search remains efficient when visual working memory is full, *Psychol. Sci.*, 12, 219–24, 2001.
128. Pashler, H., Target-distractor discriminability in visual search, *Percept. Psychophys.*, 41, 285–292, 1987.
129. Bundesen, C., A theory of visual attention, *Psychol. Rev.*, 97, 523–547, 1990.
130. Desimone, R. and Duncan, J., Neural mechanism of selective visual attention, *Annu. Rev. Neurosci.*, 18, 193–222, 1995.
131. Miller, E.K. and Cohen JD. An integrative theory of prefrontal cortex function, *Annu. Rev. Neurosci.*, 24, 167–202, 2001.
132. Stanton, G.B., Bruce, C.J., and Goldberg, M.E., Topography of projections to the frontal lobe from the macaque frontal eye fields, *J. Comp. Neurol.*, 330, 286–301, 1993.
133. White, I.M. and Wise, S.P., Rule-dependent neuronal activity in the prefrontal cortex, *Exp. Brain Res.*, 126, 315–335, 1999.
134. Wallis, J.D., Anderson, K.C., and Miller, E.K., Single neurons in prefrontal cortex encode abstract rules, *Nature,* 411, 953–956, 2001.
135. Rainer, G., Asaad, W.F., and Miller, E.K., Selective representation of relevant information by neurons in the primate prefrontal cortex, *Nature,* 393, 577–579, 1998.
136. Ferrera, V.P., Cohen, J.K., and Lee, B.B., Activity of prefrontal neurons during location and color delayed matching tasks, *NeuroReport,* 10, 1315–1322, 1999.
137. Hasegawa, R.P., Matsumoto, M., and Mikami, A., Search target selection in monkey prefrontal cortex, *J. Neurophysiol.,* 84, 1692–1696, 2000.
138. Everling, S., Tinsley, C.J., Gaffan, D., and Duncan, J., Filtering of neural signals by focused attention in the monkey prefrontal cortex, *Nat. Neurosci.*, 5, 671–676, 2002.
139. Iba, M. and Sawaguchi, T., Neuronal activity representing visuospatial mnemonic processes associated with target selection in the monkey dorsolateral prefrontal cortex, *Neurosci. Res.*, 43, 9–22, 2002.

10 Specializations of the Human Visual System: The Monkey Model Meets Human Reality

Todd M. Preuss

CONTENTS

0-8493-1243-4/04/$0.00+$1.50

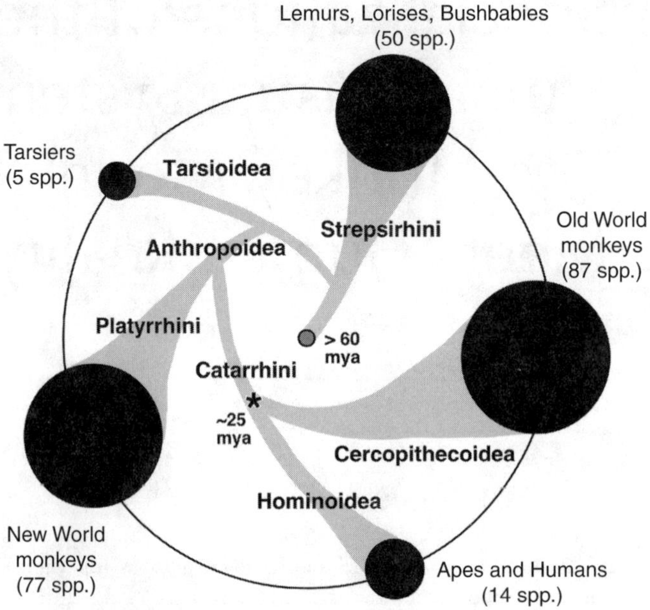

FIGURE 10.1 The phylogenetic relationships of extant primate groups. This is a top-down or end-on view, looking back in time. The area of each circle is proportional to the number of species in the group. Branching patterns and species numbers follow Fleagle.[1] The time of divergence of the hominoid and cercopithecoid branches is about 25 million years ago (mya).[2]

10.1 THE MONKEY MODEL

Much of what we think we know about the human nervous system comes from the study of nonhuman species. The emphasis on nonhuman studies is understandable, given the limitations on experimental studies in humans. But what animals should neuroscientists focus on? For many neuroscientists, the animals of choice have been the macaque monkeys, that is, the primates of the genus *Macaca*. The predominance of macaques is such that they are often referred to as "the monkey" or "the primate," notwithstanding the fact that there are over 200 different primate species (Figure 10.1). The role of macaques as proxies for humans is especially apparent in visual science: when textbooks turn from human visual function to its neural basis, they switch from human studies to macaque studies so casually that one could easily lose sight of the fact that a switch has taken place.

Why macaques? Macaques are not, as it happens, the animals most closely related to humans. That distinction falls to the apes, and in particular to the African great apes (chimpanzees and gorilla). Humans are, in fact, a recent offshoot of the African great ape group, from which they diverged 5 to 10 million years ago.[1] The next most closely related species is the orangutan, followed by the gibbons. Collectively, the ape–human group constitutes the superfamily Hominoidea. The animals most closely related to the hominoids are the Old World monkeys, the Cercopithecoidea, which is the group to which the macaque monkeys belong. The last common ancestor of the cercopithecoids and hominoids, and thus of macaques and humans, lived about 25 million years ago.[1,2] Therefore, about 50 million years of independent evolutionary history separate humans from macaques.

Although macaques are not particularly closely related to humans, Old World monkeys are in fact the animals most closely related to humans in which it is feasible to carry out invasive experimental research. Humans and apes are off limits for ethical reasons; even if it were possible to do invasive research in apes, the expense and difficulty of maintaining captive ape populations would militate against their widespread use. So we are left with Old World monkeys. Because all Old World monkeys are equally closely related to humans, the preference for macaques is a matter of practicality: macaques are hardy and breed well in captivity, and thus have been widely available for use. In addition, rhesus macaques, in particular, are renowned for the enthusiasm with which they engage in behavioral tasks.

Another factor in the transformation of macaque monkeys into the "monkey model" has been the appreciation that there are many similarities in visual function between macaques and humans. In vision research, workers have made a point of emphasizing these similarities (e.g., References 3 through 6), and as the list of functional similarities has grown, so has the apparent justification for using macaques as proxies for studying the structure of the human visual system. By the 1980s and 1990s, the accepted view was that there are no differences of consequence between the visual system organization of humans and macaques. Kiorpes et al.,[7] for example, noted that "the conventional view is that macaque and human vision are quantitatively identical." This was perhaps an overstatement: most of the workers cited above seem to have been prepared to accept that there are quantitative differences, if rather small ones. To take another example, Teller and Boothe[8] remarked that "[i]ndividual human beings and individual macaques, tested behaviourally under identical conditions, often give data sets so closely overlapping that the individual differences among members of one species are larger than the average differences between species." They concluded that the main difference between humans and macaques is simply the more rapid ontogeny of macaques. Not everyone was prepared to accept that differences in visual function are necessarily small, however. Harwerth and Smith,[9] who conducted extensive psychophysical comparisons, concluded that there are in fact "important, nontrivial differences between humans and macaques." Nevertheless, they suggested that these psychophysical differences could in many cases be explained by differences in the optics of the eye, such as its size (humans have bigger eyes than macaques) or in the amount of macular pigment through which light must pass to reach the photoreceptors. The likelihood of nontrivial differences in the neural architecture of the visual system was thus discounted. Even when evidence of neurobiological differences between macaques and humans has been uncovered, authors have tended to highlight the similarities and downplay the disparities (e.g., References 10 through 12). Few would doubt that "humans see what monkeys see."[13]

It is difficult to avoid the conclusion that visual neuroscientists have vigorously pursued an agenda that systematically emphasizes human–macaque similarities and soft-pedals the differences.[14] My intention here is to turn the tables, offering a list of human–macaque differences in visual system structure and function. I do not claim this to be a complete account, by any means. Nevertheless, it is more than enough to make the point that there are indeed nontrivial differences between macaques and humans in both the structure and function of the visual system, and that the answer to the question, "Do humans see what monkeys see?" is probably "in some ways yes, in other ways no."

10.2 STRUCTURAL VARIATIONS: RETINA AND SUBCORTICAL STRUCTURES

10.2.1 Cones and Photoreceptors

Most mammals possess two types of cones, with photopigments that respond maximally to short (S) and medium-long (M/L) wavelength light, respectively. In primate evolution, the gene coding the M/L photopigment underwent a duplication, with subsequent divergence of wavelength sensitivity, such that the catarrhine primates (i.e., the Old World monkeys and hominoids; see Figure 10.1) possess distinct S, M, and L photopigments and cones.[15,16] In most Old World monkeys that have been examined, the ratio of L:M cones is approximately 1:1,[17] and studies of chimpanzees suggest a ratio of about 1.3:1.[18] In humans, by contrast, the L:M ratio averages approximately 2:1. This conclusion is supported by evidence obtained with a variety of techniques.[14,17] Humans also exhibit considerable individual variation in color sensitivity. Some of the variation in human color vision reflects polymorphisms in the alleles coding for L and M photopigments,[19] but some is probably attributable to the remarkably wide range of individual variation in the L:M cone ratio.[20,21] By contrast, genetic and spectral sensitivity studies of chimpanzees and macaques suggest that these taxa are nearly monomorphic at the L and M opsin loci and exhibit little variation in L:M ratio.[17,18,22,23]

Another striking difference between humans and other primates that have been examined pertains to the distribution of S cones: Immunocytochemical studies in humans indicate that the central 100 μm of the fovea is nearly devoid of S cones.[24,25] By contrast, studies of macaques using immunocytochemistry and *in situ* hybridization indicate either that the S-cone-free zone is much smaller than in humans[26–28] or that S cones are actually present in the central-most fovea, albeit at low density compared to adjacent regions.[25] The New World monkeys that have been examined (*Callithrix*, *Cebus*, *Saimiri*) show no diminution of S-cone density in the foveola.[26,29] This suggests that the human condition represents a recent evolutionary modification. Unfortunately, the distribution of S cones in apes has not yet been examined, so it is not possible to state whether the presence of a 100-μm-wide S-cone-free zone in the central fovea is a human specialization or a hominoid (ape–human) specialization.

There may be additional differences in cone distribution between humans and macaques. In both, cone density is greater in the nasal than in the temporal peripheral retina, but the nasotemporal asymmetry is reported to be stronger in macaques than in humans.[30] In addition, the region of peak cone density in the fovea appears to be somewhat smaller in humans than in macaques.[31]

There appear to be substantial human–macaque differences in the distribution of rods across the retina. Humans possess a higher average ratio of rods to cones

(20:1) than do macaques (15:1).[31] This human rod advantage extends over much of the retina, exclusive of the fovea, but is particularly prominent in the rod ring, the annulus of peak rod density located at about 20° eccentricity (just eccentric to the optic disk).[31] Within the rod ring, the highest density is found dorsally (the "dorsal rod peak" described by Wikler et al.[32]); this peak appears to be much more prominent in humans than in macaques (see Figure 2 of Curcio and Hendrickson[31]). By contrast to the extrafoveal retina, where humans have the advantage in rod density, rods are said to be present but sparse within the fovea of macaques,[33] but absent within the fovea of humans.[34]

10.2.2 Retinal Ganglion Cells

Visual information originating in the photoreceptors is conveyed (after integration and transformation by the intrinsic processing network of the retina) to the brain by projections of the retinal ganglion cells (RGCs). Although a large number of morphological varieties of retinal ganglion cells have been distinguished by anatomists, current accounts of retinal organization in primates typically distinguish six main cell classes. These consist of the class of large M cells (also known as parasol cells and Y-like cells) and the smaller P cells (also known as midget cells and X-like cells) and K (W-like) cells. Each of these cell types comes in two varieties, an ON-center type and an OFF-center type, yielding a total of six cell classes. The terms M, P, and K reflect the fact that these cells project to the magnocellular, parvocellular, and koniocellular layers, respectively, of the lateral geniculate nucleus (LGN).

These types of RGCs have been recognized in all major primate groups, and homologues are likely present in other mammalian groups. Nevertheless, differences in the absolute and relative sizes of specific RGC classes of macaques and humans have been noted by Dacey and Petersen.[35] At any given eccentricity in the retina, the dendritic-field expanse of human M cells is larger than that of their macaque homologues, although the P cells of humans and macaques have very similar dimensions (Figure 10.2). The disparity between M and P cell sizes is most pronounced in the central 20° of the visual field. In humans, furthermore, Dacey and Petersen report there is an asymmetry between ON-center and OFF-center ganglion cells, which affects both the M and P cell classes: ON cells have dendritic fields that are 30 to 50% larger than those of OFF cells.[35] No comparable asymmetry has been noted in macaques,[35] although it is unclear whether the matter has been explicitly studied in these animals. No data exist concerning the relative sizes of M and P cells and ON and OFF cells in apes.

The retinal ganglion cells convey visual information to numerous structures in the diencephalon and brain stem. Little is known about differences between primate groups in the organization of most of these structures, with the exception of the suprachiasmatic nucleus (SCN) and LGN.

10.2.3 Suprachiasmatic Nucleus

Retinal projections to the SCN of the hypothalamus function to entrain the cyclicity of the circadian system. The anatomical and neurochemical organization of the SCN

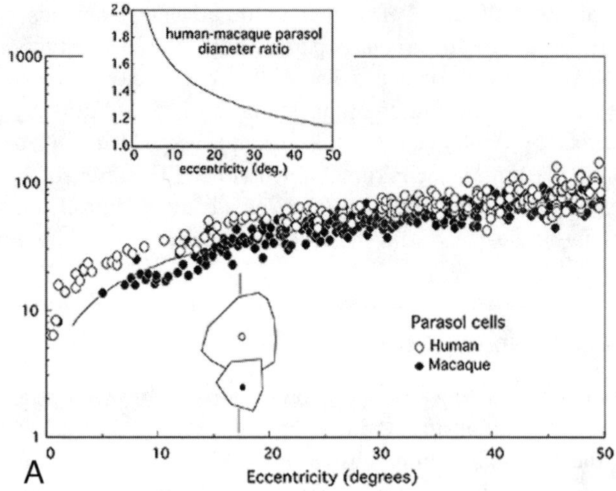

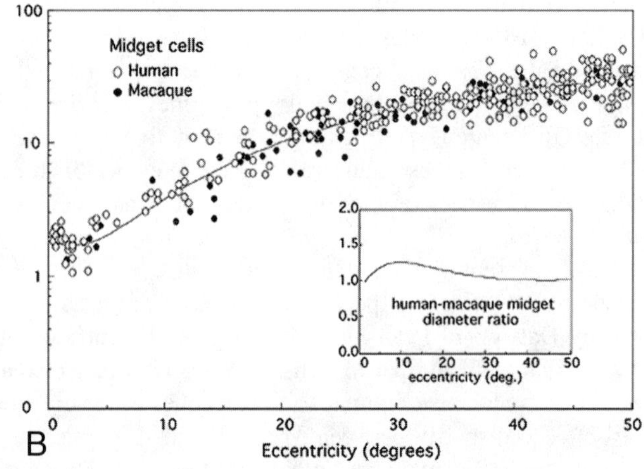

FIGURE 10.2 Dendritic field diameters of parasol (A) and midget (B) retinal ganglion cells in macaques and humans, according to Dacey and Petersen.[35] The sizes of both cell classes increase with eccentricity. At a given eccentricity, however, parasol (magno or M) cells are larger in humans than in macaques, whereas midget (parvo or P) cells are similar in size. (From Dacey, D.M. and Petersen, M.R., *Proc. Natl. Acad. Sci. U.S.A.*, 89, 9666–9670, 1992. With permission.)

is thought to be quite conservative evolutionarily.[36] Nevertheless, the neurochemical organization of the human SCN differs in some respects from that of macaques and from other mammals.[36] Human SCN possesses numerous neurons that are immunoreactive for neurotensin (NT+) and a separate population immunoreactive for neuropeptide Y (NPY+); by contrast, NT+ and NPY+ neurons are either very sparse or absent in macaques and rats. The functional consequences of these quite striking differences in SCN neurochemistry are entirely unknown. Moore's suggestion[36] that there were major modifications of the circadian system during human evolution has generated little interest among neuroscientists.

10.2.4 Lateral Geniculate Nucleus

The human LGN shares many structural characteristics with other primates, particularly other catarrhine primates. The traditional description of the LGN as "six layered" is somewhat misleading. Primate LGNs have at least three types of projection cells, each concentrated in specific layers, the magnocellular (M), parvocellular (P), and koniocellular (K) layers. There is a pair of M layers, one layer for each eye, which lie close to the optic tract. There is also a pair of P layers (one for each eye), situated superiorly to the M layers, although typically in humans and other catarrhines, and in some platyrrhines (New World monkeys) as well, each of the P layers splits and interdigitates with adjacent layers, creating the appearance that there are four P layers, rather than two.[37] Together with the two M layers, this accounting yields the conventional "six-layered" LGN. The splitting and interdigitation of the P layers presumably reflects evolutionary expansion of the P system.

The K cells have not usually figured in the enumeration of LGN layers, as they do not form very well defined layers in Nissl-stained sections from catarrhine primates. The K cells occupy the territory between the innermost M layer and the optic tract (termed the S layer) and are also situated between the main M and P layers in the so-called interlaminar zones. The K cells have only recently been recognized as a class of neurons distinct from the M and P cells; it is not even clear at present that they comprise a unitary cell class.[38,39] Owing to the difficulty of identifying K cells with routine Nissl stains, neurochemical markers have been sought that stain K cells selectively, and it is now apparent that these cells stain very well for the calcium-binding protein, calbindin (CB),[38,39] as well as for the alpha subunit of the calcium-calmodulin-dependent protein kinase 2 (CAMK2A).[40] There is, nevertheless, some variation in the expression of CB in the LGN of different species. In macaques, CB appears to be expressed almost exclusively by cells in the S layer and interlaminar zones.[38,41] In humans, by contrast, CB is also expressed by P and M cells, although the densest staining is in the interlaminar cell population.[42–44]

Humans exhibit additional distinctive features of LGN organization. Humans evidently exhibit an unusual degree of individual variability in LGN lamination (see Reference 45 and references cited therein). According to Hickey and Guillery,[45] many humans (although not all) show additional evaginations of the parvocellular layers so that certain portions of the nucleus appear to be eight layered, rather than six layered. The extent of the eight-layered segment is highly variable across individuals, although it is usually located near the presumed location of the optic disk representation. Hickey and Guillery also noted variations in the magnocellular layers, especially layer I. The magnocellular layers exhibit gaps and there are islands of M cells displaced dorsally into the parvocellular or interlaminar zones; these gaps and islands are highly variable in size and location across individuals.

The functional implications of these variations are unclear, although it is noteworthy that the caudal portion of the LGN, where foveal vision is represented, appears to be less variable in its morphology across individuals than other portions of the nucleus. On this basis, Hickey and Guillery suggested that selection pressure for good peripheral vision was relaxed during human evolution.

10.3 STRUCTURAL VARIATIONS: VISUAL CORTEX

10.3.1 Evolutionary Changes in the Laminar Organization of the Primary Visual Area

Unlike many other brain structures, we have a great deal of comparative information about the primary visual area (also known as *V1*, *striate cortex,* and *area 17*), which has been examined in each of the major primate groups.[46] In primates, area V1 is the principal target of projections from the LGN, and thus is the main (although not exclusive) conduit by which visual information reaches the cortex. The M and P geniculate layers appear to project to V1 almost exclusively, although the K projections also reach extrastriate areas. This contrasts to the situation in most nonprimate groups that have been examined, in which V2 is also a major recipient of LGN projections (reviewed by Preuss[47]).

Area V1 has been especially well studied in macaques, of course, and it is useful to give a brief overview of macaque organization. One of the distinctive features of the primary visual area in macaques, and other anthropoid primates, is its extremely well stratified appearance in sections prepared with conventional histological stains, such as Nissl and cytochrome oxidase (CO; Figure 10.3A and 10.3B). CO is especially informative, because the territories that stain most strongly for CO are known to be the targets of the projections from the LGN, i.e., the geniculostriate projections. The geniculate projections reach layer 1, the "blobs" in layer 3, layers 4A, 4C, and 6. The broad, dense band of CO staining that corresponds to layer 4C is the target of the largest projections from the P and M geniculate layers, with the P projections concentrated in the deep sublayer, 4Cβ, and the M projections in the superficial sublayer, 4Cα. Layer 4B stains more lightly for CO, and it receives no direct geniculate input, although it receives input from the M pathway via local connections with layer 4Cα. Immediately superficial to 4B, layer 4A contains a thin, irregular band of dense CO staining; this marks the location of an additional projection of the P geniculate layers.

The relationship between the M and P pathways to layers 4A and 4B has an additional element of complexity (Figure 10.3). As noted above, layer 4A receives direct input from the P geniculate layers. However, the thin sheet of P-related tissue is punctuated by bundles of apical dendrites arising from clusters or cones of pyramidal cells rooted in layer 4B; these bundles correspond to gaps in the P geniculate projections (and in the dense CO staining), which gives layer 4A a "honeycomb" appearance.[48–51] The pyramidal cells exhibit a mixture of M-like and P-like properties,[52] presumably because their cell bodies and proximal dendrites, located in layer 4B, receive M inputs, whereas their apical dendrites receive synapses from P inputs

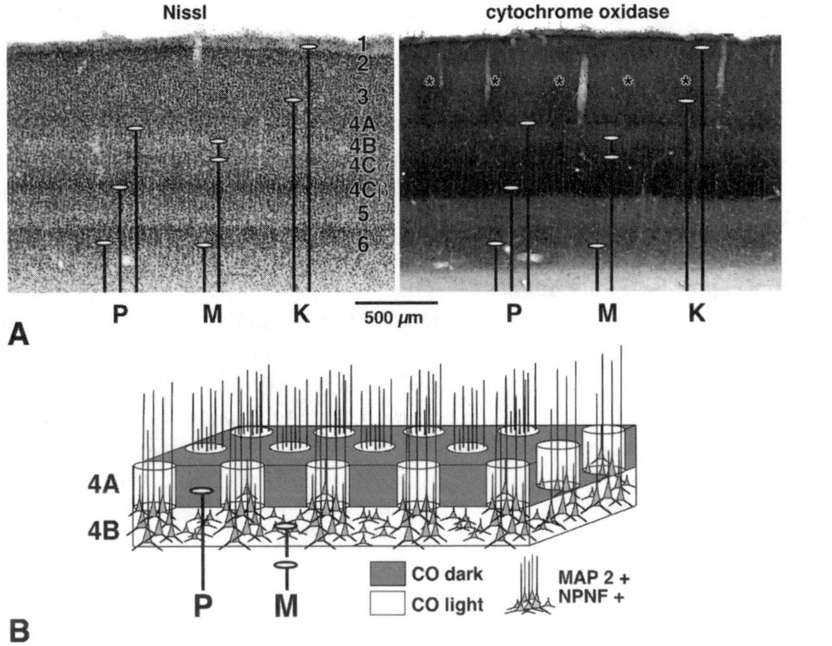

FIGURE 10.3 The laminar and compartmental organization of area V1 in macaque monkeys (*Macaca*). (A) Adjacent sections stained for Nissl substance and CO, with the locations of terminations from the P, M, and K layers of the LGN indicated schematically. CO staining is strongest where LGN projections terminate. The P layers project most strongly to layers 4Cβ and 4A, the M layers to layer 4Cα. There are no direct projections to layer 4B, but 4B receives M-related input from 4Cα. The K layers project directly to the CO-dense blobs (*) in layer 3. (B) Schematic representation of the honeycomb-like organization of layer 4A in macaques.[53] Layer 4A receives direct inputs from the P layers of the LGN, but is also punctuated by a regular array of pyramidal cell bodies and dendrites extending upward from the M-related layer 4B. (B from Preuss, T.M. et al., *Proc. Natl. Acad. Sci. U.S.A.*, 96, 11601–11606, 1999. With permission.)

in layer 4A. In this manner, layer 4A comprises two kinds of tissue compartments, one related to the P pathway, the other related to the M and P pathways.

How general is the macaque pattern of layer 4A organization? The evidence suggests that, in fact, a macaque-like organization is present in a wide variety of Old World and New World monkeys.[53,54] There is at least one conspicuous exception, however, the New World owl monkey (*Aotus* spp.). *Aotus*, the sole nocturnally active anthropoid primate, lacks a direct P geniculate projection to layer 4A, and also lacks the CO-dense band that marks this projection in other monkeys.[55,56] Nevertheless, given the widespread distribution of a macaque-like 4A organization among New World and Old World monkeys, it is quite likely that this organization was present in the last common ancestor of the anthropoid primates, as well as in the last common ancestor of the Old World monkeys and hominoids (Figure 10.4). Given that the ancestors of apes and humans probably possessed a macaque-like layer 4A, it would be a good bet that living apes and humans do as well. Such is not the case, however: layer 4A was extensively modified in ape and human evolution (Figure 10.4).

Horton and Hedley-Whyte[57] were the first to demonstrate that humans lack a band of dense staining in layer 4A, and this has been confirmed by numerous subsequent reports (reviewed by References 53 and 54). Recently, Preuss et al.[53] reported that chimpanzees, and the single orangutan they examined, also lacked a CO-dense band in layer 4A, indicating that at least some of the great apes exhibit a human-like condition. This suggests strongly that the CO-dense band was lost at some point in hominoid evolution, prior to the divergence of the lineages leading to modern chimpanzees and humans (Figure 10.4).

What is the significance of the modification of laminar histology? By analogy to the situation in owl monkeys, Horton and Hedley-Whyte[57] suggested that the lack of a CO-dense band in human layer 4A might indicate that humans lack a P geniculate projection to layer 4A. It is very difficult to evaluate this possibility, given the difficulty of studying connectivity in humans or, for that matter, in apes. There is a least one study of a human with a lesion of the geniculostriate radiation, which demonstrated projections to layer 4C, but not to 4A.[58] A transneuronal tracing study in a single chimpanzee also showed geniculate projections to 4C, but not 4A.[59] There is some additional histochemical evidence consistent with the hypothesis that hominoid primates lack a direct P geniculate projection to 4A. Muñoz and DeFelipe[11] noted that humans exhibit lower expression of $GABA_BR1a,b$ receptor in layer 4A than do macaques; otherwise, this receptor is expressed strongly in layers 4C and 6, which receive direct geniculate inputs. Similarly, Blümcke et al.[10] noted dense staining for parvalbumin, a calcium-binding protein expressed in the terminals of M and P geniculate cells, within the neuropil of layers 4C in both humans and macaques, and in layer 4A only in macaques.

These results are all consistent (if less than definitive) with the lack of a P geniculate projection to layer 4A in humans. Even if there are differences in the geniculate inputs to area V1 in apes and humans compared to macaques, one can imagine alternatives to the hypothesis that the differences in CO staining imply the loss of projections to layer 4A. For example, Preuss and Coleman[54] have noted that whereas in macaques the upper border of layer 4C is sharply demarcated from layer 4B by a marked reduction in CO staining, in apes and humans the intensity of CO staining decreases gradually from layer 4C through 4B and 4A. They suggest, therefore, that geniculate projections may be distributed in corresponding fashion in apes and humans. All that is really certain at the present time, however, is that there are histochemical differences between hominoids and other primates; these likely have connectional and functional correlates, but we can say little about those. The development of improved techniques for studying the connectivity of ape and human brains is critically important for understanding the structure and function of the human visual system.[60]

10.3.2 Unique Compartmental Organization of Layer 4A in Human Area V1

Human layer 4A is not simply macaque 4A without a CO-dense band; humans display additional evolutionary modifications. Preuss et al.[53] examined the

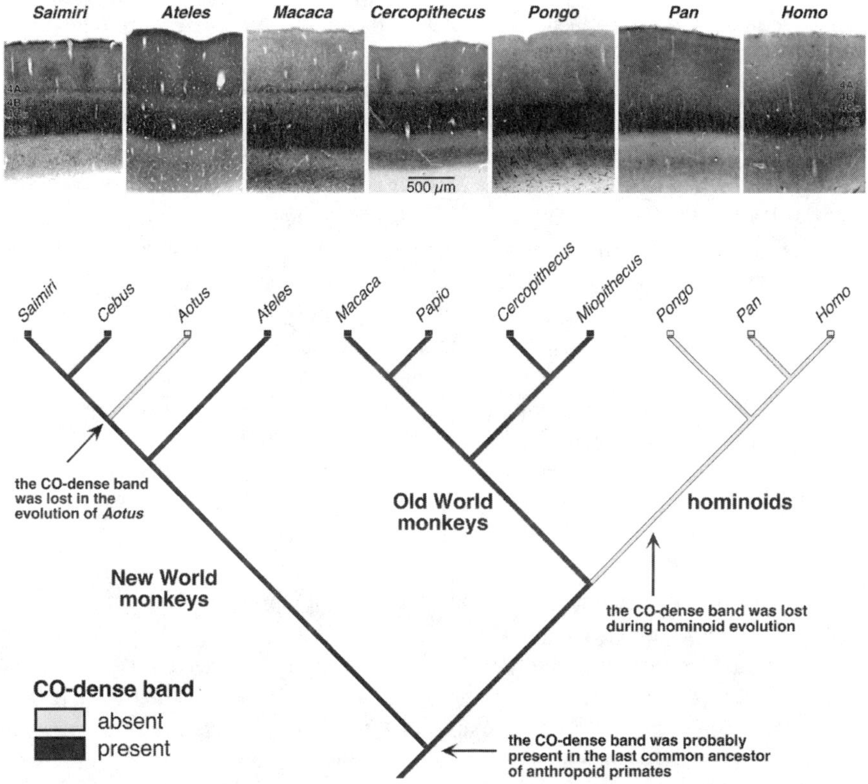

FIGURE 10.4 Laminar organization of CO staining in New World monkeys, Old World monkeys, and hominoids. In most New World and Old World monkeys that have been examined, there is a thin, dense band of CO staining corresponding to layer 4A, which is separated by the relatively lightly stained layer 4B from the thick, dense band of staining in layer 4C. In hominoids that have been examined, however, which include orangutan (*Pongo*), chimpanzees (*Pan*), and humans (*Homo*), the pattern is different: CO staining falls off gradually above layer 4C, so there is no discrete band of dense staining in layer 4A. Given this distribution of character states, and the known genealogical relationships among hominoids, Old World monkeys, and New World monkeys, it is likely that the last common ancestor of the New World and Old World monkeys possessed a distinct CO-dense band in layer 4A, similar to modern *Macaca*, and that the band was lost in the hominoid lineage at some point prior to the divergence of *Pongo*, *Pan*, and *Homo*. The absence of a CO-dense band in layer 4A of owl monkeys (*Aotus*) probably represents an independent evolutionary loss. The top panel represents material from the author's laboratory. The genera indicated in the branching diagram below represent all taxa that have been described in the literature. For details and literature citations, see Preuss et al.[53] and Preuss and Coleman.[54]

organization of layer 4A in humans, apes, and several Old World and New World monkey species using antibodies that recognize nonphosphorylated neurofilament protein (NPNF) and microtubule-associated protein 2 (MAP2), both of which are cytoskeletal proteins. In macaques and the other monkeys, these antibodies label layer 4B very densely, and reveal the clusters of apical dendrites that ascend from 4B to penetrate 4A, contributing to the honeycomb appearance of 4A.[49,51] The overall level of staining in layer 4A was very light compared to 4B, however, especially with the NPNF antibody.[53] The pattern of NPNF and MAP2 staining in humans, however, proved to be dramatically different from that in the monkeys (Figure 10.5B). Rather than vertical stalks of dendrites, humans display a geometrically complex, three-dimensional meshwork of tissue bands comprising stained cell bodies and neuropil. These bands comprise clusters of large neurons and processes that extend upward into layer 4A from 4B, curving laterally as they ascend so that they commonly cap or even encapsulate territories of unstained tissue in layer 4A.

Humans, then, have a system of dual tissue compartments in layer 4A, but the neural architecture of these compartments is very different from that in macaques. The dark-staining tissue bands in humans have some features in common with the pyramidal-cell bundles present in macaques, but these bands contain many more stained cell bodies and much more prominent dendrites in humans than in macaques (resulting in a much higher level of NPNF immunoreactivity), and the curving, woven appearance of the tissue bands in humans contrasts to the rigid, vertical dendritic bundles of macaques. Macaque 4A is like a computer punch card; human 4A is more like Swiss cheese.

One might suppose that the reorganization of human 4A seen with NPNF and MAP2 has a fairly direct causal relationship to the loss of the CO-dense band. This cannot be the case, however, because apes lack the CO-dense band, but do not show the meshwork of NPNF+ and MAP2+ tissue bands present in humans. The loss of the CO-dense band and the acquisition of the meshwork were independent evolutionary events; the meshwork is unique to the human lineage.

In their initial report, Preuss et al.[53] suggested that the human 4A tissue bands might be specifically related to the M pathway. This is because NPNF and MAP2 are enriched in neural elements related to the M pathway in macaques and other primates. There is, however, one neural marker in particular that has been strongly identified with the M pathway, namely, the monoclonal antibody Cat-301.[61–63] Given this, Preuss and Coleman[54] argued that humans might show a pattern of Cat-301 staining in human layer 4A similar to the pattern revealed with NPNF and MAP2. This proved to be the case (Figure 10.6A). Moreover, double-staining experiments indicated that Cat-301 and NPNF immunoreactivity were localized to the same tissue compartments in human layer 4A. By contrast to humans, layer 4A of apes and monkeys stained lightly and quite uniformly for Cat-301. Recent studies with another ligand thought to be enriched in the M pathway, the *Wisteria floribunda* agglutinin lectin,[64] also revealed a compartmental pattern of staining in human 4A, but not in macaques, and was found to co-localize with NPNF.[65] Results of these studies are all consistent with the hypothesis that the meshlike tissue bands of human layer 4A are allied with the M pathway, although direct confirmation by demonstration of connectivity is still lacking.

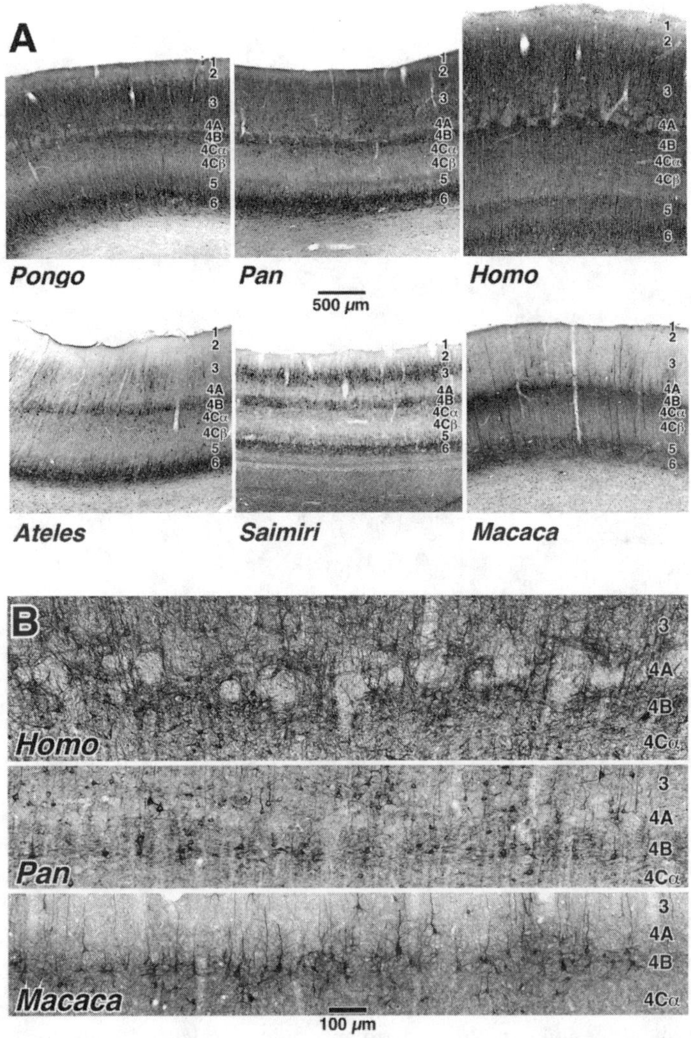

FIGURE 10.5 Immunostaining for NPNF in area V1 of humans and nonhuman primates. At low magnification (A), interspecific differences in NPNF distribution are readily apparent, as are similarities. The most eye-catching difference is the higher-level of NPNF staining, particularly in layer 3, in the hominoid primates (*Pongo*, *Pan*, *Homo*) compared to the Old World (*Macaca*) and New World (*Ateles*, *Saimiri*) monkeys. This specialization of hominoid primates is not limited to area V1, however, but extends across the cortical mantle.[53] Note, too, the more irregular appearance of layer 4A in humans compared to the apes and monkeys. As can be seen at higher magnification (B), human layer 4A possesses a network of tissue bands, consisting of NPNF+ cell bodies and dendrites that arch over and cap unstained patches of tissue. These tissue bands are apparent neither in chimpanzees (*Pan*), despite the high overall level of NPNF staining in area V1 of these animals, nor in macaques (*Macaca*). In chimpanzees and macaques, NPNF staining layer 4A is dominated by vertically oriented apical dendrites, with a few small pyramidal cell bodies, consistent with the honeycomb model.

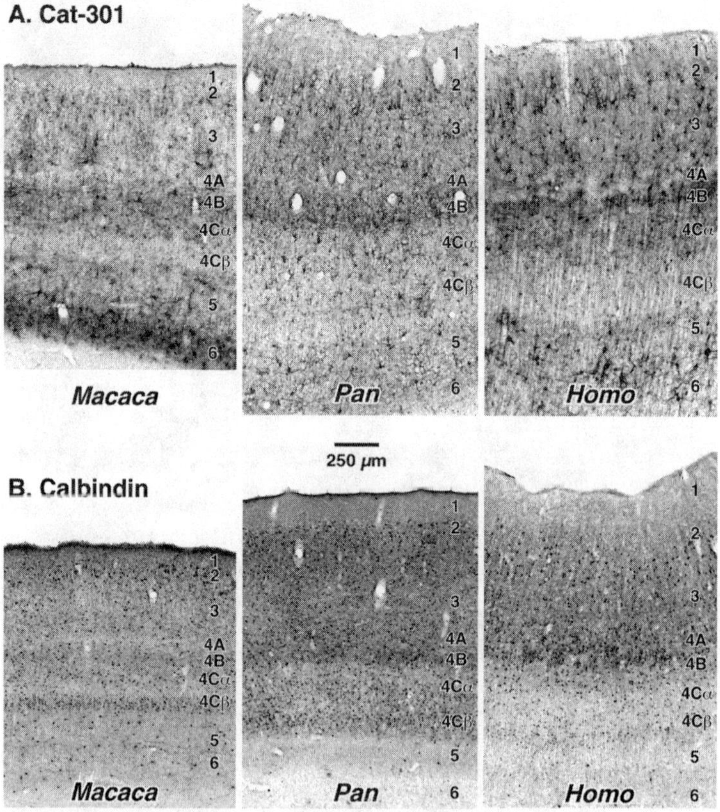

FIGURE 10.6 Cat-301 (A) and calbindin (B) immunoreactivity in area V1 of humans (*Homo*), chimpanzees (*Pan*), and macaques (*Macaca*). In both preparations, layer 4A exhibits an irregular, patchy appearance in humans that is not seen in other primates. Double-staining studies in humans[54] indicate that Cat-301 staining is coincident with NPNF staining, while the CB+ patches occupy the territories surrounded by the NPNF+ and Cat-301+ tissue bands.

What about the interstitial compartments in human layer 4A, the territories that are surrounded by the NPNF- and Cat-301-rich tissue bands? Hendry and Carder[66] described a dense, irregular pattern of staining for calbindin (CB) in layer 4A of humans, whereas they found layer 4A of macaques and squirrel monkeys to lack strong CB immunoreactivity. The CB-dense patches in human layer 4A include numerous, small immunoreactive neurons. Preuss and Coleman[54] confirmed these results and also showed that chimpanzees display an intermediate condition, with a uniform band of moderately dense CB staining in layer 4A (Figure 10.6B). Furthermore, in double-staining studies, Preuss and Coleman[54] demonstrated that dense patches of CB+ tissue fill in the gaps demarcated by the NPNF- and Cat-301-rich tissue bands. Thus, humans have a unique pattern of compartmentation within layer 4A, consisting of patches of small cells that stain strongly for CB embedded within tissue bands that are immunoreactive for NPNF, MAP2, and Cat-301.

It is unclear at present how the CB-rich patches are related to the M, P, and K pathways. It is tempting to speculate that the human layer 4A patches are related to the K pathway, because CB is enriched in the K cells of the LGN. However, the relationship of CB to the K pathway is not maintained in the cortex: the CO-dense blobs, for example, which receive direct inputs from the K cells of the LGN, stain more lightly for CB than does the tissue surrounding the blobs.

10.3.3 Extrastriate Cortex

Old World and New World monkeys possess a large set of extrastriate visual areas, areas that are distinguished by their position within the cortical mantle and by their connections, histology, retinotopy, and functional properties.[67–69] Based on functional imaging and histological studies, humans appear to possess a suite of extrastriate areas similar to the complement possessed by Old World and New World monkeys. As for specific areas, human homologues of monkey areas V1, V2, V3 (including V3d and VP), V3A (DM), and MT (V5), at a minimum, seem well established.[68–72] Moreover, as in monkeys, two broad functional territories can be distinguished within human extrastriate cortex, a dorsal stream and a ventral stream.[73] One important question that has not been resolved is whether humans, with their greatly enlarged brains, possess visual areas in addition to those possessed by Old World and New World monkeys. One might expect this to be the case, but there is at present no strong evidence to support this view. There may, however, be important differences in the relative sizes of specific visual areas in humans and macaques: several of the extrastriate areas of humans appear to be larger relative to V1 than are their macaque counterparts;[69–71] area V3A stands out in particular.[71] It also seems likely there are significant human–macaque differences in the internal organization of particular visual areas, comparable to those documented for V1. To date, there have been few detailed comparative anatomical studies of individual extrastriate areas, although Tootell and Taylor[74] report that, in humans, the CO-dense stripes in area V2 are much less distinct than in Old World and New World monkeys. Despite the current paucity of evidence for differences in the anatomical organization of the extrastriate cortex, there is evidence for functional differences, which will be considered in a subsequent section.

10.4 FUNCTIONAL VARIATIONS

10.4.1 Spectral Sensitivity

Given that humans have a relative preponderance of L cones compared to other primates examined, as well as an S-cone-free zone in the central fovea, one might expect human spectral sensitivity be somewhat "red shifted" compared to other primates. Perhaps the most comprehensive comparison of spectral sensitivity in humans and macaques was undertaken by Harwerth and Smith,[9] who examined 12 humans and 12 rhesus macaques. Harwerth and Smith found that humans showed greatest sensitivity to light in the red part of the spectrum and lowest sensitivity in the blue; macaques showed the opposite result. Their data also suggest that the interactions between L and M cones differs between humans and macaques. Despite these data, Harwerth and Smith did not conclude that these spectral sensitivity differences between humans and macaques reflect differences in the distribution or characteristics of photoreceptors, but rather that the functional differences are attributable to differences in pre-retinal light loss, possibly due to higher concentrations of macular pigment in humans than in macaques (macular pigment, which is concentrated in the fovea, attenuates short-wavelength light). Such pigment differences have been disputed, however,[17] and in any case, the observed differences in spectral sensitivity between humans and macaques are congruent with the observed differences in the distribution of L, M, and S photoreceptors.

Dobkins et al.[14] have considered possible functional and adaptive consequences of phyletic differences in L:M ratios. They suggest that the nearly equal proportions of L and M cones in macaques should result in enhanced chromatic contrast sensitivity, because of the opponent interactions of the two cones. Among humans, in fact, individuals with L:M ratios close to 1 exhibit greater chromatic contrast sensitivity than those with higher ratios.[75] Dobkins et al.[14] suggest that enhanced chromatic sensitivity could be beneficial for locating red fruit against a background of green foliage.[15] Why, then, do humans have L:M ratios averaging greater than 1, given the implied diminution of chromatic contrast sensitivity? There is some evidence that higher L:M ratios yield enhanced luminance contrast sensitivity, possibly as a result of additive input of L and M cones to magnocellular retinal ganglion cells.[75]

The relative insensitivity of humans at short wavelengths compared to macaques is plausibly related to the S-cone-free zone in the human fovea. One possible functional advantage of having an S-cone-free foveola is enhanced acuity: S cones, with peak wavelength sensitivity of about 430 nm,[15] make up only about 5% of the cone population, and eyes typically are focused for longer wavelengths,[17,18] closer to the peak sensitivities of the M and L cones, which are about 530 and 561 nm, respectively, in humans. When so focused, short-wavelength light is blurred.[76] Eliminating S cones from the foveola would eliminate this blur.[77] One might suggest, then, that compared to macaques, which have a photoreceptor array well suited for chromatic discrimination, humans have modified their photoreceptor array to enhance spatial acuity and perhaps also luminance contrast, sacrificing chromatic contrast sensitivity in the process.

10.4.2 Acuity

How, then, do humans and macaques fare on tests of acuity? There are a number of different measures of acuity, and indeed several different types of acuity; here we are most concerned with spatial resolution. One measure of this is the spatial frequency at which observers exhibit maximum luminance contrast sensitivity. Under photopic conditions, the human peak is located at a higher spatial frequency than the macaque peak: 4.0 cycles/degree (c/deg) for humans vs. 2.5 c/deg for macaques, according to Harwerth and Smith.[9] Harwerth and Smith consider this difference to be a consequence of the larger size of the human eye. Others have reported a smaller difference,[78] but with the advantage still going to humans. Thus, with low-contrast gratings, humans can distinguish finer gratings than can macaques. Humans also show advantages over macaques in discriminating line orientation and grating orientation,[79] and in measures of vernier acuity[7] (the ability to discriminate offset in the alignment of parallel line segments).

10.4.3 Luminance Contrast Sensitivity

The larger size of M ganglion cells in the human retina compared to macaques suggests that the human M pathway should show greater sensitivity for low luminance-contrast and high temporal-frequency stimuli, although at the cost of lower spatial resolution.[35] The high L:M ratio of humans might also contribute to enhanced luminance contrast, as discussed above. These expectations can be examined through comparative studies of temporal-frequency and spatial-frequency modulation sensitivity. Humans and macaques both yield qualitatively similar, inverted U-shaped functions, but humans show greater sensitivity over a broad range of temporal frequencies[5,9] and spatial frequencies[4,9] at high (photopic) and medium (mesopic) luminance levels. To state this plainly, humans can detect subtler differences between shades of gray presented in either static, sinusoidal gratings (in the spatial frequency condition) or in flickering stimuli (in the temporal frequency condition). The differences, as Harwerth and Smith[9] noted, are not small. They examined the largest samples of humans and macaques (12 individuals each) of the studies cited above, and indicated that humans exhibit about 60% greater contrast sensitivity in the spatial and temporal contrast conditions, after aligning the somewhat offset peaks of the human and macaque curves. (Differences remain even when the peaks are not aligned, however.) Human–macaque differences of comparable magnitude have been reported by De Valois et al.[4] and Merigan[5] under photopic conditions. The greater sensitivity of humans for low-contrast and high-temporal frequency gratings is consistent with the anatomical differences in the M pathway. It is not clear, however, how differences in the M pathway can explain the greater contrast sensitivity of humans compared to macaques at high spatial frequencies.

10.4.4 Perceptive Field Organization

The relatively large physical dimensions of M ganglion cells in humans might also have consequences for the receptive field dimensions of M ganglion cells. These have not been studied directly in humans, but there are psychophysical techniques that yield *perceptive fields*, which have properties that parallel to a considerable extent the antagonistic center–surround receptive field organization of retinal ganglion cells.[80] Perceptive fields have been studied using the so-called Westheimer paradigm, in which a test spot of light is presented in the center of a somewhat larger circle of light, and the size of the background light and the luminance of the test spot are varied. Thresholds for the test spot are lowest in the absence of a background stimulus; as the background stimulus is enlarged, the threshold to the spot increases, then peaks, then falls to a plateau. As discussed by Oehler[81] and Spillmann et al.,[80] the change in threshold as a function of background size has been interpreted in terms of the interplay of excitatory and inhibitory interactions within receptive fields. As the background grows within the excitatory center, the threshold for the smaller test spot increases, peaking when the center is filled; this increase is considered to reflect the elevation of neuronal adapting level due to spatial summation within the excitatory center. As the background light circle increases still further, thresholds for the test spot begin to decrease and ultimately plateau. This decrease is interpreted as reflecting the stimulation of the antagonistic surround and the lowering of the adapting level due to lateral inhibition. The background diameter that yields the peak of the threshold function is taken as the size of the perceptive field center, and the background diameter that marks the beginning of the plateau is taken as the total perceptive field size (center plus surround). Oehler[81] and Spillmann et al.[80] summarize evidence suggesting that perceptive fields measured in this way reflect the properties of M retinal ganglion cells.

Oehler[81] used this technique to map perceptive centers and surrounds in two rhesus macaques and two humans. In all subjects, not surprisingly, perceptive field size varied in an orderly fashion, increasing with eccentricity. At a given eccentricity, humans and macaques had similarly sized perceptive field centers. Perceptive field surrounds, however, were consistently larger in humans than in macaques. It should be noted, too, that the presence of a background stimulus confined to the perceptive field raised spot-detection thresholds more in humans than in macaques, prompting Oehler to suggest that "[t]ogether with differences in total perceptive field size, the differences in peak size might rather indicate a different distribution of excitation and inhibition over the perceptive field."

These findings could have important implications for human–macaque differences in visual function, particularly with respect to the spatial organization of attentional enhancement and inhibition. In macaques and humans, control of spatial attention (both of the bottom-up, stimulus-driven variety as well as the top-down variety), is dominated by the M pathway and dorsal stream,[82–86] and in macaques, attentional modulation can be detected at early stages of visual processing, including the M layers of the LGN and M-related layers 4Cα and 4B of area V1.[87] Yet, attentional modulation in area V1 has proved to be much easier to demonstrate in humans than in macaques, prompting Heeger[88] to suggest there may be real phyletic differences in the strength of attentional effects. Against this, however, Heeger notes that the differences may be an artifact of the different techniques used to demonstrate attentional modulation in humans (functional imaging) and macaques (microelectrode recording, typically). Now that functional imaging techniques have been successfully modified for use with nonhuman primates (see below), it should be possible to compare attentional modulation in humans and macaques more directly.

10.4.5 Global vs. Local Precedence

Consider a hierarchically organized visual stimulus, such as a large square or a large circle, formed by the arrangement of smaller squares or circles. Such stimuli can be consistent at global and local levels (e.g., large squares composed of small squares), or they can be compounds of different stimuli at different scales (e.g., large squares composed of small circles, large circles composed of small squares). Humans have been shown to process the global configuration of compound hierarchical stimuli faster and more accurately than the local elements, a phenomenon termed *global precedence.*[89] Moreover, even when trained to attend to local features, humans find it difficult to ignore the global configuration, so that a mismatch between the global and local features interferes with processing of the local features. The reverse is not true, however. This has been termed *global-to-local interference*.

Fagot and Deruelle[90,91] compared the performance of humans and baboons (Old World monkeys very closely related to macaques) on a variety of matching-to-sample and visual-search tasks using hierarchical stimuli. Humans showed the global advantages reported in other studies, whereas baboons exhibited strong local precedence and did not show global-to-local interference. Fagot and Tomonaga[92] carried out similar experiments comparing humans and chimpanzees, and found that chimpanzees also showed local advantage under most stimulus conditions. This advantage could be eliminated by increasing the density of the local elements, but under no conditions did chimpanzees show the global advantage exhibited by humans on the same tasks (but see Hopkins[93]).

The mechanisms that underlie global precedence in humans are unclear, but it is instructive to consider possible alternative accounts of the differences between macaques and humans. One possibility is that the global-to-local differences reflect differences in the spatial distribution of attention, as discussed above. Perhaps humans are able to focus attention on a larger region of the visual field than can macaques. Alternatively, the macaque–human behavioral differences could result from evolutionary changes in functional lateralization. In humans, there is considerable evidence that processing of global features depends on the right hemisphere and local features on the left hemisphere.[94,95] Another possibility is that humans and macaques differ in the spatial scale of binding of local features to form contours; note that there is evidence that cortical areas of the dorsal stream (including area V3A) as well as the ventral stream play a role in contour perception.[96]

10.4.6 Motion Processing

Macaque–human differences in the anatomy and function of elements of the M pathway suggest that there are differences in motion processing, to which the M pathway makes an important contribution. At present, there is little behavioral evidence bearing on this possibility, but differences in the functional characteristics of cortical areas involved in motion processing have been demonstrated. Tootell and colleagues[97] investigated motion sensitivity in dorsal extrastriate regions in humans using fMRI, and found that V3A responds strongly to motion, but V3 does not. They noted that this is the opposite of the situation reported in macaques, based on electrode recording studies. Given that the original claim of a macaque–human difference was based on data obtained with different techniques, it is significant that a recent fMRI study of awake macaques also found that V3 is strongly responsive to motion, but V3A is not, in these animals,[98] consistent with the evidence from microelectrode recording studies.

The application of fMRI to nonhuman primates opens important new avenues of comparative research.[99] Vanduffel et al.[100] recently carried out a relatively direct comparison of humans and macaques, examining patterns of cortical activation while subjects viewed a display of randomly connected lines that appear (to human eyes) to be rotating in depth (i.e., a three-dimensional structure-from-motion display). Vanduffel and colleagues found that, although macaques and humans activated a number of areas in common under this viewing condition, humans exhibited activation of several additional territories, located in the parietal cortex. Vanduffel et al. suggest the parietal areas activated in humans do not exist in macaques; alternatively, these areas might be present in both macaques and humans, but have undergone functional modification in human evolution, much like area V3A.

One possible interpretation of these results, implicit in the authors' discussion, is that macaques, unlike humans, simply do not derive information about three-dimensional structure from the visual display used in these experiments. Vanduffel et al. also suggest that the differences between humans and macaques may be related to the importance of tools in human evolution, citing evidence that viewing tool motion activates intraparietal cortex in humans.[101] It is noteworthy that some apraxic humans, who likely have parietal damage, are unable to use visual information about the spatial form of novel tools to infer the appropriate mechanical uses of those tools.[102] Furthermore, Povinelli[103] provides evidence of human–chimpanzee differences in comprehending the mechanical properties of tools: although chimpanzees readily learn how to use tools to accomplish specific tasks, their performance on transfer tasks gives little indication that they understand how the specific physical characteristics of objects relate to their usefulness as tools. These results suggest that the evolution of human tool manufacture and use was accompanied by the modification of neural systems that include parietal cortex and that support rapid assessment of the physical characteristics of objects (including their three-dimensional spatial characteristics) and the range of possible spatial and mechanical interactions of objects. Physical characteristics relevant to the manipulation of objects may also be represented in these systems.[104,105]

10.4.7 Language

It is likely that evolution modified visual cortex with respect to another distinctively human activity: language. Linguists have long appreciated that our auditory perception of language can be influenced by viewing a speaker's mouth movements. The McGurk effect[106] is a dramatic illustration of this: when subjects listen to a speaker saying one syllable ("ba") while simultaneously viewing video of the speaker making the mouth movements appropriate for another syllable ("ga"), subjects "hear" a third syllable ("da" or "tha"). It is not surprising, then, that merely viewing language-related mouth movements, without the voice-over, can activate auditory cortical areas.[107,108] There is also some evidence that when normal hearing subjects merely listen to speech, the occipitotemporal visual cortex is activated,[109,110] a cortical region that probably includes area MT. Interestingly, this region is activated even more strongly when subjects simultaneously read lips and hear speech than when they only read lips or only hear speech.[110] The activation of visual cortex by auditory stimuli, and of auditory cortex by visual stimuli, is surprising in view of the belief that the nonhuman homologues of these areas are strictly unimodal.[111] Evidently, the evolution of language involved the emergence of a new processing system that can recruit relatively low order sensory areas into transmodal functional networks.

Given the important role of vision in language, it is reasonable to suspect that evolution has also "tuned" the human visual system to be sensitive to types of movement-related stimuli associated with language.

10.5 THE HUMAN DIFFERENCE

This chapter highlights the existence of numerous differences between humans and macaques in the structure and function of the visual system. Many of these differences probably represent evolutionary specializations of humans or of the ape–human (hominoid) group, although for those features of organization for which we have information only about macaques and humans, the possibility that the differences reflect specializations of macaques or of Old World monkeys cannot be ruled out.

Human–macaque differences are found at all levels of visual processing, from the earliest, retinal levels through higher-order association cortex. This is important, because it is tempting to think that what distinguishes the human brain from that of other primates is mainly the addition of new, higher-order systems. If this were true, one could still view macaques as a faithful representation of early stages of cortical processing, at least. Clearly, however, such a view is overly simplistic: the human visual system is riddled with evolutionary modifications.

Although there is too little comparative information available at present to draw strong or general conclusions about the pattern of functional changes in human evolution, it is interesting that many of the ways in which humans differ from macaques appear to involve the M pathway and/or the dorsal stream. These changes are potentially very significant, not only for understanding the evolutionary history and normal ontology of the human visual system, but also for understanding the nature of common pathological changes. There is now very good evidence that pathological changes in the M pathway and in dorsal stream visual areas are present in at least some types of dyslexia,[112–114] an extremely common human affliction, and there is evidence that the M pathway and/or dorsal stream are more strongly affected than the P pathway or ventral stream in other developmental disorders as well, including autism,[115] schizophrenia and schizotypy,[116,117] and Williams syndrome.[118,119] Spencer et al.[115] suggest that the dorsal stream is particularly vulnerable to developmental insult. I emphasize that the likelihood that the M pathway and dorsal stream were modified in human evolution does *not* preclude the possibility that the P pathway and ventral stream were modified as well.

Another peculiarity of the human visual system that deserves comment is the apparent high degree of "normal" individual variation, best documented in the anatomy of the retina and LGN, and in color perception. Conventionally, these variations have been explained as a consequence of relaxed selection pressures on vision in the recent evolutionary history of humans. The variety of polymorphic structures and the extent of polymorphism, however, suggest that we need to consider other evolutionary accounts, including the possibility that new selection pressures have emerged in recent human evolutionary history, and that in some instances conflicting selection pressures have resulted in balanced polymorphisms.

10.6 THE MONKEY MODEL OR COMPARATIVE BIOLOGY?

How can we reconcile the existence of such a variety of differences between humans and macaques with the litany of similarities long cited by visual neuroscientists to justify the use of macaques as model humans? In fact, there is no contradiction: there *are* many similarities between humans and macaques — and there are also many differences. It is not wrong to point out that macaques and humans are similar in many respects; what is wrong is to suppose that the existence of numerous similarities rules out the existence of interesting or important differences. Logically, no list of similarities can be long enough to preclude the existence of differences, unless such a list is completely comprehensive, exhausting all the myriad features of organization that characterize complex biological systems. The problem with the monkey model, with its emphasis on similarities and soft-pedaling of differences, is that it invites us to fall into this logical trap, and encourages us to ignore one of the most obvious and central facts of biology: species differ. Little wonder that the macaque–human differences reviewed here — many of which have been in the literature for years — are so little known. It is no small irony that the institutional success of the macaque model, advocated with the best of intentions to further our understanding of human vision, has left the human visual system decidedly out of focus.

This is not a judgment on macaques: the plain fact is that *any* nonhuman species will show a variety of similarities and differences when compared to humans. This is the inevitable result of the evolutionary process, which is a process of change — if animals did not become different through time, we would not need a theory of evolution. Of course, closely related species will generally resemble each other in more ways than will distantly related species. But even closely related species will differ in some respects.

Should we then give up research on macaques and other nonhuman primates? Of course not: humans do, after all, share many features of visual system organization with other primates, and we have learned a great deal that is directly relevant to humans by studying nonhuman primates. Unquestionably, our understanding of the human nervous system would be greatly impoverished without the contribution from animal research. It is not animal research that is problematic, but rather the setting up of particular animals as general-purpose *models* (in the sense of *model organisms*).[120]

Casting animals in the role of model organisms is so much a part of modern biology it can be difficult to imagine any other way to conceptualize animal research. Nevertheless, there are alternative, comparative approaches that offer many advantages over the model-organism paradigm.[120] These approaches take as their starting point the understanding that related species will share some characteristics, by virtue of inheritance from their common ancestor, but will also exhibit differences as a consequence of the independent evolutionary history of each species. These methods encourage investigation of multiple species, so as to characterize patterns of evolutionary conservation and evolutionary change more rigorously (Figure 10.4 provides an example of such an analysis). In addition, comparative approaches offer a unique perspective on the relationship between structure and function in the nervous system, by identifying cross-species correlations between changes in the nervous system and changes in behavior. This chapter identifies a number of correlated differences in the M pathway and dorsal stream of macaques and humans that can provide starting points for understanding the neural bases of functional differences between macaques and humans (bearing in mind that any given correlation could be coincidental). Perhaps best of all, from the perspective of human visual science, by making room for the diversity of animals and adopting principled ways of dealing with their similarities and differences, comparative approaches make room for the human animal, a species that is similar to its relatives in some ways, but unique in others.

ACKNOWLEDGMENTS

The author gratefully acknowledges the continuing support of the James S. McDonnell Foundation, and the assistance provided by the New Iberia Research Center and the Northwestern University Alzheimer's Disease Center in obtaining brain tissue. The author would also like to thank Dr. Dennis Dacey for permission to use the graphs shown in Figure 10.2 and for providing the original electronic file.

REFERENCES

1. Fleagle, J.G., *Primate Adaptation and Evolution,* 2nd ed., Academic Press, San Diego, 1999.
2. Stewart, C.B. and Disotell, T.R., Primate evolution — in and out of Africa, *Curr. Biol.,* 8, R582–R588, 1998.
3. De Valois, R.L., Morgan, H.C., Polson, M.C., Mead, W.R., and Hull, E.M., Psychophysical studies of monkey vision — I. Macaque luminosity and color vision tests, *Vision Res.,* 14, 53–67, 1974.
4. De Valois, R.L., Morgan, H.C., and Snodderly, D.M., Psychophysical studies of monkey vision — III. Spatial luminance contrast sensitivity tests of macaque and human observers, *Vision Res.,* 14, 75–81, 1974.
5. Merigan, W.H., Temporal modulation sensitivity of macaque monkeys, *Vision Res.,* 20, 953–959, 1980.
6. Weiskrantz, L., Behavioural analysis of the monkey's visual nervous system, *Proc. R. Soc. Lond. B Biol. Sci.,* 182, 427–455, 1972.
7. Kiorpes, L., Kiper, D.C., and Movshon, J.A., Contrast sensitivity and vernier acuity in amblyopic monkeys, *Vision Res.,* 33, 2301–11, 1993.
8. Teller, D.Y. and Boothe, R., Development of vision in infant primates, *Trans. Ophthalmol. Soc. U.K.,* 99, 333–337, 1979.
9. Harwerth, R.S. and Smith, E.L., Rhesus monkey as a model for normal vision of humans, *Am. J. Optom. Physiol. Opt.,* 62, 633–641, 1985.
10. Blümcke, I., Hof, P.R., Morrison, J.H., and Celio, M.R., Distribution of parvalbumin immunoreactivity in the visual cortex of Old World monkeys and humans, *J. Comp. Neurol.,* 301, 417–432, 1990.
11. Muñoz, A., DeFelipe, J., and Jones, E.G., Patterns of GABA(B)R1a,b receptor gene expression in monkey and human visual cortex, *Cereb. Cortex,* 11, 104–113, 2001.
12. Yoshioka, T. and Hendry, S.H.C., Compartmental organization of layer IVA in human primary visual cortex, *J. Comp. Neurol.,* 359, 213–220, 1995.
13. Kaas, J.H., Do humans see what monkeys see? *Trends Neurosci.,* 15, 1–3, 1992.

14. Dobkins, K.R., Thiele, A., and Albright, T.D., Comparison of red-green equiluminance points in humans and macaques: evidence for different L:M cone ratios between species, *J. Opt. Soc. Am. A Opt. Image Sci. Vis.,* 17, 545–556, 2000.
15. Mollon, J.D., "Tho' she kneel'd in that place where they grew ..." The uses and origins of primate colour vision, *J. Exp. Biol.,* 146, 21–38, 1989.
16. Heesy, C.P. and Ross, C.F., Evolution of activity patterns and chromatic vision in primates: morphometrics, genetics and cladistics, *J. Hum. Evol.,* 40, 111–149, 2001.
17. Jacobs, G.H. and Deegan, J.F., II, Spectral sensitivity of macaque monkeys measured with ERG flicker photometry, *Vis. Neurosci.,* 14, 921–928, 1997.
18. Jacobs, G.H., Deegan, J.F., II, and Moran, J.L., ERG measurements of the spectral sensitivity of common chimpanzee (*Pan troglodytes*), *Vision Res.,* 36, 2587–2594, 1996.
19. Neitz, M. and Neitz, J., Molecular genetics of color vision and color vision defects, *Arch. Ophthalmol.,* 118, 691–700, 2000.
20. Carroll, J., McMahon, C., Neitz, M., and Neitz, J., Flicker-photometric electroretinogram estimates of L:M cone photoreceptor ratio in men with photopigment spectra derived from genetics, *J. Opt. Soc. Am. A Opt. Image Sci. Vis.,* 17, 499–509, 2000.
21. Brainard, D.H., Roorda, A., Yamauchi, Y., Calderone, J.B., Metha, A., Neitz, M., Neitz, J., Williams, D.R., and Jacobs, G.H., Functional consequences of the relative numbers of L and M cones, *J. Opt. Soc. Am. A Opt. Image Sci. Vis.,* 17, 607–614, 2000.
22. Jacobs, G.H., Primate photopigments and primate color vision, *Proc. Natl. Acad. Sci. U.S.A.,* 93, 577–581, 1996.
23. Jacobs, G.H. and Deegan, J.F., II, Uniformity of colour vision in Old World monkeys, *Proc. R. Soc. Lond. B Biol. Sci.,* 266, 2023–2028, 1999.
24. Curcio, C.A., Allen, K.A., Sloan, K.R., Lerea, C.L., Hurley, J.B., Klock, I.B., and Milam, A.H., Distribution and morphology of human cone photoreceptors stained with anti-blue opsin, *J. Comp. Neurol.,* 312, 610–624, 1991.
25. Bumsted, K. and Hendrickson, A., Distribution and development of short-wavelength cones differ between *Macaca* monkey and human fovea, *J. Comp. Neurol.,* 403, 502–516, 1999.
26. Martin, P.R. and Grunert, U., Analysis of the short wavelength-sensitive ("blue") cone mosaic in the primate retina: comparison of New World and Old World monkeys, *J. Comp. Neurol.,* 406, 1–14, 1999.
27. Röhlich, P., Ahnelt, P.K., Dawson, W.W., and Szel, A., Presence of immunoreactive blue cones in the fetal monkey fovea, *Exp. Eye Res.,* 58, 249–251, 1994.
28. Wikler, K.C. and Rakic, P., Distribution of photoreceptor subtypes in the retina of diurnal and nocturnal primates, *J. Neurosci.,* 10, 3390–3401, 1990.
29. Martin, P.R., Grunert, U., Chan, T.L., and Bumsted, K., Spatial order in short-wavelength-sensitive cone photoreceptors: a comparative study of the primate retina, *J. Opt. Soc. Am. A Opt. Image Sci. Vis.,* 17, 557–567, 2000.
30. Curcio, C.A., Sloan, K.R., Jr., Packer, O., Hendrickson, A.E., and Kalina, R.E., Distribution of cones in human and monkey retina: individual variability and radial asymmetry, *Science,* 236, 579–582, 1987.
31. Curcio, C.A. and Hendrickson, A., Organization and development of the primate photoreceptor mosaic, in *Progress in Retinal Research,* Vol. 11, Osborne, N.N. and Chader, G.J., Eds., Pergamon Press, Oxford, U.K., 1991, 89–120.
32. Wikler, K.C., Williams, R.W., and Rakic, P., Photoreceptor mosaic: number and distribution of rods and cones in the rhesus monkey retina, *J. Comp. Neurol.,* 297, 499–508, 1990.

33. Packer, O., Hendrickson, A.E., and Curcio, C.A., Photoreceptor topography of the retina in the adult pigtail macaque (*Macaca nemestrina*), *J. Comp. Neurol.,* 288, 165–183, 1989.
34. Curcio, C.A. and Allen, K.A., Topography of ganglion cells in human retina, *J. Comp. Neurol.,* 300, 5–25, 1990.
35. Dacey, D.M. and Petersen, M.R., Dendritic field size and morphology of midget and parasol ganglion cells of the human retina, *Proc. Natl. Acad. Sci. U.S.A.,* 89, 9666–9670, 1992.
36. Moore, R.Y., The fourth C.U. Ariens Kappers lecture. The organization of the human circadian timing system, *Prog. Brain Res.,* 93, 99–115; discussion 115–117, 1992.
37. Kaas, J.H., Huerta, M.F., Weber, J.T., and Harting, J.K., Patterns of retinal terminations and laminar organization of the lateral geniculate nucleus of primates, *J. Comp. Neurol.,* 182, 517–553, 1978.
38. Hendry, S.H.C. and Reid, R.C., The koniocellular pathway in primate vision, *Annu. Rev. Neurosci.,* 2000, 127–153, 2000.
39. Casagrande, V.A., A third parallel visual pathway to primate area V1, *Trends Neurosci.,* 17, 305–310, 1994.
40. Hendry, S.H. and Yoshioka, T., A neurochemically distinct third channel in the macaque dorsal lateral geniculate nucleus, *Science,* 264, 575–577, 1994.
41. Jones, E. and Hendry, S., Differential calcium binding protein immunoreactivity distinguishes classes of relay neurons in monkey thalamic nuclei, *Eur. J. Neurosci.,* 1, 222–246, 1989.
42. Leuba, G. and Saini, K., Colocalization of parvalbumin, calretinin and calbindin D-28k in human cortical and subcortical visual structures, *J. Chem. Neuroanat.,* 13, 41–52, 1997.
43. Leuba, G. and Saini, K., Calcium-binding proteins immunoreactivity in the human subcortical and cortical visual structures, *Vis. Neurosci.,* 13, 997–1009, 1996.
44. Münkle, M.C., Waldvogel, H.J., and Faull, R.L., The distribution of calbindin, calretinin and parvalbumin immunoreactivity in the human thalamus, *J. Chem. Neuroanat.,* 19, 155–173, 2000.
45. Hickey, T.L. and Guillery, R.W., Variability of laminar patterns in human lateral geniculate nucleus, *J. Comp. Neurol.,* 183, 221–246, 1979.
46. Casagrande, V.A. and Kaas, J.H., The afferent, intrinsic, and efferent connections of primary visual cortex in primates, in *Cerebral Cortex,* Vol. 10: *Primary Visual Cortex in Primates*, Peters, A. and Rockland, K., Eds., Plenum Press, New York, 1994, 201–259.
47. Preuss, T.M., Evolutionary specializations of primate brain systems, in *Primate Origins and Adaptations*, Ravoso, M.J. and Dagosto, M., Eds., Kluwer Academic/Plenum Press, New York, in press.
48. Hendrickson, A.E., Wilson, J.R., and Ogren, M.P., The neuroanatomical organization of pathways between the dorsal lateral geniculate nucleus and visual cortex in Old World and New World primates, *J. Comp. Neurol.,* 182, 123–136, 1978.
49. Peters, A. and Sethares, C., Layer IVA of rhesus monkey primary visual cortex, *Cereb. Cortex,* 1, 445–462, 1991.
50. Peters, A. and Sethares, C., Organization of pyramidal neurons in area 17 of monkey visual cortex, *J. Comp. Neurol.,* 306, 1–23, 1991.
51. Hendry, S.H. and Bhandari, M.A., Neuronal organization and plasticity in adult monkey visual cortex: immunoreactivity for microtubule-associated protein 2, *Vis. Neurosci.,* 9, 445–459, 1992.

52. Yabuta, N.H., Sawatari, A., and Callaway, E.M., Two functional channels from primary visual cortex to dorsal visual cortical areas, *Science,* 292, 297–300, 2001.
53. Preuss, T.M., Qi, H., and Kaas, J.H., Distinctive compartmental organization of human primary visual cortex, *Proc. Natl. Acad. Sci. U.S.A.,* 96, 11601–11606, 1999.
54. Preuss, T.M. and Coleman, G.Q., Human-specific organization of primary visual cortex: alternating compartments of dense Cat-301 and calbindin immunoreactivity in layer 4A, *Cereb. Cortex,* 12, 671–691, 2002.
55. Horton, J.C., Cytochrome oxidase patches: a new cytoarchitectonic feature of monkey visual cortex, *Philos. Trans. R. Soc. Lond. B,* 304, 199–253, 1984.
56. Kaas, J.H., Lin, C.S., and Casagrande, V.A., The relay of ipsilateral and contralateral retinal input from the lateral geniculate nucleus to striate cortex in the owl monkey: a transneuronal transport study, *Brain Res.,* 106, 371–378, 1976.
57. Horton, J.C. and Hedley-Whyte, E.T., Mapping of cytochrome oxidase patches and ocular dominance columns in human visual cortex, *Philos. Trans. R. Soc. Lond. B,* 304, 255–272, 1984.
58. Miklossy, J., Thalamocortical connections and rostral visual areas in man, in *The Functional Organization of Human Visual Cortex*, Gulyas, B., Ohoson, D., and Rowland, P.E., Eds., Pergamon Press, Oxford, U.K., 1992, 123–136.
59. Tigges, J. and Tigges, M., Ocular dominance columns in the striate cortex of chimpanzee (*Pan troglodytes*), *Brain Res.,* 166, 386–390, 1979.
60. Crick, F. and Jones, E.G., Backwardness of human neuroanatomy, *Nature,* 361, 109–110, 1993.
61. Hockfield, S. and McKay, R.D., A surface antigen expressed by a subset of neurons in the vertebrate central nervous system, *Proc. Natl. Acad. Sci. U.S.A.,* 80, 5758–5761, 1983.
62. Hendry, S.H., Jones, E.G., Hockfield, S., and McKay, R.D., Neuronal populations stained with the monoclonal antibody Cat-301 in the mammalian cerebral cortex and thalamus, *J. Neurosci.,* 8, 518–542, 1988.
63. DeYoe, E.A., Hockfield, S., Garren, H., and Van Essen, D.C., Antibody labeling of functional subdivisions in visual cortex: Cat-301 immunoreactivity in striate and extrastriate cortex of the macaque monkey, *Vis. Neurosci.,* 5, 67–81, 1990.
64. Preuss, T.M., Gray, D., and Cusick, C.G., Subdivisions of the motor and somatosensory thalamus of primates revealed with *Wisteria floribunda* agglutinin histochemistry, *Somat. Motor Res.,* 15, 211–219, 1998.
65. Bartone, S. and Preuss, T.M., Staining with *Wisteria floribunda* agglutinin (WFA) reveals human-specific compartments in primary visual cortex, *Soc. Neurosci. Abstr.,* 658.4, 2002.
66. Hendry, S.H. and Carder, R.K., Neurochemical compartmentation of monkey and human visual cortex: similarities and variations in calbindin immunoreactivity across species, *Vis. Neurosci.,* 10, 1109–1120, 1993.
67. Felleman, D.J. and Van Essen, D.C., Distributed hierarchical processing in the primate cerebral cortex, *Cereb. Cortex,* 1, 1–47, 1991.
68. Kaas, J.H., Theories of visual cortex organization in primates, in *Cerebral Cortex,* Vol. 12: *Extrastriate Cortex in Primates*, Rockland, K.S., Kaas, J.H., and Peters, A., Eds., Plenum Press, New York, 1997, 91–125.
69. Sereno, M.I., Brain mapping in animals and humans, *Curr. Opin. Neurobiol.,* 8, 188–194, 1998.
70. Tootell, R.B., Dale, A.M., Sereno, M.I., and Malach, R., New images from human visual cortex, *Trends Neurosci.,* 19, 481–489, 1996.

71. Van Essen, D.C., Lewis, J.W., Drury, H.A., Hadjikhani, N., Tootell, R.B., Bakircioglu, M., and Miller, M.I., Mapping visual cortex in monkeys and humans using surface-based atlases, *Vision Res.,* 41, 1359–78, 2001.
72. Zilles, K. and Clarke, S., Architecture, connectivity, and transmitter receptors of human extrastriate cortex, in *Cerebral Cortex,* Vol. 12: *Extrastriate Cortex in Primates*, Rockland, K.S., Kaas, J.H., and Peters, A., Eds., Plenum Press, New York, 1997, 673–742.
73. Ungerleider, L.G. and Haxby, J.V., "What" and "where" in the human brain, *Curr. Opin. Neurobiol.,* 4, 157–165, 1994.
74. Tootell, R.B. and Taylor, J.B., Anatomical evidence for MT and additional cortical visual areas in humans, *Cereb. Cortex,* 5, 39–55, 1995.
75. Dobkins, K.R., Gunther, K.L., and Peterzell, D.H., What covariance mechanisms underlie green/red equiluminance, luminance contrast sensitivity and chromatic (green/red) contrast sensitivity? *Vision Res.,* 40, 613–628, 2000.
76. Williams, D., Sekiguchi, N., and Brainard, D., Color, contrast sensitivity, and the cone mosaic, *Proc. Natl. Acad. Sci. U.S.A.,* 90, 9770–9777, 1993.
77. Sterling, P., Retina, in *Synaptic Organization of the Brain,* 4th ed., Shepherd, G.M., Ed., Oxford University Press, New York, 1998, 205–253.
78. Uhlrich, D.J., Essock, E.A., and Lehmkuhle, S., Cross-species correspondence of spatial contrast sensitivity functions, *Behav. Brain Res.,* 2, 291–299, 1981.
79. Vazquez, P., Cano, M., and Acuna, C., Discrimination of line orientation in humans and monkeys, *J. Neurophysiol.,* 83, 2639–2648, 2000.
80. Spillmann, L., Ransom-Hogg, A., and Oehler, R., A comparison of perceptive and receptive fields in man and monkey, *Hum. Neurobiol.,* 6, 51–62, 1987.
81. Oehler, R., Spatial interactions in the rhesus monkey retina: a behavioural study using the Westheimer paradigm, *Exp. Brain Res.,* 59, 217–225, 1985.
82. Steinman, B.A., Steinman, S.B., and Lehmkuhle, S., Transient visual attention is dominated by the magnocellular stream, *Vision Res.,* 37, 17–23, 1997.
83. Steinman, B.A., Steinman, S.B., and Lehmkuhle, S., Visual attention mechanisms show a center-surround organization, *Vision Res.,* 35, 1859–1869, 1995.
84. Corbetta, M., Akbudak, E., Conturo, T.E., Snyder, A.Z., Ollinger, J.M., Drury, H.A., Linenweber, M.R., Petersen, S.E., Raichle, M.E., Van Essen, D.C., and Shulman, G.L., A common network of functional areas for attention and eye movements, *Neuron,* 21, 761–773, 1998.
85. Vidyasagar, T.R., A neuronal model of attentional spotlight: parietal guiding the temporal, *Brain Res. Brain Res. Rev.,* 30, 66–76, 1999.
86. Kastner, S. and Ungerleider, L.G., Mechanisms of visual attention in the human cortex, *Annu. Rev. Neurosci.,* 23, 315–341, 2000.
87. Vanduffel, W., Tootell, R.B., and Orban, G.A., Attention-dependent suppression of metabolic activity in the early stages of the macaque visual system, *Cereb. Cortex,* 10, 109–126, 2000.
88. Heeger, D.J., Linking visual perception with human brain activity, *Curr. Opin. Neurobiol.,* 9, 474–479, 1999.
89. Navon, D., Forest before trees — precedence of global features in visual perception, *Cogn. Psychol.,* 9, 353–383, 1977.
90. Fagot, J. and Deruelle, C., Processing of global and local visual information and hemispheric specialization in humans (*Homo sapiens*) and baboons (*Papio papio*), *J. Exp. Psychol.,* 2, 429–442, 1997.
91. Deruelle, C. and Fagot, J., Visual search for global/local stimulus features in humans and baboons, *Psychonomic Bull. Rev.,* 5, 476–481, 1998.

92. Fagot, J. and Tomonaga, M., Global and local processing in humans (*Homo sapiens*) and chimpanzees (*Pan troglodytes*): use of a visual search task with compound stimuli, *J. Comp. Psychol.,* 113, 3–12, 1999.
93. Hopkins, W.D., Hemispheric specialization for local and global processing of hierarchical visual stimuli in chimpanzees (*Pan troglodytes*), *Neuropsychologia,* 35, 343–348., 1997.
94. Robertson, L.C. and Lamb, M.R., Neuropsychological contributions to theories of part/whole organization, *Cognit Psychol.,* 23, 299–330, 1991.
95. Yovel, G., Levy, J., and Yovel, I., Hemispheric asymmetries for global and local visual perception: effects of stimulus and task factors, *J. Exp. Psychol. Hum. Percept. Perform.,* 27, 1369–1385, 2001.
96. Mendola, J.D., Dale, A.M., Fischl, B., Liu, A.K., and Tootell, R.B., The representation of illusory and real contours in human cortical visual areas revealed by functional magnetic resonance imaging, *J. Neurosci.,* 19, 8560–8572, 1999.
97. Tootell, R.B., Mendola, J.D., Hadjikhani, N.K., Ledden, P.J., Liu, A.K., Reppas, J.B., Sereno, M.I., and Dale, A.M., Functional analysis of V3A and related areas in human visual cortex, *J. Neurosci.,* 17, 7060–7078, 1997.
98. Vanduffel, W., Fize, D., Mandeville, J.B., Nelissen, K., Van Hecke, P., Rosen, B.R., Tootell, R.B., and Orban, G.A., Visual motion processing investigated using contrast agent-enhanced fMRI in awake behaving monkeys, *Neuron,* 32, 565–577, 2001.
99. Tootell, R.B., Neuroimaging weighs in: humans meet macaques in "primate" visual cortex, *J. Neurosci.,* in press.
100. Vanduffel, W., Fize, D., Peuskens, H., Denys, K., Sunaert, S., Todd, J.T., and Orban, G.A., Extracting 3D from motion: differences in human and monkey intraparietal cortex, *Science,* 298, 413–415, 2002.
101. Beauchamp, M.S., Lee, K.E., Haxby, J.V., and Martin, A., Parallel visual motion processing streams for manipulable objects and human movements, *Neuron,* 34(1), 149–159, 2002.
102. Goldenberg, G. and Hagmann, S., Tool use and mechanical problem solving in apraxia, *Neuropsychologia,* 36, 581–589, 1998.
103. Povinelli, D.J., *Folk Physics for Apes: The Chimpanzee's Theory of How the World Works,* Oxford University Press, Oxford, U.K., 2000.
104. Chao, L.L. and Martin, A., Representation of manipulable man-made objects in the dorsal stream, *Neuroimage,* 12, 478–484, 2000.
105. Johnson-Frey, S.H., Cortical mechanisms of tool use, in *Taking Action: Cognitive Neuroscience Perspectives on the Problem of Intentional Movements*, Johnson-Frey, S.H., Ed., MIT Press, Cambridge, MA, in press.
106. McGurk, H. and MacDonald, J., Hearing lips and seeing voices, *Nature,* 264, 746–748, 1976.
107. Campbell, R., MacSweeney, M., Surguladze, S., Calvert, G., McGuire, P., Suckling, J., Brammer, M.J., and David, A.S., Cortical substrates for the perception of face actions: an fMRI study of the specificity of activation for seen speech and for meaningless lower-face acts (gurning), *Brain Res. Cogn. Brain Res.,* 12, 233–243, 2001.
108. Sams, M., Aulanko, R., Hamalainen, M., Hari, R., Lounasmaa, O.V., Lu, S.T., and Simola, J., Seeing speech: visual information from lip movements modifies activity in the human auditory cortex, *Neurosci. Lett.,* 127, 141–145, 1991.
109. Giraud, A.L. and Truy, E., The contribution of visual areas to speech comprehension: a PET study in cochlear implants patients and normal-hearing subjects, *Neuropsychologia,* 40, 1562–1569, 2002.

110. Kang, E., Kang, H., Kim, H., and Na, D.G., Visual modulation of speech processing in higher auditory cortices, *Soc. Neurosci. Abstr.*, 583.15, 2002.
111. Zatorre, R.J., Do you see what I'm saying? Interactions between auditory and visual cortices in cochlear implant users, *Neuron,* 31, 13–14, 2001.
112. Demb, J.B., Boynton, G.M., and Heeger, D.J., Functional magnetic resonance imaging of early visual pathways in dyslexia, *J. Neurosci.,* 18, 6939–6951, 1998.
113. Livingstone, M.S., Rosen, G.D., Drislane, F.W., and Galaburda, A.M., Physiological and anatomical evidence for a magnocellular defect in developmental dyslexia, *Proc. Natl. Acad. Sci. U.S.A.,* 88, 7943–7947, 1991; erratum, 90, 2556, 1993.
114. Stein, J., The magnocellular theory of developmental dyslexia, *Dyslexia,* 7, 12–36, 2001.
115. Spencer, J., O'Brien, J., Riggs, K., Braddick, O., Atkinson, J., and Wattam-Bell, J., Motion processing in autism: evidence for a dorsal stream deficiency, *NeuroReport,* 11, 2765–2767, 2000.
116. Doniger, G.M., Foxe, J.J., Murray, M.M., Higgins, B.A., and Javitt, D.C., Impaired visual object recognition and dorsal/ventral stream interaction in schizophrenia, *Arch. Gen. Psychiatr.,* 59, 1011–1020, 2002.
117. Richardson, A.J. and Gruzelier, J., Visual processing, lateralization and syndromes of schizotypy, *Int. J. Psychophysiol.,* 18, 227–239, 1994.
118. Atkinson, J., King, J., Braddick, O., Nokes, L., Anker, S., and Braddick, F., A specific deficit of dorsal stream function in Williams' syndrome, *NeuroReport,* 8, 1919–1922, 1997.
119. Nakamura, M., Watanabe, K., Matsumoto, A., Yamanaka, T., Kumagai, T., Miyazaki, S., Matsushima, M., and Mita, K., Williams syndrome and deficiency in visuospatial recognition, *Dev. Med. Child Neurol.,* 43, 617–621, 2001.
120. Preuss, T.M., Taking the measure of diversity: comparative alternatives to the model-animal paradigm in cortical neuroscience, *Brain Behav. Evol.,* 55, 287–299, 2000.

11 Maps of the Visual Field in the Cerebral Cortex of Primates: Functional Organization and Significance

Marcello G. P. Rosa and Rowan Tweedale

CONTENTS

11.1 INTRODUCTION: VISUOTOPIC MAPS AND VISUAL AREAS

In primates, conscious vision is particularly dependent on the projections of retinal ganglion cells to the lateral geniculate nucleus of the thalamus, and from these cells

0-8493-1243-4/04/$0.00+$1.50

to the primary visual area (also known as V1, area 17, or striate cortex). It is well established that the anatomical connections along this pathway are characterized by precise patterns of origins and terminations (Figure 11.1). For example, each geniculate neuron receives information from only a few retinal ganglion cells, and projects to a highly circumscribed region of V1.[1,2] These connections are finely organized in the spatial domain in such a way that adjacent points of the same hemiretina always project to adjacent points of V1 (e.g., Reference 3), and information coming from topographically corresponding points of the two hemiretinas converges onto single V1 cells. As a result of this anatomical specificity, cells in adjacent V1 columns always have receptive fields that represent slightly different, but overlapping, portions of the visual field.[4] This orderly relationship (Figure 11.1) creates a *visuotopic map*, whereby V1 neurons on one side of the brain collectively represent the entire contralateral half of the visual field in a gradual and predictable fashion.[5–7] The existence of a visuotopic map in V1 is one of the longest standing facts in visual neuroscience,[8,9] and the precise nature of this representation still provides lecturers with some of the best correlations among structure, function, and pathology in the cerebral cortex.

The other cortical visual areas, which in primates receive visual information mainly through V1, are collectively referred to as extrastriate cortex. Research conducted since the 1960s has suggested that visuotopic maps also exist through much of the caudal extrastriate cortex, including areas near the occipitoparietal and occipitotemporal transitions (e.g., References 10 through 13). In the "classical" view derived from these studies, neurons in each extrastriate area form a separate, more or less complete visuotopic map. As a result, visuotopic organization has become one of the key criteria for the subdivision of cortex: charting the various re-representations of the visual field should, in theory, be sufficient to define the limits of the various visual areas. However, it has gradually become clear that this is not such a simple proposition. Even now, investigators continue to debate which (if any) of the various schemata of subdivision of caudal visual cortex are valid, in the sense that they portray the full and correct extent of the cortical areas of a given species. Indeed, it may be sobering to realize that, of the many extrastriate areas described by the pioneering mapping studies in monkeys, only two have become universally accepted: the second (V2) and middle temporal (MT) areas. Other cortical fields that were initially described as forming equally well-organized visuotopic maps have since been further subdivided, or have had parts recombined with adjacent cortex to form "new" areas. These revised subdivisions of extrastriate cortex continue to be the subject of much controversy (see References 14 and 15 for reviews). Although this level of uncertainty is partly due to the different views of researchers regarding what constitutes a "cortical area," it is also a reflection of the evolution of our understanding of cortical visuotopic maps. We are now aware of more subtle and complex types of organization than initially suggested, which translate into important caveats that must be considered when one interprets the organization of the visual cortex, and the way in which visuotopic maps relate to cortical areas. With this in mind, we highlight not only examples with well-established, "classical" maps, but also those that illustrate common pitfalls in the analysis of visuotopic organization. In addition, this chapter focuses on "global" aspects: the topological properties of

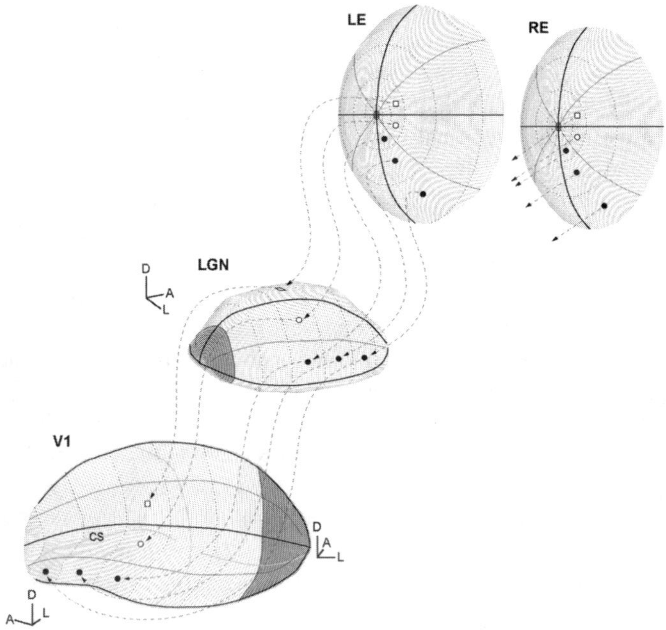

FIGURE 11.1 Cortical visuotopic organization is established by spatially precise distributions of afferent axons in the retinogeniculocortical pathway. The connections between neurons in corresponding parts of the retinas of the left and right eyes (LE and RE, respectively), the right lateral geniculate nucleus (LGN), and the right striate cortex (V1) are indicated by the arrows. Some corresponding points in these structures (i.e., points containing cells with receptive fields in the same part of visual space) are indicated by circles and squares. The foveal representations are indicated by darker gray, isoeccentricity lines by dotted lines and isopolar lines (including the vertical and horizontal meridians) by continuous lines. The anatomical orientation of the LGN and V1 diagrams is indicated (A = anterior, D = dorsal, L = lateral); because in this diagram the V1 map has been graphically "unfolded," the orientation of the calcarine sulcus (cs) is different from that of the foveal representation.

cortical maps, and their implications for understanding the evolution of sensory cortices, and for ongoing efforts to map the human visual cortex. For a more comprehensive view of other issues related to the properties of visual cortical representations (including nomenclature, homologies, local computations, and plasticity) the reader is referred to Reference 16.

11.2 CORTICAL MAPS ARE PROBABILISTIC

Simplified diagrams of the retinogeniculocortical pathway, such as the one illustrated in Figure 11.1, tend to convey the impression that visuotopic maps result from a series of point-to-point (or single cell to single cell) connections. However, such point-to-point relationships do not exist in the cerebral cortex, a structure that is characterized by extensive convergence and divergence in neuronal connections. The abundant interactions between excitatory and inhibitory cells at each stage of processing result in cortical sensory maps that are neither precise "copies" of the sensory receptor surface, nor constant over time. Thus, diagrams of visuotopic maps must be seen as depictions of the most likely location of neurons representing a given location in the visual field, at a certain point in time.

The receptive fields of cortical neurons are large in comparison with those of retinal cells, and have boundaries that change continuously. Near-instantaneous alterations in receptive field extent can be induced by variations in the spatiotemporal content of the visual image[17–20] or by its behavioral context.[21–23] Moreover, longer-term changes can occur as a result of visual experience,[24] or in response to lesions to other parts of the brain (e.g., References 25 through 29). Thus, the single-unit receptive field, as revealed by action potentials evoked by any given stimulus, reflects only a fraction of the area of the visual field capable of influencing the activity of that neuron, and therefore only a subset of its anatomical inputs.[30,31]

In addition, the receptive fields of cells belonging to the same radial column (arguably, the basic functional circuit of the cortex) vary around a mean position in an apparently random way. This variability is referred to as *receptive field scatter* (Figure 11.2A). Estimates obtained in different areas of the primate cortex indicate that angular estimates of scatter and receptive field size are positively correlated and roughly proportional, irrespective of the part of the visual field representation under consideration.[4,32–34] According to most studies, "aggregate" columnar receptive fields (i.e., the area of visual field covered by the combined receptive fields of all neurons forming a radial column) can be two to three times larger than the receptive fields of individual neurons. However, more recent studies in cat visual cortex have obtained far lower estimates of radial scatter, suggesting that a precise alignment of receptive fields may exist in some cases.[35,36] The extent to which the observed scatter reflects changes over time remains unclear, particularly in view of stimulus-linked short-term changes in receptive field extent (see above). This is not an easy issue to resolve, as we gradually realize that the very act of probing the receptive field with a given stimulus may induce its boundaries to change (an interesting parallel with the "uncertainty principle" of quantum physics). Nonetheless, more rigorous approaches, involving automated, standardized receptive field mapping and simultaneous recordings from many cells, may help to establish better confidence limits for the interpretation of cortical topographies.

Finally, it has been observed that the aggregate receptive fields of adjacent columns overlap extensively (Figure 11.2B). Studies in both V1[4] and MT[33] have concluded that the probability of finding neurons with non-overlapping receptive fields increases gradually as a function of distance in the cortex, rather than in discrete steps. Distances of several millimeters may need to separate neurons before their receptive fields are completely non-overlapping; this distance corresponds to the *point-image size*.[32,37] More recent studies have further complicated the picture by suggesting that, on a fine scale, the representation of the visual field can include significant nonlinearities, such as irregular alternations of regions of slow and fast change in receptive field position. In V1 of cats, the rate of change in visuotopic coordinates is correlated to the pattern of orientation columns: receptive field positions overlap extensively among cells located within the same iso-orientation domain, but change quickly as columns representing different orientations are

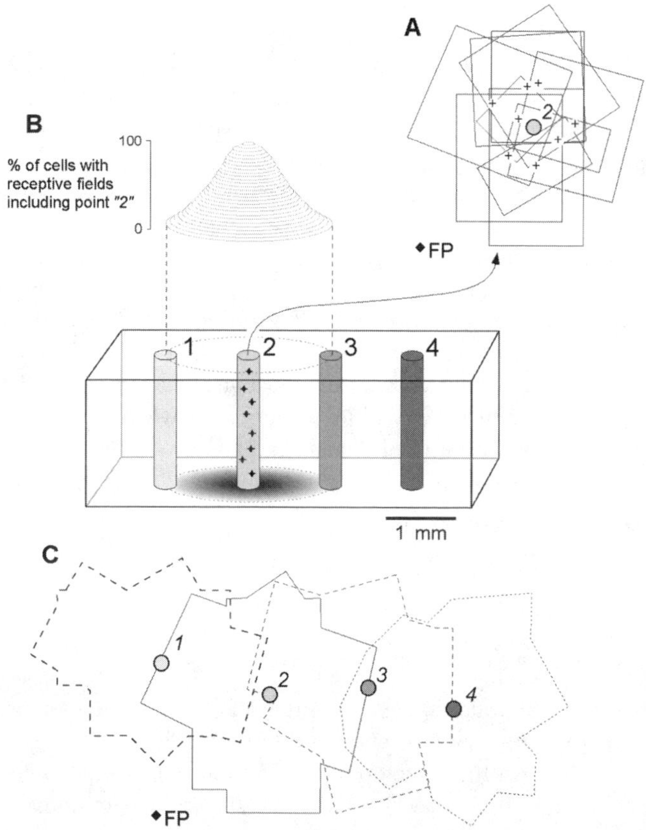

FIGURE 11.2 Statistical nature and local imprecision of cortical visuotopic organization. This diagram is largely based on the findings of Dow and collaborators,[32] in V1, and of Albright and Desimone,[33] in MT. (A) Some of the receptive fields of neurons in a hypothetical cortical column (column 2, in diagram B). Each receptive field is outlined by a rectangle, and its center indicated by a cross. Receptive fields of cells in the same column distribute around a central region (point 2, gray circle) which is included in nearly every receptive field. However, given the scatter in the receptive field distribution, the aggregate receptive field of the column is several times larger than individual neuronal fields. FP = fixation point. (B) Diagram of a small block of cerebral cortex, with four cell columns (1 to 4) indicated. Although neurons with receptive fields including point *2* are distributed over a large area (~2 mm in diameter), the probability of finding a cell with a receptive field including this point gradually decreases as a function of distance from column 2. (C) The aggregate receptive fields of neurons in columns 1 mm apart (e.g., 1 and 2) overlap significantly. Distances of 2 mm or more have to be crossed in the cortex to obtain just abutting sets of receptive fields (the actual non-overlap distance varies somewhat between areas; for example, it is smaller in V1 than in V2).[34,38]

crossed.[36] In V2 of monkeys, it has been proposed that the nonlinearities are correlated to the pattern of stripe-like cytochrome oxidase modules. For example, regions of slow change could reflect successive re-representations of each part of the visual field by adjacent cell populations, located in different stripes.[38,39] In summary, the exact amount of receptive field overlap between cells located in two radial columns is not solely dependent on the tangential distance separating them, and point images may have intricate, variable shapes.

The experimental evidence reviewed above demonstrates that only the *average* receptive field position can be seen to vary as a function of position in the cortex. Conversely, even the smallest stimulus in the visual field will stimulate an asymmetrical cluster of thousands of cortical cells distributed over a mean position. Information about the average receptive field positions in different regions of cortex is important for tasks such as defining cortical areas, or interpreting clinical signs after lesions. Yet it is likely that a better functional understanding of the computations performed by the different areas will also require knowledge about the statistical parameters of the maps within local domains.

11.3 CORTICAL MAPS CAN BE CLASSIFIED ACCORDING TO THEIR TOPOLOGICAL CHARACTERISTICS

11.3.1 First- and Second-Order Representations

We now leave the fine-scale topographic organization of the cortex, and consider the global rules governing the projection of different regions of the visual field onto cortical areas. Given that the visuotopic maps found in the different cortical layers are in register (see also References 40 and 41), it has become customary to summarize the topographic organizations of cortical areas using two-dimensional diagrams (e.g., Figure 11.3 and Figure 11.5, below). These diagrams are useful in demonstrating the differences in the spatial transformations between the retina and the cortex, in terms of features such as continuity, emphasis on the representation of different parts of the retina, and angular distortions. However, for the reasons outlined in the previous section, they should not be seen as implying the existence of point-to-point representations of the visual field.

In the simplest type of visuotopic map, all visual information being captured by the eyes at a given moment would be mapped to the cortex in a way that precisely preserves neighborhood relationships, thereby creating a continuous representation of the visual scene. Of course, there are no truly continuous visual representations in the cortex, given that the information coming from the left and right halves of the visual field projects separately to the right and left hemispheres, respectively. Nonetheless, if one restricts the analysis to the geometrical relationships between one hemifield and its contralateral cortical projection, different categories of geometrical transformation can be recognized, in terms of topological continuity.

One primary distinction, originally proposed by Allman and Kaas,[42] can be made between first- and second-order representations of the visual field. First-order representations are those in which adjacent points of the same hemifield (or hemiretina)

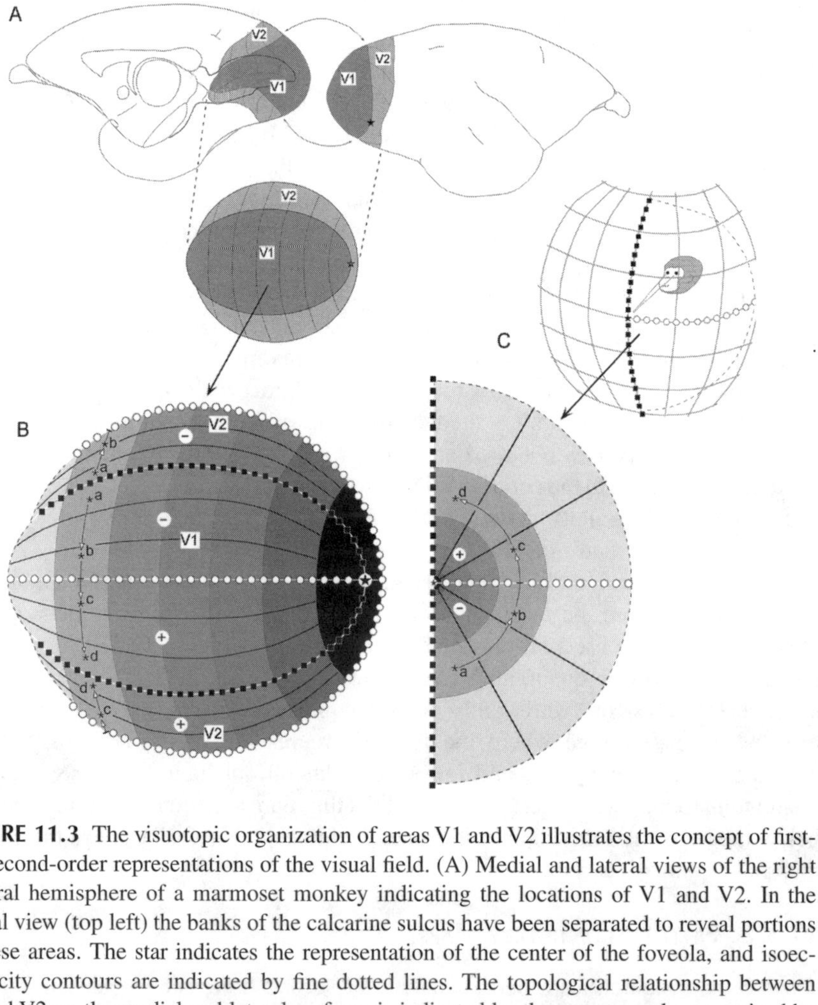

FIGURE 11.3 The visuotopic organization of areas V1 and V2 illustrates the concept of first- and second-order representations of the visual field. (A) Medial and lateral views of the right cerebral hemisphere of a marmoset monkey indicating the locations of V1 and V2. In the medial view (top left) the banks of the calcarine sulcus have been separated to reveal portions of these areas. The star indicates the representation of the center of the foveola, and isoeccentricity contours are indicated by fine dotted lines. The topological relationship between V1 and V2 on the medial and lateral surfaces is indicated by the arrows, and summarized by the "unfolded" schematic representation (dashed lines). (B) A schematic unfolded representation of V1 and V2, indicating the locations of the representations of different parts of the visual field. In this diagram, different visual field locations are illustrated by different gray shades and symbols, which are summarized in a schematic view of the animal's contralateral hemifield (C). The representation of the central 1° of the visual field is indicated in black, and progressively more eccentric parts of the visual field in gradually lighter shades of gray. Representations of the vertical meridian are outlined by the black squares, representations of the horizontal meridian by white circles, and those of the temporal perimeter of the visual field by dashed lines. The + and – signs indicate representations of the upper and lower contralateral quadrants, respectively. (C) Schematic representation of the animal's left hemifield, with four points located along an isoeccentric arc (a to d) indicated. Note that whereas the representation of this arc in V1 is continuous (B), in V2 it is not; points above and below the horizontal meridian project to nonadjacent sets of cortical columns in this area. In this and all subsequent figures, the visual hemifield is illustrated from the experimenter's point of view (see insert in C).

always map to adjacent points in the contralateral cortex. The visuotopic map in V1 is an example of this type of transformation. Consider, for example, the visual field locations indicated by points a to d, in Figure 11.3C. As with any systematic progression of points in the same hemifield, the arc a to d maps onto a continuous series of cortical columns in V1 (Figure 11.3B). In this case, the surface being mapped (the visual hemifield, or hemiretina) and the surface where the map exists (the cortex) are said to be topologically equivalent. Only two areas in the primate cortex form precisely organized first-order representations of the visual field: V1 and MT.[7,12,43–48] In both cases, the outer boundary of the map is formed by the representations of the vertical meridian* and the visual periphery (i.e., the limits of the contralateral hemifield). An example of data indicating the existence of a first-order transformation, obtained in area MT of a marmoset, is illustrated in Figure 11.4. Note that each sequence of receptive fields proceeds systematically from the lower to the upper quadrant (as the electrode moves from dorsal to ventral parts of MT) and that sequences obtained at different anteroposterior levels also gradually move from the fovea to the periphery. Despite the imperfect linearity of the sequences (reflecting receptive field scatter), each part of the visual field is only represented once, in a defined sector of MT.

In contrast, second-order representations are those in which the topological equivalence between the visual field and the cortex is violated in some way. One clear illustration of a second-order map is found in V2 (Figure 11.3B). In this area, the upper and lower quadrants of the contralateral hemifield are represented in separate locations (in ventral and dorsal occipital cortex, respectively), connected only by a small bridge where the fovea is represented.[11,42,44] Thus, despite the representations of the upper and lower contralateral quadrants in V2 as internally continuous (e.g., points a and b, or c and d), the representations of points above or below the horizontal meridian of the visual field become nonadjacent in the cortical map (e.g., points b and c).

11.3.2 Field Discontinuity Maps

The V2 map exemplifies a particular type of second-order representation, referred to as a *field discontinuity* map. As implied by the name, field discontinuity maps are organized as if a discontinuity had been introduced in the visual field. One can understand the topology of this type of map by imagining that a first-order transformation has been cut along a line, such as the representation of the horizontal meridian, and physically parted (Figure 11.5A). Although in most primates studied so far the line of field discontinuity in the V2 map corresponds to the horizontal meridian,[34,38,42,44,49] this is not necessarily the case. For example, in the prosimian *Galago* the field discontinuity corresponds to an oblique line below the horizontal

* The vertical and horizontal meridians are the primary axes of reference usually employed in describing the visuotopic organization of brain structures. The vertical meridian is an imaginary vertical line through the fixation point, which subdivides the visual field into right and left hemifields; as such, it corresponds to the line of decussation between the left and right hemiretinas of each eye. The horizontal meridian is defined as a line perpendicular to the vertical meridian, through the fixation point; it subdivides each hemifield into an upper and a lower quadrant.

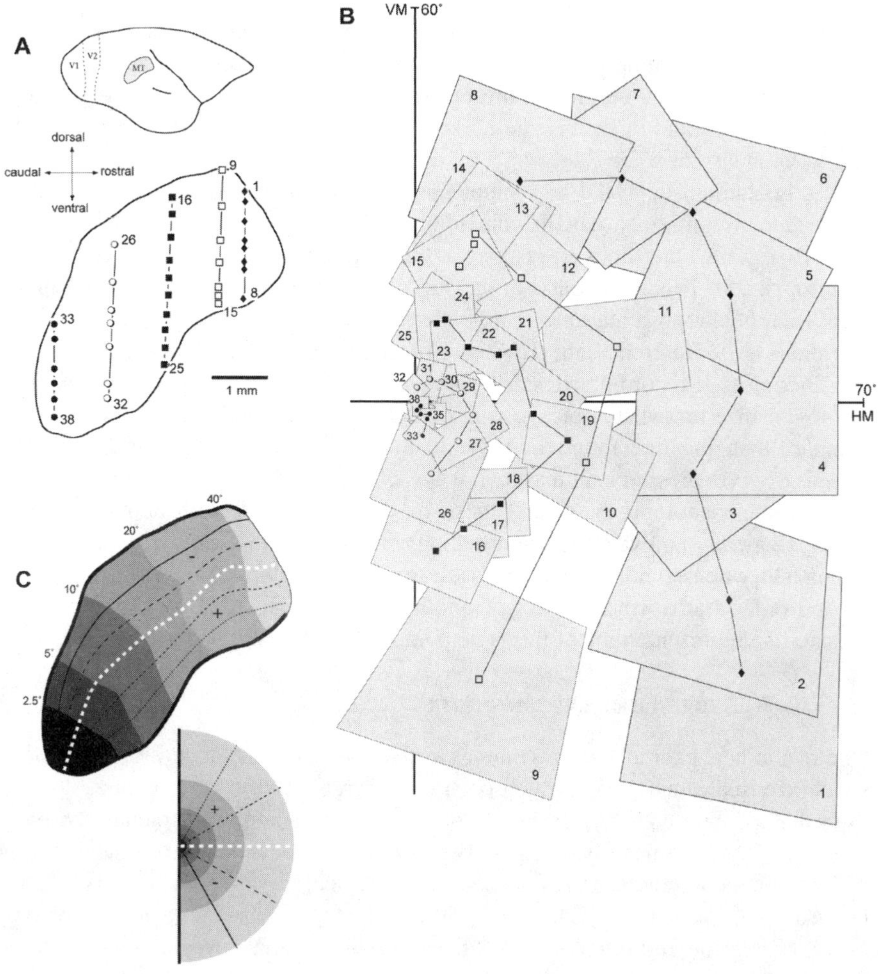

FIGURE 11.4 Visuotopic organization of MT. (A) Surface view reconstruction of MT in the right hemisphere of a marmoset monkey, with recording sites indicated by circles, squares and diamonds. The recording sites were connected to form five dorsoventral sequences, from rostral (sequence 1 to 8) to caudal (sequence 33 to 38). The diagram above illustrates the location of MT in the brain. (B) Diagram of the contralateral hemifield, with the borders and centers of the receptive fields recorded at sites 1 to 38 illustrated; centers of receptive fields of cells located at the same anteroposterior level are connected. (C) A summary map of the visuotopy of MT, based on the coordinates of the receptive field centers. The representation of the visual field is coded according to the symbols shown in the diagram below. The numbers above the map indicate the range of receptive field center eccentricities observed within the regions coded by the different shades of gray (0–2.5°, 2.5–5°, 5–10°, etc.). (Adapted from Reference 48.)

meridian.[50] As a result, the representation of the visual field in the dorsal and ventral parts of V2 is asymmetrical, with dorsal V2 representing most of the lower contralateral quadrant, and ventral V2 representing the entire upper quadrant, as well as the visual field immediately below the horizontal meridian. In fact, this pattern of representation may be more representative of mammals in general, with simian primates exceptional in having the horizontal meridian represented throughout the rostral boundary of V2.[51,52]

Early studies suggested that simple second-order transformations, involving a single field discontinuity (usually coinciding with the representation of the horizontal meridian), were ubiquitous in primate extrastriate cortex (e.g., References 13 and 53 through 55). This is the basis of the still often-made (but erroneous) assumption that visual field representations "must" alternate between the vertical and horizontal meridians (e.g., Reference 56). However, further research, based on more extensive and finer grids of recording sites, has painted a more elaborate picture. In particular, we now realize that visuotopic maps in extrastriate areas can be topologically more complex than the one found in V2, and more variable among individuals. The complexity is expressed in at least two ways: multiple field discontinuities may exist in the same visuotopic map, and these may be combined with another type of nontopologically equivalent transformation, the *map discontinuity*. It is also worth noting that, when examined on a finer scale, even the V2 map departs from a "simple" second-order transformation in yet another way: it includes locally repetitive, mosaic-like representations of the same points of the visual field (see Section 11.2).

11.3.3 Multiple Field Discontinuities

One of the best examples of a complex pattern of field discontinuities is found in the third visual area (V3),* which occupies a large fraction of the cortex located immediately anterior to V2. Primarily on the basis of anatomical tracing, the organization of V3 in primates was originally described as a reduced mirror-image version of V2, with a continuous representation of the vertical meridian forming its anterior boundary[1,59,60] (see also Reference 61). However, studies based on extensive recordings covering the entire extent of V3 have also shown that several locations along the rostral boundary of this area represent parts of the visual field away from both the vertical and the horizontal meridians.[14,62] This finding indicates that the visuotopic map in V3 includes multiple field discontinuities.[62] One can understand this type of transformation by imagining that the visuotopic map of V2 (which, as we have seen above, already includes a "primary" field discontinuity separating the upper and lower quadrant representations) has received further cuts along its perimeter. As a result (Figure 11.5B), the rostral boundary of V3 is formed in part by cells with receptive fields including the vertical meridian, and in part by cells representing other parts of the visual field. It is worth noting that this does not imply any gaps in the representation of the vertical meridian; this representation is simply discontinuous,

* The designation "V3," as used here, includes the ventral posterior (VP) and dorsolateral posterior (DLp) areas of earlier studies (e.g., References 57 and 58). It also corresponds to the ventrolateral posterior area defined in New World marmoset monkeys. It does not include the densely myelinated "V3" of other studies, which is located further dorsal and medially (see Reference 14 for a detailed discussion).

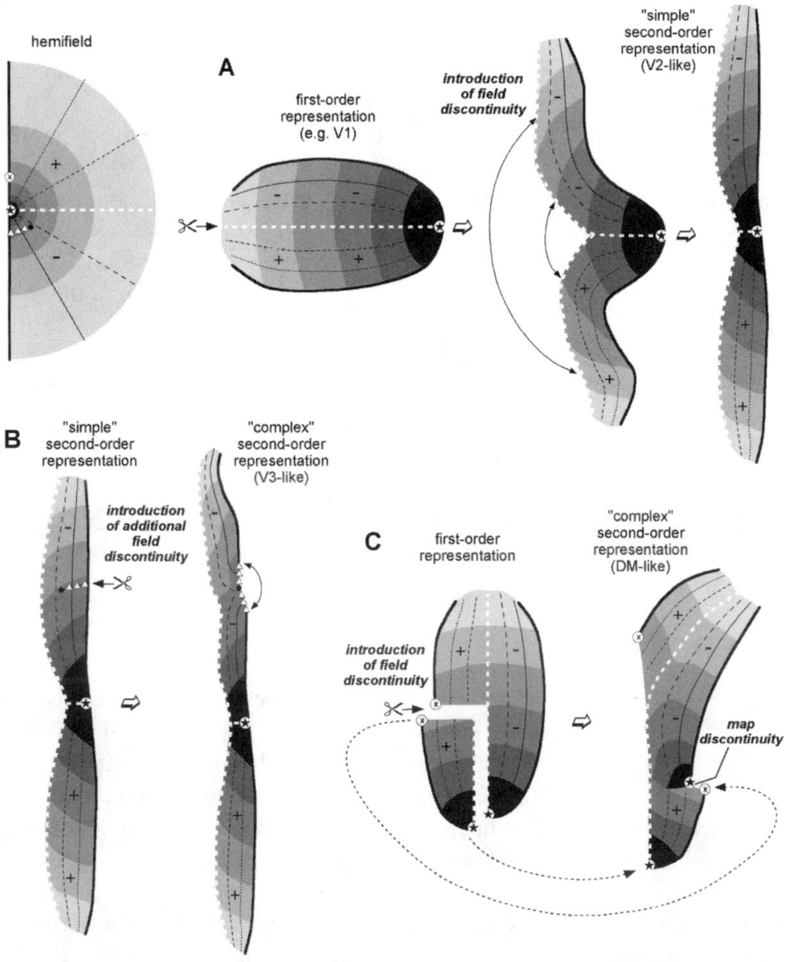

FIGURE 11.5 Different types of visuotopic maps can be seen as topological transformations of the V1 map. Top left, Schematic representation of the visual hemifield, illustrating the coding of different sections of the visual field used in the maps shown in A to C. (A) "Simple" field discontinuity maps, such as V2, can be imagined as a result of cutting and physically separating a first-order representation map. (B) Visuotopic maps can include secondary field discontinuities, such as those observed along the rostral border of primate area V3. In such cases, points located away from both the vertical and the horizontal meridians are represented along the area's outer boundary (circle and triangles). (C) More rarely, maps may include map discontinuities, regions where the receptive field coordinates change suddenly. For example, in DM of marmosets the central and peripheral parts of the upper contralateral quadrant are represented in spatially segregated sectors. This transformation can be imagined as the result of cutting off an entire "island" of cortex, and reattaching it to the map with a different orientation. As a result, there is a small region along the rostral border of DM where cells with receptive fields in the mid-peripheral region of the upper quadrant adjoin cells with receptive fields in the central lower quadrant (map discontinuity). Note, however, that this transformation preserves the continuity and the centroperipheral axis of the horizontal meridian representation (white dashed line), resulting in a congruent border between DM and V2 (see also Figure 11.6).

as it is located in two nonadjacent, but mutually complementary regions of cortex. The complex topography of this area is a robust finding, which has been observed in all three species of monkey that have been extensively studied by means of electrophysiological recordings.[14,58,62] Thus, despite the ongoing controversy regarding the precise extent of V3 (in particular, its representation of the lower peripheral visual field; see also Section 11.6), it is clear that the topographic organization of this area is more complex than initially suggested. Multiple field discontinuities also appear to exist in the visuotopic maps of the fourth visual area (V4)[63] and of the middle temporal crescent (MTc; this area includes the "V4 transitional," or V4t, region, but encompasses an upper quadrant representation as well[64]). In the MTc, it is the representation of the horizontal meridian that receives further "cuts," in such a way that receptive fields of cells near the outer boundary of this area sometimes represent the horizontal meridian, and sometimes other parts of the visual field.[48]

One final twist to the story is the suggestion that the pattern of field discontinuities may differ significantly between individuals of the same species. This has been particularly well documented by studies in carnivores,[65–68] and at least two studies have uncovered hints of variability in primate cortical areas.[14,62] However, the extent of this variability has not yet been investigated systematically in primate extrastriate cortex, and deserves further study.

11.3.4 Map Discontinuities

In the types of representation described above, adjacent cortical columns always represent adjacent or overlapping parts of the visual field. Although this local continuity in sensory maps is characteristic of the overwhelming majority of cortical representations, there are a few clear exceptions. For example, it is well established that a map discontinuity (i.e., an imaginary line separating two adjacent populations of neurons with non-overlapping receptive fields) exists at the boundary between the representations of the hand and the face in somatosensory cortex.[69] In primate visual cortex, the only well-documented example of a second-order transformation including a map discontinuity is the dorsomedial area (DM) of marmosets[70] (Figure 11.5C).

The visuotopic organization of DM is very interesting from the point of view of topological transformations, given that it includes both a field and a map discontinuity (Figure 11.5C). As a result of the field discontinuity, the representation of the upper quadrant in this area occupies two nonadjacent regions: a lateral sector, which represents the central visual field, and a medial sector, which represents the peripheral visual field. The field discontinuity between these sectors runs obliquely across the upper quadrant, in a pattern that is remarkably consistent among individuals.[70] Moreover, studies based on extensive receptive field mapping of the dorsal and medial extrastriate cortices in macaques have uncovered a similar pattern of representation in visual area 6 (V6),[71] the Old World monkey homologue of DM.[15] In both DM and V6, the medial sector of the upper quadrant representation is topologically continuous with the lower quadrant representation, forming a relatively simple map of the peripheral visual field.* In contrast, at least in the marmoset, the

* This peripheral representation is sometimes referred to as the "parietooccipital area," or PO.[72,73]

lateral sector of upper quadrant representation adjoins the remainder of DM along a line of map discontinuity. As indicated in Figure 11.5C, this discontinuity is located near the rostral border of DM, separating cells with receptive fields in parafoveal and mid-peripheral parts of the upper quadrant (laterally) and cells with receptive fields in the foveal lower quadrant (medially). It is possible that the apparent discontinuity in the visuotopic map of DM actually corresponds to a region of very rapid change in receptive field position. Studies in which neurons are studied at close intervals (50 to 100 μm) rarely observe non-overlapping receptive fields at adjacent sites.[58]

11.3.5 Coarsely Topographic Maps

The visuotopic representations discussed in the previous sections are characteristic of visual areas located in the caudal part of the brain, including V1 and most of the areas that are connected to it.[14,61,74,75] Despite a certain degree of fluctuation in the boundaries of receptive fields (as discussed above) it remains true that cells in these areas are maximally responsive to stimuli presented within relatively well defined "windows" of the visual field. Thus, it makes sense to propose that a given neuron represents a specific sector, such as the central lower quadrant or the peripheral part of the horizontal meridian. By contrast, there are other areas, located near the occipitotemporal and occipitoparietal transitions, that cannot be appropriately described in terms of the transformations discussed above. In these areas, the large single-unit receptive fields and the scatter in the representation make it more appropriate to describe their visuotopic organizations in terms of coarse gradients.

Consider, for example, the visual topography within cytoarchitectural field TEO, located at the junction between the occipital and temporal lobes (also known as the caudal subdivision of inferior temporal cortex, or ITc). Recordings in this region have revealed a topographic organization, including a gradual change of receptive field locations from the central to the peripheral visual field, with progressively more ventral recording sites, and a segregation between cells representing the upper and lower contralateral quadrants.[14,76] However, within each of these broad domains, it is impossible to predict the changes in receptive field position; their centers zigzag between the vicinities of the horizontal and vertical meridians in no apparent order. This nonsystematic mapping could be explained on the basis of a large receptive field scatter. If the proportionality rule between scatter and receptive field size (as observed in more caudal areas; see Section 11.2) is preserved, the aggregate receptive fields of each TEO column may span the distance between these meridians. Thus, a continuous representation of each quadrant can still be achieved, without a systematic mapping of polar angle. The disorder that is apparent when one considers the global visuotopic map in TEO may be further compounded by the existence of re-representations of the same part of the visual field, in side-by-side columnar aggregates.[77]

Other regions that preserve a coarse segregation between the representations of the upper, lower, central, and peripheral visual fields include visual area 6a (V6a),[78] the lateral intraparietal area (LIP),[79] and the motion-sensitive areas adjacent to MT.[48,80] In the region of inferior temporal cortex located in the ventral bank of the superior temporal sulcus, only a crude centroperipheral gradient is preserved; cells with receptive fields centered in the upper and lower quadrants are intermingled in no particular order.[81] These results indicate that there may not be a sharp division between topographically organized and non-organized areas; rather, visuotopic organization is gradually eroded between caudal and rostral visual areas.

11.3.6 Incomplete Maps

Several studies have suggested the existence of visuotopic maps lacking representations of certain parts of the visual field. However, history has demonstrated that areas with incomplete maps often prove, on more thorough examination, to be parts of larger areas representing the entire visual field. For example, the "parietooccipital area" (PO), once characterized as having no foveal representation,[73] can be more accurately described as the peripheral representation of a "complete" area, V6.[15,71] Similarly, the "ventral posterior area" (VP),[82] which was seen as representing only the upper quadrant of the visual field, is now known to correspond to the ventral half of a larger subdivision, which includes near symmetrical representations of both quadrants.[14] At present, there is no compelling evidence for areas in the primate cortex with grossly incomplete representations. A related but more difficult problem is whether or not there are visual areas lacking representations of the far periphery of the visual field. Even an area as well established as V2 was once believed not to represent the monocular crescent, a notion dismissed by studies encompassing recordings across its entire extent (see Reference 34 for a review). Certainly, the heavy emphasis on representation of central visual fields in most areas (Section 11.4) creates sampling problems, as representations of the far periphery are often highly compressed; thus, it is difficult to ascertain whether they are truly absent from a map on the basis of microelectrode recordings. The use of functional imaging techniques may be better suited to answer this question, as a result of their potential ability to reveal simultaneously all cortical fields that respond to stimulation of the far periphery. Given the caveats and cautionary tales outlined above, the evidence for "incomplete maps" in the primate cortex remains questionable.

11.3.7 Maps Are Continuous across Areal Boundaries

Descriptions of the visuotopic maps of single areas fail to convey one of the most striking characteristics of the primate visual cortex: the coherence of the visual topography across functional boundaries. As detected even by some of the earliest topographic studies of primate extrastriate cortex (cf. References 10 and 83), there are no sudden changes in receptive field position as the border between two areas is crossed. Rather, the gradient of representation usually reverts in such a way that maps in adjacent areas are mirror symmetrical (at least in the neighborhood of their common boundary). In a less common pattern, the representations of the upper and lower quadrants in two adjacent areas contact at a boundary that represents the horizontal meridian. Examples of both types of transition are illustrated in Figure 11.6, which is a composite view of the visuotopic organizations of several areas in the marmoset monkey. This illustration reveals many clear examples of "supra-areal" topography. For example, the foveal representations of several areas align near the

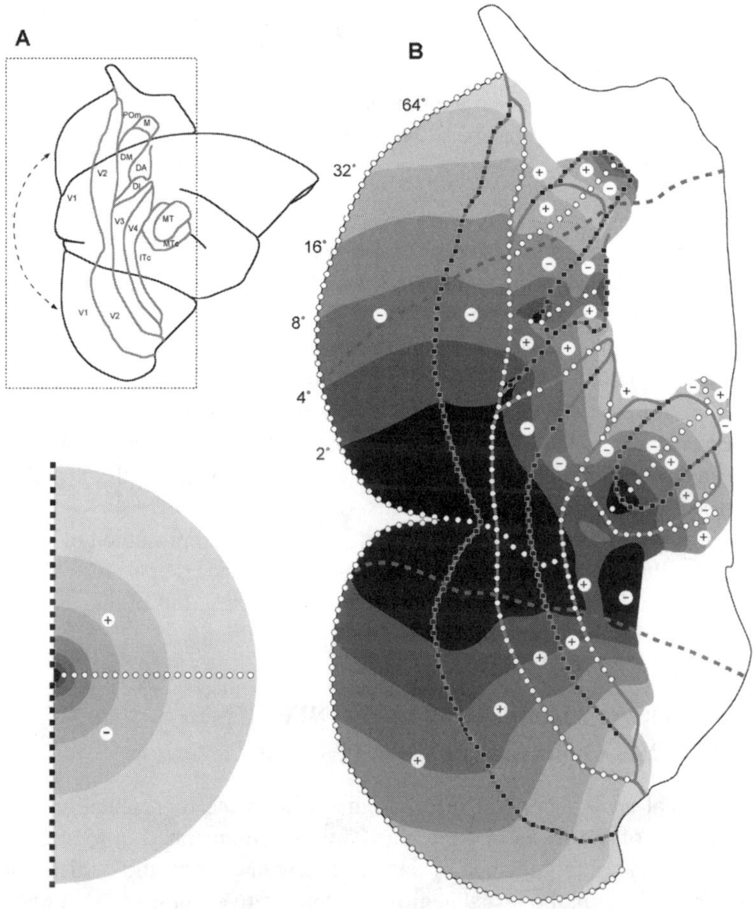

FIGURE 11.6 Supra-areal visuotopic organization of primate visual cortex. (A) Lateral view of the right hemisphere of a marmoset monkey, illustrating the anatomical relationships within the cortical region shown in part B. In generating this figure, parts of the cerebral cortex normally hidden from view (those located along the dorsal midline and ventral surface) have been graphically "unfolded"; moreover, an artificial discontinuity has been introduced along the horizontal meridian representation in V1 (arrows). The gray lines indicate the boundaries of cortical areas, according to a scheme of subdivisions based on physiological mapping and architecture.[14,48,70] (B) Unfolded map of the posterior neocortex. The thick dashed lines indicate the dorsal and ventral limits of the cortex that is normally exposed on the dorsal and lateral surfaces of the brain. The numbers to the left of V1 indicate the range of receptive field center eccentricities observed within the regions coded by the different shades of gray (0–2°, 2–4°, 4–8°, etc.). Representations of the horizontal meridian are indicated by white circles, representations of the vertical meridian by black squares, representations of the upper quadrant by the + signs, and those of the lower quadrant by the – signs. Although it is likely that visuotopy extends beyond the areas shown in this figure, the detailed organization of the maps has not yet been established in the marmoset.

lateral convexity of the occipital lobe, forming a large region that does not respect areal boundaries. Similarly, the representations of gradually more eccentric parts of the visual field form continuous bands, which occupy progressively more medial locations. Although neurons near areal boundaries often have receptive fields including the vertical and horizontal meridians, this is not always the case.

11.3.8 Summary

The evidence regarding the topology of cortical visuotopic maps, reviewed above, should be seen in the light of two basic rules:

Rule 1: Adjacent radial columns in the cortex represent adjacent regions of the visual field. This is true irrespective of whether these columns are part of the same or adjacent areas. Exceptions to this rule (map discontinuities) are very rare; in fact, it remains unclear whether, at the microscopic level, these discontinuities are real, or represent regions of unusually fast change in receptive field position.

Rule 2: Adjacent regions of the visual field *tend* to be represented in adjacent radial columns of a given area. There are numerous exceptions to this rule, as illustrated by the fact that second-order representations of the visual field are far more common than first-order representations.

11.4 CORTICAL MAPS EMPHASIZE DIFFERENT SECTORS OF THE SENSORY PERIPHERY

In most cortical areas there is a marked emphasis on the representation of central vision, and that of the fovea in particular. This is demonstrated in Figure 11.3 and Figure 11.5 by the predominance of darker shades of gray in the cortical maps, as compared with those in the visual hemifield. Quantitative studies in V1 indicate that the degree of emphasis on central vision is highly consistent among day-active primates, from marmosets (body weight 300 g) to humans: the representation of the visual field within 10° of the foveola accounts for nearly 60% of V1.[47] In contrast, nocturnal primates (both simian and prosimian) have a lesser emphasis on central representation.[7,50]

The amount of cortex dedicated to each part of the visual field can be quantified using the *magnification factor*, which is the ratio of the cortical distance between two neurons (measured parallel to the cortical layers) and the angular distance between the centers of their receptive fields. Studies in several primate visual areas have revealed that magnification factors decrease sharply and monotonically as a function of eccentricity, but are similar in symmetrical regions of the upper and lower quadrants.[6,34,50,62,63] In qualitative terms, this change resembles the variation of retinal ganglion cell densities subserving different parts of the visual field. In fact, it has been proposed that the relative magnification of the central visual field in V1 reflects the spatial density of retinal ganglion cells sending axons to the geniculostriate pathway.[2,84] According to this view, the representation in V1 is constrained by a *peripheral scaling* rule: the area of the V1 compartment devoted to the analysis of a given part of the visual field is dictated by the number of ganglion cells in the corresponding parts of the retinas. However, the possibility of peripheral scaling in V1 has been dispelled on the grounds of both quantitative analyses[50,85] and direct anatomical demonstrations.[3,86] These analyses have repeatedly demonstrated that the central visual field is even more magnified in V1 than would be expected merely on the basis of ganglion cell densities; this indicates a higher ratio of convergence to divergence in the visual pathway subserving the periphery, as compared with the fovea. In this respect, the retinogeniculocortical pathway also departs significantly from a hypothetical system of "point-to-point" connections.

Finally, it has been proposed that different visual areas have different emphases on the representation of central vision (e.g., References 54, 62, and 73). Of course, this comparison is largely dependent on the way in which the areas are defined. Nonetheless, the existence of different relative magnifications has been confirmed by direct comparisons between topographic maps in noncontroversial areas in the same individuals. For example, V2 and "ventral stream" areas devote proportionally more tissue to the representation of central vision, in comparison with V1,[14,34] whereas MT has a relatively small foveal representation.[48] Thus, even if further research reveals that peripheral scaling is manifest in the organization of some cortical areas, these will constitute special cases, rather than reflections of a rule.

11.5 DOES MAP TOPOLOGY REFLECT FUNCTION?

The existence of visuotopic maps through much of the caudal part of the primate brain reflects the importance of retinal adjacency as a determinant of correlated firing, both during development and in adulthood.[87,88] It can be argued that functional significance is implied by the parallel emergence of visuotopic maps that share several basic characteristics in many mammalian groups.[89] Perhaps maps exist primarily as a reflection of an efficient mechanism for the development of appropriate connections between neurons, during early development.[90–92] From an evolutionary point of view, visuotopy can also be interpreted as a consequence of natural selection toward configurations that minimize "wire length," thereby allowing resource-efficient processing of information.[93,94]

Both the "developmental" and the "functional" arguments above provide reasonable explanations for the fact that maps exist, as opposed to having a visual system formed primarily of random or nontopographic maps. However, the various topological categories of visuotopic maps found in primate cortex (Section 11.3) also raise the question of whether or not these reflect different functions. In some cases, it has been suggested that the particular type of map observed in an area corresponds to an efficient configuration, which facilitates certain types of neural computations (e.g., References 42, 53, 95, and 96; see Reference 16 for a review). However, there are several difficulties with these teleological arguments. First, it has become increasingly difficult to explain how the complex and variable topologies of extrastriate areas could reflect this type of "hardware optimization" by natural selection (e.g., what is the function of secondary field discontinuities along the vertical meridian?[67]). Second, it has become clear, mainly due to modeling studies, that a variety of maps could emerge simply as the logical outcome of a few developmental rules and activity-dependent refinement of neuronal receptive fields.[92] Third, as argued in detail below, several assumptions of the models proposing a functional role for map topology have been challenged, as more has become known about the temporal aspects of neuronal interconnections within and across cortical areas. Finally, it must be remembered that evolutionary "optimization" should be seen in terms of compromise solutions. Given that the object of natural selection is the individual, not the visual area, the range of possible configurations of a cortical map is necessarily limited by what can be achieved during development, without causing too much disturbance to the formation of the brain (and the organism) as a whole. Thus, theoretically "optimal" solutions may not be biologically feasible, while complex map topologies may develop and even be passed to the following generations (e.g., if they result from a mutation affecting a molecular specification step), as long as they do not interfere seriously with visual processing.

If one assumes that second-order representations arise by a combination of chance and necessity, rather than by design, then one has to ask whether there is a functional "price" to pay for having discontinuous representations of the visual field in the cortex. Many current models of cortical processing emphasize the importance of local computations involving intrinsic connections in establishing the receptive field characteristics of cortical neurons, and in creating complex "emergent" properties that may explain aspects of human perception (for recent examples, see References 94 and 97 through 99). Field discontinuities create problems in this context, as they introduce marked anisotropies in the pattern of input to single cells. Consider, for example, the inputs to neurons located at the representation of the horizontal meridian, near the rostral border of dorsal V2. As demonstrated by anatomical tracing studies, these cells have intrinsic connections with parts of V2 representing neighboring parts of the lower quadrant, but little access to information coming from the upper quadrant representation of this area.[100–102] One possibility worth investigating is that certain perceptual tasks involving integration across different parts of the visual field may be affected by this partial "disconnection." In particular, tasks that depend on computations occurring in V2[103–105] could be more efficiently performed in situations where integration within the same quadrant is sufficient, as opposed to those requiring integration between the upper and lower quadrants. Nonetheless, it is likely that any functional correlates of second-order representation architectures in extrastriate cortex will be translated into rather slight perceptual effects, if indeed these are present at all. The correlates of a far more radical segregation (that of left and right hemifield representations between the two cerebral hemispheres) remain unnoticed in daily life, requiring controlled experiments and quantitative analyses to be evidenced (e.g., References 106 and 107).

Thus, visual perception is not seriously affected by the additional "cable length" required to link neurons across field discontinuities, at least not in any way that could affect an organism's survival. Clues to the reason this is the case can be found in recent studies that have explored the temporal aspects of neuronal interactions in visual cortex, both from the physiological and the anatomical/developmental points of view. It is now clear that the timing of stimulus-evoked neuronal responses in different areas is largely overlapping, with the processing of any stimulus involving multiple re-entrant loops of feedforward and feedback connections (for reviews, see

References 108 and 109). At the same time, the conduction velocity of intrinsic connections has been reported to be far lower than that of extrinsic connections.[110] This may result in feedback, feedforward, and intrinsic connections acting simultaneously to determine the activity of a cell,[109] thereby raising the possibility that feedback from areas with larger receptive fields may help bridge the gaps introduced by field discontinuities. Moreover, the study of callosal connections has also demonstrated that separate branches of the same axon may take quite different paths in the white matter, some branches going directly to the target, and others following tortuous trajectories. It has been suggested that this may represent another mechanism whereby temporal matching across spatially segregated neuronal populations can be achieved during development, circumventing apparent connection length mismatches.[111,112] In summary, physical distance between points in the adult cortex is not directly proportional to functional "distance." The evidence now available indicates that activity-dependent processes during development can circumvent possible discrepancies introduced by the separation of neuronal populations representing adjacent parts of the visual field.

11.6 RELATIONSHIP BETWEEN MAPS AND AREAS

The current dominant view in the scientific literature is reflected in a recent statement that "the cerebral cortex is divided into morphologically distinct, functionally dedicated and stereotypically connected cortical areas."[113] Essentially, cortical areas continue to be seen as "organs of the brain,"[114] well defined units that interact via orderly feedforward and feedback connections. Reflecting this view, charts of the monkey cerebral cortex, in which the boundaries of cortical areas are indicated by sharp lines (e.g., Figure 11.6), have become common textbook illustrations. Yet it is important to recognize that the picture painted by our current knowledge of cortical organization is far from complete. Moreover, even in the best-studied animal models, the degree of precision (and confidence) with which one can delimit visual areas varies immensely, from pencil-sharp limits (e.g., V1) to tenuous, gradual transitions in response properties and architecture. In fact, it could be argued that the notion that the cortex is entirely formed by well-segregated areas remains an extrapolation, based on a few examples (mainly, primary sensory/motor areas). Intermediate configurations, including partially overlapping or interdigitating fields, are also likely to exist (e.g., Reference 115).

Our knowledge of visual areas is particularly instructive in this respect, given that they represent the most intensively studied part of the primate neocortex. At one extreme, there are three areas (V1, V2, and MT) that can be unambiguously identified by any investigator with access to the proper histological tools. At the same time (or perhaps for this reason), the visuotopic map correlates of these areas are now established. However, at the other end of the scale there are regions that, despite intensive investigation, are still not well characterized in terms of their constituent areas. For example, it is clear that the inferior temporal cortex is functionally heterogeneous, given that recordings from different regions yield neurons with distinct response properties, and that different lesions result in distinct constellations of symptoms (e.g., References 116 through 119). Yet histological transitions in this region are gradual, the patterns of interconnections are difficult to assign as clearly "intrinsic" or "extrinsic" (e.g., References 120 and 121), and visuotopic maps are coarse (Section 11.3.5). All evidence considered, the organization of the inferior temporal cortex can be described just as adequately in terms of gradual changes, without the implication of a mosaic of sharply defined areas.

Between these extremes, several topographically organized visual areas exist for which we know the broad outlines, but not the complete picture. Perhaps the best example is reflected in the long-standing controversy (e.g., References 13 and 122) regarding the organization of the cortical belt immediately anterior to V2 in monkeys. Although a consensus has gradually emerged that a V3-like area (i.e., similar to area 19 of other mammals[68,89]) occupies a large fraction of this belt, the exact extent of primate V3 remains a subject of debate (Figure 11.7). As seen above, progress is being made; for example, it is now clear that "ventral V3," or VP, is part of a larger area that represents both the upper and lower quadrants.[14,29,61,123] Yet results of different investigations on the extent and visuotopy of dorsal V3 (the lower quadrant representation) remain incompatible. According to one point of view, which has emerged from electrophysiological mapping studies in New World monkeys, there is a cortical area (DM) representing the upper quadrant immediately adjacent to dorsal V2 (Figure 11.5C and Figure 11.7A). In this proposal, the V3-like area is restricted to ventral and lateral cortices (Figure 11.6).[14,15,70] Another point of view, reflecting the results of anatomical tracing experiments (Figure 11.7B), is that dorsal V3 extends all the way to the dorsal midline, separating V2 from DM.[11,61,123] Either way, it is fair to say that, at this stage, no one can be certain that an electrode, tracer injection, or lesion is located in dorsal V3 (at least not with the same degree of confidence that one can know that it is located in V1, V2, or MT).

It is also unlikely that any single technique will be able to resolve this type of controversy. For example, given the discrete nature of the sampling in electrophysiological mapping experiments, it could be argued that such studies have so far "missed" the evidence for a thin, continuous lower quadrant V3 separating V2 from DM in New World monkeys, due to the separation between recording sites. However, the use of discrete injections of anatomical tracers is also problematic, because it allows the study only of the representations of a few points of the visual field at a time. Thus, complex topographies such as those suggested by Figure 11.5B and Figure 11.7A are not revealed particularly well by this technique (see also Reference 58). Combining either method with histological analysis is also not always the answer, given that the gradual nature of histological transitions in extrastriate cortex often precludes unambiguous decisions. For example, both schemes shown in Figure 11.7 have been validated by architectural analyses, depending on whether the distribution of myelin[70,124–126] or cytochrome oxidase[61,123] was charted.

Perhaps the main lesson to be learned from this chapter is that, while it is possible that many of the current controversies will be solved by future studies, one must also be prepared to face the prospect that the cerebral cortex is not entirely formed by neatly defined, homogeneous fields. One possibility worth investigating is that borders that are sharply defined and methodologically robust in adults are those that are specified by the sharpest transitions in the expression of cell surface molecules during development (e.g., V1).[113] These early-specified boundaries are likely to

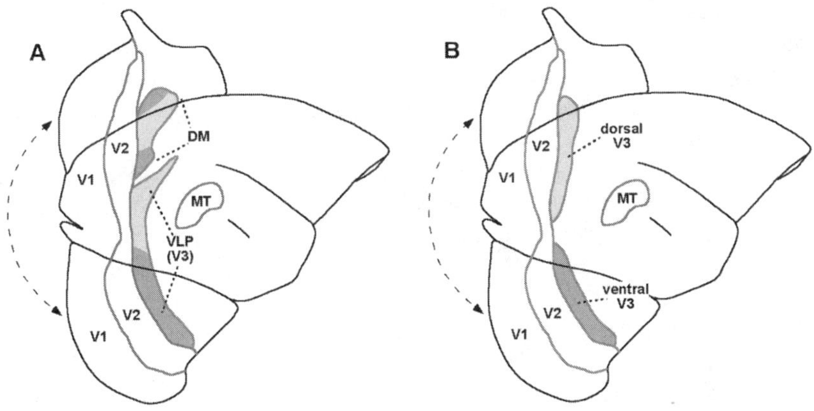

FIGURE 11.7 Uncertainty regarding the organization of the third visual complex. Lateral views of the brain of a New World monkey, illustrating two current hypotheses on the organization of the cortex immediately anterior to the second visual area (V2). These diagrams include graphically "unfolded" views of the cortex hidden in the medial and ventral surfaces of the brain (see also legend to Figure 11.6 for details). Dark gray indicates representations of the upper contralateral quadrant (or parts thereof), and light gray indicates representations of the lower contralateral quadrant. (A) Hypothesis based on the results of electrophysiological recordings. According to this view, the dorsolateral and ventral regions of extrastriate cortex include a field homologous to area 19 of most mammals, named the ventrolateral posterior area (VLP), or "ventral V3."[129] Because neurons with receptive fields in the upper quadrant have been recorded immediately anterior to V2 in dorsomedial cortex,[13,70,130,131] the extension of V3 to the dorsal midline is regarded as unlikely. Instead, a dorsomedial area (including the "dorsal V3" of Zeki[55]) is proposed. The dorsomedial and ventrolateral components of the "third visual complex" also differ in terms of myeloarchitecture.[125,132] (B) Alternative hypothesis, based on the results of anatomical tracing.[61] As in the first diagram, the cortex anterior to ventral V2 forms a relatively simple representation of the upper quadrant. However, it is hypothesized that the dorsomedial cortex anterior to dorsal V2 lacks upper quadrant representations. Instead, a simple strip-like dorsal V3 is proposed to exist in the corresponding location, representing only the lower contralateral quadrant. This view is also supported by architectural analyses; in particular, both dorsal and ventral V3 are characterized by a coarse system of cytochrome oxidase bands. Although a similar controversy exists regarding the organization of Old World monkey cortex,[15] the visuotopic configuration of the areas is different; in particular, the DM homologue does not have an upper visual field representation adjacent to V2.[62]

characterize phylogenetically older, "core" areas, which can be reliably identified across many species.[52] In evolutionarily newer cortices, less precise boundaries (e.g., interdigitating areas and gradual transition zones) as well as more complex visuotopic maps and larger individual variability are to be expected, particularly if activity-dependent processes are responsible for a large fraction of the formation of the circuitry.[92] These regions are also likely to be more variable among species, thereby compounding the inherent difficulty in the identification of homologous fields (e.g., References 56 and 127). Thus, clarification of the developmental mechanisms underlying the formation of cortical maps, and in particular the degree of genetic specification involved in the establishment of different areas, will also have a significant impact on our understanding of the evolution of visual cortex.

Finally, the realization that there may exist different "correct" ways of subdividing the brain will probably require a more pragmatic view of what constitutes a cortical area. In particular, many areas could become operational definitions, based on a certain criterion; for example, one could use the designation V3 to refer to a part of the human brain that can be reliably defined by visuotopic analysis of functional MRI.[49] This would ensure a common frame of reference for clinicians or experimenters without necessarily implying that this is an area validated by multiple, converging criteria, or that it corresponds in all respects to V3 defined by a different combination of methods in a monkey species (e.g., Reference 128). Indeed, given the relative histological uniformity of the neocortex, visuotopic organization is likely to remain one of the main anatomical–functional criteria for identification of cortical fields in humans. In this context, high-resolution data on the precision, topology, and variability of visuotopic maps in nonhuman primates, as well as their relationship to cortical histology or connectivity, will become important in establishing boundaries for the scientific interpretation of noninvasive functional mapping or clinical data in humans.

ACKNOWLEDGMENTS

The authors thank Dmitri Chklovskii for comments on sections of this manuscript, and acknowledge the Australian Research Council and the National Health and Medical Research Council for funding our research. Equipment support from the Clive and Vera Ramaciotti Foundation and the ANZ Charitable Trust is also gratefully acknowledged.

REFERENCES

1. Blasdel, G.G. and Lund, J.S., Termination of afferent axons in macaque striate cortex, *J. Neurosci.*, 3, 1389, 1983.
2. Schein, S.J. and de Monasterio, F.M., Mapping of retinal and geniculate neurons onto striate cortex of macaque, *J. Neurosci.*, 7, 996, 1987.
3. Azzopardi, P. and Cowey, A., Preferential representation of the fovea in the primary visual cortex, *Nature*, 361, 719, 1993.
4. Hubel, D.H. and Wiesel, T.N., Uniformity of monkey striate cortex: a parallel relationship between field size, scatter, and magnification factor, *J. Comp. Neurol.*, 158, 295, 1974.
5. Talbot, S.A. and Marshall, W.H., Physiological studies on neural mechanisms of visual localization and discrimination, *Am. J. Ophthalmol.*, 24, 1255, 1941.
6. Daniel, P.M. and Whitteridge, D., The representation of the visual field on the cerebral cortex in monkeys, *J. Physiol.* (London), 159, 203, 1961.
7. Allman, J.M. and Kaas, J.H., Representation of the visual field in striate and adjoining cortex of the owl monkey (*Aotus trivirgatus*), *Brain Res.*, 35, 89, 1971.
8. Inouye, T., *Die Sehstörungen bei Schußverletzungen der kortikalen Sehsphäre*, Engelmann, Leipzig, 1909.

9. Holmes, G. and Lister, W.T., Disturbances of vision from cerebral lesions with special reference to the cortical representation of the macula, *Brain,* 39, 34, 1916.
10. Cowey, A., Projection of the retina on to striate and prestriate cortex in the squirrel monkey, *Saimiri sciureus*, *J. Neurophysiol.*, 27, 366, 1964.
11. Zeki, S.M., Representation of central visual fields in prestriate cortex of monkey, *Brain Res.*, 14, 271, 1969.
12. Allman, J.M. and Kaas, J.H., A representation of the visual field in the caudal third of the middle temporal gyrus of the owl monkey (*Aotus trivirgatus*), *Brain Res.*, 31, 85, 1971.
13. Allman, J.M. and Kaas, J.H., The dorsomedial cortical visual area: a third tier area in the occipital lobe of the owl monkey (*Aotus trivirgatus*). *Brain Res.*, 100, 473, 1975.
14. Rosa, M.G.P. and Tweedale, R., Visual areas in lateral and ventral extrastriate cortices of the marmoset monkey, *J. Comp. Neurol.*, 422, 621, 2000.
15. Rosa, M.G.P. and Tweedale, R., The dorsomedial visual areas in New World and Old World monkeys: homology and function, *Eur. J. Neurosci.*, 13, 421, 2001.
16. Rosa, M.G.P., Visuotopic organisation of primate extrastriate cortex, in *Cerebral Cortex,* Vol. 12: *Extrastriate Cortex in Primates*, Rockland K.S., Kaas, J.H., and Peters, A., Eds., Plenum Press, New York, 1997, 127.
17. Worgotter, F., Suder, K., Zhao, Y., Kerscher, N., Eysel, U.T., and Funke, K., State-dependent receptive-field restructuring in the visual cortex, *Nature*, 396, 165, 1998.
18. Kapadia, M.K., Westheimer, G., and Gilbert, C.D., Dynamics of spatial summation in primary visual cortex of alert monkeys, *Proc. Natl. Acad. Sci. U.S.A.*, 96, 12073, 1999.
19. Sceniak, M.P., Ringach, D.L., Hawken, M.J., and Shapley, R., Contrast's effect on spatial summation by macaque V1 neurons, *Nat. Neurosci.*, 2, 733, 1999.
20. Suder, K., Funke, K., Zhao, Y., Kerscher, N., Wennekers, T., and Worgotter, F., Spatial dynamics of receptive fields in cat primary visual cortex related to the temporal structure of thalamocortical feedforward activity: experiments and models, *Exp. Brain. Res.,* 144, 430, 2002.
21. Duhamel, J.-R., Colby, C.L., and Goldberg, M.E., The updating of the representation of visual space in parietal cortex by intended eye movements, *Science*, 255, 90, 1992.
22. Tolias, A.S., Moore, T., Smirnakis, S.M., Tehovnik, E.J., Siapas, A.G., and Schiller, P.H., Eye movements modulate visual receptive fields of V4 neurons, *Neuron,* 29, 757, 2001.
23. Ben Hamed, S., Duhamel, J.R., Bremmer, F., and Graf, W., Visual receptive field modulation in the lateral intraparietal area during attentive fixation and free gaze, *Cereb. Cortex*, 12, 234, 2002.
24. Eysel, U.T., Eyding, D., and Schweigart, G., Repetitive optical stimulation elicits fast receptive field changes in mature visual cortex, *NeuroReport*, 9, 949, 1998.
25. Kaas, J.H., Krubitzer, L.A., Chino, Y.M., Langston, A.L., Polley, E.H., and Blair, N., Reorganization of retinotopic cortical maps in adult mammals after lesions of the retina, *Science,* 248, 229, 1990.
26. Heinen, S.J. and Skavenski, A.A., Recovery of visual responses in foveal V1 neurons following bilateral foveal lesions in adult monkey, *Exp. Brain Res.*, 83, 670, 1991.
27. Darian-Smith, C. and Gilbert, C.D., Topographic reorganization in the striate cortex of the adult cat and monkey is cortically mediated, *J. Neurosci.*, 15, 1631, 1995.
28. Schmid, L.M., Rosa, M.G.P., and Calford, M.B., Retinal detachment induces massive immediate reorganization in visual cortex, *NeuroReport*, 6, 1349, 1995.
29. Rosa, M.G.P., Tweedale, R., and Elston, G.N., Visual responses of neurons in the middle temporal area of New World monkeys after lesions of striate cortex, *J. Neurosci.*, 20, 5552, 2000.

30. Bringuier, V., Chavane, F., Glaeser, L., and Fregnac, Y., Horizontal propagation of visual activity in the synaptic integration field of area 17 neurons, *Science*, 283, 695, 1999.
31. Li, W., Thier, P., and Wehrhahn, C., Neuronal responses from beyond the classic receptive field in V1 of alert monkeys, *Exp. Brain Res.*, 139, 359, 2001.
32. Dow, B.M., Snyder, A.Z., Vautin, R.G., and Bauer, R., Magnification factor and receptive field size in foveal striate cortex of the monkey, *Exp. Brain Res.*, 44, 213, 1981.
33. Albright, T.D. and Desimone, R., Local precision of visuotopic organization in the middle temporal area (MT) of the macaque, *Exp. Brain Res.,* 65, 582, 1987.
34. Rosa, M.G.P., Fritsches, K.A., and Elston, G.N., The second visual area in the marmoset monkey: visuotopic organisation, magnification factors, architectonical boundaries, and modularity, *J. Comp. Neurol.*, 387, 547, 1997.
35. Sherk, H. and Mulligan, K.A., Retinotopic order is surprisingly good within cell columns in the cat's lateral suprasylvian cortex, *Exp. Brain Res.*, 91, 46, 1992.
36. Das, A. and Gilbert, C.D., Distortions of visuotopic map match orientation singularities in primary visual cortex, *Nature*, 387, 594, 1997.
37. McIlwain, J.T., Large receptive fields and spatial transformations in the visual system, *Int. Rev. Physiol.*, 10, 223, 1976.
38. Rosa, M.G.P., Sousa, A.P.B., and Gattass, R., Representation of the visual field in the second visual area in the *Cebus* monkey, *J. Comp. Neurol.*, 275, 326, 1988.
39. Roe, A.W. and Ts'o, D.Y., Visual topography in primate V2: multiple representation across functional stripes, *J. Neurosci.,*15, 3689, 1995.
40. Tootell, R.B.H., Switkes, E., Silverman, M.S., and Hamilton, S.J., Functional anatomy of macaque striate cortex. II. Retinotopic organization, *J. Neurosci.*, 8, 1531, 1988.
41. Tootell, R.B.H. and Hamilton, S.L., Functional anatomy of the second visual area (V2) in the macaque, *J. Neurosci.*, 9, 2620, 1989.
42. Allman, J.M. and Kaas, J.H., The organization of the second visual area (V II) in the owl monkey: a second order transformation of the visual hemifield, *Brain Res.*, 76, 247, 1974.
43. Allman, J.M., Kaas, J.H., and Lane, R.H., The middle temporal area (MT) in the bushbaby, *Galago senegalensis*, *Brain Res.*, 57, 197, 1973.
44. Gattass, R., Gross, C.G., and Sandell, J.H., Visual topography of V2 in the macaque, *J. Comp. Neurol.,* 201, 519, 1981.
45. Gattass, R., Sousa, A.P.B., and Rosa, M.G.P., Visual topography of V1 in the *Cebus* monkey, *J. Comp. Neurol.,* 259, 529, 1987.
46. Fiorani, M., Jr., Gattass, R., Rosa, M.G.P., and Sousa, A.P.B., Visual area MT in the *Cebus* monkey: location, visuotopic organization, and variability, *J. Comp. Neurol.*, 287, 98, 1989.
47. Fritsches, K.A. and Rosa, M.G.P., Visuotopic organisation of striate cortex in the marmoset monkey (*Callithrix jacchus*), *J. Comp. Neurol.*, 372, 264, 1996.
48. Rosa, M.G.P. and Elston, G.N., Visuotopic organisation and neuronal response selectivity for direction of motion in visual areas of the caudal temporal lobe of the marmoset monkey (*Callithrix jacchus*): middle temporal area, middle temporal crescent, and surrounding cortex, *J. Comp. Neurol.*, 393, 505, 1998.
49. Sereno, M.I., Dale, A.M., Reppas, J.B., Kwong, K.K., Belliveau, J.W., Brady, T.J., Rosen, B.R., and Tootell, R.B., Borders of multiple visual areas in humans revealed by functional magnetic resonance imaging, *Science*, 268, 889, 1995.
50. Rosa, M.G.P., Casagrande, V.A., Preuss, T., and Kaas, J.H., Visual field representation in striate and prestriate cortices of a prosimian primate (*Galago garnetti*), *J. Neurophysiol.*, 77, 3193, 1997.

51. Rosa, M.G.P., Schmid, L.M., and Pettigrew, J.D., Organization of the second visual area in the megachiropteran bat *Pteropus*, *Cereb. Cortex*, 4, 52, 1994.
52. Rosa, M.G.P. and Krubitzer, L.A., The evolution of visual cortex: where is V2? *Trends Neurosci.*, 22, 242, 1999.
53. Allman, J.M. and Kaas, J.H., A crescent-shaped cortical visual area surrounding the middle temporal area (MT) in the owl monkey (*Aotus trivirgatus*), *Brain Res.*, 81, 199, 1974.
54. Allman, J.M. and Kaas, J.H., Representation of the visual field on the medial wall of occipital-parietal cortex in the owl monkey, *Science*, 191, 572, 1976.
55. Zeki, S.M., The third visual complex of rhesus monkey prestriate cortex, *J. Physiol.* (London), 277, 245, 1978.
56. Pigarev, I.N., Nothdurft, H.C., and Kastner, S., Neurons with radial receptive fields in monkey area V4A: evidence of a subdivision of prelunate gyrus based on neuronal response properties, *Exp. Brain Res.*, 145, 199, 2002.
57. Newsome, W.T., Maunsell, J.H., and Van Essen, D.C., Ventral posterior visual area of the macaque: visual topography and areal boundaries, *J. Comp. Neurol.*, 252, 139, 1986.
58. Sereno, M.I., McDonald, C.T., and Allman, J.M., Analysis of retinotopic maps in extrastriate cortex, *Cereb. Cortex*, 4, 601, 1994.
59. Cragg, B.G. and Ainsworth, A., The topography of the afferent projections in the circumstriate visual cortex of the monkey studied by the Nauta method, *Vision Res.*, 9, 733, 1969.
60. Zeki, S.M., Simultaneous anatomical demonstration of the representation of the vertical and horizontal meridians in areas V2 and V3 of rhesus monkey visual cortex, *Proc. R. Soc. Lond. B*, 195, 517, 1977.
61. Lyon, D.C. and Kaas, J.H., Connectional and architectonic evidence for dorsal and ventral V3, and dorsomedial area in marmoset monkeys, *J. Neurosci.*, 21, 249, 2001.
62. Gattass, R., Sousa, A.P.B., and Gross, C.G., Visuotopic organization and extent of V3 and V4 of the macaque, *J. Neurosci.*, 8, 1831, 1988.
63. Piñon, M.C.G.P., Gattass, R., and Sousa, A.P.B., Area V4 in *Cebus* monkey: extent and visuotopic organization, *Cereb. Cortex*, 8, 685, 1998.
64. Kaas, J.H. and Morel, A., Connections of visual areas of the upper temporal lobe of owl monkeys: the MT crescent and dorsal and ventral subdivisions of FST, *J. Neurosci.*, 13, 534, 1993.
65. Tusa, R.J., Rosenquist, A.C., and Palmer, L.A., Retinotopic organization of areas 18 and 19 in the cat, *J. Comp. Neurol.*, 185, 657, 1979.
66. Albus, K. and Beckmann, R., Second and third visual areas of the cat: interindividual variability in retinotopic arrangement and cortical location, *J. Physiol.* (London), 299, 247, 1980.
67. Sherk, H. and Mulligan, K.A., A reassessment of the lower visual field map in striate-recipient lateral suprasylvian cortex, *Vis. Neurosci.*, 10, 131, 1993.
68. Manger, P.R., Kiper, D., Masiello, I., Murillo, L., Tettoni, L., Hunyadi, Z., and Innocenti, G.M., The representation of the visual field in three extrastriate areas of the ferret (*Mustela putorius*) and the relationship of retinotopy and field boundaries to callosal connectivity, *Cereb. Cortex*, 12, 423, 2002.
69. Manger, P.R., Woods, T.M., Munoz, A., and Jones, E.G., Hand/face border as a limiting boundary in the body representation in monkey somatosensory cortex, *J. Neurosci.*, 17, 6338, 1997.
70. Rosa, M.G.P. and Schmid, L.M., Visual areas in the dorsal and medial extrastriate cortices of the marmoset, *J. Comp. Neurol.*, 359, 272, 1995.

71. Galletti, C., Fattori, P., Gamberini, M., and Kutz, D.F., The cortical visual area V6: brain location and visual topography, *Eur. J. Neurosci.*, 11, 3922, 1999.
72. Colby, C.L., Gattass, R., Olson, C.R., and Gross, C.G., Topographical organization of cortical afferents to extrastriate visual area PO in the macaque: a dual tracer study, *J. Comp. Neurol.,* 269, 392, 1988.
73. Neuenschwander, S., Gattass, R., Sousa, A.P.B., and Piñon, M.C.G.P., Identification and visuotopic organization of areas PO and POd in *Cebus* monkey, *J. Comp. Neurol.*, 340, 65, 1994.
74. Perkel, D.J., Bullier, J. and Kennedy, H., Topography of the afferent connectivity of area 17 in the macaque monkey: a double-labelling study, *J. Comp. Neurol.*, 253, 374, 1986.
75. Sousa, A.P.B., Piñon, M.C.G.P., Gattass, R., and Rosa, M.G.P., Topographic organization of cortical input to striate cortex in the *Cebus* monkey: a fluorescent tracer study, *J. Comp. Neurol.*, 308, 665, 1991.
76. Boussaoud, D., Desimone, R., and Ungerleider, L.G., Visual topography of area TEO in the macaque, *J. Comp. Neurol.*, 306, 554, 1991.
77. Felleman, D.J., Xiao, Y., and McClendon, E., Modular organization of occipito-temporal pathways: cortical connections between visual area 4 and visual area 2 and posterior inferotemporal ventral area in macaque monkeys, *J. Neurosci.*, 17, 3185, 1997.
78. Galletti, C., Fattori, P., Kutz, D.F., and Gamberini, M., Brain location and visual topography of cortical area V6A in the macaque monkey, *Eur. J. Neurosci.*, 11, 575, 1999.
79. Ben Hamed, S., Duhamel, J.R., Bremmer, F., and Graf, W., Representation of the visual field in the lateral intraparietal area of macaque monkeys: a quantitative receptive field analysis, *Exp. Brain Res.*, 140, 127, 2001.
80. Komatsu, H. and Wurtz, R.H., Relation of cortical areas MT and MST to pursuit eye movements, I. Localization and visual properties of neurons, *J. Neurophysiol.*, 60, 580, 1988.
81. Hikosaka, K., Representation of foveal visual fields in the ventral bank of the superior temporal sulcus in the posterior inferotemporal cortex of the macaque monkey, *Behav. Brain Res.*, 96, 101, 1998.
82. Burkhalter, A., Felleman, D.J., Newsome, W.T., and Van Essen, D.C., Anatomical and physiological asymmetries related to visual areas V3 and VP in macaque extrastriate cortex, *Vision Res.*, 26, 63, 1986.
83. Woolsey, C.N., Comparative studies on cortical representation of vision, *Vision Res.*, 3, 365, 1971.
84. Wässle, H., Grünert, U., Röhrenbeck, J., and Boycott, B.B., Cortical magnification factor and the ganglion cell density of the primate retina, *Nature*, 341, 643, 1989.
85. Silveira, L.C.L, Perry, V.H., and Yamada, E.S., The retinal ganglion cell distribution and the representation of the visual field in area 17 of the owl monkey, *Aotus trivirgatus*, *Vis. Neurosci.*, 10, 887, 1993.
86. Azzopardi, P. and Cowey, A., The overrepresentation of the fovea and adjacent retina in the striate cortex and dorsal lateral geniculate nucleus of the macaque monkey, *Neuroscience*, 72, 627, 1996.
87. Arnett, D. and Spraker, T.E., Cross-correlation analysis of the maintained discharge of rabbit retinal ganglion cells, *J. Physiol.* (London), 317, 29, 1981.
88. Wong, R.O., Meister, M., and Shatz, C.J., Transient period of correlated bursting activity during development of the mammalian retina. *Neuron*, 11, 923, 1993.
89. Rosa, M.G.P., Topographic organisation of extrastriate areas in the flying fox: implications for the evolution of mammalian visual cortex, *J. Comp. Neurol.*, 411, 503, 1999.

90. Cowey, A., Why are there so many visual areas? in *The Organization of the Cerebral Cortex*, Schmitt, F.O., Worden, F.G., Adelman, G., and Dennis, S.G., Eds., MIT Press, Cambridge, MA, 1981, 395.
91. Weinberg, R.J., Are topographic maps fundamental to sensory processing? *Brain Res. Bull.*, 44, 113, 1997.
92. Rosa, M.G.P., Visual maps in the adult primate cerebral cortex: some implications for brain development and evolution, *Braz. J. Med. Biol. Res.*, 35, 1485, 2002.
93. Cherniak, C., Neural component placement, *Trends Neurosci.*, 18, 522, 1995.
94. Koulakov, A.A. and Chklovskii, D.B., Orientation preference patterns in mammalian visual cortex: a wire length minimization approach, *Neuron*, 29, 519, 2001.
95. Schwartz, E.L., Spatial mapping in the primate sensory projection: analytic structure and relevance to perception, *Biol. Cybern.*, 25, 181, 1977.
96. Gattass, R., Rosa, M.G.P., Sousa, A.P.B., Piñon, M.C.G.P., Fiorani, M., Jr., and Neuenschwander, S., Cortical streams of visual information processing in primates, *Braz. J. Med. Biol. Res.*, 23, 375, 1990.
97. Dragoi, V. and Sur, M., Dynamic properties of recurrent inhibition in primary visual cortex: contrast and orientation dependence of contextual effects, *J. Neurophysiol.*, 83, 1019, 2000.
98. Lauritzen, T.Z., Krukowski, A.E., and Miller, K.D., Local correlation-based circuitry can account for responses to multi-grating stimuli in a model of cat V1, *J. Neurophysiol.*, 86, 1803, 2001.
99. Adini, Y., Sagi, D., and Tsodyks, M., Context-enabled learning in the human visual system, *Nature*, 415, 790, 2002.
100. Stepniewska, I. and Kaas, J.H., Topographic patterns of V2 cortical connections in macaque monkeys, *J. Comp. Neurol.*, 371, 129, 1996.
101. Gattass, R., Sousa, A.P.B., Mishkin, M., and Ungerleider, L.G., Cortical projections of area V2 in the macaque, *Cereb. Cortex,* 7, 110, 1997.
102. Collins, C.E., Stepniewska, I., and Kaas, J.H., Topographic patterns of V2 cortical connections in a prosimian primate (*Galago garnetti*), *J. Comp. Neurol.,* 431, 155, 2001.
103. Heider, B., Meskenaite, V., and Peterhans, E., Anatomy and physiology of a neural mechanism defining depth order and contrast polarity at illusory contours, *Eur. J. Neurosci.*, 12, 4117, 2000.
104. Marcar, V.L., Raiguel, S.E., Xiao, D., and Orban, G.A., Processing of kinetically defined boundaries in areas V1 and V2 of the macaque monkey, *J. Neurophysiol.*, 84, 2786, 2000.
105. Ramsden, B.M., Hung, C.P., and Roe, A.W., Real and illusory contour processing in area V1 of the primate: a cortical balancing act, *Cereb. Cortex*, 11, 648, 2001.
106. Chaudhuri, A. and Glaser, D.A., Metastable motion anisotropy, *Vis. Neurosci.*, 7, 397, 1991.
107. Pillow, J. and Rubin, N., Perceptual completion across the vertical meridian and the role of early visual cortex, *Neuron*, 33, 805, 2002.
108. Lamme, V.A. and Roelfsema, P.R., The distinct modes of vision offered by feedforward and recurrent processing, *Trends Neurosci.*, 23, 571, 2000.
109. Bullier, J., Integrated model of visual processing, *Brain Res. Rev.*, 36, 96, 2001.
110. Girard, P., Hupé, J.M., and Bullier, J., Feedforward and feedback connections between areas V1 and V2 of the monkey have similar rapid conduction velocities, *J. Neurophysiol.*, 85, 1328, 2001.
111. Houzel, J.C., Milleret, C., and Innocenti, G., Morphology of callosal axons interconnecting areas 17 and 18 of the cat, *Eur. J. Neurosci.*, 6, 898, 1994.

112. Innocenti, G.M., Lehmann, P., and Houzel, J.C., Computational structure of visual callosal axons, *Eur. J. Neurosci.*, 6, 918, 1994.
113. Donoghue, M.J. and Rakic, P., Molecular gradients and compartments in the embryonic primate cerebral cortex, *Cereb. Cortex,* 9, 586, 1999.
114. Brodmann, K., *Localisation in the Cerebral Cortex*, translated by L.J. Garey, 1994, Smith-Gordon, London, 1909.
115. Krubitzer, L., The organization of neocortex in mammals: are species differences really so different? *Trends Neurosci.*, 18, 408, 1995.
116. Baylis, G.C., Rolls, E.T., and Leonard, C.M., Functional subdivisions of the temporal lobe neocortex, *J. Neurosci.*, 7, 330, 1987.
117. Kobatake, E. and Tanaka, K., Neuronal selectivities to complex object features in the ventral visual pathway of the macaque cerebral cortex, *J. Neurophysiol.*, 71, 856, 1994.
118. Cowey, A., Heywood, C.A., and Irving-Bell, L., The regional cortical basis of achromatopsia: a study on macaque monkeys and an achromatopsic patient, *Eur. J. Neurosci.*, 14, 1555, 2001.
119. Tamura, H. and Tanaka, K., Visual response properties of cells in the ventral and dorsal parts of the macaque inferotemporal cortex, *Cereb. Cortex*, 11, 384, 2001.
120. Kondo, H., Hashikawa, T., Tanaka, K., and Jones, E.G., Neurochemical gradient along the monkey occipito-temporal cortical pathway, *NeuroReport*, 5, 613, 1994.
121. Suzuki, W., Saleem, K.S., and Tanaka, K., Divergent backward projections from the anterior part of the inferotemporal cortex (area TE) in the macaque, *J. Comp. Neurol.*, 422, 206, 2000.
122. Van Essen, D.C. and Zeki, S.M., The topographic organization of rhesus monkey prestriate cortex, *J. Physiol.* (London), 277, 193, 1978.
123. Lyon, D.C. and Kaas, J.H., Evidence for a modified V3 with dorsal and ventral halves in macaque monkeys, *Neuron*, 33, 453, 2002.
124. Krubitzer, L.A. and Kaas, J.H., Cortical connections of MT in four species of primates: areal, modular, and retinotopic patterns, *Vis. Neurosci.,* 5, 165, 1990.
125. Beck, P.D. and Kaas, J.H., Cortical connections of the dorsomedial visual area in New World owl monkeys (*Aotus trivirgatus*) and squirrel monkeys (*Saimiri sciureus*), *J. Comp. Neurol.,* 400, 18, 1998.
126. Beck, P.D. and Kaas, J.H., Cortical connections of the dorsomedial visual area in Old World macaque monkeys, *J. Comp. Neurol.,* 406, 487, 1999.
127. Tootell, R.B.H. and Hadjikhani, N., Where is "dorsal V4" in human visual cortex? Retinotopic, topographic and functional evidence, *Cereb. Cortex*, 11, 298, 2001.
128. Tootell, R.B.H., Mendola, J.D., Hadjikhani, N.K., Ledden, P.J., Liu, A.K., Reppas, J.B., Sereno, M.I., and Dale, A.M., Functional analysis of V3A and related areas in human visual cortex, *J. Neurosci.*, 17, 7060, 1997.
129. Rosa, M.G.P., Piñon, M.C., Gattass, R., and Sousa, A.P.B., "Third tier" ventral extrastriate cortex in the New World monkey, *Cebus apella*, *Exp. Brain Res.*, 132, 287, 2000.
130. Krubitzer, L.A. and Kaas, J.H., The dorsomedial visual area of owl monkeys: connections, myeloarchitecture, and homologies in other primates, *J. Comp. Neurol.*, 334, 497, 1993.
131. Allman, J., Jeo, R., and Sereno, M., The functional organization of the visual cortex in owl monkeys, in *Aotus*: *The Owl Monkey,* Baer, J.F., Weller, R.E., and Kakoma, I., Eds., Academic Press, San Diego, 1994, 287.
132. Rosa, M.G.P., Soares, J.G.M., Fiorani, M., Jr., and Gattass, R., Cortical afferents of visual area MT in the *Cebus* monkey: possible homologies between New and Old World monkeys, *Vis. Neurosci.*, 10, 827, 1993.

12 Face Expertise and Category Specialization in the Human Occipitotemporal Cortex

Isabel Gauthier

CONTENTS

12.1 INTRODUCTION

Much has been written about face recognition, including reviews addressing what, if anything, may be "special" about face processing.[1–3] This chapter takes a different approach and instead asks what we can learn from the case of face processing about how object recognition is generally supported by the visual system. Interestingly, if face recognition were truly special, answering to rules different from the recognition of other objects and supported by a neural network functionally independent from that involved in object recognition, then what we would learn about face recognition would be of little relevance to object recognition. Therefore, the extent to which we

0-8493-1243-4/04/$0.00+$1.50

can learn about object recognition using the particular case of face processing is by itself an answer to the question of whether faces are special.

First, it is important to define the concepts of modularity and of specialization as used in this chapter. When authors refer to the possibility that faces are "special," they often refer to the concept of modularity as defined by Fodor[4] rather than to the more general concept of specialization. Specialization of face processing, broadly speaking, refers to the undisputed fact that there are parts of the visual system that respond more to faces than to other objects. How this specialization should be interpreted, rather than its existence, is a source of debate. One interpretation is provided by the concept of modularity. The original Fodorian definition of a module includes several characteristics, including the idea of cognitive encapsulation (a process independent of the rest of the system, our goals, desires, and knowledge of the world), that of automaticity and speed of processing, and the suggestions that a module is localized and innate.[4] It is important to note that Fodor did not believe that all of his proposed properties of modularity would apply in every case, and different authors have since proposed various definitions of modularity that do not include all of Fodor's original criteria. However, one of Fodor's criteria has been argued to embody the very nature of modularity and by itself defines modularity according to a neo-Fodorian approach.[5] This criterion is domain specificity, the requirement that modules deal exclusively with a single information type, albeit one of enormous relevance to the species. For face processing, the domain would be the geometry proper to faces, so that any visual input that does not possess this geometry would not be processed by a face module. The idea of domain specificity is at the heart of a debate in cognitive neuroscience regarding the specialization in the visual system for faces: whereas some emphasize evidence for face-specific visual processes,[6,7] others argue that putatively face-selective effects, both behavioral and neural, can be engaged by nonface objects under specific circumstances.[8–10] According to this latter view, there is much to learn from those circumstances under which object processing becomes facelike. In particular, these studies offer a window onto the forces that may control the specialization of face processing and can serve as valuable case studies in the functional plasticity of the mechanisms responsible for object processing.

This chapter focuses on the processing of faces in terms of their identity at the individual level. This applies not only to the recognition of those faces with which we are familiar, but also to unfamiliar faces, which can also be encoded as individuals for later recognition. The face is certainly a source of much more information, including age, sex, race, and emotion, dimensions that have been the object of much work reviewed elsewhere.[11–13] Whereas the most important neural substrates for processing face identity are in the lateral occipital and fusiform gyri of the occipitotemporal pathway, the processing of facial emotions recruits other areas such as the amygdala and the superior temporal sulcus[12,14] — also engaged by neutral facial movements.[14] Despite these differences, and despite neuropsychological evidence for double dissociations between deficits in face recognition[15] and those in emotion recognition,[16,17] it remains uncertain whether these two dimensions (as well as others, such as lip reading or sex judgments) are supported by independent (i.e., modular) systems.[18]

12.2 IS THERE AN INNATE BIAS FOR FACE PROCESSING?

It is much easier to find evidence for face selectivity than to explain its origins. Developmental psychologists have long been fascinated by this question, and infants' face perception has been studied extensively.[19] It is important to remember that even newborns can learn very rapidly. For example, in one study[20] newborns who were shown four briefly flashed faces (400 ms each) for only a few repetitions preferred to look at a composite of these four faces than at a composite of four new faces. Such learning, within less than 1 min of exposure to the entire set of four faces, reminds us that any preference observed in newborns with even the shortest of opportunity to learn can be difficult to interpret unequivocally as an innate bias. That said, several studies have uncovered a preference in newborns for facelike over nonfacelike stimuli.[21–23] Johnson and Morton[22] proposed that faces are special in the newborn, who may possess a template-matching device termed "Conspec," which would contain information about the general structure of a face (similar to the "*config*" stimulus shown in Figure 12.1A). Conspec would lead newborns to attend to facelike stimuli, until a mechanism that would come into play later and called "Conlern" would allow children to learn to distinguish individual faces. The Conspec mechanism is thought to be supported subcortically (whereas Conlern relies on cortical mechanisms). This is supported by findings that the preference is present only when objects are presented in the temporal hemifield,[23] an index of extrageniculate mediation, as well as the finding that the preference disappears by 6 weeks of age,[24] consistent with the timing of a switch from subcortical to cortical influences.

An interesting characteristic of infants' face processing is that it changes rapidly during the first few months of life. This was demonstrated by a study in which preference for different types of face and control stimuli was explored in newborns (mean age of 53 min), 6-week-olds, and 12-week-olds.[24] Newborns preferred the config stimuli over its inverted version (Figure 12.1A), but this preference was not present in the older infants. When a stimulus consisting of the phase spectrum of a face but the amplitude spectrum of a lattice was contrasted with a stimulus containing the phase of a lattice but the amplitude of a face (Figure 12.1B), most newborns preferred the image containing the amplitude of a face (a stimulus that looks to us less facelike) but all infants in the older groups preferred the image containing the phase of a face. Finally, when positive and negative contrast faces are compared (Figure 12.1C), only 12-week-olds show a preference for the positive contrast. In light of such complex changes in visual preference early in life, only the earliest preference, that for a config stimulus over its inverted version, has been proposed to be innate.[22,24] This conclusion, however, is seriously challenged by more recent studies. In an elegant series of experiments with newborns who were a few days old (about 60 h old), the preference for the up-down asymmetry of many different patterns was investigated. The idea was to test whether the fact that config presents more elements in its upper than lower part compared to its inverted version was crucial in determining the newborns' preference. Newborns were found to prefer nonfacelike stimuli when more high-contrast elements were in the upper part than the inverted versions of these patterns; this was true for several patterns, including

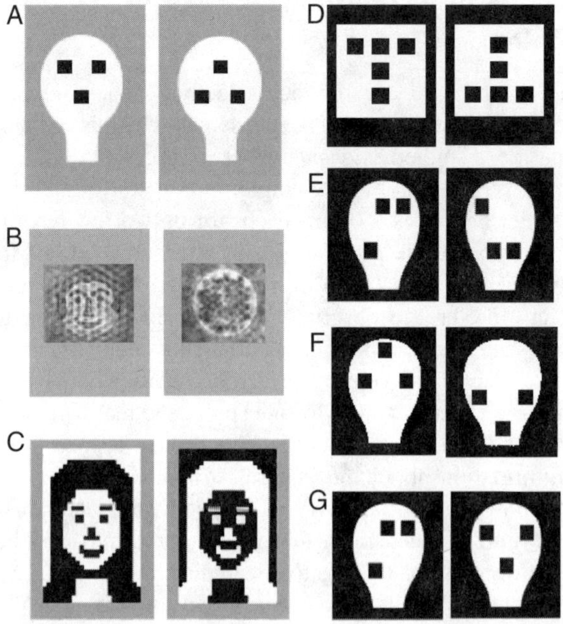

FIGURE 12.1 Different stimuli used to assess newborns' visual preferences. (A, B) Stimuli used by Mondloch et al.,[24] including *Config* and its inversion (A), a combination of the phase spectrum of a face and amplitude spectrum of a lattice and the reversed combination (B, left and right, respectively), and a positive and negative contrast face (C). (Stimuli courtesy of Catherine Mondloch.) Stimuli used by Simion et al.[25] (D) and by Turati et al.[26] (E to G) to study the role of an up-down asymmetry in newborns' preference for facelike and nonface like stimuli.

a T (Figure 12.1D) and an inverted U shape.[25] This suggests that the preference for config may not depend on the facelike arrangement of its elements, which is further supported by newborns' preference of the stimulus with more upper elements when the face arrangement is broken.[26] Perhaps most striking is that the up-down bias seems to win over the facelike arrangement: newborns prefer the left-most stimuli in Figure 12.1F, even though the right-most stimulus has the more facelike structure.[26] When two stimuli have an equal up-down asymmetry, newborns no longer show a significant preference for the one that has a face structure (Figure 12.1G).

There is certainly much to learn from studying the development of face perception. One recent study demonstrated that visual experience during the first months of life is necessary for the normal development of face processing. Children who had bilateral cataracts removed early in life were found to show deficits for the processing of configural (but not featural) facial information.[27] This is consistent with the prior findings of rapid changes in face processing early in life. Although the mechanisms behind this effect remain to be discovered, early deprivation[28] poses limits on what can be learned later in life. However, parsimony suggests that the preference for a face configuration present in newborns reflects a more general up-down bias that is not specific to facelike objects. The source of this preference in the visual system and its function in the general development of pattern vision remain to be determined. More work is needed that focuses on the up-down bias, now clearly demonstrated in the newborn across a range of stimuli, rather than on the perhaps more captivating preference for a facelike structure (which is apparently the result of a confound).

12.3 DESCRIPTIVE STUDIES OF FACE SELECTIVITY

There is very little support for an innate Conspec mechanism supporting face processing, at least one that would be specific to a facelike geometry. However, this is not sufficient grounds to reject the idea that face perception is modular. Following Fodor himself, some authors have postulated that an adult mechanism specialized for face processing could be modular without necessarily having an innate basis. For example, on the basis of intracranial event-related potentials measured with scalp electrodes from the cortical surface of patients with intractable epilepsy, Puce and colleagues[29] proposed a modular, but perhaps not innate, basis for adult face processing. They recorded from sites on the ventral occipitotemporal cortex — most of them on the fusiform gyrus — where faces elicited a negative response peaking 200 ms poststimulus onset (the N200). No similar response was observed on these sites for nonface objects such as cars, flowers, or scrambled faces. Stimulation at face-specific sites in some cases led to a temporary face recognition deficit or to face hallucinations. One indication that the N200 may not reflect an innate preference is that other electrode sites revealed selectivity for letter strings compared to faces and objects. Puce et al.[29] suggested that the neuronal population responding to faces reflects the activity of an encapsulated system (one of Fodor's modularity criteria, which describes a system that is immune to outside influences such as desires or beliefs). This is supported by the finding that the N200 is not primed semantically and is unaffected by face familiarity.[29] However, it should be noted that although the lack of top-down influences on the N200 may speak to the level of analysis performed in this neuronal population, strictly speaking the N200 can only be interpreted as modular if demonstrated to be domain specific, that is, part of a system that accepts only faces as input.[5,30] Although the intracranial recording method offers a very powerful combination of high temporal resolution together with good spatial resolution (although poor coverage compared to functional magnetic resonance imaging (fMRI) or even scalp recordings), there are reasons to doubt that the issue of domain specificity can be addressed definitively with this technique. These reasons include the limited availability of this method, the possible functional reorganization taking place in the epileptic brain, and the fact that the patients are generally under medication, which prevents the use of difficult tasks. As will become clear by the end of this chapter, challenges to the domain-specificity claims come from experiments that require subjects to process nonface objects to the same depth as they seem to process faces by default, or from experiments with experts for nonface categories who seem to automatically process nonface objects with which they are expert in a similar manner as they process faces. The domain specificity of face processing is currently best evaluated using other methods such as neuropsychological studies, fMRI, and scalp ERPs, techniques that have all successfully demonstrated face-selective effects.

12.3.1 Neuropsychology

The first evidence for face selectivity in the human brain came from the most easily accessible technique to cognitive neuroscientists: the study of brain damage. Neuropsychology is often limited by a lack of experimental control and the large variability in symptoms as well as in the localization of lesions, but it is nonetheless one of the most fertile grounds for theory.

Bodamer[31] coined the term prosopagnosia to describe a failure in the visual recognition of faces. It is a rare type of visual agnosia, diagnosed after amodal deficits have been ruled out (e.g., amnesia, aphasia). Material-specific visual agnosias are generally considered associative agnosias,[32] which means that early visual deficits can be ruled out: these patients can generally draw, copy, or match images that they cannot name or recognize. Evidence from prosopagnosic patients has been used to make several different claims regarding face processing, and is used here to introduce many of the questions that motivate experimental studies conducted with other methods.

Early on, the study of prosopagnosic patients allowed researchers to implicate the occipitotemporal pathway in face processing, and in particular a right hemisphere lesion has generally been thought necessary for the deficit.[33] What has been debated is whether a bilateral lesion is necessary or whether a right hemisphere lesion is sufficient,[34] as some rare patients appear to develop the deficit following unilateral right hemisphere damage.[35]

In a meta-analysis of 99 cases of visual agnosias, Farah[36] noticed a general pattern in the co-occurrence of material-specific visual deficits: object agnosia was always associated with either prosopagnosia or with alexia (visual deficit specific for words) and all cases of prosopagnosia or alexia show some degree of object agnosia. This led Farah to suggest that two processes rather than three (one for each domain) may be sufficient to explain object, face, and word agnosia, The first would be necessary to process faces, the other necessary to process words, and both would contribute to some extent to object recognition. Hypothetically, the two processes could be identifiable with left vs. right hemisphere functions, part-based vs. whole-based processing, namable vs. non-namable parts, etc. Since this original proposal, however, Farah's interpretation has been challenged by the report of case studies that are not consistent with the original pattern of co-occurrence. This includes the case of patients with an isolated deficit for objects with preserved skill for words and faces,[37,38] as well as a patient who can recognize objects, but not faces and words.[39]

Although prosopagnosia may be dissociable from both object and word agnosia, pure cases of prosopagnosia are very rare and it is often doubtful that testing was sensitive enough to demonstrate problems with object recognition. For this reason, Farah and colleagues[40] reasoned that although prosopagnosia may not be dissociable from object agnosia, it may be possible to demonstrate that the deficit with faces is disproportionate relative to the deficits with objects and, as such, cannot be explained by it. These authors compared LH, a patient with prosopagnosia, with normal controls for recognition of different exemplars of nonface object categories. Although LH was not as good as normal subjects at recognizing nonface objects, he was even more impaired when it came to the recognition of faces. Farah et al.[40] argued that LH's deficit with faces could not simply be a deficit in discriminating among highly similar exemplars of the same object category because his performance was relatively better when discriminating between different eyeglass frames. However, Gauthier et al.[41] argue that LH's good performance with eyeglass frames may have been due to much slower responses than normals (RTs were not reported). They also demonstrated the importance of taking response bias into account in a signal detection framework[42] when testing patients (as patients often adopt different response criteria for different object categories, making accuracy difficult to interpret). Using this approach they showed that patients similar to LH were disproportionately sensitive to the visual similarity between stimuli, for both faces as well as nonface objects, suggesting that a general deficit in making fine discriminations may be at the root of at least some cases of prosopagnosia.

Perhaps the most radical claim originating from the study of prosopagnosia was argued on the basis of a single case study. The patient, Adam, suffered from prosopagnosia acquired as a result of damage to ventral temporo-occipital cortex at 1 day of age (although the exact regions damaged at that time remain unknown). He showed profound impairments in tasks with pictures of faces and a more moderate deficit with common objects, when tested as a 16-year-old. Farah and colleagues[28] concluded, "The distinction between face and object recognition, and the anatomical localization of face recognition, are explicitly specified in the genome." However, Adam's ability to process visually similar nonface objects was not assessed and his response times were not considered.[41] In addition, because there is clear evidence for specialization for faces in a consistent region of the adult brain, it should be expected that damaging this region of the brain would impair the development of face recognition, if only because this is the best possible part of the visual system for the processing necessary to discriminate faces. An analogy is that abnormalities in parts of the brain that are important for acquiring reading skills may cause developmental dyslexia, even if reading is not explicitly specified in the genome.[43]

12.3.2 Brain Imaging (fMRI and ERPs)

Building on a long history of neuropsychological findings indicating specialization in the primate visual system for face processing, brain imaging techniques such as positron emission tomography (PET) and fMRI have localized face-selective regions in the ventral temporal lobe.[44–47] Kanwisher and colleagues[48–50] have focused on one face-selective region in the human fusiform gyrus, in a location consistent with lesions resulting in prosopagnosia, and have explored its responses to many different object categories. They have named this region the fusiform face area (FFA) and have emphasized the generality of its response to faces in many different formats (front and profile views, two-tone version, inverted face, cartoon faces, animal faces) as well as the specificity of its response (relative to faces, it responds poorly to objects, eyes only, bodies without faces, hands, etc.). The FFA is relatively small (about the size of a pea) and although its location averaged across a group of subjects is surprisingly similar across studies, its location is variable enough that it is often

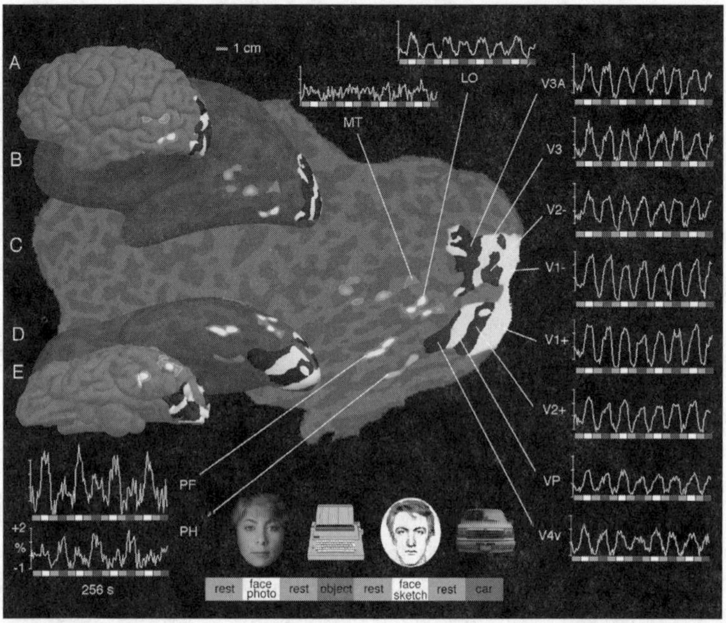

FIGURE 12.2 FMRI responses in a single subject in lateral (A) and ventral (E) views, inflated lateral (B) and ventral (D) views, and in a flattened view (C). The time course of activation in response to gray-scale stimuli of different categories (faces, varied objects, and cars) and fixation is shown in retinotopic areas of the inferior (V1–, V2–, V3, V3A) and superior (V1+, V2+, VP, V4v) visual fields, as well as area MT, area PF (equivalent to FFA in Reference 44), area LO,[100] and area PH (equivalent to PPA in Reference 84). (From Halgren, E. et al., *Hum. Brain Mapping,* 7, 29–37, 1999. With permission.)

judged necessary to define it functionally within individual subjects, using a comparison such as passive viewing of faces minus objects.

When earlier visual areas are likewise functionally defined within a single subject,[51] the FFA can be shown to lie beyond the classically defined retinotopic cortex, approximately 1 to 2 cm anterior to area V4v (see Figure 12.2 and Reference 52). However, when retinotopy is explored using stimuli that contain images of objects in different parts of the visual field, an eccentricity map is defined beyond the classical retinotopic areas.[53] Levy et al.[53] have demonstrated that the FFA lies within an area that has a relatively center-biased representation (in terms of retinal eccentricity), and suggested that this may be because face recognition relies on detailed central scrutiny.

In parallel with PET and fMRI studies on this area, techniques with poorer spatial precision but higher temporal resolution such as EEG and magnetoencephalography (MEG) have also isolated face-selective responses. Similar to the N200 obtained in intracranial electrodes, the N170 is the first posterior negative deflection obtained at the scalp that follows the visual presentation of a face (or of an object, see Figure 12.5f, below). It peaks at occipitotemporal sites at around 170 ms.[54,55] Although this potential was originally suggested to be specific to faces[55,56] several EEG/ERP studies have described large and clear negative N170 components following the presentation of nonface objects as well as faces (e.g., References 10 and 57 through 59). Recording of the magnetic field on the scalp using MEG reveals a "M170" component with similar response properties as the N170 (e.g., References 60 and 61). N170 responses are recorded for objects (e.g., houses, chairs, cars, birds, dogs, shoes, or novel objects such as Greebles or two-dimensional "blob" shapes) often at latencies comparable to faces but always at lower amplitudes. Interestingly, N170 responses of large amplitude are also recorded in response to visually presented words.[62]

The N170 can vary widely in amplitude for different object categories (e.g., chairs vs. cars; see Reference 59), suggesting that it could be generally related to object categorization (for a similar argument in fMRI, see Reference 63) and that the processes reflected by the N170 are not uniquely engaged by faces (see Reference 2). The N170 increases in amplitude when the same object is categorized at the subordinate vs. the basic level (e.g., beagle vs. dog[64] and when dog and bird experts — who recognize their objects of expertise automatically at the subordinate level — are presented with their objects of expertise (dog/birds)).[10] Together with other recent studies (e.g., References 65 through 67) this suggests that the level of categorization and degree of visual expertise are critical in explaining at least part of the difference in the N170 between faces and objects.[68]

The neural source of the N170 is unclear, as is its relationship to the face-sensitive regions identified in fMRI, even though ERPs reflect essentially postsynaptic activity that is likely to be the source of the BOLD signal used in fMRI.[69] The fMRI signal is measured over several seconds whereas the N170 reflects transient changes of electrical potential on the order of a few tenths of milliseconds. More than a dozen areas are likely to show activity between 100 and 200 ms following the presentation of a visual complex stimulus such as a face, with several showing differential activation to faces and objects. Therefore, the N170 is likely to reflect the combination of several generators with different orientations, active around this latency. Several local electrophysiological components responsive to faces (or to objects) such as the N200 are recorded on a multitude of sites in intracranial recordings along the ventral and lateral parts of occipitotemporal cortex.[70] Most of these local components are likely to contribute to the far-field N170 scalp potential. For this reason, functional links between the scalp ERPs such as the N170 and the activity of face-sensitive regions in fMRI may be the only meaningful approach.

12.4 HYPOTHESES ACCOUNTING FOR FACE SPECIALIZATION

Because the FFA can be so reliably defined either in a group or in individual subjects, many fMRI studies have used faces and/or focused on this brain region as a means to study more general questions about the ventral visual pathway.[71,72] In the last few years, however, several researchers have used the response properties of the FFA as well as other putatively category-selective regions to account for the phenomenon of category selectivity and its implications for theories of object recognition. Although these hypotheses have been mainly discussed in the context of fMRI experiments, they offer potential frameworks for our general understanding of how object recognition is performed in the occipitotemporal cortex. These different hypotheses are reviewed next.

12.4.1 Module for Detection of Faces

According to one type of model, faces are processed in a separate module, reasonably enough, given their sociobiological relevance, and all other objects (perhaps with the exception of letters, which also often get their own module) are processed by a "general" system. Support for this model is obtained from neuroimaging experiments in which activation to faces is compared with another condition using a variety of nonface objects.[44,73] However, such designs cannot reveal possible differences between categories of nonface objects. Many experiments have now revealed differences in occipitotemporal cortex between nonface object categories.[74–76] Therefore, the idea that faces are special but most every other category is processed in a "hodgepodge" system can be refuted.

In one version of this hypothesis, Kanwisher[6] suggests that face selectivity in the FFA as demonstrated by fMRI reflects a special-purpose mechanism for detecting the geometry of faces, that is, a mechanism that determines that a face is a face, but not the individual identity of the face (see also Reference 77). Evidence for this detection role comes from stimulus manipulations that affect face recognition performance but have little impact on activation in FFA. One example is face inversion, which only minimally influences activity in the FFA.[49,78,79] Kanwisher et al.[49] found that although the inversion effect in the FFA was small for gray-scale faces, it is much larger for two-tone faces. Such two-tone faces are typically not recognized as faces when upside-down. These authors argued that FFA activity is influenced by manipulations that influence the detection of a face rather than identification. However, as can be appreciated from Figure 12.3, a stimulus that is not detected as a face is also unlikely to be encoded as an individual face: inversion of two-tone faces influences face detection as well as face identification, and thus it fails to dissociate

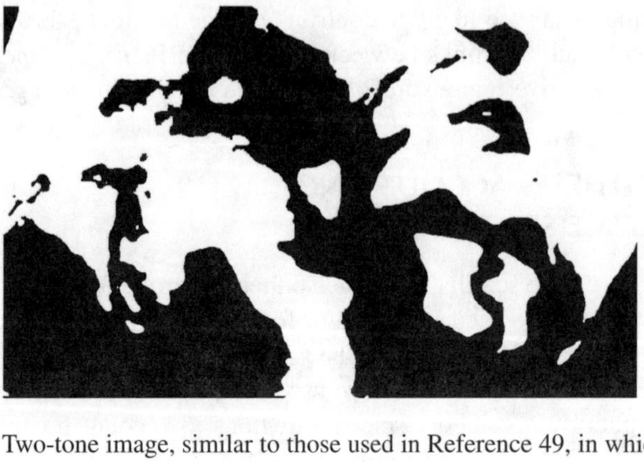

FIGURE 12.3 Two-tone image, similar to those used in Reference 49, in which it is difficult to detect faces unless it is turned upside-down.

the two effects. One argument against models that postulate a stage of face detection that precedes that of identification comes from the work of Tanaka,[80] who showed that faces are recognized at the individual level ("Brian") as fast as they are recognized as faces. This is thought to be due to our expertise with faces: similarly, bird and dog experts identify stimuli outside of their domain of expertise fastest using basic categorical level (e.g., "bird" or "car," more similar to "face") but for objects within their domain of expertise, they can recognize objects as fast at the subordinate or individual level as at the basic level (e.g., a bird expert could recognize an object equally fast as a "jay" than as a "bird").[81] This argues against an early stage of processing for detection of faces. Neuroimaging results also suggest that the FFA, as well as the "occipital face area" (OFA) in the lateral occipital lobe, processes faces at the individual level.[82] Even the earliest neurophysiological signature of face selectivity in the human brain, the N170 potential, which peaks around 170 ms poststimulus onset, is influenced by manipulations that limit face identification (e.g., inversion)[59] and is sensitive to expertise at identifying nonface objects individually.[10,67] At this point, clear evidence for a separate face detection stage is lacking, which is problematic for the suggestion that a given face-selective area or face potential is primarily involved in this processing stage.

12.4.2 Module for Each Object Category

One hypothesis that does not find many supporters and is generally only mentioned as a "straw man" (e.g., Reference 83) is the possibility that there is a module for every object category (e.g., a system that only accepts chairs as an input, another for houses, another for plants, etc.) Such a category-specific model receives some support from rare patients with selectively impaired (or preserved) recognition of a category, such as cows or sheep (see Reference 36 for review). One issue is whether such a solution would lead to a serious storage problem: Is there enough cortex to support all the possible categories? Even more problematic is that modules are supposed to be domain specific, which means that they need a clear definition of the kinds of appropriate inputs (their "domain"). Most categories do not have sharp definitions and it would seem inappropriate to implement our loose and flexible representations of objects in a compartmentalized "one category/one module" system.

12.4.3 Topography of Object Features

Another account of category selectivity proposes that object-recognition cortex may be "featurotopic" in the same way that primary visual cortex is retinotopic and somato/motor cortex is somatotopic. This was suggested on the basis of an fMRI experiment in which faces, houses, and chairs were presented to subjects.[74] Within the ventral temporal cortex and consistently across all subjects, an area was found to respond more strongly to chairs than to houses and faces, laterally to the FFA, and another area medial to the FFA responded more to houses than to faces and chairs (see also Reference 84). Apart from the relative selectivity of each of these three regions, each area gave a significant and discriminative response to the other two categories (see Rossion et al.[67] for similar results based on ERPs). Therefore, an argument can be made that each area reflects enough information to discriminate faces from chairs and from houses, and that the best computational strategy to categorize objects at this level of abstraction would be a system that could read out all three regions. The idea is simple and elegant: if extrastriate cortex includes a map in which similar visual features are found near one another, then visually similar objects should activate similar regions of this map. Tanaka and colleagues[85] have championed a similar model on the basis of single-cell recording evidence. However, the cortical organization that Tanaka and colleagues have observed in inferior temporal cortex (IT) consists of very small regions of cortex in which cells within a tangential distance of 0.2 to 0.7 mm respond to similar stimuli. Beyond this distance, cells no longer respond to those stimuli, although another small patch of neurons may be found that responds to similar features a distance of 0.4 to 1.0 mm away. Although there is some degree of clustering according to visual features in IT, there is no evidence for a larger-scale map of features whereby two of Tanaka's modules near to each other are more likely to respond to similar features than are two other modules farther apart. Therefore, it is difficult to map the much larger areas described by Ishai et al.[74] (each at least over a centimeter wide) onto the organization described by Tanaka and colleagues.

In a recent experiment, Haxby et al.[63] obtained fMRI results that are even more suggestive of a topography of features. They presented subjects with objects from different object categories (faces, houses, cats, bottles, scissors, shoes, and chairs) and divided the fMRI data into two halves. The results revealed a distributed pattern of selectivity over the entire occipitotemporal gyrus, which was replicable over time within a given subject (unlike the areas in the Ishai et al. 1999 study;[74] however, the details of this distributed pattern was found to be the same across subjects). To further demonstrate the power of a distributed representation, Haxby et al.[63] also analyzed their results by taking out "hot spots" of selectivity, such as the FFA, and found that the rest of the distributed pattern was sufficient by itself to show a response that discriminated one category (e.g., faces) from the others. This analysis illustrates that, although from our point of view it may often be easier to think of hot spots of selectivity as most informative, a population coding over a wide expanse of cortex is a viable possibility.

Certain issues remain problematic for a model based on a topography of features. First, we do not know at this point how the information present over a very large part of occipitotemporal cortex would be combined to lead to a categorization response. Second, certain parts of the putative feature-map do not respond in a manner that is consistent with any specific feature set. For example, Epstein and Kanwisher[84] found that the house-selective region (which they call the "parahippocampal place area," or PPA) responded strongly to a picture of a house, and to a furnished room or the same room when empty, but did not respond to an image of the furniture from the same room floating in a rearranged configuration that did not reflect spatial information about the environment. This result is difficult to explain by the feature-map account because it suggests that there is not a set of features common to all objects that activate the PPA, which are not present in objects that fail to recruit the PPA. Third, the model at this point cannot explain how objects are discriminated within a category. Even more problematic is the evidence that once subjects become very skilled at discriminating objects of a given category, the FFA becomes active regardless of the objects' geometry.[86]

12.4.4 Process-Map Model

An alternative model arises from consideration of how face-specific results (i.e., any difference obtained between faces and nonface objects) may be interpreted, regardless of the technique. Tanaka and Gauthier[3] suggested that such effects can have many possible explanations, because faces fall at the extreme of many dimensions relevant to object recognition. Some of the dimensions often confounded with category domain (face/nonface) include categorization level (basic/subordinate) and expertise (novice/expert). These dimensions (as well as other factors, many of which may not have been considered yet) can interact to specify a large number of different object recognition situations (see Figure 12.4). For example, differences obtained in an experiment comparing face recognition to processing of many different objects (e.g., a mix of varied artificial and living objects) could be due to at least three factors (domain, level of categorization, expertise) or any of their interactions.

Although the prevailing interpretation finding that faces and objects are not processed in the same way has been in terms of domain specificity (i.e., faces being "special"), an alternative approach has been to manipulate other relevant dimensions one at a time in the domain of nonface objects so that if putative "face-specific" effects could be replicated, the domain-specific hypothesis would be refuted. A

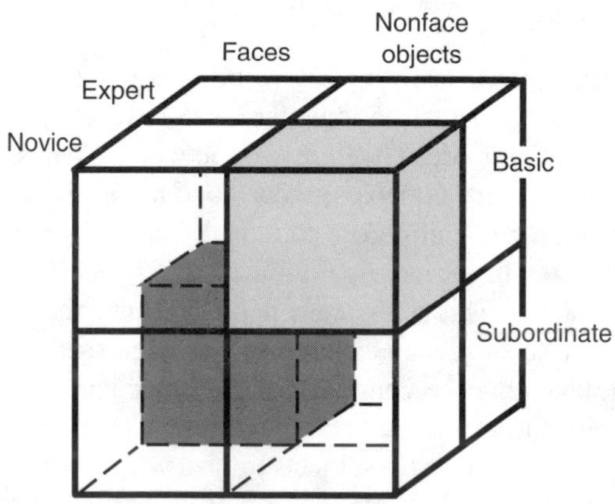

FIGURE 12.4 Simplified representation of the multidimensional space arising from all possible combinations of factors (here three; in reality, an infinite number) that may constrain object recognition, such as stimulus-class membership, expertise, and categorization level. Tanaka and Gauthier[3] and Gauthier et al.[41] argued that such factors are likely to be confounded and are likely to interact (e.g., that it is very difficult to argue having "equated" a manipulation of level of categorization for faces and for objects). Therefore, the best strategy to test whether an effect is domain specific is to see whether it can be obtained through a manipulation of other dimensions *within one domain* (in particular, manipulations that are likely to be confounded with category membership in most natural situations). (From Gauthier, I. et al., *J. Cogn. Neurosci.*, 11(4), 349–370, 1999. With permission.)

landmark study was that by Diamond and Carey,[8] who demonstrated that dog show judges showed an inversion effect with dog breeds with which they were experts. This inversion effect in experts, similar to the one obtained before for faces but not objects,[87] suggested the possibility that given enough expertise (in this case over 10 years) behavioral effects once thought to be face specific could be obtained with nonface objects. A practical question is how much expertise is necessary for this to happen? Does it require several years of training,[8] or can expertise be acquired relatively rapidly in the laboratory? This question was addressed in the first study of expertise created in the laboratory, in which subjects were trained to become "experts" with novel objects called Greebles.[88] These objects were created so that they would be very similar to each other and would all possess the same number of parts in the same general configuration, something that is thought to make faces especially difficult to discriminate. The rationale was that becoming very good at discriminating among Greebles would recruit the same mechanisms (and brain areas) that are useful to discriminate among faces. Greeble experts are trained to a criterion, which models what is considered to be a hallmark of expertise in the natural world. Whereas we are generally slower to recognize a picture as a Ferrari or a Subaru than as a car, car experts can access the more specific (or subordinate) level of recognition equally fast.[81] This is also true of our expertise with faces, as we can recognize faces at the individual level (e.g., Tom Cruise) as fast as at a more categorical level (e.g., human).[80] Greeble experts are therefore trained to be able to categorize Greebles at the individual level as fast and accurately as they can make an easier discrimination (e.g., categorize them according to whether all parts point up or down).

Figure 12.5 summarizes many of the results obtained with Greebles using a variety of techniques. In general, the logic has been to identify behavioral or neural effects that represented important differences between objects and faces, and to investigate these effects with Greebles under conditions of expertise. Tanaka and Sengco[89] found that the recognition of individual face features (e.g., "Jim's nose") is disrupted by changes in the configuration of other face features (e.g., moving Jim's eyes apart slightly). This is the case neither in inverted faces or houses, nor with Greebles for novice subjects. However, in Greeble experts recognizing a single part of the object, changing the configuration of other parts (e.g., the top parts in Figure 12.5A) significantly increased reaction time.[88] This effect was also found for upright but not inverted Greebles (in subjects trained with upright Greebles). Another transformation that impairs face recognition is brightness reversal.[90,91] Greeble experts are also more sensitive to this transformation (Figure 12.5B) than are novices.[79] The recognition of a composite face's top or bottom half is slower when the other half (from a different individual) is aligned with the other part rather than misaligned.[92] This effect, found with upright but not inverted faces, reflects the fact that we find it very difficult to ignore any part of a face when the parts are in the correct configuration. Whereas novices have little difficulty ignoring Greeble parts as instructed, Greeble experts have similar difficulties when told to ignore a Greeble part. In addition, when told to make a judgment about the top of composite Greebles (Figure 12.5C), Greeble experts cannot ignore irrelevant information from the bottom part, and just as in the same effect for faces, misaligned irrelevant information is not as disruptive. Greeble expertise also leads to an increase of activity for Greebles in the FFA, both in a comparison with nonface objects or when upright Greebles are compared to inverted ones.[93] In Figure 12.5E, an inversion effect for faces in the FFA is replicated in five subjects viewing Greebles, but only once they reached Greeble expertise. With ERPs, the face inversion effect leads to an increase in amplitude and a delay of 10 ms in the face-selective potential called the N170, which is not found for other objects in novice observers.[59] However, the same effect can be obtained in Greeble experts, as shown in Figure 12.5F.[67] In a recent study, changes on a behavioral measure of holistic processing using composite Greebles (Figure 12.5C) were found to correlate with the right FFA activity for upright Greebles (but not inverted Greebles) during the acquisition of Greeble expertise.[9] This suggests that holistic processing may offer a window into the computations taking place in the FFA. Finally, Greebles have also been used to test patients with prosopagnosia. Patients SM and CR are prosopagnosic patients who report difficulties mainly with faces. These patients are fast at discriminating Greebles from other objects, but are dramatically slowed by the visual similarity of Greebles when making simultaneous matching judgments (Figure 12.5D).[41] This suggests that in at least some cases of visual agnosia for faces, the problem may not be face specific but rather a general impairment for the discrimination of visually similar objects. Greeble studies have provided converging evidence from several techniques showing that putative face-specific effects can be obtained with visually similar nonface objects once subjects have acquired expertise with this category. It is worth emphasizing that all these expertise effects have been obtained using novel Greebles that the subjects were never exposed to during the training: they are therefore not mere effects of practice with specific exemplars but generalize to the entire category.

Some authors have suggested that the Greebles recruit similar mechanisms as faces because they possess a facelike configuration of parts.[6] However, facelike expertise effects have also been obtained with dogs[8] and birds (Tanaka and Curran, 2002), as well as cars,[86] casting doubt on the hypothesis that any specific geometrical configuration is necessary for these effects. Rather, the expertise framework has emphasized the processing biases that experts may have acquired with a given category, rather than its geometry, as a principle that could explain most cases of category selectivity in the visual system.

The expertise account is therefore an instantiation of the more general "process-map" framework[68,94] that also offers an explanation for why we observe category selectivity for different object categories in occipitotemporal cortex.[68] The explanation is based on two assumptions. First, through experience we learn to process object categories in different manners, leading to biases in processing (e.g., we may have different recognition goals, learn that a given dimension is particularly diagnostic, come to expect a certain level of similarity between different exemplars of a category. and so forth). Second, the visual system is not homogeneous in its ability to support different recognition processes. Examples of processing biases orthogonal to category preferences include regions that are important for the processing of identity vs. location,[46] within-category discriminations,[95,96] processing of implied motion[97] or "natural" object motion,[98] grouping,[99] holistic processing,[3] and eccentricity bias.[53] Such processing biases could intersect so that each part of

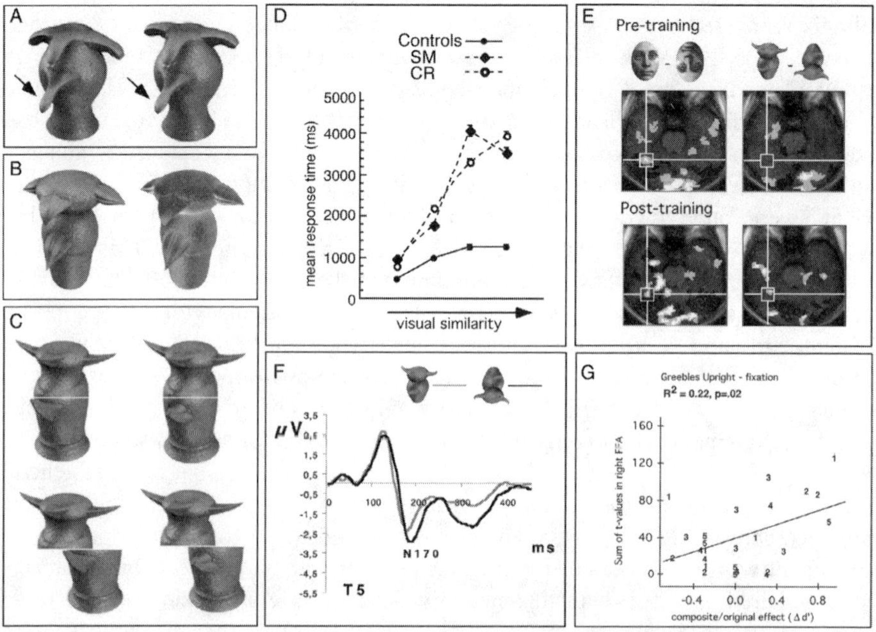

FIGURE 12.5 Summary of many expertise effects, replicating putative face-specific effects, obtained with Greeble stimuli. (A) Gauthier and Tarr[88] found that when identifying a single part of a Greeble, novices can ignore changes in the configuration of the top parts, whereas Greeble experts cannot. (B) Gauthier et al.[79] found that Greeble experts, but not novices, are disrupted by brightness inversion. (C) Gauthier and Tarr[9] found that Greeble novices can make a judgment about half of a Greeble while ignoring the other half, whereas Greeble expertise leads to an increase in the effect of the identity and the configuration of the to-be-ignored distractor part. (D) Gauthier et al.[41] found that prosopagnosic patients, compared to controls, can be dramatically sensitive to a manipulation of visual similarity with Greebles. (E) Gauthier et al.[82] found a face inversion effect in the FFA, which could be replicated with Greebles only after expertise training. Each brain image shows an average of five subjects before (top) and after (bottom) training, with brain areas showing more activity for upright than for inverted images in white to pale gray, and those showing more activity for inverted than for upright images in dark gray to black, for faces (left) and Greebles (right). (F) Rossion et al.[67] found that the effect of stimulus inversion on the N170 ERP component (with faces, an increase in amplitude and delay of 10 ms for inverted relative to upright images) can be obtained with Greebles but only after expertise training. (G) Gauthier and Tarr[9] found a relationship between an increase in the composite effect (as measured with stimuli shown in C) and the response to Greebles in the right FFA during the acquisition of expertise. Numbers represent the five subjects tested; each was tested five times.

occipitotemporal cortex displays a unique combination. Through experience the geometry of an object category may become a trigger for activity in part of the visual system (e.g., experts who develop a holistic processing strategy for a category will later demonstrate this bias when it is irrelevant and even inconsistent with the task).[3,88] In this framework, although neurons in extrastriate cortex display preferences for certain visual features, such preferences are not fixed or intrinsic to their functional role. Imagine, for example, moving to a new planet where shelters look entirely different from what we know as houses. After a while, all of the neurons that previously responded strongly to images of houses might now respond to images of the new habitations. According to the process-map model, this would not reflect a change in the functional role of these neurons (or the neuronal assemblies they constitute); rather, they would be performing very similar computations, only using different inputs.

12.5 CONCLUSIONS

Face recognition has proved to be a rich case study in the cognitive neuroscience of object recognition. The traditional modularist framework, which considers face and object processing to be performed by functionally independent systems, has led to a large number of studies describing many important differences between these two domains. An expertise framework, in contrast, rejects the idea that face and object processing can truly be studied separately and seeks unifying principles that can account for the more general phenomenon of category selectivity. The next challenge for this approach may be to apply some of what was learned from the case of faces, including the importance of processing over geometry, to other examples of specialization in occipitotemporal cortex (e.g., specialization for printed material or for scenes).

ACKNOWLEDGMENTS

The author thanks many colleagues, too many to name but several of them from the Perceptual Expertise Network, for the many discussions that led to the work summarized here, and thanks Kim Curby for comments on the manuscript. This work was supported by the National Science Foundation, the National Eye Institute, and the James S. McDonnell Foundation.

REFERENCES

1. Farah, M.-J., Wilson, K.-D., Drain, M., and Tanaka, J.-N., What is "special" about face perception? *Psychol. Rev.,* 105(3), 482–498, 1998.
2. Tovee, M.J., Is face processing special? *Neuron,* 21, 1239–1242, 1998.
3. Tanaka, J.W. and Gauthier, I., Expertise in object and face recognition, in *Mechanisms of Perceptual Learning*, Goldstone, R.L. and Schyns, P.G., Eds., Academic Press, San Diego, CA, 1997, 83–125.
4. Fodor, J.A., *Modularity of Mind,* MIT Press, Cambridge, MA, 1983.

5. Coltheart, M., Modularity and cognition, *Trends Cogn. Sci.,* 3(3), 115–120, 1999.
6. Kanwisher, N., Domain specificity in face perception, *Nat. Neurosci.,* 3(8), 759–763, 2000.
7. Carmel, D. and Bentin, S., Domain specificity versus expertise: factors influencing distinct processing of faces, *Cognition,* 83(1), 1–29, 2002.
8. Diamond, R. and Carey, S., Why faces are and are not special: an effect of expertise, *J. Exp. Psychol. Gen.,* 115(2), 107–117, 1986.
9. Gauthier, I. and Tarr, M.J., Unraveling mechanisms for expert object recognition: bridging brain activity and behavior, *J. Exp. Psychol. Hum. Percept. Performance,* 28(2), 431–446, 2002.
10. Tanaka, J.W. and Curran, T., A neural basis for expert object recognition, *Psychol. Sci.,* 12(1), 43–47, 2001.
11. Haxby, J.V., Hoffman, E.A., and Gobbini, M.I., Human neural systems for face recognition and social communication, *Biol. Psychiatr.,* 51(1), 59–67, 2002.
12. Adolphs, R., Recognizing emotion from facial expressions: psychological and neurological mechanisms, *Behav. Cogn. Neurosci. Rev.,* 1(1), 21–62, 2002.
13. Valentine, T. and Bruce, V., The effect of race, inversion and encoding activity upon face recognition, *Acta Psychol.,* 61(32), 259–273, 1986.
14. Puce, A., Allison, T., Bentin, S., Gore, J.C., and McCarthy, G., Temporal cortex activation in humans viewing eye and mouth movements, *J. Neurosci.,* 18(6), 2188–2199, 1998.
15. DeRenzi, E., Current issues in prosopagnosia, in *Aspects of Face Processing*, Eliss, H.D., Jeeves, M.A., Newcombe, F., and Young, A., Eds., Martinus Nijhoff, Dordrecht, the Netherlands, 1986.
16. Parry, F.M., Young, A.W., Saul, J.S., and Moss, A., Dissociable face processing impairments after brain injury, *J. Clin. Exp. Neuropsychol.,* 13(4), 545–558, 1991.
17. Rapcsak, S.Z., Kaszniak, A.W., and Rubens, A.B., Anomia for facial expressions: evidence for a category specific visual-verbal disconnection syndrome, *Neuropsychologia,* 27(8), 1031–1041, 1989.
18. Walker, S., Bruce, V., and O'Malley, C., Facial identity and facial speech processing: familiar faces and voices in the McGurk effect, *Percept. Psychophys.,* 57(8), 1124–1133, 1995.
19. Gauthier, I. and Nelson, C.A., The development of face expertise, *Curr. Opin. Neurobiol.,* 11(2), 219–224, 2001.
20. Walton, G.E. and Bower, T.G.R., Newborns form "prototypes" in less than 1 minute, *Psychol. Sci.,* 4(3), 203–205, 1993.
21. Goren, C., Sarty, M., and Wu, P., Visual following and pattern discrimination of face-like stimuli by newborn infants, *Pediatrics* 56, 544–549, 1975.
22. Johnson, M.H. and Morton, J., *Biology and Cognitive Development: The Case of Face Recognition,* Blackwell, Oxford, U.K., 1991.
23. Simion, F., Valenza, E., Umilta, C., and Dalla Barba, B., Preferential orienting to faces in newborns: a temporal-nasal asymmetry, *J. Exp. Psychol. Hum. Percept. Perform.,* 24(5), 1399–1405, 1998.
24. Mondloch, C.J., Lewis, T.L., Budreau, D.R., Maurer, D., Dannemiller, J.L., Stephens, B.R., and Kleiner-Gathercoal, K.A., Face perception during early infancy, *Psychol. Sci.,* 10, 419–422, 1999.
25. Simion, F., Valenza, E., Macchi Cassia, V., Turati, C., and Umiltà, C., Newborns' preference for up-down asymmetrical configurations, *Dev. Sci.,* 4, in press.
26. Turati, C., Simion, F., Milani, I., and Umilta, C., Newborns' preference for faces: what is crucial? *Dev. Psychol.*, in press.

27. Le Grand, R., Mondloch, C.J., Maurer, D., and Brent, H.P., Early visual experience and face processing, *Nature,* 410, 890, 2001.
28. Farah, M.J., Rabinowitz, C., Quinn, G.E., and Liu, G.T., Early commitment of neural substrates for face recognition, *Cogn. Neuropsychol.,* 17(1/2/3), 2000.
29. Puce, A., Allison, T., and McCarthy, G., Electrophysiological studies of human face perception. III: Effects of top-down processing on face-specific potentials, *Cereb. Cortex,* 9(5), 445–458, 1999.
30. Faulkner, T.F., Rhodes, G., Palermo, R., Pellicano, E., and Ferguson, D., Recognizing the un-real McCoy: priming and the modularity of face recognition, *Psychonomic Bull. Rev.,* 9(2), 327–334, 2002.
31. Bodamer, J., Die Prosop-Agnosie, *Arch. Psychiatr. Nervenkr.,* 179, 6–54, 1947.
32. Lissauer, H., Ein Fall von Seelenblindheit nebst einem Beitrage zur Theorie derselben, *Arch. Pychiatr. Nervenkr.,* 21, 222–270, 1890.
33. Benton, A.L. and Van Allen, M.W., Prosopagnosia and facial discrimination, *J. Neurol. Sci.,* 15, 167–172, 1972.
34. Damasio, A.R., Damasio, H., and Van Hoesen, G.W., Prosopagnosia: anatomical basis and behavioral mechanisms, *Neurology,* 32, 331–341, 1982.
35. DeRenzi, E., Perani, D., Carlesimo, G.A., Silveri, M.C., and Fazio, F., Prosopagnosia can be associated with damage confined to the right hemisphere — an MRI and PET study and a review of the literature, *Neuropsychologia,* 32, 893–902, 1994.
36. Farah, M.J., *Visual Agnosia: Disorders of Object Recognition and What They Tell Us about Normal Vision,* MIT Press, Cambridge, MA, 1990.
37. Rumiati, R.I. and Humphreys, G.W., Visual object agnosia without alexia or prosopagnosia: arguments for separate knowledge stores, *Vis. Cognit.,* 4(2), 207–217, 1997.
38. Humphreys, G.W. and Rumiati, R.I., Agnosia without prosopagnosia or alexia: evidence for stored visual memories specific to objects, *Cogn. Neuropsychol.,* 15(3), 243–277, 1998.
39. Buxbaum, L.J., Glosser, G., and Coslett, H.B., Impaired face and word recognition without object agnosia, *Neuropsychologia,* 37(1), 41–50, 1999.
40. Farah, M.J., Levinson, K.L., and Klein, K.L., Face perception and within-category discrimination in prosopagnosia, *Neuropsychologia,* 33(6), 661–674, 1995.
41. Gauthier, I., Behrmann, M., and Tarr, M.J., Can face recognition really be dissociated from object recognition? *J. Cogn. Neurosci.,* 11(4), 349–370, 1999.
42. Green, D.M. and Swets, J.A., *Signal Detection Theory and Psychophysics,* Wiley, New York, 1966.
43. Jenner, A.R., Rosen, G.D., and Galaburda, A.M., Neuronal asymmetries in primary visual cortex of dyslexic and nondyslexic brains, *Ann. Neurol.,* 46(2), 189–196, 1999.
44. Kanwisher, N., McDermott, J., and Chun, M.M., The fusiform face area: a module in human extrastriate cortex specialized for face perception, *J. Neurosci.,* 17, 4302–4311, 1997.
45. Puce, A., Allison, T., Asgari, M., Gore, J.C., and McCarthy, G., Differential sensitivity of human visual cortex to faces, letterstrings, and textures: a functional magnetic resonance imaging study, *J. Neurosci.,* 16(16), 5205–5215, 1996.
46. Haxby, J.V., Grady, C.L., Horwitz, B., Ungerleider, L.G., Mishkin, M., Carson, R.E., Herscovitch, P., Schapiro, M.B., and Rapoport, I., Dissociation of object and spatial visual processing pathways in human extrastriate cortex, *Proc. Natl. Acad. Sci. U.S.A.,* 88, 1621–1625, 1991.
47. Sergent, J., Ohta, S., and MacDonald, B., Functional neuroanatomy of face and object processing: a positron emission tomography study, *Brain,* 115, 15–36, 1992.

48. Kanwisher, N., McDermott, J., and Chun, M.M., A module for the visual representation of faces, *NeuroImage,* 3(Suppl.), S361, 1996.
49. Kanwisher, N., Tong, F., and Nakayama, K., The effect of face inversion on the human fusiform face area, *Cognition,* 68(1), B1–11, 1998.
50. Kanwisher, N., Stanley, D., and Harris, A., The fusiform face area is selective for faces not animals, *NeuroReport,* 10(1), 183–187, 1999.
51. Engel, S.A., Glover, G.H., and Wandell, B.A., Retinotopic organization in human visual cortex and the spatial precision of functional MRI, *Cereb. Cortex,* 7(2), 181–192, 1997.
52. Halgren, E., Dale, A.M., Sereno, M.I., Tootell, R.B.H., Marinkovic, K., and Rosen, B.R., Location of human face-selective cortex with respect to retinotopic areas, *Hum. Brain Mapping,* 7, 29–37, 1999.
53. Levy, I., Hasson, U., Avidan, G., Hendler, T., and Malach, R., Center-periphery organization of human object areas, *Nat. Neurosci.,* 4(5), 533–539, 2001.
54. Botzel, K. and Grusser, O.J., Electric brain potentials evoked by pictures of faces and non-faces: a search for "face-specific" EEG potentials, *Exp. Brain Res.,* 77(2), 349–360, 1989.
55. Bentin, S., Allison, T., Puce, A., Perez, E. et al., Electrophysiological studies of face perception in humans, *J. Cogn. Neurosci.,* 8(6), 551–565, 1996.
56. Eimer, M., Does the face-specific N170 component reflect the activity of a specialized eye processor? *NeuroReport,* 9(13), 2945–2948, 1998.
57. Eimer, M., Event-related brain potentials distinguish processing stages in face perception and recognition, *Clin. Neurophysiol.,* 111, 694–705, 2000.
58. Eimer, M., The face-specific N170 component reflects late stages in the structural encoding of faces, *NeuroReport,* 11, 2319–2324, 2000.
59. Rossion, B., Gauthier, I., Tarr, M.J., Despland, P., Bruyer, R., Linotte, S., and Crommelinck, M., The N170 occipito-temporal component is delayed and enhanced to inverted faces but not to inverted objects: an electrophysiological account of face-specific processes in the human brain, *NeuroReport,* 11(1), 69–74, 2000.
60. Linkenkaer-Hansen, K., Palva, J.M., Sams, M., Hietanen, J.K., Aronen, H.J., and Ilmoniemi, R.J., Face-selective processing in human extrastriate cortex around 120 ms after stimulus onset revealed by magneto- and electroencephalography, *Neurosci. Lett.,* 253(3), 147–150, 1998.
61. Liu, J., Higuchi, M., Marantz, A., and Kanwisher, N., The selectivity of the occipitotemporal M170 for faces, *NeuroReport,* 11(2), 337–341, 2000.
62. Bentin, S., Mouchetant-Rostaing, Y., Giard, M.H., Echallier, J.F., and Pernier, J., ERP manifestations of processing printed words at different psycholinguistic levels: time course and scalp distribution, *J. Cogn. Neurosci.,* 11(3), 235–260, 1999.
63. Haxby, J.V., Gobbini, M.I., Furey, M.L., Ishai, A., Schouten, J.L., and Pietrini, P., Distributed and overlapping representations of faces and objects in ventral temporal cortex, *Science,* 293(5539), 2425–2430, 2001.
64. Tanaka, J., Luu, P., Weisbrod, M., and Kiefer, M., Tracking the time course of object categorization using event-related potentials, *NeuroReport,* 10(4), 829–835, 1999.
65. Curran, T., Tanaka, J.W., and Weiskopf, D.M., Differentiating object categorization and recognition processes: a high density ERP study, poster presentation, 7th Human Brain Mapping Conference, Brighton, U.K., 2001.
66. de Haan, M., Pascalis, O., and Johnson, M.H., Specialization of neural mechanisms underlying face recognition in human infants, *J. Cogn. Neurosci.,* 14(2), 199–209, 2002.

67. Rossion, B., Gauthier, I., Goffaux, V., Tarr, M.J., and Crommelinck, M., Expertise training with novel objects leads to left lateralized face-like electrophysiological responses, *Psychol. Sci.,* 13(3), 250–257, 2002.
68. Tarr, M.J. and Gauthier, I., FFA: a flexible fusiform area for subordinate-level visual processing automatized by expertise, *Nat. Neurosci.,* 3(8), 764–769, 2000.
69. Logothetis, N.K., Pauls, J., Augath, M., Trinath, T., and Oeltermann, A., Neurophysiological investigation of the basis of the fMRI signal [see comments.], *Nature,* 412(6843), 150–157, 2001.
70. Allison, T., Puce, A., Spencer, D.D., and McCarthy, G., Electrophysiological studies of human face perception. I: Potentials generated in occipitotemporal cortex by face and non-face stimuli, *Cereb. Cortex,* 9, 415–430, 1999.
71. O'Craven, K.M., Downing, P.E., and Kanwisher, N., fMRI evidence for objects as the units of attentional selection, *Nature,* 401(6753), 584–587, 1999.
72. Tong, F., Nakayama, K., Vaughan, J.T., and Kanwisher, N., Binocular rivalry and visual awareness in human extrastriate cortex, *Neuron,* 21(4), 753–759, 1998.
73. McCarthy, G., Puce, A., Gore, J.C., and Allison, T., Face-specific processing in the human fusiform gyrus, *J. Cogn. Neurosci.,* 9, 605–610, 1997.
74. Ishai, A., Ungerleider, L.G., Martin, A., Schouten, J.L., and Haxby, J., Distributed representation of objects in the human ventral visual pathway, *Proc. Natl. Acad. Sci. U.S.A.,* 96, 9379–9384, 1999.
75. Kreiman, G., Koch, C., and Fried, I., Category-specific visual responses of single neurons in the human medial temporal lobe [see comments], *Nat. Neurosci.,* 3(9), 946–953, 2000.
76. Polk, T.A. and Farah, M.J., The neural development and organization of letter recognition: evidence from functional neuroimaging, computational modeling, and behavioral studies, *Proc. Natl. Acad. Sci. U.S.A.,* 95(3), 847–852, 1998.
77. Bentin, S. and Deouell, L., Structural encoding and identification in face processing: ERP evidence for separate mechanism, *Cogn. Neuropsychol.* 17, 35–54, 2000.
78. Haxby, J.V., Ungerleider, L.G., Clark, V.P., Schouten, J.L., Hoffman, E.A., and Martin, A., The effect of face inversion on activity in human neural systems for face and object perception, *Neuron,* 22, 189–199, 1999.
79. Gauthier, I., Williams, P., Tarr, M.J., and Tanaka, J., Training "Greeble" experts: a framework for studying expert object recognition processes, *Vision Res.,* 38(15/16), 2401–2428, 1998.
80. Tanaka, J.W., The entry point of face recognition: evidence for face expertise, *J. Exp. Psychol. Gen.,* 130(3), 534–543, 2001.
81. Tanaka, J.W. and Taylor, M., Object categories and expertise: is the basic level in the eye of the beholder? *Cogn. Psychol.,* 23, 457–482, 1991.
82. Gauthier, I., Tarr, M.J., Moylan, J., Anderson, A.W., Skudlarski, P., and Gore, J.C., The fusiform "face area" is part of a network that processes faces at the individual level, *J. Cogn. Neurosci.,* 12(3), 495–504, 2000.
83. Ishai, A., Ungerleider, L.G., Martin, A., and Haxby, J.V., The representation of objects in the human occipital and temporal cortex, *J. Cogn. Neurosci.,* 12(Suppl. 2), 35–51, 2000.
84. Epstein, R. and Kanwisher, N., A cortical representation of the local visual environment, *Nature,* 392(6676), 598–601, 1998.
85. Tanaka, K., Saito, H., Fukada, Y., and Moriya, M., Coding visual images in the inferotemporal cortex of the macaque monkey, *J. Neurophysiol.,* 66(1), 170–189, 1991.

86. Gauthier, I., Skudlarski, P., Gore, J.C., and Anderson, A.W., Expertise for cars and birds recruits brain areas involved in face recognition, *Nat. Neurosci.,* 3(2), 191–197, 2000.
87. Yin, R.K., Looking at upside-down faces, *J. Exp. Psychol.,* 81(1), 141–145, 1969.
88. Gauthier, I. and Tarr, M.J., Becoming a "Greeble" expert: exploring mechanisms for face recognition, *Vision Res.,* 37(12), 1673–1682, 1997.
89. Tanaka, J.W. and Sengco, J.A., Features and their configuration in face recognition, *Memory Cognit.,* 25(5), 583–592, 1997.
90. Galper, R.E., Recognition of faces in the photographic negative, *Psychonomic Sci.,* 19, 207–208, 1970.
91. Johnston, A., Hill, H., and Carman, N., Recognising faces: effects of lighting direction, inversion, and brightness reversal, *Perception,* 21, 365–375, 1992.
92. Young, A.W., Hellawell, D., and Hay, D., Configural information in face perception, *Perception,* 10, 747–759, 1987.
93. Gauthier, I., Tarr, M.J., Anderson, A.W., Skudlarski, P., and Gore, J.C., Activation of the middle fusiform "face area" increases with expertise in recognizing novel objects, *Nat. Neurosci.,* 2(6), 568–573, 1999.
94. Gauthier, I., What constrains the organization of the ventral temporal cortex? *Trends Cogn. Sci.,* 4(1), 1–2, 2000.
95. Gauthier, I., Tarr, M.J., Moylan, J., Anderson, A.W., and Gore, J.C., Does subordinate-level categorization engage the functionally-defined fusiform face area? *Cogn. Neuropsychol.,* 17(1/2/3), 143–163, 2000.
96. Gauthier, I., Anderson, A.W., Tarr, M.J., Skudlarski, P., and Gore, J.C., Levels of categorization in visual recognition studied using functional magnetic resonance imaging, *Curr. Biol.,* 7(9), 645–651, 1997.
97. Kourtzi, Z. and Kanwisher, N., Activation in human MT/MST by static images with implied motion, *J. Cogn. Neurosci.,* 12(1), 48–55, 2000.
98. Downing, P.E., Jiang, Y., Shuman, M., and Kanwisher, N., A cortical area selective for visual processing of the human body, *Science,* 293(5539), 2470–2473, 2001.
99. Hasson, U., Hendler, T., Ben Bashat, D., and Malach, R., Vase or face? A neural correlate of shape-selective grouping processes in the human brain, *J. Cogn. Neurosci.,* 13(6), 744–753, 2001.
100. Malach, R., Reppas, J.B., Benson, R.R., Kwong, K.K., Jiang, H., Kennedy, W.A., Ledden, P.J., Brady, T.J., Rosen, B.R., and Tootell, R.B., Object-related activity revealed by functional magnetic resonance imaging in human occipital cortex, *Proc. Natl. Acad. Sci. U.S.A.,* 92(18), 8135–8139, 1995.

13 Motion Processing in Human Visual Cortex

Randolph Blake, Robert Sekuler, and Emily Grossman

CONTENTS

13.1 INTRODUCTION

During the past few decades our understanding of the neural bases of visual perception and visually guided actions has advanced greatly, and the chapters in this volume amply document these advances. It is fair to say, however, that no area of visual neuroscience has enjoyed greater progress than the physiology of visual motion perception. Publications dealing with visual motion perception and its neural bases would fill volumes, and the topic remains central within the visual neuroscience literature even today.

0-8493-1243-4/04/$0.00+$1.50

This chapter aims to provide an overview of brain mechanisms underlying human motion perception. The chapter comprises four sections. We begin with a short review of the motion pathways in the primate visual system, emphasizing work from single-cell recordings in Old World monkeys; more extensive reviews of this literature appear elsewhere.[1,2] With this backdrop in place, the next three sections concentrate on striate and extrastriate areas in the human brain that are implicated in the perception of motion. These sections draw on experimental work from brain imaging studies, from studies employing transcranial magnetic stimulation, and from psychophysical studies in patients with localized brain damage. These three methodologies have generated complementary characterizations of multiple neural mechanisms underlying different aspects of human motion perception: perception of translational motion (including direction and speed), perception of three-dimensional (3D) shape from motion, perception of heading from optic flow and perception of biological activity. To isolate these different aspects of motion perception, researchers have utilized a variety of different motion stimuli, each designed to challenge the visual system in some characteristic, revealing fashion. Examples of some of those stimuli are illustrated schematically in Figure 13.1.

Because of space limitations, this chapter does not cover the extensive psychophysical literature on motion perception, except for selected phenomena that are related directly to brain activation studies or lesion studies. For up-to-date reviews of the psychophysical literature, readers may consult either of two recent chapters dealing with human motion perception.[3,4] Nor do we deal with visual motion processing as it applies to the generation and control of eye movements; for work on this topic, readers are referred to other sources (for example, Horowitz and Newsome[5]). Also omitted from coverage are studies of motion processing that use visual evoked responses[6] or magnetoencephalography[7] as the dependent variable. Finally, our coverage of deficits in motion perception is limited to neuropsychological studies of patients with restricted brain lesions — we will not cover the important and potentially revealing literature on deficits in motion perception that can accompany psychopathology,[8] reading disorders,[9] amblyopia,[10] Alzheimer's disease,[11] and developmental disorders including William's syndrome[12] and autism.[13]

13.2 VISUAL MOTION PATHWAYS IN NONHUMAN PRIMATES

It is generally agreed that explicit analysis of visual motion is inaugurated in the primary visual cortex. Although it is true that cells in the retina and in the thalamus respond vigorously to moving contours, spots of light, and gratings, those responses are *not* selective for the direction or the speed of motion, and, hence, those cells are not representing anything about velocity (a fundamental value when specifying information about motion). The foundations for motion processing in the primate visual system are the direction-selective (DS) cells first encountered in primary visual cortex, here termed V1. First described by Hubel and Wiesel[14] in the anesthetized monkey, DS cells in V1 are characterized by their vigorous response to stimulus motion in a given direction, with other directions of motion giving little or no

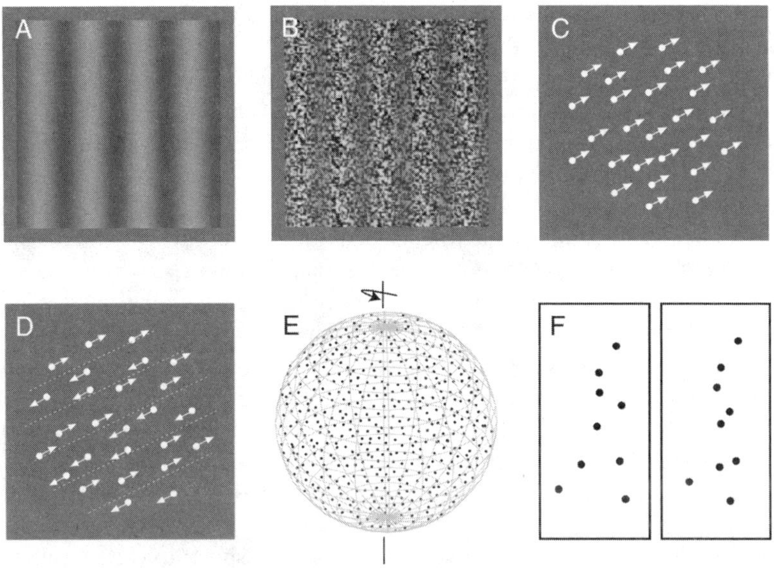

FIGURE 13.1 Schematic examples of some of the animation displays used in many brain imaging and neuropsychological studies of human motion perception. As discussed in this chapter, these different types of visual motion engage different neural mechanisms in the human brain. (A) One-dimensional luminance grating (sinusoidal variations in luminance in the horizontal direction). When shifted in phase in small steps over time, the grating appears to drift in one of the two directions orthogonal to its contour orientation; this is one example of a first-order motion stimulus. (B) One-dimensional, contrast-defined grating. The contrast of this random array of dots is sinusoidally modulated in the horizontal direction, producing visible "bars" defined by contrast modulation. The contrast envelope can be shifted over time in small steps in either direction orthogonal to its contour orientation, thereby creating an impression of motion; this is one example of a second-order motion stimulus. (C) Array of small, randomly positioned dots that can move in any direction from frame to frame of the animation. Called a random-dot cinematogram (RDC), the "strength" of motion can be varied by manipulating the percentage of total dots that moves in a given direction. In this example, 100% of dots move up and to the right. (D) Subsets of dots within virtual regions of the array move in different directions, creating "boundaries" defined by differential optic flow. In this example, dots within alternating "bars" of a grating move in opposite directions, creating the vivid impression of oriented contours. This so-called kinetic grating (and other figures like it) can be employed to isolate regions of the human brain responsive to motion-defined form. (E) Array of small, randomly positioned dots whose frame to frame motion corresponds to the optic flow associated with a real, 3D object, a sphere that rotates about its virtual vertical axis in this example; complex flow fields of this sort create shape from motion (SFM). (F) Two frames from a special type of SFM animation in which the motions of a small cluster of dots track the changing positions of the joints of a person engaged in some activity. In this example, the two frames (nonsuccessive) depict a human walking in a rightward direction. Called "point-light animation" this technique was devised by Johansson[162] as a means for isolating the kinematics associated with biological motion. Brain imaging studies have compared activations associated with viewing examples of biological motion to activations associated with viewing scrambled versions of these same dot motions wherein the perception of biological motion is destroyed; results from those studies point to a region on the superior temporal sulcus (STS) as a reliable source of activation.

response.* It is estimated that about one third of cells in primary visual cortex of the macaque monkey fall in the DS category,[18] and these cells tend to congregate in the upper portions of cortical layer 4 (which receives strong input from the magnocellular layers of the lateral geniculate nucleus) and in layer 6.[19] Within large samples of DS cells there is no obvious bias toward a given direction of motion, and the breadth of directional tuning varies considerably among cells. DS cells are also found in the superior colliculus of the primate visual system, but here their responses depend importantly on whether moving stimuli are presented against a background of stationary contours or against a background in which contours move in a direction different from that of the "target" stimulus — the latter condition produces activity in many more superior colliculus cells than does the former.[20]

Although varying in the details of their behavior, DS cells in V1 have in common a receptive field structure that allows them to register the presence of a pattern moving through the receptive field. These cells can be modeled as spatiotemporal filters that extract motion energy within local regions of the retinal image.[21,22] One useful way to portray the receptive field of a DS cell is by depicting its preferred stimulus in a plot of the variation of that stimulus in spatial position over time. Portrayed in this way, it is easy to see that DS cells respond to events embodying visual energy oriented in space/time. And just as one can model spatial orientation selectivity based on the interplay between excitatory and inhibitory regions arrayed spatially, one can model direction selectivity as the product of an interplay between excitation and inhibition arrayed in space and time.

Much more could be written about DS cells, including their vulnerability to early visual deprivation, their selectivity for spatial frequency, and their relative insensitivity to the wavelength composition of moving stimuli. For our purposes, however, we concentrate on an inherent limitation in the responses of DS cells, a limitation arising from the "local" nature of the motion extraction performed by DS cells. It is important to understand this limitation and the subsequent computational steps needed to overcome it, because this understanding sets the stage for appreciating the neural operations performed in extrastriate visual areas.

As a point of departure for considering this limitation of DS cells, we can construe their receptive fields as spatially restricted "windows" within which motion is "measured." When an extended contour, such as a bar, moves through that window, the receptive field of the DS cell may be too small to "see" the ends of the bar, and, hence, the cell cannot unambiguously respond to the bar's direction of motion independent of its orientation. Instead, the cell will respond as though the contour were moving perpendicular to the orientation imaged within the receptive field of the cell.[23] Thus, for example, a vertical bar moving smoothly up and to the right may traverse a series of small DS receptive fields over time. Those cells, however, can only signal the motion component perpendicular to the orientation of the bar,

* The idea of individual brain cells responsive to given directions of motion can be traced to Reichardt's seminal chapter[15] in which he employs simple logic devices to construct a circuit for registering the presence of motion in a given direction. For that matter, an even earlier version of a "motion detector" can be found in the monograph by Exner,[16] who proposed a simple neural network involving comparisons of successive stimulation arising at different points of the retina at different points in time. (See Smith and Snowden[17] for a more detailed explication of this earliest of neural models of motion detection.)

which, in the case of a vertical bar, means the rightward motion; the upward component of motion of the bar produces no measurable change within the receptive fields of those cells. In other words, the DS cells "see" only what transpires within their local receptive fields, which means that their responses are inherently ambiguous with respect to the actual direction of motion of an extended contour or border. Think about it this way: for any given DS receptive field there exists a family of directions of motion that could generate the same spatiotemporal luminance changes within the local receptive field of the cell and, hence, generate the same response in the cell. How, then, can the actual direction of motion be recovered from these ambiguous responses?

For reasons just explained, the local signals generated by V1 DS cells must be processed further by subsequent stages of the nervous system, where overall, global direction of flow specified by local, overlapping motion can be extracted. And it is widely believed that this second stage of processing transpires within the brain area known as MT (middle temporal area). Some of the DS cells in V1 project to the so-called "thick" stripes in V2, and from here projections innervate MT, or V5 as it is sometimes termed. Other DS cells in V1 project to area V3 and then on to MT,[24] and still other DS cells in V1 project directly to MT.[25] Moreover, some believe that a fraction of MT cells can be activated even when area V1 has been surgically removed, implicating subcortical inputs to this extrastriate area.[26]

13.2.1 Middle Temporal Area

In the monkey, the middle temporal gyrus (MT) is a visual area near the parietal-occipital juncture. Similar to primary visual cortex, MT contains a retinotopically organized representation of visual space, and the receptive field sizes of cells increase with eccentricity.[27] It is estimated that the latency between photoreceptor stimulation and MT activation may be as brief as 20 to 25 ms,[28] which befits MT's role in registering information about events that may require immediate behavioral reaction. In MT, unlike V1, essentially all neurons are direction selective,[29] with this receptive field property topographically organized in a columnar system.[30] The distribution of preferred directions is isotropic among a large sample of cells,[31] which dovetails with the uniform sensitivity to all directions of motion found psychophysically in humans.[32,33]

Several lines of evidence imply that MT plays an important role in motion perception. For one thing, some MT neurons are as sensitive to weak motion in random dot cinematograms (RDC; see Figure 13.1C) as is the alert, behaving monkey from whom MT recordings are being made — there exists a close correspondence between cell firing rates and psychophysical performance.[34,35] For another, microstimulation applied to MT neurons can bias the perceived direction of motion reported by the animal.[36]

But why is MT thought to embody the second-stage process discussed above, i.e., the stage that disambiguates direction of motion? The answer is because a substantial fraction of MT neurons respond to a given direction of motion independently of the orientation of the moving stimulus pattern.[25] These cells behave as if their receptive fields are combining multiple, appropriate outputs from V1 to provide an explicit representation of velocity disambiguated from contour orientation. The discovery of such cells represents a remarkable advance in our understanding of motion computations, and these findings have had a marked influence on computational modeling of human motion perception.[37–39]

In addition to their sensitivity to motion, some MT cells are also disparity selective, i.e., selectively responsive to the depth plane in which motion appears.[40] Moreover, activity in this class of cells fluctuates coincident with reversals in the perceived depth ordering of motion-defined surfaces,[41] as predicted by Nawrot and Blake's[42] neural model of the kinetic depth effect. There is also evidence that activity evoked by stimulation within the receptive fields of MT neurons can be modulated by stimulation falling outside those "classic receptive fields," i.e., within areas of the visual field where stimulation on its own does not evoke neural responses.[43] This response property could provide a mechanism for registering motion-defined boundaries within complex optic flow fields. Finally, it has been shown that MT responses can be modulated by the attentional state of the animal.[44]

Closely allied with MT is a companion area, the medial superior temporal area (MST), that receives input from MT. Neurons in area MST have even larger receptive fields organized in ways that endow them with selectivity for complex optic flow such as expansion or contraction (e.g., Reference 45). MST neurons are also sensitive to nonretinal information about eye movements.[46] This information could be especially important to an organism that is trying to extract its direction of heading from optic flow, while its eyes are moving about.[47] In fact, Bradley et al.[48] demonstrated that neurons in the dorsal part of MST behave in just the manner needed to accomplish this extraction: a large fraction of dorsal MST neurons responded to optic flow centered at a given region of the visual field independent of the direction in which the eyes were pointed. These cells, in other words, can signal the direction in which the animal was headed even as it turned its eyes from side to side while maintaining a constant direction of self-movement.

13.2.2 Superior Temporal Polysensory Region

Anatomically situated just anterior to MT and its satellites on the superior temporal sulcus (STS) of the monkey brain is the anterior superior temporal polysensory region (STPa); for a review on the debate regarding the exact location of this region, see Cusick.[49] Unlike cells in MT, cells in STPa are characterized by their multimodal responses. This region receives inputs from visual, auditory, and somatosensory cortices, but the most dominant modality is vision.[50] Receptive fields of the visually responsive neurons usually cover the entire visual field, and nearly all cells prefer moving stimulation over stationary stimulation.[51] While most cells have no size or shape preferences for the object in motion and half have no preference for direction of motion, many STPa cells prefer particular types of motion such a movement in depth, radial motion, or optic flow.[52] Generally STPa cells respond well to smooth motion over a range of speeds, but many prefer jerky movements and respond maximally at the onset of each discrete step.[51]

Cells in STPa, which appear to respond indiscriminately to form but very selectively to movement, have been implicated in perception of body movement.

Hietanen and Perrett[53] measured responses of cells in the dorsal bank of the anterior STS of the macaque brain, the region corresponding to STPa. They found a majority of the cells to be sensitive to viewing external body movements; only a few cells responded when the animal viewed movements of its own body. Others have found cells in this area that respond selectively to a human form that is walking, rotating, moving in and out of occlusion, and flexing the limbs.[54] Oram and Perrett[55] found a small population of cells (about one third of all cells sampled) in the upper bank of STPa that responded to whole-body movement (such as walking) portrayed in animations consisting of a dozen or so "points" of light attached to the major joints of the body (i.e., the "biological motion" animation technique illustrated schematically in Figure 13.1F). Many of these cells were selective for a given direction of movement implied by the gait of the biological figure, even when there was no net translational motion because the figure was walking in place (as if on a treadmill). Most of the cells did not respond well to "jumbled motion" consisting of the same component motion vectors as biological motion but scrambled to disrupt their spatial coherence. Also found were cells that responded when the biological motion depicted some form of articulation (e.g., bending of the arm). Oram and Perrett noted that cells with similar response selectivity (such as a preference for head rotations toward the animal) tended to be found in close proximity to each other, with these cells clustered according to motion and viewpoint preference.

With this overview of the nonhuman primate visual system, we now are ready to turn to three categories of research studies whose results bear on the issue of brain mechanisms underlying human motion perception. As will become evident, the stimulation conditions used in these studies have been guided by the monkey physiology just reviewed.

13.3 HUMAN BRAIN AREAS RESPONSIVE TO MOTION REVEALED BY BRAIN IMAGING

Based on functional magnetic resonance imaging (fMRI), it is estimated that the human brain contains as many as 17 anatomically distinct areas that can be activated by visual motion of one sort or another.[56] Figure 13.2 illustrates the approximate locations of most of the motion-responsive areas that we focus on in this section. Unless otherwise noted, results described here were obtained from normal adult humans in whom brain activity was inferred either from metabolic maps obtained from positron emission tomography (PET) or from the blood oxygen level-dependent (BOLD) signals recorded while the observers viewed visual animation sequences inside the bore of a magnetic resonance image scanner (fMRI); details of the scanning procedures (field strength, sequence type, image plans, spatial resolution, concurrent task) are not included in the discussion unless specifically relevant for a given finding.

13.3.1 Visual Area V1

In the human brain, V1 can be readily identified using well-established retinotopic mapping procedures that clearly delineate the boundaries and layout of this primary

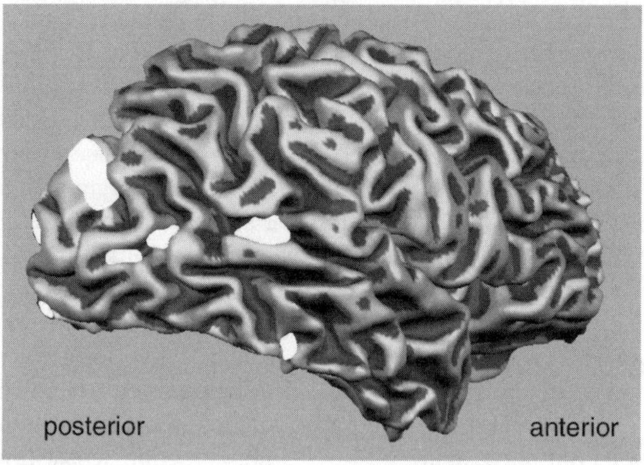

FIGURE 13.2 (Color figure follows p. 142.) Human brain with colored regions denoting approximate locations of several well-studied motion-responsive areas of the human brain, areas that are discussed in this chapter. These regions are shown superimposed on a high-resolution surface map of the right hemisphere, reconstructed at the level of the white matter/gray matter junction. These motion-responsive visual areas are color-coded as follows: purple = V1 (shown here is just the small part of V1 visible in lateral view; the bulk of V1 is located on the medial surface of the hemisphere and, therefore, is largely hidden in this lateral view); orange = V3a; cyan = kinetic occipital region (KO); green = MT+ complex (sometimes referred to as V5); pink = posterior part of the STS (STSp); red = fusiform face area (typically localized by contrasting two viewing conditions: faces vs. objects). There are additional motion-responsive areas that have been identified using fMRI, located in more dorsal and more anterior regions of the brain and, as well, in the cerebellum (not pictured). For a comprehensive map of these multiple motion areas, see Sunaert et al.[56]

sensory cortex located on the medial bank of the calcarine fissure.[57] Targeting this region of interest, it is well known that strong, reliable signals can be evoked by a variety of forms of dynamic stimulation, including flickering or drifting gratings (Figure 13.1A), rotating wedges, expanding and contracting rings (including "second-order" rings defined exclusively by contrast modulation similar to that illustrated in Figure 13.1B), moving dots (Figure 13.1C), and flashing lights. These responses to dynamic stimulation, however, are not unequivocal evidence for explicit representation of motion information in the activity of DS neurons. Indeed, the indiscriminate responses of V1 to a wide variety of dynamic visual stimuli can all be accounted for by the amount of spatiotemporal energy in the display sequence, irrespective of the motion quality per se.[58,59] So, to obtain unequivocal evidence for DS neurons in V1, one must employ strategies for producing BOLD responses attributable to DS cells. One such strategy, described in the next several paragraphs, was developed by Heeger and colleagues.[60]

A hallmark property of nearly all models of motion processing is "motion opponency." In a neural network exhibiting motion opponency, DS cells with opposing preferred directions of motion suppress one another's firing rate, the aim being to generate a net motion signal for given local regions of the visual field. Thus, when opposite directions of motion of equal strength are present simultaneously at a given location in the visual field, motion opponency produces equal, relatively low levels of activity within DS cells tuned to those two directions. But when one direction of motion is stronger than the other, the balance of activity within the opponent DS cells tips in favor of those cells whose preferred direction corresponds to the stronger motion stimulus.

Now there is good psychophysical evidence for the existence of motion opponency in human vision,[61] and Heeger and colleagues used fMRI to identify specific brain areas exhibiting BOLD responses consistent with motion opponency. In the condition we focus on, observers viewed two different motion sequences that alternated over time: a single grating drifting in a given direction interleaved with presentation of two identical gratings drifting in opposite directions ("counterphase" grating). In V1 the fMRI BOLD signal increased during presentation of the counterphase grating and decreased during presentation of the single direction of motion. This outcome implies that the two opposite directions of motion produced greater neural activity than did either grating on its own; there was not strong evidence for motion opponency, in other words. The opposite pattern of BOLD signal responses was found in MT: counterphase gratings produced weaker responses than did single moving gratings, implying strong motion opponency.

From that study it would be premature to conclude that DS cells are not present in human V1, for the interpretation of the findings depends on assumptions about the nature of the inhibitory interactions among DS cells. A more direct brain imaging test for DS cells in V1 was recently performed by the Heeger laboratory using an adaptation procedure, which is described in more detail in the next section. In brief, the procedure produces estimates of the strength of direction-selective adaptation by comparing (1) BOLD responses to repetitive stimulation with visual motion in a single direction to (2) BOLD responses to repetitive stimulation by visual motion that changes direction from moment to moment. The extent to which BOLD signals are smaller in the former case provides an index of the extent of DS processing in a region of interest. Using this technique, Huk et al.[62] found evidence for relatively weak but reliable direction-selective adaptation in V1, reminiscent of the relatively modest proportion of DS cells in V1.

Before moving to subsequent extrastriate cortical areas, we should say a word about the V1 response to apparent motion of an illusory figure created by rotating the sectors in an array of sectored disks ("pacmen") forming an illusory square. When appropriately sequenced, this kind of animation can create the vivid impression that the illusory square is moving back and forth over a set of black disks.[63] When observers view these sequences in the fMRI scanner, robust activation is evoked in V1 as well as in other visual areas including MT.[64] This does not mean, however, that V1 directly computes motion from illusory contours, for the V1 responses could reflect feedback from extrastriate areas. (For a discussion of the interpretation of V1 BOLD signals, see Logothetis et al.[65])

13.3.2 Visual Area 3A

A portion of the output from V1 destined for dorsal portions of the brain, in the parietal lobe, innervates a retinotopically organized brain region located near the junction of the transverse occipital sulcus and the intraparietal sulcus (Figure 13.2). This dorsal extrastriate area, called V3a, was shown by Tootell et al.[66] to be motion sensitive. In fact, subsequent studies have shown that V3a exhibits the strongest direction selectivity of any of the retinotopically defined areas,[62] and it responds much more strongly to coherent motion of random dots than to incoherent, random motion.[58] Culham et al.[67] discuss possible homologies between human V3a and monkey visual area V3.

13.3.3 Visual Area MT+

Everyone agrees that the human homology to monkey MT is located within a relatively large expanse of tissue anterior to early, retinotopically organized visual areas. This putative MT is situated posterior and inferior to the STS on the ascending branch of the inferior temporal sulcus and lateral temporal sulcus.[67–70] In neuroimaging studies, human MT is most commonly isolated by comparing activity levels while observers view a moving pattern (often, moving dots) to activity levels while observers view a stationary pattern or one in which stationary elements flicker on and off (Braddick et al.;[71] but see Orban et al.[72]). Simply comparing moving to stationary conditions, however, lacks the specificity needed to distinguish MT and its surrounding satellite areas, and so the entire motion-responsive region in humans has often been referred to as MT+; in some publications, this brain area is identified as V5. Recently, Huk et al.[73] used novel fMRI mapping techniques to show that human MT+ can be subdivided into two regions. One is retinotopically organized and is activated almost exclusively by stimulation presented in contralateral portions of the visual field; the other is nonretinotopically organized (meaning that activation sites are not dependent on the location of visual stimulation within the visual field) and responds to both contralateral and ipsilateral stimulation. By analogy to receptive field properties from single neurons in macaque monkeys, Huk and colleagues speculate that the retinotopically organized area corresponds to MT and the nonretinotopically organized area to MST. Using somewhat different stimulus conditions, including pursuit eye movements, Dukelow et al.[74] have also partitioned MT+ into functionally distinct subregions that map nicely onto the findings of Huk and colleagues. (For a somewhat different perspective on the MT/MST distinction, see the recent fMRI paper by Morrone et al.[75] in which different types of optic flow were used to differentiate subregions of MT+.)

BOLD signal levels from MT+, unlike signals from any other motion-responsive area, are linearly proportional to the degree of motion coherence present in random-dot cinematograms evoking those BOLD signals (Rees et al.;[76] but see McKeefry et al.[77] for conflicting results obtained using PET); with a few simplifying assumptions, this linear relation between BOLD signal and motion coherence can be nicely modeled based on population responses from single neurons in area MT of alert, behaving monkeys.[78] When activated using gratings varying in contrast, MT+ shows reliable responses to very low contrasts, below the threshold for evoking activation in V1. Those MT+ responses, however, saturate at relatively low contrast values.[70,79] In addition to luminance-defined gratings and RDCs, MT+ is also strongly activated by expanding and contracting rings whose borders are defined exclusively by contrast-modulation, which constitutes a form of second-order motion.[59] As an aside, no fMRI study has uncovered a brain area that responds exclusively to second-order motion (recall Figure 13.1B), although visual area V3 and its ventral counterpart VP respond more strongly to second-order motion than they do to first-order motion.[59]

There is disagreement concerning the extent to which MT+ responds selectively to more complex fields of motion depicting spatial structure (e.g., curvature) or motion-defined forms. Some laboratories report evidence for shape specificity,[80] whereas others find no differential responsiveness to simple translation vs. structured flow fields (e.g., Reppas et al.[81] and Grossman and Blake[82]). It remains to be learned whether these seemingly conflicting results arise because different laboratories are measuring BOLD signals from different subregions within the MT+ complex, subregions containing neurons that differ in their functional properties. The recent work of Huk and colleagues[73] raises some hope that it may be possible to develop MT localization techniques comparable in spatial precision to those used to identify boundaries between early visual areas V1, V2, and V3a.

Activity in MT+ varies in several interesting ways, all of which point to the involvement of this region in the actual perception of motion. For one thing, signal levels from this area are reliably higher when observers genuinely experience motion and not just flicker. This result was found by Muckli et al.[83] who performed an event-related study in which observers viewed repetitive, two-frame animation sequences, which sometimes yielded the strong impression of apparent motion (AM) and other times an impression of on/off flicker with no hint of AM. During brain scanning, observers tracked transitions between these perceptual states, and the resulting BOLD signals were analyzed contingent on the observers' perceptual reports. Activation levels from MT+ covaried over time, with perceptual fluctuations strongest when observers reported seeing motion and weakest when flicker was seen. In a related study, Sterzer et al.[84] utilized a bistable motion stimulus — a "spinning wheel" — whose apparent direction of rotation alternated irregularly over time between clockwise and counterclockwise. While in the scanner, observers tracked these changes in direction of perceived rotation, which allowed the experimenters to identify brain regions in which activation coincided with switches in perceptual state. Sterzer and colleagues found reliable event-related activations in MT+, as well as in several brain regions in parietal and frontal areas; no activations coincident with perceptual fluctuations were found anywhere in temporal lobe areas generally thought to be involved in object recognition. Based on these two studies, then, it seems clear that neural events underlying perceptual fluctuations when viewing bistable motion displays arise within the same brain region — MT+ — that responds reliably to physical changes in the motion characteristics of the retinal image.

Also consistent with the involvement of MT+ in the actual experience of motion, Blakemore et al.[85] found that simple animations implying causation produced greater activation in MT+ than did essentially the same animations in which the perception of causality was absent. While in the scanner, observers viewed brief animations in which a colored circle resembling a "ball" moved horizontally across the visual field. On some trials this ball appeared to collide with another, previously stationary ball, with the collision "launching" this second ball into motion (the "causal" event);

on other trials, the initially moving ball appeared to move underneath the stationary ball without collision, thereby continuing its path of motion ("noncausal" event). These animations are similar to those devised by Michotte[86] in his classic work on perception of causality. In the study by Blakemore et al., the actual motion trajectories in both conditions were essentially identical, with only a change in the color of the balls controlling whether or not collision was perceived. Still, the "causal" event produced significantly greater MT+ activation than did the "noncausal" event, even on trials where attention was directed to the motion trajectory itself and not to the implied causality. Incidentally, this same pattern of activations was also observed in portions of the STS (a brain region we discuss in the next subsection). Blakemore and colleagues interpret these results as evidence that sensory areas of the human brain registering the presence of visual motion are responsive to the causal relations implied by that motion.

Another strategy for studying the relation between MT+ activation and perception exploits the well-known motion aftereffect (MAE), wherein prolonged viewing of motion in one direction causes a subsequently viewed, stationary stimulus to appear to move in the opposite direction. Several laboratories have reported that BOLD signals recorded from MT+ are larger when observers view a stationary figure undergoing illusory motion, compared with when they view the same stimulus after adaptation conditions that do not produce illusory motion.[87–90] Moreover, several of these studies have found that the initially elevated BOLD signal associated with the MAE drops to its baseline level over the same period of time that the MAE itself dissipates (e.g., Culham et al.[88]). This temporary elevation in MT+ activation during MAE but not during control trials was interpreted as the neural concomitant of illusory motion. However, that conclusion was questioned by Huk et al.,[62] who showed that differential levels of attention during MAE and control trials contribute to the differences in BOLD signal differences between these two conditions. The compelling nature of illusory motion during the test phases, in other words, may have unwittingly boosted MT+ activation, a likely possibility in view of the well-established finding that explicit attentional manipulations modulate MT+ BOLD signals (e.g., Buchel et al.[91]).

Huk et al.[62] went on to show that the MAE could be attributable to temporary reductions in activity among neurons stimulated during the adaptation phase. To arrive at this conclusion, they compared BOLD signals measured in blocks of trials during which a single direction of motion was repeatedly presented ("same" direction trials) to BOLD signals measured in blocks of trials during which motion directions varied from presentation to presentation ("mixed" trials). During both trial types, observers performed a challenging speed discrimination task requiring sustained attention. MT+ signals were considerably smaller during "same" trials compared to "mixed" trials, presumably because repeated stimulation of DS cells tuned to a given direction of motion depressed their responsiveness when there was no opportunity for recovery, which the "mixed" trials on the other hand did afford. Based on these and other results, Huk et al. concluded that the MAE results from a temporary imbalance in activation within populations of DS cells. This theory of the MAE, incidentally, was advanced by Mather[92] two decades ago based on psychophysical evidence. Huk and colleagues were able to index the "strength" of direction selectivity by computing the extent to which adaptation reduced BOLD signals, relative

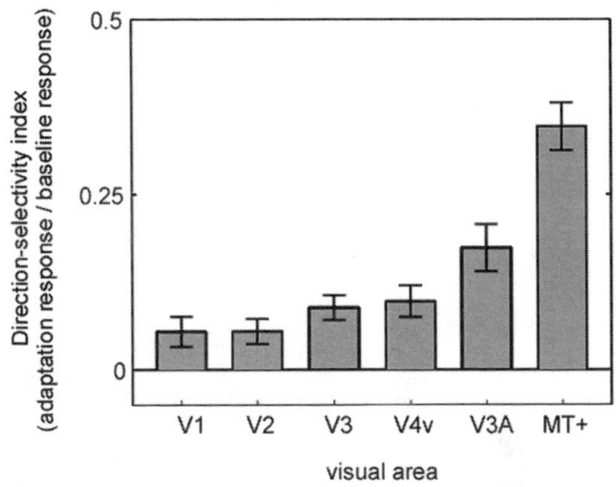

FIGURE 13.3 Histogram showing values of a "direction selectivity" index measured from six different areas of the human brain responsive to motion. These indices were obtained using an fMRI adaptation procedure in which brief episodes of motion were interleaved over closely spaced trials; details of the procedure are given in the text. (From Huk, A.C. et al., *Neuron,* 32, 161–172, 2002. With permission.)

to baseline response; the resulting "direction selectivity" indices for the six different brain areas studied are shown in Figure 13.3. Obviously, this index is highest in MT+, befitting its putatively strong role in motion perception.

Recall that one hallmark property of many cells in monkey MT is their selectivity for given directions of motion independent of the orientations of the moving contours — these cells respond to the overall *pattern* of motion within their receptive fields, not to a given component of motion. Huk and Heeger[93] found compelling evidence for the existence of such "pattern-motion" responses in human visual cortex, including MT+. In their study, activation levels in multiple visual areas were measured while observers viewed a pair of superimposed gratings, differing in orientation and drift direction. In all cases, the pair of gratings cohered to form a "plaid" that appeared to drift in a direction dependent on the orientations of the components. During one block of trials ("same direction" trials), the component directions of the two gratings varied from trial to trial, but the resulting plaid itself appeared always to drift in the same direction throughout the block of trials. During another block of trials, the same component gratings were arranged in pairings in which the direction of coherent motion of the plaid varied from trial to trial ("mixed direction" trials).* Huk and Heeger reasoned that the "same direction" condition, wherein the plaid always moved in a given direction, should produce strong adaptation in "pattern motion" cells, compared to adaptation of those cells in the "mixed direction" condition. Thus in a brain area containing "pattern motion" cells, the fMRI response

* The perceived direction of a plaid composed of two superimposed component gratings is governed by the orientations of the two components, by their drift directions, and by their relative spatial frequencies. Thus, it is possible to produce plaids from different components all of which yield the same global direction of drift, and it is possible to produce plaids from different combinations of the same components each of which yields a different direction of global motion.

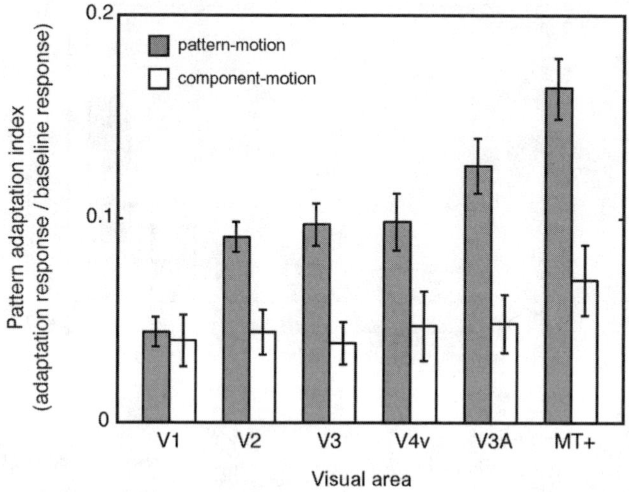

FIGURE 13.4 Histogram showing the extent to which brain activation in different visual areas fluctuates in coincidence with presentation of pairs of moving gratings that stimuluate "pattern motion" neurons vs. "component motion" neurons. Larger index values denote greater pattern motion responsivness. (From Huk, A.C. and Heeger, D.J., *Nat. Neurosci.,* 5, 72–75, 2001. With permission.)

should be periodically weaker and stronger as viewing conditions alternated between "same" and "mixed" trials. Brain areas with only component-motion cells would respond the same to both conditions. Huk and Heeger calculated a "pattern motion" index reflecting the extent to which the fMRI response was modulated, and the values of that index for different visual areas are shown in Figure 13.4. Here it can be seen that MT+ generates the highest degree of "pattern motion" adaptation, implying the existence of a substantial proportion of "pattern motion" cells. V1, on the other hand, shows evidence for component motion cells only, with intermediate visual areas showing intermediate amounts of pattern motion responsiveness. So it appears that the human brain, like that of the macaque monkey, solves the ambiguity problem inherent in V1 cell responses by the addition of a second stage of processing in which "pattern motion" neurons extract the global direction of motion independent of the local contours signaling that direction.

MT+ also appears to receive input from other sources besides earlier visual areas. Two studies[94,95] report that human MT+ can also be activated when observers view stationary photographs of an object in motion (e.g., a person engaged in an athletic maneuver), with activation greater than that produced by viewing the same object without implied motion (e.g., the same person standing still). Activation associated with implied motion could be construed as evidence for "top-down" influences on MT+. Consistent with the existence of such top-down influences, activity levels in human MT+ are significantly boosted when observers voluntarily attend to a set of moving dots seen against a background of stationary dots; shifting attention to the stationary dots yields weakened activation even though the physical stimulation conditions remain identical.[96] Even a stationary object (e.g., an arrow pointing in a given direction) can produce significant activation in MT+ when that object cues the direction of motion to be portrayed in a subsequently viewed animation.[97] For that matter, simply imagining an array of moving dots with eyes closed is reported to yield activation in MT+ compared with a condition that involves imagination of stationary dots.[98] Also pointing to top-down influences, Zacks et al.[99] found transient spikes in MT activation time-locked to perception of transitions between two discrete activities portrayed in an extended video sequence (e.g., a person ironing a shirt and then playing a musical instrument). This finding can be interpreted as the consequence of knowledge-based monitoring of a sequence of events, with the "boundaries" between events being particularly salient. Tong[100] provides additional speculations on the implications of these intriguing results. Considered together, these findings imply that BOLD signals originating within MT+ are not modulated solely by the physical characteristics of visual motion stimulation: mental set, imagination and intentions all influence activity within this intermediate stage of the motion pathway.

Before leaving MT and moving into neighboring visual areas responsive to motion, it is worth saying a word about the responsiveness of MT+ to motion defined by color exclusively. As discussed elsewhere in this volume, an overarching model in visual neuroscience assumes the existence of two visual pathways coursing throughout the human brain, the so-called P-pathway originating in the parvocellular layers of the lateral geniculate nucleus and the M-pathway originating in the magnocellular layers of the lateral geniculate nucleus. When this dual pathways theory was first being developed,[101,102] great emphasis was placed on the functional segregation of color processing (P-pathway) and motion processing (M-pathway). In more recent years, however, it has become abundantly clear that vivid impressions of motion can be created by moving chromatic contours in the absence of luminance contrast (e.g., ffytche et al.,[103] Dougherty et al.[104]), although perceived speed can appear slower for chromatically defined stimuli. Similarly, single-unit recording experiments have documented the existence of neurons in monkey MT responsive to chromatic motion (e.g., Thiele et al.,[105] Seidemann et al.[106]), again undermining the notion that color and motion are strictly segregated visual processes. It is not surprising to learn, then, that human MT+ can be activated by chromatic contours defined exclusively by chromatic stimuli that produce only S-cone signals.[79] These activations are relatively weak, mimicking those generated by very low contrast luminance contours; this contrast equivalence may account for the apparent slowing of motion defined exclusively by color. It remains to be learned just how these color signals are transmitted to MT+, with possibilities including the koniocellular pathway, which receives "on" signals from S cones, or the parvocellular pathway, which receives "off" signals from S cones (see Chapter 1 in this volume by Casagrande and Royal). Whatever their route to MT, it is important to keep in mind that the existence of color-driven responses in MT+ does not necessarily mean that those signals influence color appearance, a point underscored by Wandell et al.[79]

13.3.4 Kinetic Occipital Area

Posterior to the MT+ region and lateral to V3a is another putative motion-selective area, named the kinetic occipital area (KO) by Orban and colleagues;[80,107] it is thought by some that KO may overlap with visual areas LOC/LOP identified and named by Malach et al.,[108] although Dupont et al.[107] argue otherwise. In brain imaging studies using PET and fMRI, KO has been distinguished by its enhanced responsiveness to contours defined by differences in velocity.[80,107,109] Figure 13.1D shows a schematic example of a "kinetic" grating in which alternating "bars" are defined solely by differences in direction of motion of dots. Area KO responds more vigorously to kinetic gratings than it does to coherent dot motion in a single direction or to multiple, superimposed planes of dots. In addition to kinetic gratings, KO also responds selectively to curved motion-defined contours and to alphabetic letters defined by velocity differences between figure and background.[80] One wonders whether the neural interactions endowing KO with this ability to signal form from motion might arise from center/surround antagonism of the sort that has been described for monkey MT neurons.[43]

Whatever the underlying neural architecture, we know that the responsiveness of KO to motion-defined forms does not extend to point-light animations depicting a human body engaged in some form of biological activity.[110] For brain areas responsive to this class of novel, compelling motion stimuli, we must shift our attention to yet another brain area located in the dorsal portion of the STS, a location that could be construed as the confluence of the so-called parietal lobe pathway situated in dorsal portions of the cortex and the temporal lobe pathway situated in more ventral regions of the cortex.

13.3.5 Superior Temporal Sulcus

Using PET and fMRI, several different laboratories have sought to identify brain regions selectively activated when one views someone else engaged in various familiar activities, a form of motion perception termed biological motion. Results from this work point to the unique activation of a neural area in the STS, adjacent to human MT+. We start by summarizing a study from the Vanderbilt laboratory, as those results are highly representative of other findings.

Grossman et al.[110] had observers view alternating presentations of point-light animations (see Figure 13.1F) and spatially scrambled versions of the same animations. The biological motion sequences differentially activated a region on the posterior end of the STS, just anterior to the dorsal bend (see Figure 13.2). This region is anterior and superior to MT+, which itself was localized by Grossman et al. on the ascending limb of the lateral occipital sulcus. Neural activity in MT+ increased relative to resting baseline at the onset of the point-light animations, indicating that MT+ certainly registers the presence of motion contained in these animations. However, the levels of MT+ activation were not different for the point-light motion compared to the scrambled motion; MT+, in other words, does not respond selectively to biological motion. Grossman and colleagues also compared neural activity associated with biological motion to that evoked by kinetic gratings (see Figure 13.1D), the stimulus used to reveal visual area KO (e.g., Van Oostende et al.[80]). They found reliable activation to kinetic gratings in regions posterior and superior to MT+ along the lateral occipital sulcus, which dovetails closely with the published landmarks of KO. However, this region, like MT+, did not respond selectively to biological motion sequences. Nor did STS — the biological motion area — respond to the kinetic gratings. Thus, STS is anatomically and functionally distinct from other brain regions putatively involved in registering shape information from motion.

In a similar demonstration of unique patterns of brain activity associated with point-light stimuli, Howard et al.[111] compared neural activity while viewing coherent motion, optic flow, and point-light biological motion. Observers in this study first viewed a high-contrast checkerboard either stationary or moving coherently. The resulting areas with higher neural activity during the coherent motion condition consisted of a large region extending over the ascending limb of the lateral occipital sulcus and into posterior parietal cortex, the region corresponding to area MT+. Observers also viewed optic flow displays and point-light biological motion displays, and the resulting neural activity was compared to that measured in a condition involving the same individual dot motions only now spatially randomized to destroy their coherence. Howard and colleagues found that viewing optic flow resulted in increased neural activity in a region overlapping and adjacent along the superior boundary of MT+. Biological motion sequences also produced increased neural activity superior to MT+, and this region overlapped with the cortical regions active during viewing the coherent motion and optic flow animations. Howard and colleagues interpreted their results as evidence for specialization of motion perception within MT+. Outside of the MT+ region, they also found activation on the superior temporal gyrus (STG) during the biological motion and optic flow conditions. Anterior regions on the STG are known to be involved in auditory perception,[112,113] and the site of activation on STG during biological motion perception overlapped with neural areas active during listening to a book-on-tape. Howard and colleagues suggest either that MT+ is connected to auditory cortex associated with interpretation of language, or that the overlapping region on STG corresponds to a polysensory area active during visual or auditory stimulation.

This is not the only study to find neural activity associated with auditory perception in regions quite close to those associated with visual perception. An fMRI study on voice-selective regions in cortex found two loci in human STS that had higher levels of activation when listeners heard speech sounds than when they heard energy-matched nonspeech sounds.[114] Speech sounds included isolated words and strings of words connected together and presented in various languages. Energy-matched nonspeech sounds included natural sounds, mechanical noises, and animal cries. The most posterior area of activation on STS corresponds quite closely to the reported site of activation during viewing biological motion.[110,111] Thus, it is quite possible that the posterior region of STS, which is situated between primarily visual- and primarily auditory-responsive cortex, may in fact be a polymodal region.

Point-light animations not only convey movements such as running and walking, but can also convey specific goal-oriented actions, such as picking up a box or grasping a glass.[115] Bonda et al.[116] compared neural activity while observers viewed point-light perception of goal-oriented hand actions with that found when observers viewed whole-body movement. A point-light hand and arm conveyed goal-oriented activity by reaching for an invisible glass and bringing it to the (invisible) mouth. Body movements comprised video segments of a point-light walker moving in various directions. Like the previous studies, Bonda and colleagues[116] found that viewing body movements resulted in higher activity levels on posterior STS than when observers viewed random motion or when they viewed meaningless dot patterns moving rigidly. Viewing hand movements also produced increased activity levels on posterior STS. Based on these findings, plus those from single-unit recordings (discussed earlier), Bonda and colleagues concluded that posterior STS may be involved in the planning and comprehension of goal-directed reaching. Interestingly, during the body movement condition, significantly higher levels of activity were measured in the amygdala than during the hand movement conditions, as if observers experienced emotional responses to the whole-body sequences.

It is interesting to note, too, that brain activity during perception of biological motion depends on whether the perceived visual motion is biologically possible or not. In a study by Stevens et al.[117] observers viewed successive images of whole-body motion that gave the impression of smooth "apparent" motion of limbs either around the body (possible motion) or passing though the body (impossible motion). Following the lead of Shiffrar and Freyd,[118] Stevens and colleagues manipulated the perceived path of the limb by changing the interframe interval (IFI): observers see the biologically plausible path of motion when IFI is long (approximately 450 ms) but see the impossible motion path when IFI is short. Stevens and colleagues identified a region on STG that responded preferentially to the possible movement over the impossible movement, as did primary motor, premotor cortex, and the cerebellum.

Even simple changes in facial features — movement of the eyes or opening and closing of the mouth — are sufficient to produce increased activity levels in posterior STS.[119] BOLD signal levels in posterior STS are higher during both the eye and mouth movement animations than during conditions in which the eyes or mouth are replaced with a low-contrast checkerboard that gives the impression of random movement. For that matter, Beauchamp et al.[120] have shown that STS responds more strongly when observers view moving manipulable objects (e.g., a hammer moving up and down) compared to moving gratings; still, the response of STS to manipulable objects is significantly less than to animations depicting biological activity, implying that it is a region especially designed for the complex kinematics associated with these activities.

In addition to activating posterior STS, viewing biological motion sequences in point-light animations also activates four brain areas thought to be involved in object recognition: the extrastriate body area (EBA), the lateral occipital complex (LOC), and the occipital and fusiform face areas (OFA and FFA).[121] In only two of these areas, however, the FFA and OFA, were the evoked neural signals capable of differentiating biological from nonbiological motion. It seems likely that activity in FFA and OFA is consequent to the "shape" generated by the motion sequences, although this speculation remains to be tested. In any event, these results imply that the perception of biological motion engages a network of distributed neural areas in visual pathways specialized for form perception (the ventral stream) and for motion perception (the dorsal stream).

As mentioned above, STS responds more strongly to biological motion than to the characteristic movements of common objects such as a hammer. However, located inferior to STS in the posterior middle temporal gyrus (MTG) the opposite patterns of activation are found: viewing motion of manipulable objects produces stronger activation in MTG than does viewing of humans performing different behaviors.[120] This functional segregation of motion processing streams for human activity and for manipulable objects is not too surprising when we consider that the two classes of events entail quite different patterns of motion (fluid, hierarchical pendular motions vs. rigid, translational motions).

Before moving to the next section, we should point out that very recent brain imaging studies have begun to focus on visual perception of "intention" and object "affordance."[122,123] Although not explicitly manipulating motion information, these studies seek to identify brain areas that become active when people plan activities (e.g., grasping) in response to presentation of pictures of objects or when people think about imitating the activity of another person. Results from these studies highlight possible links between the visual perception of actions and the execution of those actions; these studies, thus, are extending the scope of "motion perception" to the realm of goal-directed behavior.

13.4 MOTION PROCESSING REVEALED BY TRANSCRANIAL MAGNETIC STIMULATION

A very promising technique for relating motion perception and brain activity is provided by transcranial magnetic stimulation (TMS), wherein a brief magnetic pulse applied to a restricted region of the scalp induces a local electric field in the immediately underlying brain tissue.[124] Because of its high temporal resolution and moderately good spatial resolution, TMS provides a unique means for producing a temporary, reversible "lesion" (functional disruption) within the brain area near the cortical surface. Rather than destruction of neural tissue, this "lesion" actually results from the brief disruption of ongoing, coherent neural activity; TMS, in other words, transiently introduces "noise" into the underlying neural processes.

Several research groups have examined the consequence of TMS applied to the scalp region overlying visual area MT while observers view visual animation sequences.[125–128] In general, TMS directed to this brain area disrupts motion perception, as evidenced by disruptions in performance on a direction discrimination task and/or a speed discrimination task; performance on control tasks (e.g., form acuity) is unaffected by TMS-induced disruption of MT activity. There is some disagreement about exactly how this degradation in motion perception varies with the relative timing of TMS and visual stimulation (e.g., compare Hotson et al.[127] with Beckers and Zeki[129]), but these disagreements could arise from methodological differences, such as the exact placement of the coil used to induce TMS.

It is illuminating to compare the effects on motion perception of TMS applied to visual areas V1 and MT. Beckers and Zeki[129] applied brief TMS pulses just before, during, or just after presentation on a video monitor of a small array of dots moving coherently either leftward or rightward, with direction varying randomly over trials. (The dots were imaged to one side of fixation, so that the TMS pulse could be directed to the contralateral hemisphere in which those dots would normally evoke activity.) Observers were instructed to report the direction of motion, with TMS pulses applied to regions of the scalp coinciding as closely as possible with area V1 or area MT (Beckers and Zeki acknowledge the imprecision of targeting these areas exclusively).

When delivered within ±50 ms of the visual stimulus, TMS targeted on V1 had essentially no effect on motion perception, as evidenced by accurate performance on the direction discrimination task. Only when delivered about 60 ms after the onset of the visual stimulus did TMS of V1 disrupt performance on this task, and then by only a modest amount. TMS targeting MT, however, produced pronounced masking, lowering task performance to chance levels when TMS and visual stimulation fell within about 20 ms of one another. Based on this pattern of results, Beckers and Zeki concluded that at least some motion-related neural signals arise within MT before reaching V1, which they take as evidence for a visual input to MT that bypasses V1. This possibility is discussed further in a subsequent section of this chapter, when we consider the consequences of selective brain damage on motion perception.

In another, related study, Matthews et al.[128] applied brief TMS pulses to the scalp region overlying V1 or the region overlying MT, and they required observers to judge perceived speed or perceived direction of motion associated with brief presentations of translational RDCs. TMS at both sites produced reliable slowing of perceived speed but no shift in perceived direction of motion. In a second experiment, these differential consequences of TMS applied at these two locations were replicated using forced-choice testing procedures (although there were intriguing differences in the timing relationship between RDC presentation and TMS for speed discrimination and speed perception). The failure of TMS to alter perceived direction of motion is not necessarily in conflict with the results of Beckers and Zeki,[129] because that study did not directly measure perceived direction of motion but, instead, asked observers to indicate in which one of four cardinal directions an array of semicoherent dots was moving.

In the studies just summarized, TMS was used to disrupt motion processing instigated by presentation of real motion signals. But through appropriate adjustment of stimulation levels, TMS can also be used to generate visual sensations ("phosphenes") of motion in the absence of external stimulation; deployed in this way, TMS transiently induces illusory motion when applied to visual area MT.[130,131] Two experiments using TMS to trigger phosphenes have produced intriguing results bearing on the question of the role of V1 in motion perception.

Cowey and Walsh[130] created TMS-induced phosphenes in normally sighted individuals, in an adult male with complete blindness owing to bilateral optic nerve damage (peripheral blindness), and in an adult male missing his left striate cortex (blindsight patient GY, who is discussed in more detail in the following section). In all observers — sighted and blind — magnetic pulses were applied over the occipital scalp at points targeting primary visual cortex as well as at more dorsolateral points targeting MT+. The normal, sighted individuals all experienced phosphenes, sometimes colored but never moving, when stimulation was directed to regions presumably activating neural tissue around the calcarine sulcus and, by inference, the primary visual cortex. The visual field location of the phosphenes depended on the specific site of stimulation and, of course, on the hemisphere stimulated (phosphenes always appeared in the visual field contralateral to TMS stimulation). When the location of pulse delivery was moved to the more dorsolateral site (presumptive area MT+), all normally sighted observers experienced moving phosphenes in the visual field contralateral to the stimulated hemisphere. The overwhelming majority of phosphene descriptions characterized them as grayish and sparkly; faint color sensations were rarely reported. When TMS stimulation was applied to the same regions of the occipital lobe of the man blind from optic nerve damage, he described the same two categories of phosphenes (stationary, colored vs. uncolored, moving). Incidentally, this individual also expressed surprise and pleasure upon experiencing visual phosphenes. Patient GY experienced normal phosphenes — stationary and moving — when TMS was delivered to the appropriate sites over his right hemisphere. But when pulses targeted the equivalent sites in his left hemisphere, GY experienced no phosphenes whatsoever. The absence of stationary phosphenes with more posterior stimulation is not surprising, since GY's damage encompasses nearly all of the striate cortex in the left hemisphere.[132] However, the absence of visual phosphenes when area MT+ is stimulated is rather surprising, because this region is intact in GY as evidenced by successful activation of MT+ in both hemispheres in fMRI studies.[133] From this pattern of results, Cowey and Walsh conclude that activity in area MT+ alone is necessary but not sufficient for generation of visual motion perception.

This same conclusion has been reached based on results from a clever experiment in which TMS was used both to evoke motion sensations and to disrupt them. Pascual-Leone and Walsh[134] applied TMS to both V1 and MT in close temporal proximity; importantly, the MT pulse was set to a level yielding perception of moving phosphenes when delivered on its own and the V1 pulse was set to a level too weak to yield phosphenes of any sort. Results from this intriguing study are reproduced in Figure 13.5. When TMS was applied to V1 before MT, moving phosphenes were reliably reported by observers, as if the V1 TMS pulse had not been delivered at all. But when TMS was delivered to V1 5 to 45 ms after TMS stimulation to MT, perception of moving phosphenes was abolished. This outcome is remarkable because the disrupting V1 event occurred later in time than the evoking MT event. The authors interpret their findings to indicate that V1 activation is necessary for conscious awareness of visual motion, even though the signals critical for signaling that motion arise in MT. Obviously, this conclusion dovetails nicely with the findings of Cowey and Walsh.[130]

TMS applied to the scalp overlying MT+ can also neutralize the effect of prior exposure to a motion stimulus that, ordinarily, would facilitate or "prime" subsequent performance on a motion task.[135] In this study, observers were administered a series of brief trials separated by short intervals. During each trial the observer viewed four arrays of moving dots presented around a central fixation point. In three of the arrays all dots moved in the same direction and in the fourth array the dots moved in a direction orthogonal to the other three; observers indicated by manual response which one of the four arrays portrayed the "odd" direction. Ordinarily, observers are faster at making this judgment when the direction of motion of the "odd" target is the same on consecutive trials, regardless of whether it appears in the same location. This perceptual priming effect implies that visual information during one exposure is briefly retained in visual memory and utilized during processing in the following exposure. Campana et al.[135] showed that a brief TMS pulse delivered during the intertrial interval, when no motion stimulus was present, was sufficient to neutralize this priming effect. This disruption of processing was specific to motion stimulation: when the same task structure required the use of color for the "odd" discrimination, TMS applied to MT had no deleterious effect on priming. These results add weight to the idea that visual memory engages some of the same areas

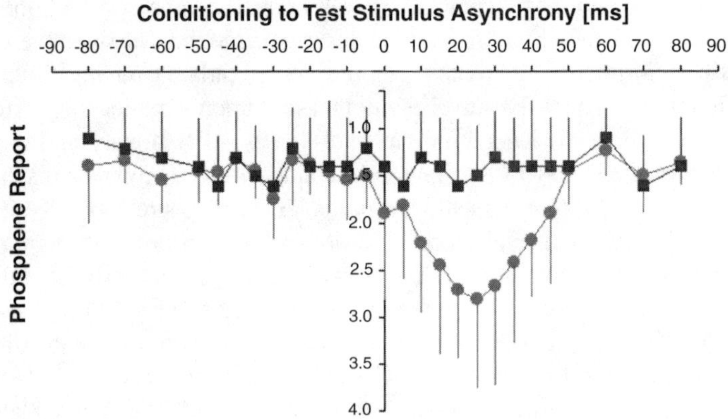

FIGURE 13.5 Graph showing the incidence of moving phosphenes elicited by the application of a brief TMS pulse to a region of the scalp overlying visual area MT in the human brain. Human volunteers judged the quality of the phosphenes using a four-category scale (1 = phosphenes present and moving; 2 = phosphenes present but uncertain whether they moved; 3 = stationary phosphenes; 4 = no phosphene observed). A second, brief TMS pulse was also applied at varying times relative to the MT pulse, with the interval between the two pulses plotted on the abscissa. For data shown with squares, both pulses were applied to area MT — this pairing had no effect on the high incidence of perceived motion regardless of the interval between pulses. For the data shown with circles, one TMS pulse of the pair was applied over visual area V1 at varying times relative to the MT pulse — when the V1 pulse closely followed the MT pulse, the incidence of moving phosphenes was reduced significantly. (From Pascual-Leone, A. and Walsh, V., *Science,* 292, 510–512, 2001. With permission.)

involved in actual visual perception, and they further highlight the potentially powerful marriage of psychophysics and TMS.

13.5 VISUAL MOTION PERCEPTION AND BRAIN DAMAGE

In this final section we focus on deficits in motion perception consequent to accident-induced or disease-induced lesions of restricted regions of the brain, another complementary strategy for identifying motion-responsive areas of the human brain. Our survey of this literature is highly selective, and readers seeking a more comprehensive overview will want to consult any of several available sources.[136,137]

13.5.1 CEREBRAL MOTION BLINDNESS

Certainly one of the most dramatic examples of motion impairment is the "motion blindness" or "akinetopsia" suffered by LM,* a woman who sustained bilateral

* LM is almost certainly not the first case of akinetopsia to be described in the neurological literature. Zeki[140] describes several other, similar cases dating from the early part of the 20th century, and he also describes cases in which motion perception is preserved in the face of losses in visual object recognition and color vision.

damage to the occipitoparietal region, consequent to cerebral infarction.[138] Later brain imaging confirmed that the damage included (but was not limited to) visual area MT in both cerebral hemispheres.[139] This fascinating individual has been extensively studied by a number of laboratories. While maintaining serviceable spatial vision, reasonably good object recognition, and normal color vision, LM denies experiencing any sense of visual motion. Video clips of LM reveal that she pours water into a glass past the point where it overflows, presumably because she cannot see the rising water level, and she experiences anxiety when attempting to cross the street or navigate environments replete with moving objects. Even with forced-choice testing in the laboratory, LM experiences great difficulty telling which of two objects is moving at a higher speed[141] or the direction in which a single stimulus moves.[139,142] For example, LM sometimes confuses upward motion and downward motion, especially when coherent, translational motion is presented against a background of static, random noise. When asked to identify letters defined by moving dots presented against a stationary background (motion-defined form), LM experiences great difficulty.[142] LM has also been tested on the "Ternus" ambiguous motion display in which three evenly spaced bars shift positions back and forth by one cycle from frame to frame.[143] Normal observers perceive apparent motion with this display, with the particulars of the motion depending on the interval separating successive frames. On viewing the Ternus display, LM denies *any* experience of motion and, instead, describes the outer bars as blinking on and off regardless of the timing of the frames.[142]

It is perhaps not so surprising that LM, because of her brain damage, which encompasses visual area MT, suffers impaired motion perception. After all, monkeys with lesions in the homologous region of the brain show pronounced deficits in detection and discrimination of motion,[144–146] although recovery can lessen the perceptual impact of such lesions.[147] (LM's impaired motion perception, by contrast, showed little recovery over the 15 years or so following her stroke.) Furthermore, the neural consequences of LM's damaged MT most likely adversely affected other brain regions innervated by MT and normally involved in motion processing (e.g., the frontal eye field).* Finally, we must keep in mind that LM's lesion encompasses tissue outside area MT, and it encroaches on white matter as well. In fact, some of her perceptual difficulties may be related to the large cerebellar lesion that extends to some midline structures. Consistent with this possibility, Nawrot and Rizzo[148] have shown that patients with midline cerebellar lesions exhibit elevated thresholds for detecting coherent motion in random-dot cinematograms. These impairments were measured at least 2 years after the incident leading to cerebellar damage, and they are not attributable to unstable fixation or poor tracking eye movements, as both of these oculomotor functions were normal in this sample of ten patients. Nawrot and Rizzo[148] speculate that the cerebellum may play a crucial role in the

---

* In the Zihl et al.[138] paper, LM's pursuit movements were characterized as normal when tracking a relatively slowly moving target, but at higher speeds the tracking movements were jerky and saccade-like. To our knowledge, no other, detailed eye movement records were published on LM (she died in the spring of 2002).

perceptual interpretation of retinal image motion, which can arise from object motion, head movements, and postural adjustments.

So, given the large extent of LM's bilateral damage, it is simply impossible to attribute her motion disorders to the loss of any single visual area such as MT. There are, however, other neuropsychological studies that point to the crucial involvement of MT in normal motion perception. In one large sample of 50 patients suffering stroke with unilateral brain damage, deficits in direction discrimination using RDCs were reliably observed in those individuals whose damage included the MT+ complex but not in patients whose lesions were confined to more ventral structures in the temporal lobe.[137] Two studies[149,150] have examined motion perception in patients with parietal lobe lesions, finding no deficits in speed perception or detection of moving gratings. On the other hand, patients with damage to the right parietal lobe do evidence deficits in their ability to track the motion of a subset of targets within an array of moving elements,[151] a deficit that probably involves an inability to attend to transient events, not necessarily a deficit in registration of motion information per se.

As well, there are some single case studies in which specific motion deficits are associated with highly restricted, unilateral lesions, and these results also bear on the role of MT in motion perception. Several of those studies are summarized in the next section.

13.5.2 Brain Damage and Perception of First-Order vs. Second-Order Motion

An intriguing selective deficit in motion perception has been identified in patient FD, an adult male with a small, shallow lesion located dorsal to, but not including, area MT. Unlike LM's lesion, FD's damage is confined to one hemisphere, making it possible to compare visual performance in the two visual fields, contralateral and ipsilateral to the brain damage. Patient FD has been described in some detail by Vaina and Cowey,[152] whose findings can be summarized as follows. FD has normal visual acuity, normal visual fields, good color vision and stereopsis, and normal contrast sensitivity. He can readily judge two-dimensional (2D) and 3D shape from motion (recall Figure 13.1E), and he has no trouble discriminating the direction or speed of moving stimuli. His coherence thresholds for detection of translational motion in RDCs (Figure 13.1C) are within the normal range, regardless in which visual field the moving dots are imaged. So in these respects, FD's motion perception is very different from LM's. However, FD does show pronounced impairments in perception of "second-order" motion in which movement is signaled by displacements of boundaries defined solely by contrast variations or by texture discontinuities, not by luminance differences (compare Figure 13.1A and Figure 13.1B). This deficit in perception of second-order motion was found on several different tasks.

In a related study, Vaina et al.[153] identified a male patient, TF, with a very small, unilateral lesion that involved visual area V2 only, with MT completely spared. In contrast to FD, TF experienced difficulties on motion tasks involving "first-order" motion (e.g., direction and speed discrimination in random-dot cinematograms), but he exhibited normal motion sensitivity to "second-order" motion. This distinction between the consequences of brain damage on first-order vs. second-order motion is buttressed by Greenlee and Smith's[154] study of a large sample of patients with unilateral brain damage to regions of the temporal, lateral occipital, and/or posterior parietal cortex. All patients exhibited essentially normal performance on a grating orientation task, but all performed below normal individuals on several motion discrimination tasks; the magnitude of the motion deficits were dependent on the nature of the task (direction vs. speed discrimination) and the nature of the motion stimuli (first-order vs. second-order). This pattern of results, too, supports the idea that motion perception is not a unitary ability arising from a single set of neural mechanisms. Considered together, these three neuropsychological studies, along with others by Plant and colleagues,[155,156] lead to the conclusion that the human brain contains multiple motion pathways that implement different computational operations, pathways whose components are differentially vulnerable to brain damage.

13.5.3 Brain Damage and Perception of Biological Motion

Next we turn to neuropsychological studies relating brain damage to performance on "shape from motion" tasks, with particular emphasis on perception of biological shapes defined by motion. Vaina et al.[157] examined an individual, patient AF, with bilateral lesions of the temporal-parietal-occipital junction extending into the posterior parietal cortex of the left hemisphere (this pattern of damage was consequent to a series of strokes that occurred over an unspecified period of time). AF's ability to see shape from luminance or from color was intact, but he was unable to see shapes defined by retinal disparity or by motion. AF also had elevated thresholds for detecting coherent motion signals in RDCs. Nonetheless, AF could easily identify biological motion in Johansson's point-light displays (Figure 13.1F). He readily described animations of a man climbing the stairs or riding a bicycle, and he had no problem reporting the direction in which the "walker" was headed. So despite obvious difficulties on more basic motion perception tasks, AF experienced no problem seeing the more complex forms of motion specifying human activity in point-light animations.

Another interesting patient, AL, suffered two strokes that affected her temporal lobe and posterior parietal cortex.[158] Immediately following her second stroke AL was unable to recognize friends and family members by looking at their faces or watching their gaits, but she could readily identify them by their voices. She soon regained her ability to recognize familiar faces when people remained still, but complained that she could not recognize those same faces when the people were moving. During psychophysical testing AL was almost perfect at discriminating luminance-defined geometric shapes, but she could not detect objects defined by flicker nor could she discriminate objects defined by motion. She failed to tell the difference between a virtual cylinder created from moving dots and a spatially unstructured display of the same moving dots. AL also utterly failed to identify any human activities depicted in point-light animations, even after having been told that the display showed a person engaged in a routine action. So unlike patient AF, AL's motion deficits included perception of biological motion.

Schenk and Zihl[159,160] measured motion and form perception in two patients with bilateral lesions in the parietal lobe and extensive damage to the underlying white matter. These patients had normal coherent motion thresholds and were able to segregate sketches of objects superimposed on top of each other. These patients also recognized human activities depicted in point-light animations when a single biological actor was animated; however, when the actor appeared embedded in a field of static or randomly moving dots, the patients, unlike normal observers, were unable to recognize the point-light figure.

Considered together, results from these four studies imply that perception of one particular form of shape from motion — the human form defined by the kinematics in point-light animations — can be spared in the face of deficits in other aspects of motion perception and, conversely, can be impaired even when other aspects of motion perception are normal. In the neuropsychological field, these kinds of double dissociations are typically taken as evidence for the involvement of multiple, anatomically distinct neural mechanisms.

13.6 CONCLUSIONS

It was the noted comparative anatomist Gordon Walls[161] who poetically described the significance of visual motion perception to behaving organisms:

> If asked what aspect of vision means the most to them, a watchmaker may answer "acuity," a night flier, "sensitivity," and an artist, "color." But to the animals which invented the vertebrate eye, and hold the patents on most of the features of the human model, the visual registration of movement was of the greatest importance.

As Walls knew, the ability to detect and recognize visual events in our environment is crucial to our survival, and motion perception is the key to event perception. Motion specifies where objects are headed (direction) and how soon they will get there (speed). Motion aids in the perception of the shapes of moving objects (shape from motion) and their relative positions in 3D space (depth from motion). In fact, motion can reveal the presence of an otherwise camouflaged object when that object moves. Motion also allows mobile organisms to move about in their environments, avoiding obstacles while intercepting objects of interest (heading from optical flow). And motion portrays important information concerning other people's body movements and facial expressions, allowing us to recognize who those people are and what their intentions might be.

Befitting this set of crucial visual abilities, it is not surprising to learn that considerable neural territory in the human brain is devoted to the registration and analysis of motion information. Moreover, damage to portions of that neural hardware can produce serious perceptual problems that require substantial compensatory modifications to behavior. Our understanding of brain mechanisms underlying motion perception has advanced greatly during the last 15 years, in part because of exciting technological advances in brain imaging and in brain stimulation. It is also worth noting that those advances were grounded in a solid psychophysical literature on motion perception and on sophisticated computational models that often guided the search for underlying neural mechanisms.

ACKNOWLEDGMENTS

The authors thank David Heeger, Alex Huk, Nestor Matthews, and Mark Nawrot for comments on sections of this chapter. David Bloom helped greatly with manuscript formatting and references. During preparation of this chapter R.B. and E.G. were supported by research grants from the National Institutes of Health (EY07760) and the National Science Foundation (BCS0121962) and by a Core Grant from the National Institutes of Health (EY01826) awarded to the Vanderbilt Vision Research Center.

REFERENCES

1. Logothetis, N.K., Physiological studies of motion inputs, in *Visual Detection of Motion*, Smith, A.T. et al., Eds., Academic Press, London, 1994, 177.
2. Andersen, R.A., Neural mechanisms of visual motion perception in primates, *Neuron,* 18, 865, 1997.
3. Shiffrar, M., Movement and event perception, in *Blackwell Handbook of Perception*, Goldstein, E.B., Ed., Blackwell Publishers, Oxford, U.K., 2001, 237.
4. Sekuler, R., Watamaniuk, S.N.J., and Blake, R., Motion Perception, in *Stevens' Handbook of Experimental Psychology,* Vol. 1: *Sensation and Perception*, 3rd ed., Yantis, S., Ed., John Wiley & Sons, New York, 2002, 121.
5. Horowitz, G.D. and Newsome, W.T., Target selection for saccadic eye movements: direction-selective visual responses in the superior colliculus, *J. Neurophysiol.,* 86, 2527, 2001.
6. Kubova, Z. et al., Is the motion system relatively spared in amblyopia? Evidence from cortical evoked responses, *Vision Res.,* 36, 181, 1996.
7. Ahlfors, S.P. et al., Spatiotemporal activity of a cortical network for processing visual motion revealed by MEG and fMRI, *J. Neurophysiol.,* 82, 2545, 1999.
8. Chen, Y. et al., Psychophysical isolation of a motion-processing deficit in schizophrenics and their relatives and its association with impaired smooth pursuit, *Proc. Natl. Acad. Sci. U.S.A.,* 96, 4724, 1999.
9. Demb, J., Boynton, G., and Heeger, D., Brain activity in visual cortex predicts individual differences in reading performance, *Proc. Natl. Acad. Sci. U.S.A.,* 94, 13363, 1997.
10. Hess, R.F. and Anderson, S.J., Motion sensitivity and spatial undersampling in amblyopia, *Vision Res.,* 33, 881, 1993.
11. Rizzo, M. and Nawrot, M., Perception of movement and shape in Alzheimer's disease, *Brain,* 121, 2259, 1998.
12. Atkinson, J. et al., A specific deficit of dorsal stream function in Williams' syndrome, *NeuroReport,* 8, 1919, 1997.
13. Blake, R. et al., Visual recognition of biological motion is impaired in children with autism, *Psychol. Sci.,* 14, 151, 2003.
14. Hubel, D.H. and Wiesel, T.N., Receptive fields and functional architecture of monkey striate cortex, *J. Physiol.,* 195, 215, 1968.
15. Reichardt, W., Autocorrelation, a principle for the evaluation of sensory information by the central nervous system, in *Sensory Communication*, Rosenblith, W.A., Ed., MIT Press, Cambridge, MA, 1961, 303.
16. Exner, S., Über optische Bewegungsempfindungen, *Biol. Zentralbl.,* 8, 437, 1888.

17. Smith, A.T. and Snowden, R.J., Motion detection: an overview, in *Visual Detection of Motion*, Smith, A.T. et al., Eds., Academic Press, London, 1994, 3.
18. Schiller, P.H., Finlay, B.L., and Volman, S.F., Quantitative studies of single-cell properties in monkey striate cortex, *J. Neurophysiol.,* 39(6), 1288, 1976.
19. Hawken, M.J., Parker, A.J., and Lund, J.S., Laminar organization and contrast sensitivity of direction-selective cells in the striate cortex of the Old World monkey, *J. Neurosci.,* 8, 3541, 1988.
20. Davidson, R.M. and Bender, D.B., Selectivity for relative motion in the monkey superior colliculus, *J. Neurophysiol.,* 65, 1115, 1991.
21. Adelson, E.H. and Bergen, J.R., Spatiotemporal energy models for the perception of motion, *J. Opt. Soc. Am.,* 2, 284, 1985.
22. van Santen, J.P.H. and Sperling, G., Temporal covariance model of human motion perception, *J. Opt. Soc. Am.,* 1(5), 451, 1984.
23. Fennema, C.I. and Thompson, W.B., Velocity determination via scenes containing several moving images, *Comput. Graph. Image Process.,* 9, 301, 1979.
24. Gegenfurtner, K., Kiper, D., and Levitt, J., Functional properties of neurons in macaque area V3, *Neurophysiology,* 77, 1906, 1997.
25. Movshon, J.A. and Newsome, W.T., Visual response properties of striate cortical neurons projecting to area MT in macaque monkeys, *J. Neurosci.,* 23, 7733, 1996.
26. Rodman, H.R. and Albright, T.D., Single-unit analysis of pattern-motion selective properties in the middle temporal visual area (MT), *Exp. Brain Res.,* 75, 53, 1989.
27. Gattass, R. and Gross, C.G., Visual topography of striate projection zone (MT) in posterior superior temporal sulcus of the macaque, *J. Neurophysiol.,* 46, 621, 1981.
28. Bair, W. et al., The timing of response onset and offset in macaque visual neurons, *J. Neurosci.,* 22(8), 3189, 2002.
29. Maunsell, J.H.R. and Van Essen, D.C., Functional properties of neurons in middle temporal visual area of the macaque monkey. I. Selectivity for stimulus direction, speed, and orientation, *J. Neurophysiol.,* 49, 1127, 1983.
30. Albright, T.D., Desimone, R., and Gross, C.G., Columnar organization of directionally selective cells in visual area MT of the macaque, *J. Neurophysiol.,* 51, 16, 1984.
31. Albright, T.D., Direction and orientation selectivity of neurons in visual area MT of the macaque, *J. Neurophysiol.,* 52, 1106, 1984.
32. Ball, K. and Sekuler, R., A specific and enduring improvement in visual motion discrimination, *Science,* 218(4573), 697, 1982.
33. Gros, B.L., Blake, R., and Hiris, E., Anisotropies in visual motion perception: a fresh look, *J. Opt. Soc. Am.,* 15(8), 2003, 1998.
34. Britten, K.H., Newsome, W.T., and Saunders, R.C., Effects of inferotemporal cortex lesions on form-from-motion discrimination in monkeys, *Exp. Brain Res.,* 88, 292, 1992.
35. Britten, K.H. et al., A relationship between behavioral choice and the visual responses of neurons in macaque MT, *Vis. Neurosci.,* 13(1), 87, 1996.
36. Salzman, C.D., Britten, K.H., and Newsome, W.T., Cortical microstimulation influences perceptual judgements of motion direction, *Nature,* 346, 174, 1990.
37. Wilson, H.R., Ferrera, V.P., and Yo, C., A psychophysically motivated model for two-dimensional motion perception, *Vis. Neurosci.,* 9, 79, 1992.
38. Simoncelli, E.P. and Heeger, D.J., A model of neuronal responses in visual area MT, *Vision Res.,* 38, 743, 1998.
39. Simoncelli, E.P. and Heeger, D.J., Representing retinal image speed in visual cortex, *Nat. Neurosci.,* 4(5), 461, 2001.

40. Maunsell, J.H.R. and Van Essen, D.C., Functional properties of neurons in middle temporal visual area of the macaque monkey. II. Binocular interactions and sensitivity to binocular disparity, *J. Neurophysiol.,* 49, 1148, 1983.
41. Bradley, D.C., Chang, G.C., and Andersen, R.A., Encoding of three-dimensional structure-from-motion by primate area MT neurons, *Nature,* 392, 714, 1998.
42. Nawrot, M. and Blake, R., A neural network model of kinetic depth, *Vis. Neurosci.,* 6, 219, 1991.
43. Allman, J., Miezin, F., and McGuinness, E., Direction- and velocity-specific responses from beyond the classical receptive field in the middle temporal visual area (MT), *Perception,* 14, 105, 1985.
44. Treue, S. and Maunsell, J.H., Attentional modulation of visual motion processing in cortical areas MT and MST, *Nature,* 382(6591), 539, 1996.
45. Saito, H. et al., Integration of direction signals of image motion in the superior temporal sulcus of the macaque monkey, *J. Neurosci.,* 6, 145, 1986.
46. Komatsu, H. and Wurtz, R.H., Relation of cortical areas MT and MST to pursuit eye movements. I. Localization and visual properties of neurons, *J. Neurophysiol.,* 60, 580, 1988.
47. Barinaga, M., Researchers find neurons that may help us navigate, *Science,* 273(5281), 1489, 1996.
48. Bradley, D.C. et al., Mechanisms of heading perception in primate visual cortex, *Science,* 273, 1544, 1996.
49. Cusick, C.G., The superior temporal polysensory region in monkeys, in *Extrastriate Cortex in Primates*, Rockland, K. et al., Eds., Plenum Press, New York, 1996, 435.
50. Desimone, R. and Gross, C.G., Visual areas in the temporal cortex of the macaque, *Brain Res.,* 178, 363, 1979.
51. Bruce, C., Desimone, R., and Gross, C.G., Visual properties of neurons in a polysensory area in superior temporal sulcus of the macaque, *J. Neurophysiol.,* 46, 369, 1981.
52. Anderson, K.C. and Siegel, R.M., Optic flow selectivity in the anterior superior temporal polysensory area, STPa, of the behaving monkey, *J. Neurosci.,* 19(7), 2681, 1999.
53. Hietanen, J.K. and Perrett, D.I., Motion sensitive cells in the macaque superior temporal polysensory area, *Exp. Brain Res.,* 93, 117, 1993.
54. Perrett, D.I. et al., Visual cells in the temporal cortex sensitive to face view and gaze direction, *Proc. R. Soc. Lond. B Biol. Sci.,* 223, 293, 1985.
55. Oram, M.W. and Perrett, D.I., Responses of anterior superior temporal polysensory (STPa) neurons to "biological motion" stimuli, *J. Cogn. Neurosci.,* 6, 99, 1994.
56. Sunaert, S. et al., Motion-responsive regions of the human brain, *Exp. Brain Res.,* 127, 355, 1999.
57. Wandell, B.A., Computational neuroimaging of human visual cortex, *Annu. Rev. Neurosci.,* 22, 145, 1999.
58. Braddick, O.J. et al., Brain areas sensitive to coherent visual motion, *Perception,* 30, 61, 2001.
59. Smith, A.T. et al., The processing of first- and second-order motion in human visual cortex assessed by functional magnetic resonance imaging (fMRI), *J. Neurosci.,* 18(10), 3816, 1998.
60. Heeger, D., Boynton, G., and Demb, J., Motion opponency in visual cortex, *J. Neurosci.,* 19(16), 7162, 1999.
61. Levinson, E. and Sekuler, R., Inhibition and disinhibition of direction-specific mechanisms in human vision, *Nature,* 254(5502), 692, 1975.

62. Huk, A.C., Ress, D., and Heeger, D.J., Neuronal basis of the motion aftereffect reconsidered, *Neuron,* 32, 161, 2001.
63. Bravo, M., Blake, R., and Morrison, S., Cats see subjective contours, *Vision Res.,* 28(8), 861, 1988.
64. Seghier, M. et al., Moving illusory contours activate primary visual cortex: an fMRI study, *Cereb. Cortex,* 10, 663, 2000.
65. Logothetis, N.K. et al., Neurophysiological investigation of the basis of the fMRI signal, *Nature,* 412, 150, 2001.
66. Tootell, R.B. et al., Functional analysis of V3A and related areas in human visual cortex, *J. Neurosci.,* 17(18), 7060, 1997.
67. Culham, J. et al., Visual motion and the human brain: what has neuroimaging told us? *Acta Psychol.* (Amsterdam), 107, 69, 2001.
68. Zeki, S. et al., A direct demonstration of functional specialization in human visual cortex, *J. Neurosci.,* 11, 641, 1991.
69. Watson, J.D.G. et al., Area V5 of the human brain: evidence from a combined study using positron emission tomography and magnetic resonance imaging, *Cereb. Cortex,* 3, 79, 1993.
70. Tootell, R.B.H. et al., Functional analysis of human MT and related visual cortical areas using magnetic resonance imaging, *J. Neurosci.,* 15(4), 3215, 1995.
71. Braddick, O.J. et al., Brain areas differentially activated by coherent visual motion and dynamic noise, *NeuroImage,* 7, S322, 1998.
72. Orban, G.A. et al., Human cortical regions involved in extracting depth from motion, *Neuron,* 24, 929, 1999.
73. Huk, A.C., Dougherty, R.F., and Heeger, D.J., Retinotopy and functional subdivision of human areas MT and MST, *J. Neurosci.,* 22(16), 7195, 2002.
74. Dukelow, S.P. et al., Distinguishing subregions of the human MT+ complex using visual fields and pursuit eye movements, *J. Neurophysiol.,* 86, 1991, 2001.
75. Morrone, M.C. et al., A cortical area that responds specifically to optic flow, revealed by fMRI, *Nat. Neurosci.,* 3(12), 1322, 2000.
76. Rees, G., Friston, K., and Koch, C., A direct quantitative relationship between the functional properties of human and macaque V5, *Nat. Neurosci.,* 3(7), 716, 2000.
77. McKeefry, D. et al., The activity in human areas V1/V2, V3, and V5 during the perception of coherent and incoherent motion, *NeuroImage,* 5, 1, 1997.
78. Britten, K.H. et al., Responses of neurons in macaque MT to stochastic motion signals, *Vis. Neurosci.,* 10, 1157, 1993.
79. Wandell, B.A. et al., Color signals in human motion-selective cortex, *Neuron,* 24, 901, 1999.
80. Van Oostende, S. et al., The kinetic occipital (KO) region in man: an fMRI study, *Cereb. Cortex,* 7, 690, 1997.
81. Reppas, J. et al., Representation of motion boundaries in retinotopic human visual cortical areas, *Nature,* 388, 175, 1997.
82. Grossman, E.D. and Blake, R., Brain activity evoked by inverted and imagined biological motion, *Vision Res.,* 41, 1475, 2001.
83. Muckli, L. et al., Apparent motion: event-related functional magnetic resonance imaging of perceptual switches and states, *J. Neurosci.,* 22(9), RC219, 2002.
84. Sterzer, P. et al., Neural correlates of spontaneous direction reversals in ambiguous apparent visual motion, *NeuroImage,* 15, 908, 2002.
85. Blakemore, S.-J. et al., How the brain perceives causality: an event-related fMRI study, *NeuroReport,* 12(17), 3741, 2001.
86. Michotte, A., *The Perception of Causality,* Basic Books, New York, 1963.

87. Tootell, R.B.H. et al., Visual motion aftereffect in human cortical area MT revealed by functional magnetic resonance imaging, *Nature,* 375, 139, 1995.
88. Culham, J.C. et al., Recovery of fMRI activation in motion area MT following storage of the motion aftereffect, *J. Neurophysiol.,* 81, 388, 1999.
89. He, S., Cohen, E., and Hu, X., Close correlation between activity in brain areas MT/V5 and the perception of a visual motion aftereffect, *Curr. Biol.,* 8, 1215, 1998.
90. Taylor, J.G. et al., The network of brain areas involved in the motion aftereffect, *NeuroImage,* 11, 257, 2000.
91. Buchel, C. et al., The functional anatomy of attention to visual motion: a functional MRI study, *Brain,* 121, 1281, 1998.
92. Mather, G., The movement aftereffect and a distribution-shift model of coding the direction of visual movement, *Perception,* 9, 379, 1980.
93. Huk, A.C. and Heeger, D.J., Pattern-motion responses in human visual cortex, *Nat. Neurosci.,* 5, 72, 2001.
94. Senior, C. et al., The functional neuroanatomy of implicit motion perception or "representational momentum," *Curr. Biol.,* 10, 16, 2000.
95. Kourtzi, Z. and Kanwisher, N., Activation in human MT/MST by static images with implied motion, *J. Cogn. Neurosci.,* 12(1), 48, 2000.
96. O'Craven, K.M. et al., Voluntary attention modulates fMRI activity in human MT-MST, *Neuron,* 18, 591, 1997.
97. Shulman, G.L. et al., Areas involved in encoding and applying directional expectations to moving objects, *J. Neurosci.,* 19(21), 9480, 1999.
98. Goebel, R. et al., The constructive nature of vision: direct evidence from functional magnetic resonance imaging studies of apparent motion and motion imagery, *Eur. J. Neurosci.,* 10, 1563, 1998.
99. Zacks, J.M. et al., Human brain activity time-locked to perceptual event boundaries, *Nat. Neurosci.,* 4(6), 651, 2001.
100. Tong, F., Brain at work: play by play, *Nat. Neurosci.,* 4(6), 560, 2001.
101. Livingstone, M. and Hubel, D., Segregation of form, color, movement, and depth: anatomy, physiology and perception, *Science,* 240, 740, 1988.
102. Zeki, S., *A Vision of the Brain,* Blackwell Scientific Publications, Oxford, U.K., 1993.
103. ffytche, D.H., Skidmore, B.D., and Zeki, S., Motion-from-hue activates area V5 of human visual cortex, *Proc. R. Soc. Lond. B Biol. Sci.,* 260, 353, 1995.
104. Dougherty, R.F., Press, W.A., and Wandell, B.A., Perceived speed of colored stimuli, *Neuron,* 24, 893, 1999.
105. Thiele, A., Dobkins, K.R., and Albright, T.D., Neural correlates of chromatic motion perception, *Neuron,* 32, 351, 2001.
106. Seidemann, E. et al., Color signals in area MT of the macaque monkey, *Neuron,* 24, 911, 1999.
107. Dupont, P. et al., The kinetic occipital region in human visual cortex, *Cereb. Cortex,* 7, 283, 1997.
108. Malach, R. et al., Object-related activity revealed by functional magnetic resonance imaging in human occipital cortex, *Proc. Natl. Acad. Sci. U.S.A.,* 92, 8135, 1995.
109. Orban, G.A. et al., Processing of kinetically defined boundaries in the cortical motion area MT of the macaque monkey, *J. Neurophysiol.,* 74, 1258, 1995.
110. Grossman, E. et al., Brain areas involved in perception of biological motion, *J. Cogn. Neurosci.,* 12, 711, 2000.
111. Howard, R. et al., A direct demonstration of functional specialization within motion-related visual and auditory cortex of the human brain, *Curr. Biol.,* 6, 1015, 1996.

112. Berry, I. et al., Activation of association auditory cortex demonstrated with functional MRI, *NeuroImage,* 2(3), 215, 1995.
113. Cansino, S., Williamson, S.J., and Karron, D., Tonotopic organization of human auditory association cortex, *Brain Res.,* 17, 305, 1994.
114. Belin, P. et al., Voice-selective areas in human auditory cortex, *Nature,* 403, 309, 2000.
115. Hill, H. and Pollick, F.E., Exaggerating temporal differences enhances recognition of individuals from point light displays, *Psychol. Sci.,* 11(3), 223, 2000.
116. Bonda, E. et al., Specific involvement of human parietal systems and the amygdala in the perception of biological motion, *J. Neurosci.,* 16, 3737, 1996.
117. Stevens, J.A. et al., New aspects of motion perception: selective neural encoding of apparent human movements, *NeuroReport,* 11(1), 109, 2000.
118. Shiffrar, M. and Freyd, J.J., Apparent motion of the human body, *Psychol. Sci.,* 1, 257, 1990.
119. Puce, A. et al., Temporal cortex activation in humans viewing eye and mouth movements, *J. Neurosci.,* 18(6), 2188, 1998.
120. Beauchamp, M.S. et al., Parallel visual motion processing streams for manipulable objects and human movements, *Neuron,* 34, 149, 2002.
121. Grossman, E. and Blake, R., Brain areas active during visual perception of biological motion, *Neuron,* 35, 1167, 2002.
122. Grèzes, J. and Decety, J., Does visual perception of object afford action? Evidence from a neuroimaging study, *Neuropsychologia,* 40, 212, 2002.
123. Chaminade, T., Meltzoff, A.N., and Decety, J., Does the end justify the means? A PET exploration of the mechanisms involved in human imitation, *NeuroImage,* 15, 318, 2002.
124. Walsh, V. and Cowey, A., Magnetic stimulation studies of visual cognition, *Trends Cogn. Sci.,* 2, 103, 1998.
125. Walsh, V., Ashbridge, E., and Cowey, A., Cortical plasticity in perceptual learning demonstrated by transcranial magnetic stimulation, *Neuropsychologia,* 36, 45, 1998.
126. Beckers, G. and Homberg, V., Cerebral visual motion blindness: transitory akinetopsia induced by transcranial magnetic stimulation of human area V5, *Proc. R. Soc. Lond. B Biol. Sci.,* 249(1325), 173, 1992.
127. Hotson, J. et al., Transcranial magnetic stimulation of extrastriate cortex degrades human motion direction discrimination, *Vision Res.,* 34, 2115, 1994.
128. Matthews, N. et al., Transcranial magnetic stimulation differentially affects speed and direction judgments, *Exp. Brain Res.,* 140, 397, 2001.
129. Beckers, G. and Zeki, S., The consequences of inactivating areas V1 and V5 on visual motion perception, *Brain,* 118, 49, 1995.
130. Cowey, A. and Walsh, V., Magnetically induced phosphenes in sighted, blind and blindsighted observers, *NeuroReport,* 11(14), 3269, 2000.
131. Battelli, L., Black, K.R., and Wray, S.H., Transcranial magnetic stimulation of visual area V5 in migraine, *Neurology,* 58, 1066, 2002.
132. Baseler, H.A., Morland, A.B., and Wandell, B.A., Topographic organization of human visual areas in the absence of input from primary cortex, *J. Neurosci.,* 19(7), 2619, 1999.
133. Barbur, J.L. et al., Conscious visual perception without V1, *Brain,* 116, 1293, 1993.
134. Pascual-Leone, A. and Walsh, V., Fast backprojections from the motion to the primary visual area necessary for visual awareness, *Science,* 292, 510, 2001.
135. Campana, G., Cowey, A., and Walsh, V., Priming of motion direction and area V5/MT: a test of perceptual memory, *Cereb. Cortex,* 12, 663, 2002.

136. Nawrot, M., Disorders of motion and depth, in *Neurologic Clinics of North America*, Barton, J. et al., Eds., W.B. Saunders, Philadelphia, in press.
137. Vaina, L.M. et al., Regional cerebral correlates of global motion perception: evidence from unilateral cerebral brain damage, *Brain,* 124, 310, 2001.
138. Zihl, J., von Cramon, D., and Mai, N., Selective disturbance of movement vision after bilateral brain damage, *Brain,* 106, 313, 1983.
139. Shipp, S. et al., The brain activity related to residual motion vision in a patient with bilateral lesions of V5, *Brain,* 117, 1023, 1994.
140. Zeki, S., Cerebral akinetopsia (visual motion blindness), *Brain,* 114, 811, 1991.
141. Hess, R.F., Baker, C.L., and Zihl, J., The "motion blind" patient: low level spatial and temporal filters, *J. Neurosci.,* 9, 1628, 1989.
142. Rizzo, M., Nawrot, M., and Zihl, J., Motion and shape perception in cerebral akinetopsia, *Brain,* 118, 1105, 1995.
143. Pantle, A. and Petersik, J.T., Factors controlling the competing sensations produced by a bistable stroboscopic motion display, *Vision Res.,* 19, 143, 1979.
144. Rudolph, K. and Pasternak, T., Transient and permanent deficits in motion perception after lesions of cortical areas MT and MST in the macaque monkey, *Cereb. Cortex,* 9(1), 90, 1999.
145. Cowey, A. and Marcar, V.L., The effect of removing superior temporal cortical motion areas in the macaque monkey: I. Motion discrimination using simple dots, *Eur. J. Neurosci.,* 4, 1219, 1992.
146. Wurtz, R.H. and Roy, J.P., The role of disparity-sensitive cortical neurons in signalling the direction of self-motion, *Nature,* 348, 160, 1990.
147. Newsome, W.T. and Pare, E.B., A selective impairment of motion perception following lesions of the middle temporal visual area (MT), *J. Neurosci.,* 8(6), 2201, 1988.
148. Nawrot, M. and Rizzo, M., Chronic motion perception deficits from midline cerebellar lesions in human, *Vision Res.,* 38, 2219, 1998.
149. Greenlee, M.W. et al., Visual short-term memory of stimulus velocity in patients with unilateral posterior brain damage, *J. Neurosci.,* 15, 2287, 1995.
150. Spinelli, D. and Zoccolotti, P., Perception of moving and stationary gratings in brain damaged patients with unilateral spatial neglect, *Neuropsychologia,* 30(4), 393, 1992.
151. Battelli, L. et al., Unilateral right parietal damage leads to bilateral deficit for high-level motion, *Neuron,* 32, 985, 2001.
152. Vaina, L.M. and Cowey, A., Impairment of the perception of second order motion but not first order motion in a patient with unilateral focal brain damage, *Proc. R. Soc. Lond. B Biol. Sci.,* 263, 1225, 1996.
153. Vaina, L.M. et al., A lesion of cortical area V2 selectively impairs the perception of the direction of first-order visual motion, *NeuroReport,* 11(5), 1039, 2000.
154. Greenlee, M.W. and Smith, A.T., Detection and discrimination of first- and second-order motion in patients with unilateral brain damage, *J. Neurosci.,* 17, 804, 1997.
155. Plant, G.T. and Nakayama, K., The characteristics of residual motion perception in the hemifield contralateral to lateral occipital lesions in humans, *Brain,* 116(6), 1337, 1993.
156. Plant, G.T. et al., Impaired visual motion perception in the contralateral hemifield following unilateral posterior cerebral lesions in humans, *Brain,* 116(6), 1303, 1993.
157. Vaina, L.M. et al., Intact "biological motion" and "structure from motion" perception in a patient with impaired motion mechanisms: a case study, *Vis. Neurosci.,* 5, 353, 1990.
158. Cowey, A. and Vaina, L.M., Blindness to form from motion despite intact static form perception and motion perception, *Neuropsychologia,* 38, 566, 2000.

159. Schenk, T. and Zihl, J., Visual motion perception after brain damage: I. Deficits in global motion perception, *Neuropsychologia,* 35, 1289, 1997.
160. Schenk, T. and Zihl, J., Visual motion perception after brain damage: II. Deficits in form-from-motion perception, *Neuropsychologia,* 35, 1299, 1997.
161. Walls, G., *The Vertebrate Retina and Its Adaptive Radiation,* Cranbrook Press, Bloomfield Hills, MI, 1942.
162. Johansson, G., Visual perception of biological motion and a model for its analysis, *Percept. Psychophys.,* 14, 195, 1973.

14 The Functional Organization of Monkey Inferotemporal Cortex

Manabu Tanifuji

CONTENTS

14.1 INTRODUCTION

The inferotemporal (IT) cortex in monkey is a large portion of the temporal lobe ventral to the superior temporal sulcus (STS). The dorsal and posterior border of the IT cortex is located next to area V4 at around the inferior occipital sulcus (IOS), and its ventral and anterior border reaches the perirhinal cortex a few millimeters ventral to the anterior middle temporal sulcus (AMTS). The IT cortex is situated in the ventral visual pathway: Area V4 is the major source of visual information relayed to the posterior part of the IT cortex, designated as area TEO, and area TEO is the major source of visual information relayed to the anterior part of the IT cortex,

0-8493-1243-4/04/$0.00+$1.50

designated as area TE. On the other hand, the IT cortex projects to multiple areas that are not specific to visual information processing (for further subdivisions and extrinsic connections of the IT cortex, see Reference 1).

Early physiological and lesion studies have shown that the IT cortex is essential for visual perception and recognition of objects by their visual images (for a historical review, see Reference 2). Results of these studies together with the position of the IT cortex in the ventral visual pathway have shown that the IT cortex is considered as the site that represents object images essential for perception and recognition. The major purpose of this chapter is to reveal the current view of object representation in the IT cortex based on recent findings obtained by electrophysiological and functional imaging studies.

14.2 VISUAL IMAGES ACTIVATING IT NEURONS

In general, neurons in the IT cortex respond to visual images of natural objects.[3] However, this does not necessarily mean that neurons are specific to object images. For example, some IT neurons are also activated by artificially generated two-dimensional shapes, such as Fourier descriptors.[4] Extensive investigations have revealed two aspects concerning visual images that activate IT neurons.[5] First, there are neurons responding specifically to images of behaviorally important objects, such as faces, hands, and other body parts. Second, efforts to determine essential features that activate IT neurons revealed that optimal visual stimuli of some IT neurons are visual features less complex than the images of natural objects.

14.2.1 Neurons Specific for Behaviorally Important Objects

Visual images of behaviorally important objects, such as faces and hands, activate some IT neurons. In particular, those responding to faces have been well investigated. To examine their specificity for faces, these neurons were examined for facial parts such as eyes and mouth, stimuli from which facial parts were removed, scrambled faces, and other visual stimuli unrelated to faces.[5–8] These investigations showed that some neurons (face neurons) are specifically activated by "faces." These neurons also respond to "facial parts" or "faces lacking some facial parts," but the responses are not as strong as those to "faces" (Figure 14.1).[5,7] Selectivity among different faces was also examined in several investigations.[9–12] Some of the face neurons are selective for a particular face, but in general, they are broadly tuned to different faces.[10–12] One unique property of face neurons is that most of these neurons are tuned to a particular view of faces, whereas only a small number of neurons are view invariant (Figure 14.1).[5,13] According to Perrett and colleagues,[13] 92% of face neurons in STS are tuned to certain view angles, only 4% of cells respond to faces regardless of view angles, and another 4% of cells respond to any views with preference for particular view angles. Face neurons are widely distributed in area TE including the fundus and ventral bank of STS, and also in the upper bank of STS (for the distribution of face neurons see Reference 14). Face neurons, however, may be distributed more densely in the dorsal bank, fundus, and ventral bank of STS, than other regions. In fact, the face neurons reported in most studies are located in STS.

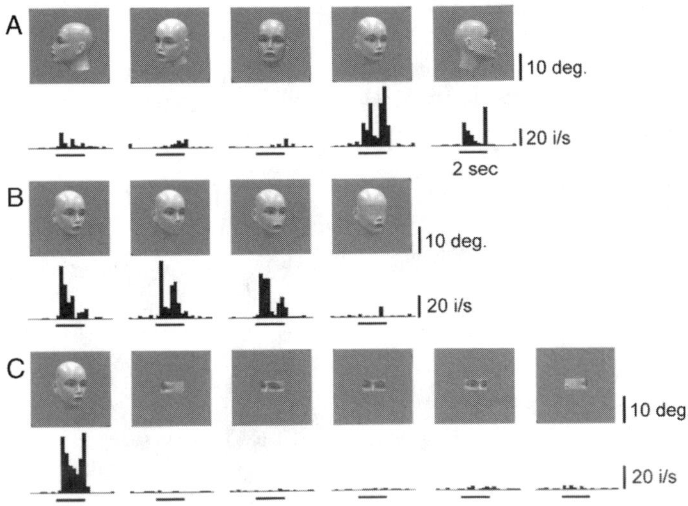

FIGURE 14.1 Response properties of a face neuron. Each histogram shows neuronal responses elicited by the above stimulus. The recording consists of three sessions. In the first session (A), the selectivity of the neuron for different views of a face is demonstrated. In the second and third sessions (B, C), selectivity of the same neuron for a face and facial parts were examined. Note that facial parts activated the cell, but the response was weaker than that to a complete face (B), and that no activation by either a face without eyes (B) or an eye alone (C) was observed. (From Fukuda and Tanifuji, unpublished observations.)

There are also neurons specific for hands, although the response properties of these neurons have not been systematically investigated as have those of face neurons. Neurons responding to the entire human body were found in the dorsal and fundus of STS.[15] Some of these neurons had response properties similar to face neurons; namely, the neurons were tuned to a particular view and maximally activated by the entire body, but not by component parts such as the head or body alone.[15]

14.2.2 Neurons Responding to Visual Features Less Complex Than Objects

Many neurons in the IT cortex responding to objects are also activated by visual features that are less complex than objects. There are a number of attempts to search for features essential for activating these neurons using complex and simple visual stimuli.[5,8,16] Tanaka and colleagues[8,16] systematically simplified visual stimuli to determine the simplest visual feature that maximally activates each cell ("critical feature"). First, for each cell, they searched for the most effective stimulus among more than 100 three-dimensional object stimuli, including stuffed animals, plastic fruits and vegetables, and experimenter's hand and body. These stimuli were presented to the subject from various viewpoints to maximize the chance for finding the most effective stimulus. Then, they generated modifications of the most effective stimulus and examined responses evoked by the simplified stimuli. If the cell responded to one of the simplified stimuli equally well as compared to that to the

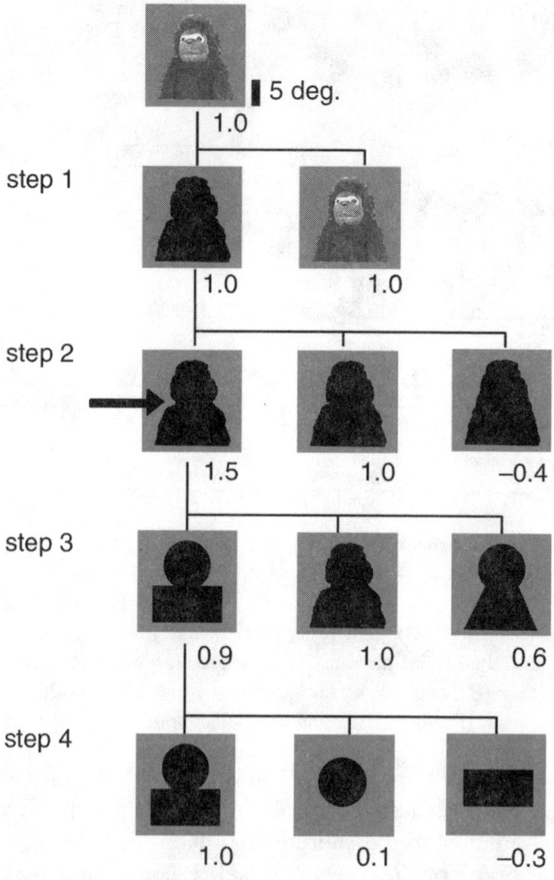

FIGURE 14.2 Systematic stimulus simplification of an object image. The most effective object stimulus (top) was simplified step by step to determine the simplest stimulus that maximally activates the cell. Step 1 shows that neuronal activity elicited by the most effective object and by its silhouette are the same. The number below each picture indicates the response amplitudes normalized to the response to the reference stimulus. Step 2 examines the effect of the "sharpness" of the corner at the junction of the upper and lower parts (arrow) and shows that the silhouette with the sharpest corners (left) is the most effective stimulus. Step 3 shows that activities elicited by the two leftmost stimuli are not significantly different. Step 4 shows that neither the upper nor the lower part activates the cell. In this case, the critical feature was determined to be the combination of a circle and a rectangle (leftmost image in step 4). (From Yamane and Tanifuji, unpublished observations.)

original object stimulus, this new stimulus was further simplified. This procedure was repeated until the experimenter encountered a drastic decrease in responsiveness due to simplification. Figure 14.2 shows an example of the stimulus simplification procedure, where a combination of the circle and rectangle was essential for the maximal activation of the cell. Tanaka and colleagues examined many neurons in this way and identified visual features essential for activating individual cells

FIGURE 14.3 Critical features determined for IT cells. (Adapted from Reference 63.)

(Figure 14.3). Many of these features are combinations of simple shapes, colors, luminance gradient/contrast, and textures. These features are more complex than the optimal visual stimuli for cells in areas V1, V2, and V4, but still less complex than natural objects.

Neurons responding to these moderately complex visual features are not evenly distributed in the IT cortex. Kobatake and Tanaka[16] found that the proportion of cells that prefer these features vs. simple visual stimuli such as oriented bars was higher in area TE than in area TEO.

14.3 SPATIAL ORGANIZATION OF NEURONS IN THE IT CORTEX

Some of the above investigations of neurons selective for faces or moderately complex visual features revealed that neurons with similar response properties tend to cluster together, suggesting the existence of columnar organizations in the IT cortex.[8,9,17,18] The columnar organization, such as orientation and ocular dominance columns in area V1, is characterized by two specific spatial arrangements of neurons:

1. Neurons with similar response properties are clustered in a vertical strip extending from the cortical surface to the white matter.
2. These strips are systematically arranged along the cortical surface: adjacent clusters form a distinct border with respect to responsiveness as in the case of ocular dominance columns, or the distribution of neurons with particular responsiveness changes gradually in the tangential direction such that response properties averaged over local neurons change continuously as in the case of orientation columns.

In addition to these criteria, we have to consider that (1) the cortical area may not be entirely occupied by these columns, and (2) depending on the property of the stimuli, the degree of response similarity may be different, and thus the spatial organization of neurons may appear different.

Fujita and colleagues[19] were the first to examine systematically the columnar organization in the IT cortex. First, they penetrated electrodes perpendicular to the cortical surface and determined the critical feature of one neuron within the track. Then, they prepared a set of visual stimuli including this critical feature and its modifications, and examined the response selectivity of other neurons within the track using the set of visual stimuli. This experiment showed that the most effective stimulus within the set for individual neurons in the track is the critical feature or visual features similar to it. Next, they examined neuronal responses along a tangential electrode penetration and found that neurons with similar responsiveness were localized within about 0.4 mm. These results confirm the first two criteria and thus provide strong evidence of the presence of columnar organization in the IT cortex at least when the response similarity is measured using a set of simple visual features.

Intrinsic signal imaging has enabled further characterization of the spatial organization of columns in the IT cortex.[20–22] This technique measures the decrease in the degree of light reflection elicited by neural activation from the exposed cortical surface using a CCD camera.[23] The reflection changes are due to metabolic changes elicited by neural activation including changes in deoxygenation of hemoglobin in capillaries.[23] Although these reflection changes are not a direct measure of neural activation, intrinsic signals coincide well with the activity of neurons examined by conventional extracellular recordings.[20–22] Thus, if nearby neurons are simultaneously activated and elicit detectable changes in reflection, the spatial organization of the activated neurons can be investigated by this technique. Intrinsic signal imaging in area TE revealed multiple spots elicited by visual stimulation (Figure 14.4).[20–22] The mean size of the "active spot" was 0.50 ± 0.13 mm along the longer axis and 0.35 ± 0.09 mm along the shorter axis.[22] These dimensions agreed well with the size previously reported for a column of cells with similar responsiveness in this area.[18,19] Strictly speaking, "active spots" do not necessarily correspond to columns because intrinsic signals may reflect neural activity only in superficial layers. Thus, we use the term *spots* instead of *columns* for cellular clustering revealed by intrinsic signal imaging. The typical distribution pattern of active spots showed that some spots overlap partially (Figure 14.5). This observation suggests that response properties averaged over local neurons change continuously across the cortical surface. Another example of continuous changes in response properties across the cortical surface is shown in Figure 14.6, where active spots produced by images of faces from different vantage points shift along the cortical surface as the face rotates from the left profile to the right profile, through the front face. Intrinsic signal imaging also showed that there are regions unoccupied by any spots. One explanation is that these regions are not activated by the stimulus set used in this experiment but by other stimuli. Alternatively, neurons responding to different stimuli may be intermingled in these regions, and thus any of these stimuli could not elicit spots. A conservative view of spatial organization of the IT cortex at present would be that some part of the IT cortex consists of columns where

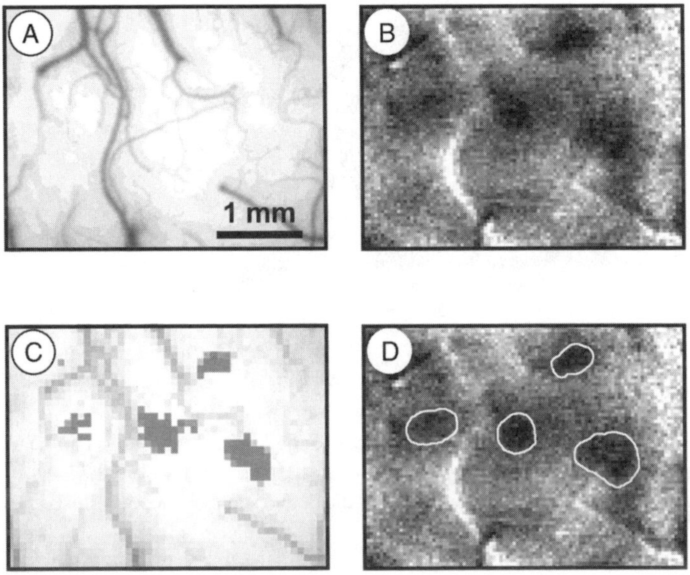

FIGURE 14.4 Active spots revealed by intrinsic signal imaging. (A) The surface view of area TE where intrinsic signals were recorded. (B) A differential image showing local increases in absorption. (C) Active regions where the degree of reflection change evoked by the stimulus was significantly greater than that in the case of no stimulus presentation ($p < 0.05$). (D) Extracted active spots outlined by connecting pixels using 1/2 of the peak absorption value. (Modified from Reference 22.)

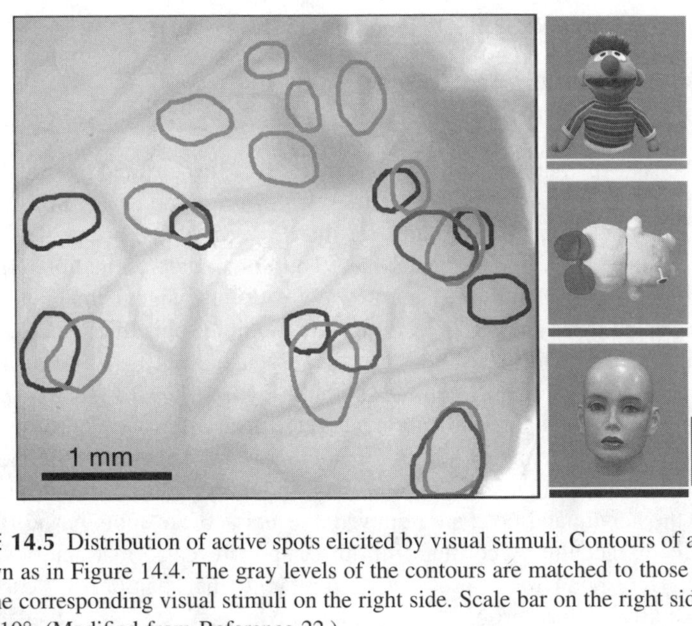

FIGURE 14.5 Distribution of active spots elicited by visual stimuli. Contours of active spots are drawn as in Figure 14.4. The gray levels of the contours are matched to those of the bars below the corresponding visual stimuli on the right side. Scale bar on the right side of visual stimuli, 10°. (Modified from Reference 22.)

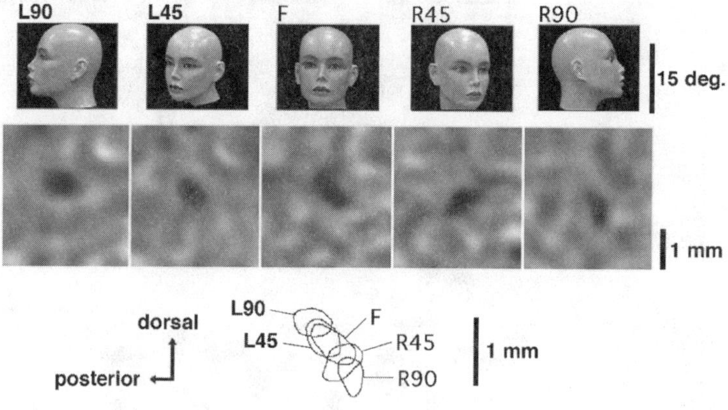

FIGURE 14.6 Systematic shift in distribution of active spots with rotation of the face. Images of the same cortical area (middle panels) obtained from five different views of the same mannequin face (top panels). The contours of the active spots are superimposed at the bottom. (Adapted from Reference 20.)

neurons with similar response properties are distributed continuously, and the other parts of the cortex consist of a mixture of neurons with different responsiveness. By taking into account that neurons with different orientation preference are intermingled at the close vicinity of the orientation singularity points,[24] the organization of orientation columns in area V1 appears similar to the proposed organization in area TE.

14.4 OBJECT REPRESENTATION IN THE IT CORTEX

14.4.1 Representation by Combined Activation of IT Neurons

As described above, there are neurons that are maximally activated by visual features less complex than images of objects. These neurons respond to any objects as long as the objects contain critical features for the cells. Similarly, a face neuron is not activated exclusively by one particular face. Thus, object representation requires the combined activation of multiple neurons. With intrinsic signal imaging, we have recently revealed that complex objects activate multiple spots distributed across the cortex (Figure 14.5) and that simplified stimuli activate only a subset of the spots that were activated by the more complex stimuli (Figure 14.7).[22] Figure 14.7A shows one in which each step of simplification led to the disappearance of a subset of spots activated by the more complex stimuli. In another example, spots A, B, and C were activated by a "fire extinguisher" (*Stimulus 1*), and the simplified stimulus (*Stimulus 3*), where the handle and hose are removed, did not activate spots A and B (Figure 14.7B). The extracellular recording from these spots revealed that neurons in spots A, B, and D respond to a "protrusion" similar to the "handle," a "curved line segment" similar to the "hose," and an "elongated shape" similar to the "cylinder" of a "fire extinguisher," respectively.[22] These results suggest that each spot represents

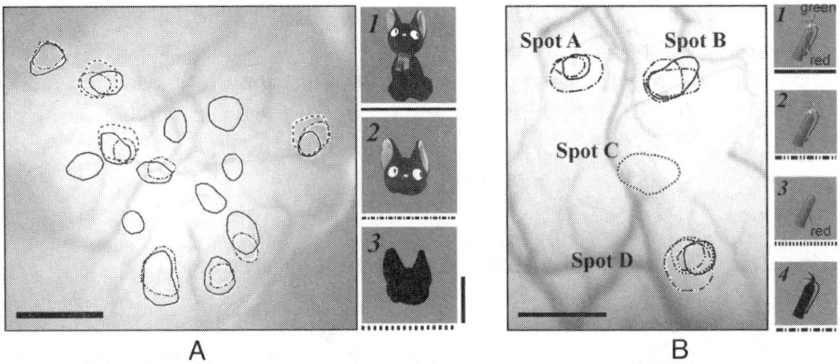

FIGURE 14.7 Representation of complex object images and their simplification in area TE. (a) A case in which simplified stimuli elicited only a subset of spots evoked by more complex stimuli. (b) Cases in which new spots appeared when the original stimulus was simplified. Spots A and B were activated by stimuli 1, 2, and 4, but spot C was activated only by stimulus 3. Spot D was activated by all four stimuli. Scale bars on the activation patterns indicate 1 mm, and those on the right side of visual stimuli indicate 10°. (Adapted from Reference 22.)

particular visual features within an object image, and that an object image is specifically represented by a combination of these spots (distributed representation of objects). In some cases, new spots also emerged by apparent simplification of an object image. For example, spot C appeared when the handle and hose were removed from the fire extinguisher (Figure 14.7B). Extracellular recordings revealed that neurons in spot C were activated by a "rectangle" but were inhibited by a "circle." Because there was no activation by an "ellipse," the balance between these excitatory and inhibitory inputs seems to determine the responses of neurons in spot C.[22] These results could account for the lack of activation by the "fire extinguisher," in which the hose (circular shape) attached to the cylinder makes the entire shape elliptical. Some spots representing a particular feature are inactive when other features are presented together with it.

Based on these experiments, we proposed an extended scheme of distributed representation, in which objects are represented not by the simple sum of feature columns but by combinations of active and inactive columns for individual features. According to this scheme, the fire extinguisher in Figure 14.7B is supposed to be represented not only by the active spots representing the handle (spot A), hose (spot B), and cylinder (spot D), but also by the absence of spot C representing the entire elliptical structure. Combinations of inactive as well as active columns increase the number of available activation patterns and thus, in general, can increase the number of objects that are specifically represented by neuronal population.

From a physiological point of view, the relationship between increased activation patterns and object representation is considered as follows. In our scheme, it is assumed that there is an optimal visual feature for a set of columns. However, the tuning specificities of these columns are not exactly the same. In Figure 14.7B, for example, both spots C and D were maximally activated by a "rectangle," but the tuning specificities of these spots were different: spot C was sharply tuned to a "rectangle," but spot D responded to both an "ellipse" and a "rectangle." Because of this difference in tuning specificity, slight changes in object images cause the disappearance or appearance of spots in the set. Thus, these changes can elicit different activation patterns. One case is the representation of objects consisting of the same set of visual features in a different arrangement, and another case is the representation of the same object that appears different under different vantage points, with partial occlusion, and shading.

14.4.2 Representation of Spatial Arrangement of Parts in Object Images

Results of intrinsic signal imaging suggest that some of the active spots represent "local features" in object images (Figure 14.7). Because information about the spatial arrangement of "local features" is necessary for the specific representation of an object image, some neurons may represent visual features in relation to the spatial arrangement of local features ("configurational information"). Here, we refer to "local features" as visual features that occupy part of an object image and are distinguishable from other parts of the image by their particular shapes, colors, or textures. On the other hand, we refer to "configurational information" as information about the spatial arrangement of local features per se or of parts of the images containing local features. Intrinsic signal imaging with extracellular recording suggests that the presence or absence of activity in spot C constrains the spatial arrangement of the hose and the cylinder (local features) of the fire extinguisher (Figure 14.7B). This spot was activated when the hose is attached to the side of the cylinder making the entire shape elliptical; however, it may not be activated if the hose is secured above the handle and the rectangular shape of the cylinder is exposed. Thus, we consider that the activity in spot C provides information about the configuration of the two parts to some extent, even though this may just be one particular way of representing "configurational information."

To further examine the neuronal representation of configurational information, we searched for neurons representing a typical spatial arrangement among "parts."[25] We found some neurons that respond equally well to object images consisting of different colors, textures, or local shapes, as long as they consist of two vertically aligned parts. These cells responded to combinations of the two vertically aligned parts activated as well as the object images, but did not respond to either part presented alone (Figure 14.2 and Figure 14.8). These cells were highly selective to a particular position of the upper part relative to that of the lower part (Figure 14.8).

These characteristic response properties enable these neurons to respond to visual stimuli regardless of the local features embedded in either part, but only when these local features are aligned in a particular orientation, as shown in Figure 14.8. The results from these neurons, as well as those from neurons in spot C in Figure 14.7B, suggest that neurons in area TE not only represent local features but also configurational information of an object image. Object images could be specifically represented by a combination of spots representing local features and those representing configurational information.

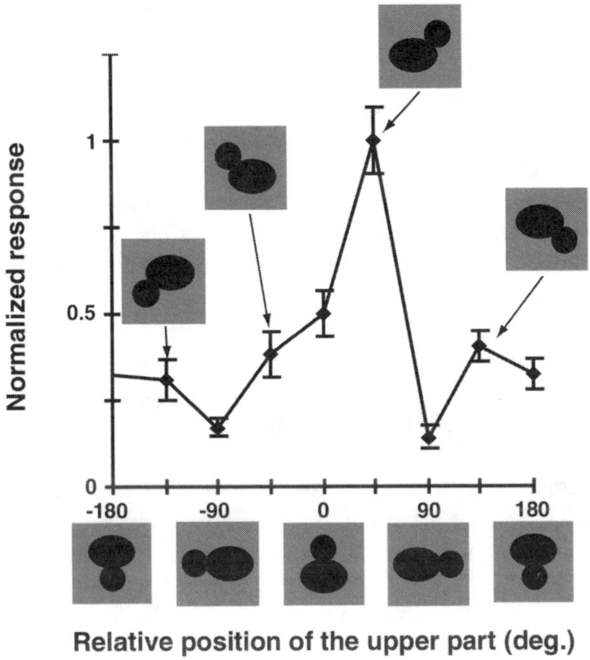

FIGURE 14.8 Response specificity of a cell representing the spatial arrangement of parts. The upper part of the critical feature of the cell was rotated relative to the lower part. The horizontal axis indicates the angle between the line connecting the center of the upper and lower parts of each stimulus and that of the critical feature. The vertical axis indicates normalized value of stimulus-evoked responses. (From Yamane and Tanifuji, unpublished observations.)

14.4.3 Spatial Organization of Face-Specific Neurons

Intrinsic signal imaging revealed that images of a face, as well as of nonface objects, activate multiple spots (Figure 14.5).[21,22] Most of these spots represent visual features that are common among objects including faces. For example, Figure 14.5 shows that spots activated by faces overlap with spots activated by nonface objects. However, there were regions activated exclusively by faces.[21] By using different views of a face as visual stimuli, we found that face-specific regions have a unique spatial organization as previously described (Figure 14.6).[20,21]

14.5 ANATOMICAL SUBSTRATES FOR SHAPING RESPONSIVENESS OF IT NEURONS

The visual features represented by IT neurons cover a wide range of complexity: some of them are relatively simple, such as a combination of two orthogonal line segments, whereas others are rather complex, such as a combination of circles filled with orthogonal grating patterns (see Figure 14.3). If we consider that face neurons represent a particular configuration specific for faces, a face would be the most complex visual feature represented by IT neurons. The neuronal mechanisms that shape particular visual features for individual cells and allow various visual features to be associated with IT neurons are an important issue to be investigated. Several recent findings provide clues to these mechanisms.

First, the IT cortex receives dense anatomical projections from area V4. Thus, visual attributes represented by V4 neurons would provide elements necessary to generate visual features represented by IT neurons. Recently, for example, neurons responding to particular curvatures[26] and those responding to texture have been found in area V4.[27] Second, intrinsic horizontal connections of excitatory neurons have been demonstrated in the IT cortex.[28,29] Recently, we have compared the spatial patterns of intrinsic connections with those of active spots activated by various visual stimuli. One example suggests that spots responding to slightly different visual features are interconnected by intrinsic horizontal connections.[30] Thus, the various degrees of responsiveness of IT neurons may be due to the degree of convergence of axons carrying different visual attributes. Third, Wang and colleagues[31] found that inhibitory connections are involved in shaping the responsiveness of neurons in area TE. They examined the stimulus selectivity of TE neurons before and after the local blockade of inhibitory inputs by bicuculline focal injections, and found that responses to some visual stimuli are augmented when inhibitory inputs are blocked. Finally, several reports have shown that neuronal responses are different in the early and late phases of stimulus presentation.[32–34] These results suggest that late feedback projections from other areas, such as the perirhinal cortex, are involved in the further modification of response properties.

14.6 EFFECTS OF VISUAL EXPERIENCES ON RESPONSE PROPERTIES OF IT NEURONS

A number of studies have shown that extensive training of adult monkeys with particular visual stimuli changes the response properties of IT neurons to the visual stimuli used for training.[35–41] The changes in the response properties of IT neurons could play a role in shaping the response properties suitable for subjects to perceive and recognize objects efficiently.

Logothetis and colleagues[38] trained monkeys with consecutive multiple views of computer-generated objects, paper clips, and spheroidal objects. They found that some visually responsive neurons in the trained animals are broadly tuned to particular views of these objects.[38] These cells responded to objects used for training, but did not respond to computer-generated objects not used for training. Because the objects used and those not used for training have similar local features, these neurons seem to represent the configuration of the images of computer-generated objects rather than that of local features.[38] This study provides evidence that extensive training can broadly tune some neurons, such as face neurons, to a particular view of an object.

Some investigations showed that when monkeys are trained in a visual association task, a single neuron responding to the stimulus used for training also tends to be activated by stimuli presented in succession.[35,42] These results suggest that visual features and images of objects can be associated with each other depending on the strength of their temporal correlation, and broadness of tuning to a particular view of complex objects can be generated through this process.[43]

14.7 SELECTIVITY OF IT NEURONS FOR SIZE AND POSITION

Visual recognition of an object is not strongly affected by its size and position in the visual field. It is also known that visual recognition of familiar objects is not influenced by vantage points. Apart from the above-mentioned visual stimuli maximally activating IT neurons, the sensitivity of IT neurons to these properties can explain the neuronal mechanism of invariance in visual recognition. As described above, most face neurons selectively respond to a particular view of faces. Thus, some mechanism associating different views is required for invariant recognition of faces. Continuous mapping of faces at different view angles (Figure 14.6) would help in associating different views. On the other hand, compared with neurons in early visual areas, IT neurons respond invariantly to changes in stimulus size[4,44,45] and have large receptive fields (RFs).[16,17,46,47]

Ito and colleagues[45] examined the selectivity of neurons in area TE for optimal visual stimuli of different sizes using anesthetized monkeys. They found that the size eliciting the maximal response ranged from 1.7 to 30° and that the full-width at half-maximum (FWHM) of the response was more than two octaves (twofold in size) in nearly 60% of the cells studied. Similarly, Rolls and Baylis[44] recorded from face neurons in STS and reported that the FWHM of the response to faces was more than two octaves for 95% of the cells studied. Although these studies showed IT neurons broadly tuned to visual stimuli of different size, there were still 40% of neurons responding to stimuli with FWHM less than two octaves in the former study.[45]

The RF sizes of IT neurons examined by Kobatake and Tanaka in anesthetized monkeys are $16.5 \pm 6.1°$ and $5.4 \pm 2.8°$ in areas TE and TEO, respectively.[16] Because the optimal stimulus is large, the RF size evaluated based on the outer edge of the stimulus and that evaluated based on the inner edge of the stimulus can be very different. They determined the borders of the RFs at the center of effective stimuli at the most peripheral location to minimize the effect of stimulus size. These RFs, particularly those in area TE, are larger than the RFs in area V4 ($4.8 \pm 2.5°$) examined by the same group. Neurons in area TE are also unique in that their RFs usually include foveal representation. Although these studies showed large RFs particularly in area TE, the responses to visual stimuli at different locations within the RFs are different.[45,47] A recent study has quantitatively measured neuronal responses to a small visual stimulus (3.3°) presented at one position within a grid of seven by eight locations in the visual field (spacing between nearby locations was 4°) in awake monkeys. The study showed that the RF profiles of TE neurons could be fitted well with a two-dimensional Gaussian function and that the RF size, defined as the square root of the area where neuronal responses were more than 50% of the maximum response, was $10.3 \pm 9.6°$.[47] In relation to the RFs, retinotopic representation exists in area TEO but not in area TE.[48]

The broad tuning to the stimulus size and large RF size of IT neurons are consistent with size and translation invariance in object recognition, by which we can recognize objects regardless of their size and position in the visual field. However, we cannot immediately extend these response properties as the neuronal mechanisms underlying size and translation invariance in object recognition. In the real world, for example, an object appears in complex visual environments where many other objects are present simultaneously, whereas the studies mentioned above examined the responsiveness of neurons to stimulus size and position when a visual stimulus is presented in isolation. All studies that examined the interaction between two objects in the visual field show that neural responses to a stimulus are suppressed when the second stimulus appears in the visual field.[49–52] Thus, in a natural complex environment, the neuronal response to a target object can be greatly reduced. There is evidence that stimulus selectivity is not influenced by interacting stimuli in the environment so that neurons can still detect a target object even with weak responses.[52,53] To understand the neuronal mechanisms of size and translation invariance, the receptive field size and neuronal responses to objects in complex environments should be investigated in relation to attention and awareness of visual objects in the scene.[53,54]

14.8 DISPARITY TUNED NEURONS IN INFEROTEMPORAL CORTEX

According to the view that visual information is processed separately in the ventral and dorsal visual pathways,[55] binocular disparity information seems to be processed mainly in the dorsal pathway where the visual perception of object location in three-dimensional space is mediated. However, recent investigations have provided evidence that disparity information is also processed in the IT cortex.[56–61]

Uka and colleagues[56] found that about 60% of IT neurons are sensitive to horizontal disparity added to visual stimuli. Most of these disparity-sensitive neurons were tuned to visual stimuli presented either nearer or farther than the fixation point. Because the investigators used flat visual stimuli on a front parallel plane at different depths, the neurons may not be related to surface representation of three-dimensional objects. It is more plausible that the disparity sensitivity of these neurons provides a cue for defining the shapes of object images. In fact, the same group showed that the shape selectivities of about 40% of these disparity-sensitive neurons are similar regardless of whether the shapes are defined by luminance contrast or by difference in disparity.[61]

In a separate series of studies, Janssen and colleagues[57–60] examined the disparity sensitivity of neurons using three-dimensional surfaces generated by random-dot stereograms. These surfaces included flat surfaces tilted in depth, concave and convex surfaces, and protrusions in front or at the back from the center of a front parallel plane. They found that neurons in the lower bank of STS of area TE respond selectively to a particular three-dimensional surface with a particular sign of curvature. For example, some neurons responded to a concave but not to a convex surface, and vice versa. For a substantial number of these neurons, the selectivity is not affected by the position of the stimuli in depth relative to the fixation point.[58] Thus, these neurons do not represent the local surface position relative to the fixation point; instead, they represent the spatial variation of disparity gradient.[58,59]

The above-mentioned experiments demonstrated that there are indeed disparity-tuned neurons in the IT cortex, which may provide neuronal substrates for the visual representation of objects that are three dimensional in the real world or may use disparity information to define the shapes of visual images.

14.9 SUMMARY AND CONCLUSIONS

A number of investigations searching for essential visual features activating IT neurons revealed that many neurons are effectively activated by visual features less complex than natural objects, and thus suggest that an object image is represented by the combined activation of neurons. The stimulus simplification technique shown in Figure 14.2 enables us to systematically identify such visual features, which range from simple visual features to relatively complex ones.[8]

On the other hand, extensive investigations of the response properties of IT neurons revealed that there are neurons maximally activated by biologically important objects, such as faces. Taking a face neuron as a representative, the author has described three characteristic properties of these neurons: (1) their maximum activation requires an image of a complete face, (2) they are tuned to a particular view of faces, and (3) they are not tuned to a face of a particular individual. These properties suggest that the role of face neurons is to represent a certain visual feature common in faces. The feature is uniquely defined by a view, and is not a local feature of faces. As previously described, some cells in the IT cortex of trained monkeys are tuned to a particular view of the objects repeatedly used during training.[38] In a similar way, IT neurons are tuned to biologically important objects because monkeys frequently interact with visual features of these objects in daily life or during development.

Taken together, it would not be appropriate to classify neurons into those specific to biologically important objects and those specific to moderately complex visual features. Instead, neurons in the IT cortex seem to represent various visual features that can be as simple as line segments or as complex as a configuration specific for faces. There are visual features common among many objects, such as that representing two vertically aligned parts (see Figure 14.2 and Figure 14.8). The configuration of faces is also a feature common to many individual faces. Frequency for monkeys to encounter a certain visual feature, in general, depends on how common the feature appears in objects and how often monkeys encounter the objects containing the feature. Depending on this frequency, IT neurons could be tuned to particular visual features. The wide range of complexity in critical features of IT neurons is the consequence of this process. As a result of shaping visual features this way, the representation of objects by combinations of IT neurons is tuned to necessary visual tasks or it facilitates perception of objects consisting of common features.

Intrinsic signal imaging allows us to investigate directly how these feature neurons are used to represent objects. This technique demonstrated that a complex object is represented by a combination of columns in area TE, each of which represents the visual features of the object image. The combination is not based simply on summing up of the columns, but may instead rely on a combination of active and inactive columns to represent objects. Furthermore, we showed evidence that some visual features are related to "configurational information" rather than to local features.

Visual features represented by IT neurons have been considered to be two dimensional because the IT cortex is situated in the ventral visual pathway and early studies did not reveal significant differences between responses to three-dimensional objects and responses to two-dimensional images of these objects. However, recent findings that some neurons are tuned to a three-dimensional surface suggest that these visual features are three dimensional.[57–60] It should be kept in mind that neurons tuned to a three-dimensional surface are localized in the ventral bank of STS.[59] The functional roles of neurons in the ventral bank and the other regions of area TE can be separately considered.

In the above discussion on object representation, only visual images eliciting maximum responses are taken into account. However, neurons also respond significantly well to suboptimal stimuli. For example, a face neuron can be maximally activated by a complete face, but it can also be significantly activated by an incomplete face (see Figure 14.1). Thus, if there is no mechanism to differentiate suboptimal activation from optimal activation, the selectivity of neurons decreases, and thus the specific representation of objects becomes difficult. On the other hand, suboptimal responses play an important role in some stimulus properties, such as size and position in the visual field as described above. How the brain manages suboptimal responses is an issue for future investigation. It has been shown that the activities of IT neurons in an awake monkey depend on whether or not it is aware of the objects in its visual field.[62] Control of neuronal activity as shown in this study may be a key to resolve this issue.[54]

REFERENCES

1. Logothetis, N.K. and Sheinberg, D.L., Visual object recognition, *Annu. Rev. Neurosci.*, 19, 577–621, 1996.
2. Gross, C.G., How inferior temporal cortex became a visual area, *Cereb. Cortex*, 5, 455–469, 1994.
3. Gross, C.G., Bender, D.B., and Gerstein, G.L., Activity of inferior temporal neurons in behaving monkeys, *Neuropsychology*, 17, 215–229, 1979.
4. Schwartz, E.L., Desimone, R., Albright, T.D., and Gross, C.G., Shape recognition and inferior temporal neurons, *Proc. Natl. Acad. Sci. U.S.A.*, 80(18), 5776–5778, 1983.
5. Desimone, R., Albright, T.D., Gross, C.G., and Bruce, C., Stimulus-selective properties of inferior temporal neurons in the macaque, *J. Neurosci.*, 4(8), 2051–2062, 1984.
6. Bruce, C., Desimone, R., and Gross, C.G., Visual properties of neurons in a polysensory area in superior temporal sulcus of the macaque, *J. Neurophysiol.*, 46(2), 369–384, 1981.
7. Perrett, D.I., Rolls, E.T., and Caan, W., Visual neurones responsive to faces in the monkey temporal cortex, *Exp. Brain Res.*, 47(3), 329–342, 1982.

8. Tanaka, K., Saito, H., Fukada, Y., and Moriya, M., Coding visual images of objects in the inferotemporal cortex of the macaque monkey, *J. Neurophysiol.,* 66(1), 170–189, 1991.
9. Perrett, D.I., Smith, P.A., Potter, D.D., Mistlin, A.J., Head, A.S., Milner, A.D., and Jeeves, M.A., Neurones responsive to faces in the temporal cortex: studies of functional organization, sensitivity to identity and relation to perception, *Hum. Neurobiol.,* 3(4), 197–208, 1984.
10. Baylis, G.C., Rolls, E.T., and Leonard, C.M., Selectivity between faces in the responses of a population of neurons in the cortex in the superior temporal sulcus of the monkey, *Brain Res.,* 342(1), 91–102, 1985.
11. Yamane, S., Kaji, S., and Kawano, K., What facial features activate face neurons in the inferotemporal cortex of the monkey? *Exp. Brain Res.,* 73(1), 209–214, 1988.
12. Young, M.P. and Yamane, S., Sparse population coding of faces in the inferotemporal cortex, *Science,* 256(5061), 1327–1331, 1992.
13. Perrett, D.I., Oram, M.W., Harries, M.H., Bevan, R., Hietanen, J.K., Benson, P.J., and Thomas, S., Viewer-centred and object-centred coding of heads in the macaque temporal cortex, *Exp. Brain Res.,* 86(1), 159–173, 1991.
14. Perrett, D.I., Hietanen, J.K., Oram, M.W., and Benson, P.J., Organization and functions of cells responsive to faces in the temporal cortex, *Philos. Trans. R. Soc. Lond. B Biol. Sci.,* 335(1273), 23–30, 1992.
15. Wachsmuth, E., Oram, M.W., and Perrett, D.I., Recognition of objects and their component parts: responses of single units in the temporal cortex of the macaque, *Cereb. Cortex,* 4(5), 509–522, 1994.
16. Kobatake, E. and Tanaka, K., Neuronal selectivities to complex object features in the ventral visual pathway of the macaque cerebral cortex, *J. Neurophysiol.,* 71(3), 856–867, 1994.
17. Gross, C.G., Rocha-Miranda, C.E., and Bender, D.B., Visual properties of neurons in inferotemporal cortex of the macaque, *J. Neurophysiol.,* 35(1), 96–111, 1972.
18. Gochin, P.M., Miller, E.K., Gross, C.G., and Gerstein, G.L., Functional interactions among neurons in inferior temporal cortex of the awake macaque, *Exp. Brain Res.,* 84(3), 505–516, 1991.
19. Fujita, I., Tanaka, K., Ito, M., and Cheng, K., Columns for visual features of objects in monkey inferotemporal cortex, *Nature,* 360(6402), 343–346, 1992.
20. Wang, G., Tanaka, K., and Tanifuji, M., Optical imaging of functional organization in the monkey inferotemporal cortex, *Science,* 272(5268), 1665–1668, 1996.
21. Wang, G., Tanifuji, M., and Tanaka, K., Functional architecture in monkey inferotemporal cortex revealed by in vivo optical imaging, *Neurosci. Res.,* 32(1), 33–46, 1998.
22. Tsunoda, K., Yamane, Y., Nishizaki, M., and Tanifuji, M., Complex objects are represented in macaque inferotemporal cortex by the combination of feature columns, *Nat. Neurosci.,* 4(8), 832–838, 2001.
23. Grinvald, A., Shoham, D., Shmuel, A., Glaser, D., Vanzetta, I., Shtoyerman, E., Slovin, H., Wijnbergen, C., Hildesheim, R., and Arieli, A., *In-vivo* optical imaging of cortical architecture and dynamics, in *Modern Techniques in Neuroscience Research*, Windhorst, U. and Johansson, H., Eds., Springer, New York, 1999, 893–970.
24. Maldonado, P.E., Godecke, I., Gray, C.M., and Bonhoeffer, T., Orientation selectivity in pinwheel centers in cat striate cortex, *Science,* 276(5318), 1551–1555, 1997.
25. Yamane, Y., Tsunoda, K., Matsumoto, M., Phillips, A., and Tanifuji, M., Decomposition of object images by feature columns in macaque inferotemporal cortex, *Soc. Neurosci. Abstr.,* 27, 399.6, 2001.

26. Pasupathy, A. and Connor, C.E., Responses to contour features in macaque area V4, *J. Neurophysiol.,* 82(5), 2490–2502, 1999.
27. Hanazawa, A. and Komatsu, H., Influence of the direction of elemental luminance gradients on the responses of V4 cells to textured surfaces, *J. Neurosci.,* 21(12), 4490–4497, 2001.
28. Fujita, I. and Fujita, T., Intrinsic connections in the macaque inferior temporal cortex, *J. Comp. Neurol.,* 368(4), 467–486, 1996.
29. Tanigawa, H., Fujita, I., Kato, M., and Ojima, H., Distribution, morphology, and gamma-aminobutyric acid immunoreactivity of horizontally projecting neurons in the macaque inferior temporal cortex, *J. Comp. Neurol.,* 401(1), 129–143, 1998.
30. Tanifuji, M., Li, H., Yamane, Y., and Rockland, K.S., Horizontal intrinsic connections as the anatomical basis for the functional columns in macaque inferior temporal cortex, *Soc. Neurosci. Abstr.,* 27, 620.7, 2001.
31. Wang, Y., Fujita, I., and Murayama, Y., Neuronal mechanisms of selectivity for object features revealed by blocking inhibition in inferotemporal cortex, *Nat. Neurosci.,* 3(8), 807–813, 2000.
32. Sugase, Y., Yamane, S., Ueno, S., and Kawano, K., Global and fine information coded by single neurons in the temporal visual cortex, *Nature,* 400(6747), 869–873, 1999.
33. Naya, Y., Yoshida, M., and Miyashita, Y., Backward spreading of memory-retrieval signal in the primate temporal cortex, *Science,* 291(5504), 661–664, 2001.
34. Tamura, H. and Tanaka, K., Visual response properties of cells in the ventral and dorsal parts of the macaque inferotemporal cortex, *Cereb. Cortex,* 11(5), 384–399, 2001.
35. Miyashita, Y., Neuronal correlate of visual associative long-term memory in the primate temporal cortex, *Nature,* 335(6193), 817–820, 1988.
36. Miyashita, Y. and Chang, H.S., Neuronal correlate of pictorial short-term memory in the primate temporal cortex, *Nature,* 331(6151), 68–70, 1988.
37. Sakai, K. and Miyashita, Y., Neural organization for the long-term memory of paired associates, *Nature,* 354(6349), 152–155, 1991.
38. Logothetis, N.K., Pauls, J., and Poggio, T., Shape representation in the inferior temporal cortex of monkeys, *Curr. Biol.,* 5(5), 552–563, 1995.
39. Kobatake, E., Wang, G., and Tanaka, K., Effects of shape-discrimination training on the selectivity of inferotemporal cells in adult monkeys, *J. Neurophysiol.,* 80(1), 324–330, 1998.
40. Sigala, N., Gabbiani, F., and Logothetis, N.K., Visual categorization and object representation in monkeys and humans, *J. Cogn. Neurosci.,* 14(2), 187–198, 2002.
41. Baker, C.I., Behrmann, M., and Olson, C.R., Impact of learning on representation of parts and wholes in monkey inferotemporal cortex, *Nat. Neurosci.,* 5(11), 1210–1216, 2002.
42. Yakovlev, V., Fusi, S., Berman, E., and Zohary, E., Inter-trial neuronal activity in inferior temporal cortex: a putative vehicle to generate long-term visual associations, *Nat. Neurosci.,* 1(4), 310–317, 1998.
43. Stryker, M.P., Neurobiology. Temporal associations [comment], *Nature,* 354(6349), 108–109, 1991.
44. Rolls, E.T. and Baylis, G.C., Size and contrast have only small effects on the responses to faces of neurons in the cortex of the superior temporal sulcus of the monkey, *Exp. Brain Res.,* 65(1), 38–48, 1986.
45. Ito, M., Tamura, H., Fujita, I., and Tanaka, K., Size and position invariance of neuronal responses in monkey inferotemporal cortex, *J. Neurophysiol.,* 73(1), 218–226, 1995.

46. Gross, C.G., Bender, D.B., and Rocha-Miranda, C.E., Visual receptive fields of neurons in inferotemporal cortex of the monkey, *Science,* 166(910), 1303–1306, 1969.
47. Op De Beeck, H. and Vogels, R., Spatial sensitivity of macaque inferior temporal neurons, *J. Comp. Neurol.,* 426(4), 505–518, 2000.
48. Boussaoud, D., Desimone, R., and Ungerleider, L.G., Visual topography of area TEO in the macaque, *J. Comp. Neurol.,* 306(4), 554–575, 1991.
49. Sato, T., Interactions of visual stimuli in the receptive fields of inferior temporal neurons in awake macaques, *Exp. Brain Res.,* 77(1), 23–30, 1989.
50. Miller, E.K., Gochin, P.M., and Gross, C.G., Suppression of visual responses of neurons in inferior temporal cortex of the awake macaque by addition of a second stimulus, *Brain Res.,* 616(1–2), 25–29, 1993.
51. Rolls, E.T. and Tovee, M.J., The responses of single neurons in the temporal visual cortical areas of the macaque when more than one stimulus is present in the receptive field, *Exp. Brain Res.,* 103(3), 409–420, 1995.
52. Missal, M., Vogels, R., Li, C.Y., and Orban, G.A., Shape interactions in macaque inferior temporal neurons, *J. Neurophysiol.,* 82(1), 131–142, 1999.
53. DiCarlo, J.J. and Maunsell, J.H., Form representation in monkey inferotemporal cortex is virtually unaltered by free viewing, *Nat. Neurosci.,* 3(8), 814–821, 2000.
54. Sheinberg, D.L. and Logothetis, N.K., Noticing familiar objects in real world scenes: the role of temporal cortical neurons in natural vision, *J. Neurosci.,* 21(4), 1340–1350, 2001.
55. Ungerleider, L.G. and Mishkin, M., Two cortical visual systems, in *Analysis of Visual Behavior,* Ingle, D.G., Goodale, M.A., and Mansfield, R.J.Q., Eds., MIT Press, Cambridge, MA, 1982, 549–586.
56. Uka, T., Tanaka, H., Yoshiyama, K., Kato, M., and Fujita, I., Disparity selectivity of neurons in monkey inferior temporal cortex, *J. Neurophysiol.,* 84(1), 120–132, 2000.
57. Janssen, P., Vogels, R., and Orban, G.A., Macaque inferior temporal neurons are selective for disparity-defined three-dimensional shapes, *Proc. Natl. Acad. Sci. U.S.A.,* 96(14), 8217–8222, 1999.
58. Janssen, P., Vogels, R., and Orban, G.A., Three-dimensional shape coding in inferior temporal cortex, *Neuron,* 27(2), 385–397, 2000.
59. Janssen, P., Vogels, R., and Orban, G.A., Selectivity for 3D shape that reveals distinct areas within macaque inferior temporal cortex, *Science,* 288(5473), 2054–2056, 2000.
60. Janssen, P., Vogels, R., Liu, Y., and Orban, G.A., Macaque inferior temporal neurons are selective for three-dimensional boundaries and surfaces, *J. Neurosci.,* 21(23), 9419–9429, 2001.
61. Tanaka, H., Uka, T., Yoshiyama, K., Kato, M., and Fujita, I., Processing of shape defined by disparity in monkey inferior temporal cortex, *J. Neurophysiol.,* 85(2), 735–744, 2001.
62. Sheinberg, D.L. and Logothetis, N.K., The role of temporal cortical areas in perceptual organization, *Proc. Natl. Acad. Sci. U.S.A.,* 94(7), 3408–3413, 1997.
63. Tanaka, K., Neuronal mechanisms of object recognition, *Science,* 262(5134), 685–688, 1993.

15 Comparative Studies of Pyramidal Neurons in Visual Cortex of Monkeys

Guy N. Elston

CONTENTS

15.1 INTRODUCTION

The visual cortex is perhaps the most studied of all cortical regions. Prior to the 1970s, most studies focused on the striate, or primary (V1), visual area. However, with the advent of better techniques, increasing attention has been devoted to the extrastriate cortex and many new areas have been defined. As a result, it is now known that a large part of the brain may be dedicated to visual processing. In the monkey, visual cortex may include up to half of the cortex and contain more than 30 different areas[1–5] (Figure 15.1). In this chapter I focus on recent findings of

0-8493-1243-4/04/$0.00+$1.50

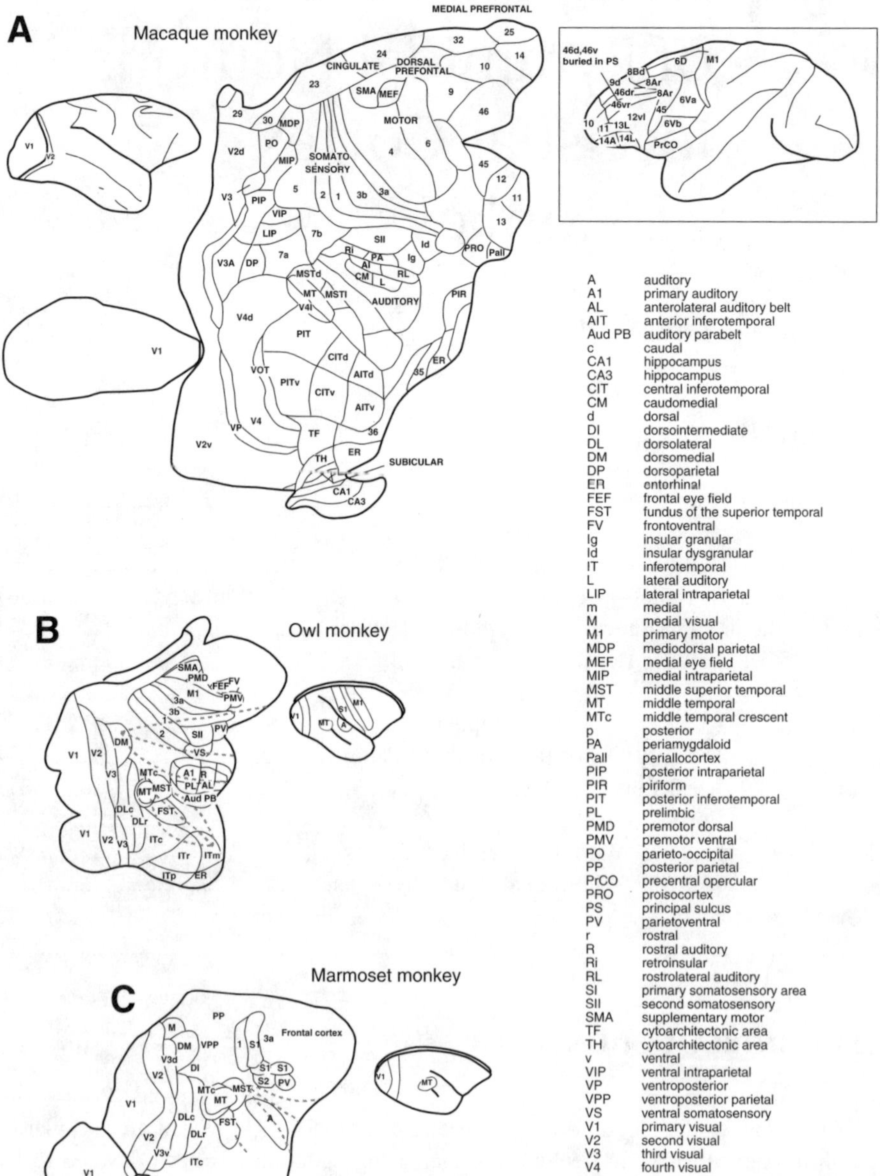

FIGURE 15.1 Schematic showing summary maps of cortical organization in macaque (A), owl (B), and marmoset (C) monkeys. Nomenclature, the number of distinguishable areas and the borders between areas may vary between reports. (Modified from Felleman and Van Essen,[2] Preuss and Goldman-Rakic,[129] and Lyon and Kaas.[130,131])

regional specialization in cortical circuitry in these different visual areas and speculate on how this may influence visual processing.

15.2 ORGANIZATION OF VISUAL CORTEX

From Figure 15.1 it is clear that a large proportion of the monkey cerebral cortex is involved in visual processing. Furthermore, there is a degree of similarity in the number and organization of cortical areas in different species, despite appreciable differences in the absolute size of their brains. However, not all mammals have the same complement of visual areas. Visual cortex in rats, for example, is thought to include fewer cortical areas than higher primates.[6] Consequently, it has been proposed that some visual areas, such as V1, were common in ancestral mammals whereas other areas have evolved with the expansion of the brain (see References 6 through 11 for reviews). This expansion was widely believed to occur through cortical folding and the addition of more of the same basic repeated unit of cortex.[12–20] New methods of quantification have revealed that this is not the case. A closer look reveals marked, systematic differences in cortical circuitry across different visual areas.

15.2.1 Cellular Composition of Cortex

Cortex is composed of two basic types of neurons, spiny and nonspiny. Spiny cells that contain the neurotransmitter glutamate are widely believed to be excitatory (see References 21 through 24 for reviews). Pyramidal cells are the most common type of spiny neuron, comprising approximately 70% of all neurons in cortex.[25] Arguably, pyramidal cells are the principal neurons of the cerebral cortex, generating nearly all cortically initiated excitation. They include many phenotypes, but are distinguished by their prominent apical dendrite and basal dendritic arbor (Figure 15.2). Nonspiny neurons include many different morphological types, most of which contain GABA (see References 24, 26, and 27 for reviews). They are characterized by different receptor profiles and project to different postsynaptic targets.[28] In general, these cells are thought to be important in modulating the activity of excitatory neurons (see References 29 and 30 for reviews).

15.2.2 Regional Specialization in Cortical Organization

Recent studies have revealed significant and systematic variation in circuit structure in different visual areas (Figure 15.3). In particular, pyramidal cells in visual areas of the parietal and temporal lobes of the macaque monkey are larger, more branched, and more spinous than those in the occipital lobe.[31–41] In addition, pyramidal cells in the frontal eye field (FEF) in prefrontal cortex, which are thought to control purposive eye movements,[42–45] are even more branched than those in the aforementioned lobes.[46] These differences in pyramidal cell structure can be dramatic; pyramidal cells in the superior temporal polysensory area (STP), which contains neurons that integrate visual, somatosensory, and auditory inputs,[47–51] have as many as 13-fold more dendritic spines than those in V1.[39] Marked differences

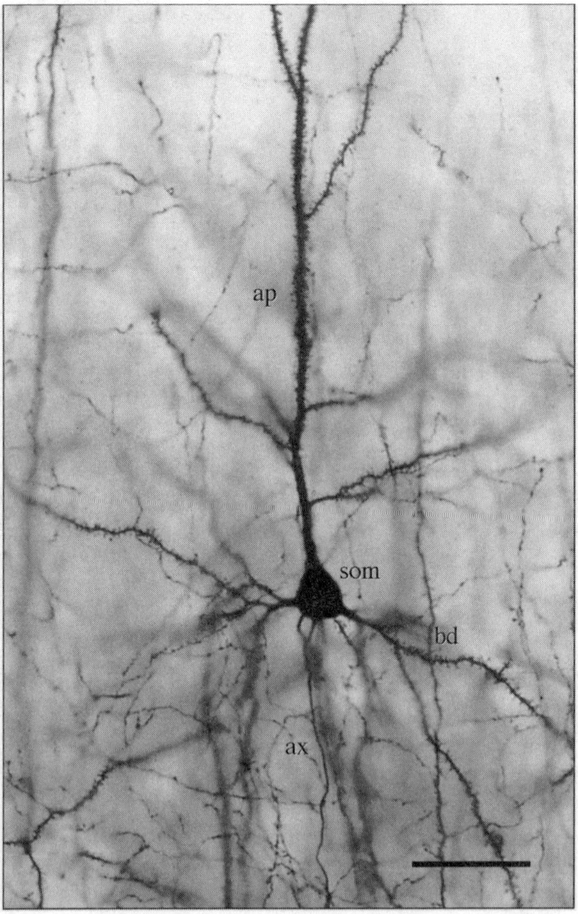

FIGURE 15.2 Golgi-impregnated pyramidal cell as seen in the traditional transverse plane, illustrating the apical dendrite (ap), basal dendrites (bd), pyramidal soma (som), and axon (ax). Scale bar = 20 μm. (Courtesy of Professor Fecundo Valverde, from his own macaque monkey V1 preparations.)

in pyramidal cell structure have also been reported in visual cortex of marmoset and owl monkeys[34,40,52,53] (Figure 15.4).

Recent studies have also revealed systematic variation in the density, distribution, and connectivity of inhibitory interneurons in different cortical areas.[28,54–57] For example, there are measurable differences in the number of parvalbumin (PV), calbindin (CB), and calretinin (CR) immunoreactive (ir) interneurons in many visual areas (Figure 15.5). As inhibitory neurons project to different regions of the pyramidal cell, including the dendrites, soma, and axon, they may modulate different aspects of pyramidal cell activity (see References 29 and 30 for reviews). More specifically, axon terminals of PV-ir chandelier cells form synapses primarily with the axon initial segment of pyramidal cells, and are thought to provide a powerful inhibitory influence over pyramidal cell discharge.[58] CB-ir and CR-ir interneurons,

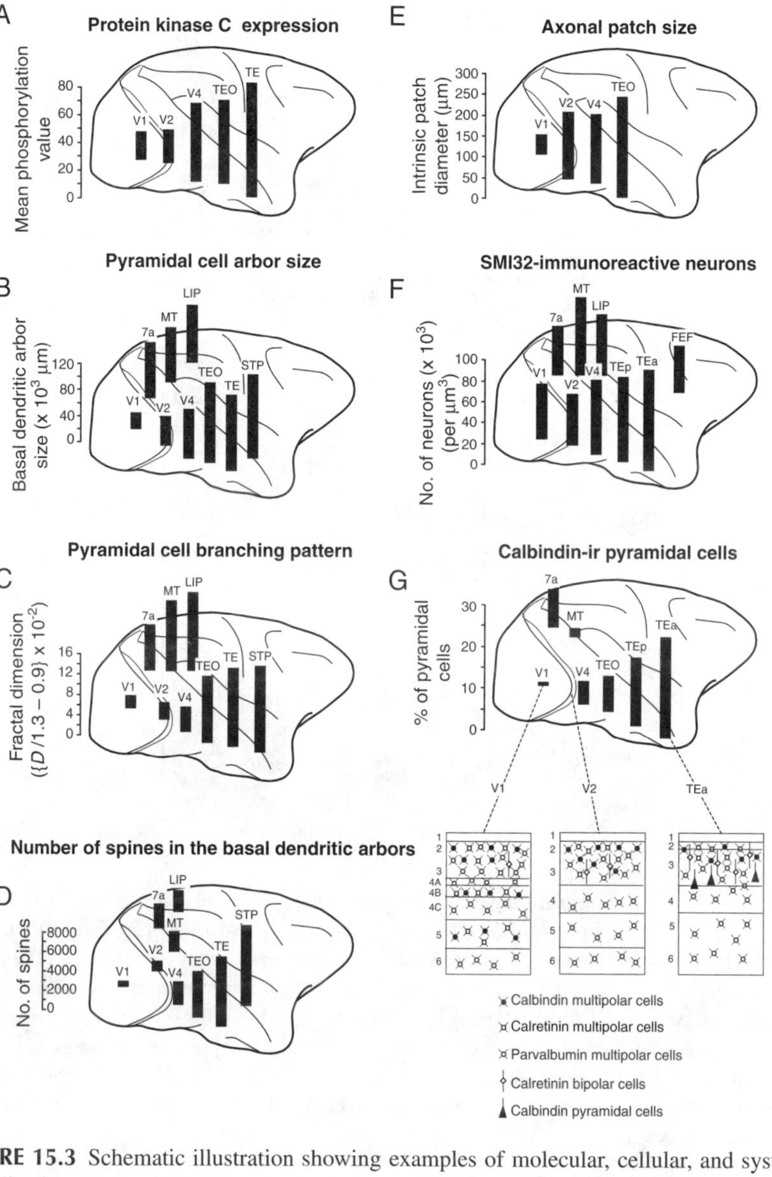

FIGURE 15.3 Schematic illustration showing examples of molecular, cellular, and systems specializations in visually related cortical areas of the macaque monkey. (A) Mean phosphorylation values of protein kinase C (F1/50 kDa) expressed through all cortical layers.[127] Size[36,37,39] (B), fractal dimension of branching patterns[128] (C), and number of spines[36,37,39] (D) in the basal dendritic arbors of layer III pyramidal cells. (E) Diameter of intrinsic axonal patches in supragranular cortical layers. (From Lund[40] and Fujita and Fujita.[70]) (F) Density of layer III cells immunoreactive for SMI32. (From Hof and Morrison.[132]) (G) Number of cells immunoreactive for calbindin, which are of the pyramidal type (from Kondo and colleagues[133]), and differences in the distribution of cells immunoreactive for calcium-binding proteins (from DeFelipe and colleagues[55]). (Modified from Elston.[60])

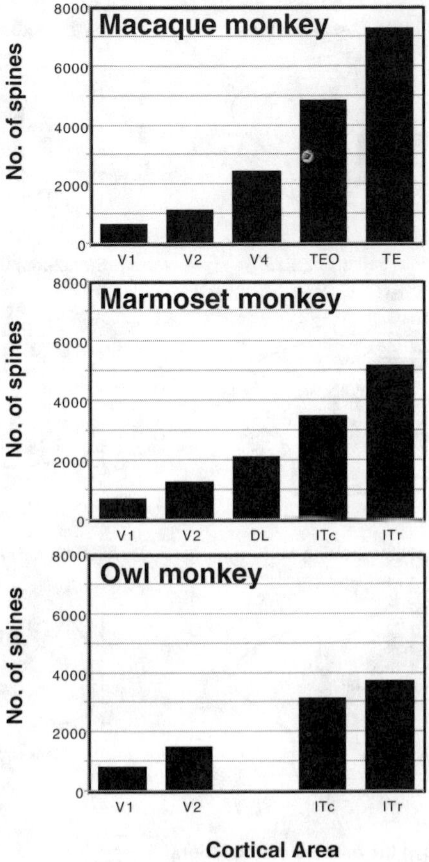

FIGURE 15.4 Graphs illustrating estimates of the total number of spines in the basal dendritic arbor of the "average" pyramidal cell in the primary (V1), second (V2), fourth (V4; corresponding to the dorsolateral, DL, area) visual areas and cytoarchitectonic areas TEO and TE (corresponding to the caudal, ITc, and rostral, ITr, subdivisions of inferotemporal cortex) in the macaque, marmoset, and owl monkeys. These reveal a trend for successively more spinous pyramidal cells with anterior progression through occipitotemporal visual areas. (Data from References 37, 39, 40, and 53.) Note that data obtained from the owl monkey were obtained from an animal older than the animals from which data were obtained for macaque and marmoset monkeys.

on the other hand, project primarily to the dendrites and somata of the pyramidal cells and are believed to modulate activity within the dendritic arbors via local vetoing, or shunting inhibition.[59] Thus, differences in the density and distribution of these different types of interneurons are likely to underlie the regional variation reported in their patterns of connectivity.[54,55]

The magnitude of these regional differences in cortex and their consistency across species make it implausible that they might be "accidental," as suggested by some.[15,19] Instead, it is likely that the structural differences in cortical circuitry in

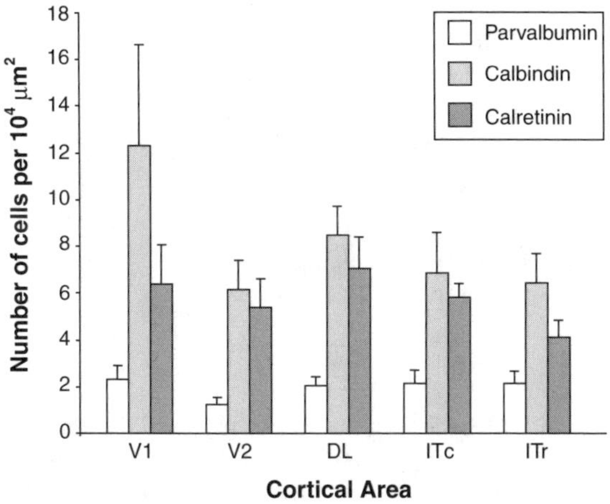

FIGURE 15.5 Areal differences in the density of neurons immunoreactive for parvalbumin, calbindin, and calretinin in layer III of the primary (V1), second (V2), dorsolateral (DL), and caudal and rostral inferotemporal (ITc and ITr, respectively) areas of the owl monkey. Numbers of immunoreactive cells are reported per 10,000 μm². (Modified from Elston and González-Albo.[28])

different visual areas are fundamental in determining the functional characteristics of the neurons therein (see References 60 through 62 for reviews).

15.3 STRUCTURE AND FUNCTION OF PYRAMIDAL CELLS

15.3.1 Cellular Level

The function of any neuron is to receive, integrate, and relay inputs to other cells. Three basic aspects of the neuronal structure may influence how inputs are received and integrated: size of the dendritic arbor, number and distribution of inputs, and dendritic branching pattern. Each of these aspects of cell structure can influence different aspects of the input and integration of synapses in various ways. In addition, the relay of information depends on axonal structure and postsynaptic targets. These too have been shown to vary between different cortical areas.

15.3.1.1 Pyramidal Cell Arbor Size

Cortical neurons characterized by a small dendritic arbor may integrate inputs over a smaller region of cortex than larger cells.[63] In topographically organized visual cortex, the effect of this is that a greater portion of the sensory map is sampled. In the macaque monkey, the increase in arbor size, coupled with topographic map compression (see Reference 3 for a review), results in more than a 100-fold difference in the region of the visual map sampled by individual cells in areas V1 and TEO.[37] The region of cortex (or the proportion of the topographic representation) from which individual cells may sample is further increased by the nature of intrinsic circuitry. For example, supragranular visual cortex is organized into a lattice of intrinsic axonal patches,[64–66] the axons of which originate from supragranular pyramidal neurons.[67–69] As the extent of the patches is reportedly greater in areas such as the fourth visual area (V4) and TEO, as compared with V1,[41,70] a greater portion of the visual map may be sampled by cells in the former cortical areas. In addition, the increase in the size of these intrinsic patches is directly correlated with the increase in the size of the pyramidal cell arbors,[37,41] potentially influencing their sampling strategies (see Reference 71 for a review).

15.3.1.2 Number of Inputs

As well as the regional differences in the size of pyramidal cells, there are appreciable differences in the number of inputs they sample. As each dendritic spine receives at least one asymmetrical synapse,[72–74] the presynaptic terminals of which contain the excitatory neurotransmitter glutamate,[75,76] differences in the number of dendritic spines found on pyramidal cells are likely to reflect differences in the total number of excitatory inputs they integrate. Thus, larger, more spinous pyramidal cells such as those in area STP are likely to receive more excitatory inputs than smaller, less spinous cells such as those in V1 (Figure 15.6). Most postsynaptic targets of GABAergic inhibitory inputs to spiny neurons are also found on the dendrites.[77–81] As a result, interareal differences in pyramidal cell structure could influence the number of inhibitory inputs to their arbors. Differences in both the density of interneurons and the size of pyramidal cell dendritic arbors in different cortical areas are likely to affect the number and diversity of interneurons sampled by pyramidal cells.[82]

15.3.1.3 Dendritic Branching Patterns

Differences in the branching patterns of pyramidal cells potentially allow varying degrees of compartmentalization within their dendritic arbors (see References 83 through 88 for reviews). The significance of this was demonstrated by Poirazi and Mel,[89] who showed that compartmentalization acts as a mechanism for boosting cellular potential in learning paradigms. For example, scaling the size, or changing the branching patterns of a cell's dendritic arbor results in a logarithmic increase in the input-output functions in nonlinear cells, but only a relatively minor increase in linear cells (Figure 15.7). Thus, pyramidal cells in areas such as TE, STP, and FEF, which are characterized by highly branched dendritic arbors,[39,46] may compartmentalize processing to a greater extent than smaller, less complex cells such as those in areas V1 and V2. Although the relationship between dendritic branching and functional capacity has yet to be demonstrated in cortical pyramidal cells, compartmentalization is reportedly instrumental in determining function in retinal ganglion cells.[90]

15.3.2 Areal Level

Functional implications of structural differences that can be distinguished at the cellular level must necessarily be amplified in circuits. Cortical circuits composed of fundamentally different pyramidal cell phenotypes, which receive different

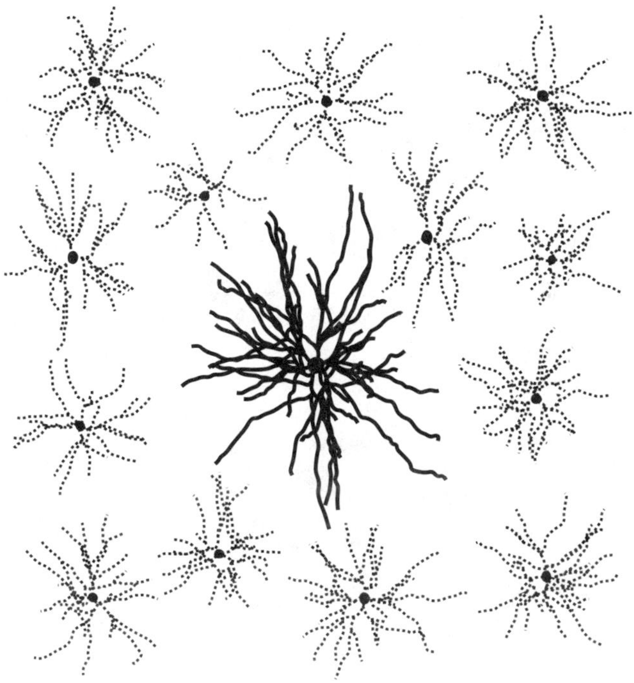

FIGURE 15.6 Scale drawings of the basal dendritic arbors of layer III pyramidal cells (in the tangential plane) taken from the superior polysensory area (STP, central black cell) and the primary visual area (V1, surrounding gray cells). A single STP cell is, on average, 13 times more spinous than those in V1.[39] If each spine received at least one excitatory input, cells in STP putatively integrate, on average, at least 13 times more excitatory inputs than those in V1.

numbers of inputs and project to different numbers of neurons, will be characterized by different patterns of connectivity. These differences in connectivity are likely to be reflected in the functional properties of the circuits. Thus, it is reasonable to suppose that two visual areas composed of pyramidal cells with different phenotypes will have different functional signatures.

There are several direct demonstrations of how the functional characteristics of individual neurons, and circuits, differ in the various visual cortical areas. For example, single-unit recording has revealed a fundamental difference in the discharge properties of pyramidal cells in the occipital and temporal lobes: cells in V1 are characterized by phasic discharge properties, whereas cells in inferotemporal cortex are distinguished by tonic discharge properties.[91] The former response characteristics are thought to allow quick "reset time" and real-time saccadic processing of retinal inputs,[60] whereas the sustained tonic activity of cells in visual areas in the temporal lobe[91–94] is thought to be necessary for memory (see Reference 95 for a review).

In addition, the shape-specific direction selectivity reported for cells in STP[96] could be a feature endowed by their large spinous dendritic arbors. Clustering of like inputs throughout the dendritic arbor may allow individual cells to detect

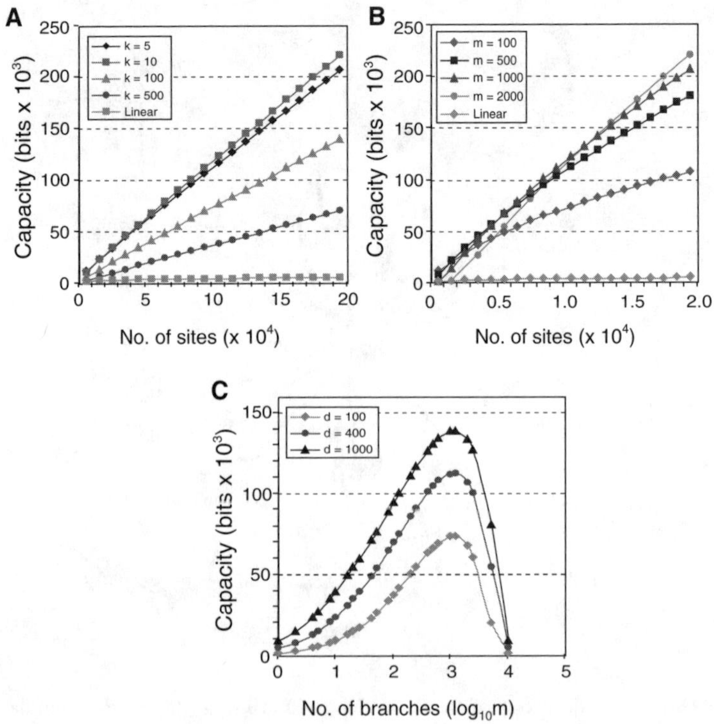

FIGURE 15.7 The *capacity* (measured in bits) of a given cell of fixed branch length (A) and branch count (B) increases with an increase in the number of sites. Cells of different branch length (k) and branch count (m) are illustrated in A and B, respectively. An increase in either of these variables results in increasing *capacity* with an increase in the number of sites. The *capacity* of a cell of fixed size (C) increases to a peak with an increase in the number of branches, before decreasing to values expected by linear models. (From Poirazi, P. and Mel, B., *Neuron*, 29, 779, 2000. With permission.)

different direction selectivities, which due to the high degree of dendritic branching may be juxtaposed (or *linked*) with specific objects. Polysensory processing by cells in STP[97,98] may also depend on their structure. The integration of different sensory modalities necessitates sampling of multiple sensory inputs within the same dendritic arbor, while maintaining the distinction between inputs. In addition, differences in pyramidal cell structure may be important in determining the diametrically opposed response characteristics of cortical circuits in the occipital and temporal lobes: tetanic stimulation in TE causes long-term potentiation, but the same stimulus applied to V1 results in long-term depression.[99]

15.3.3 Cortical Level

We have seen how structure may influence neuronal function at the cellular level, and how this could be important in determining the functional characteristics of the circuitry in different visual areas. However, visual areas do not function autonomously. Normal visual processing involves the interaction of ensembles of neurons in many cortical areas, acting in concert to orchestrate the visual percept. Although objective analysis fails to determine any overall trends in connectivity and functional relationships between visual areas,[100,101] it is clear that not all of aspects of visual processing are performed by all neurons in visual cortex. Instead, neurons in different visual areas reportedly process different aspects of vision. Based on varying neuronal response properties, these areas have been grouped loosely into those that process for motion detection and those that process for object recognition.[102] How then do areal specializations in cortical circuitry affect these aspects of visual processing?

15.3.3.1 Within Hierarchies

In monkeys, most retinal inputs project via the thalamus to V1, where the inputs are processed before being sent to other cortical areas (see References 3, 103, and 104 for reviews). Based on this flow of information it has been proposed that visual processing occurs through a quasi-hierarchy of various cortical areas. In this system, visual information is fed forward primarily by supragranular pyramidal cells, whereas information is fed back primarily by infragranular pyramidal cells.

Based on this type of organization, supragranular pyramidal neurons provide a route for the rapid flow of information through successive visual areas. For example, axonal projections of pyramidal neurons in V1 that stream to extrastriate areas arborize mainly in layers III and IV,[105,106] where other supragranular pyramidal cells may sample the inputs and, in turn, send them to another cortical area. By virtue of the systematic differences in arbor structure of pyramidal cells in different areas, such feedforward flow of processing may result in the reconstruction of progressively larger regions of the visuotopic representation (Figure 15.8). The progressive increase in the proportion of the visual scene sampled by populations of cells in "higher" visual areas may then be considered instrumental in reconstructing a single global visual percept from the relatively fragmented sampling in V1. Such a rapid transmission route may also be important for rapid motion detection, and recognition of specific stimuli in "real" time.

In contrast, feedback projections arise primarily from infragranular pyramidal cells and preferentially arborize in supra- and infragranular layers (see Reference 3 for a review). Thus, top-down modulation of visual processing within such a hierarchy is also subject to putative differences in the integrative ability of cells in different cortical areas. For example, as can be seen in Figure 15.9, there is *potential* for greater serial summation, or convergence, to individual cells through areas in the forward direction as compared with the feedback direction. This basic difference in patterns of connectivity may be offset, to some degree, by the arborization patterns of feedback projections given that feedforward axons terminate in tight, high-density axon clusters[107–111] whereas feedback axonal projections are more divergent.[105,110,112–116] Further quantitative data are required to determine the relative convergence and divergence in forward and back projections if we are to gain a better understanding of the role of regional pyramidal cell specialization in the reconstruction and modulation of visual processing.

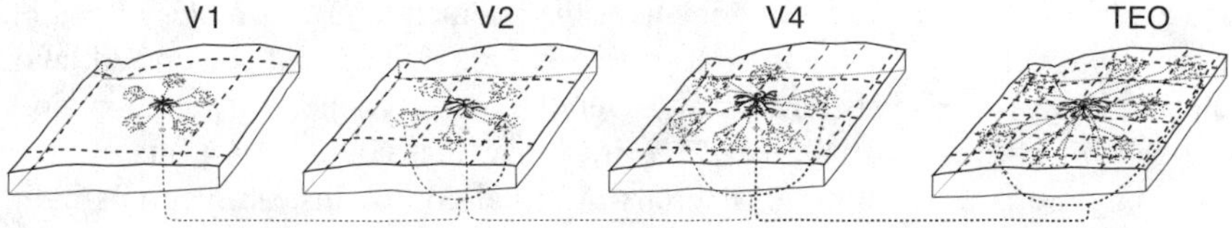

FIGURE 15.8 Schematic showing a serial feedforward passage of cortical projections in visuotopically organized occipitotemporal visual areas, including the primary (V1), second (V2), and fourth (V4) visual areas and cytoarchitectonic area TEO. Supragranular pyramidal neurons in V1 project to layers III and IV in V2. Supragranular pyramidal neurons in V2 may sample these feedforward inputs directly, or indirectly through a layer IV cell relay. Projections from cells in V2, which arborize in layers III/IV may then be sampled directly by supragranular cells in V4 and sent on to area TEO. Pyramidal neurons in V2 are larger and more spinous than those in V1, allowing them to integrate a greater number of more diverse inputs. Supragranular pyramidal neurons in V4 and TEO are progressively larger and more spinous, allowing them to sample from an increasing number of inputs over a wider proportion of the visuotopic representation. Coupled with the fact that supragranular pyramidal cells in the four areas form an intrinsic lattice of connections with progressively more of the visuotopic representation, an individual neuron can potentially sample a greater number of more diverse inputs with successive progression through cortical areas V1, V2, V4, and TEO. At the same time, feedforward projections become more divergent. The combination of the diverging feedforward projection and the converging intrinsic connectivity may allow for the progressive reconstruction of the visual scene through cortical areas, with the divergent feedforward projections increasing the extent of visuotopic connectivity through cortical areas and the intrinsic connections acting in concert to ensure the integration of larger portions of the visuotopic representation by populations of cells. Such a sampling strategy may also be important for the integration of different sensory modalities in individual cortical neurons (within networks) in visual association areas.

15.3.3.2 Within Distributed Systems

Visual processing appears to be more complex than just the anterior and posterior flow of information through a few selected cortical areas arranged into hierarchies. Data from multiple converging lines of research suggest that a strict hierarchical interpretation is not tenable. For example, not all retinothalamic inputs project to V1, not all projections adhere to the feedforward/feedback hierarchical scheme, the response latencies of neurons in different cortical areas do not necessarily comply with the proposed hierarchies, imaging studies reveal large ensembles of neurons may be recruited across putative hierarchies for a particular visual task, and ablation of V1 does not necessarily result in complete blindness (see References 117 through 122 for reviews). Instead, it appears as though the flow of information through visual areas operates within a much wider distributed system.[3,12,123] Nonetheless, the selection of a few visual areas serves as a useful illustration of how regional variation in pyramidal cell structure may influence a particular aspect of visual processing among cortical areas.

Within the context of a distributed system, specializations in pyramidal cell structure may serve to maintain topographic and/or temporal coherence in different

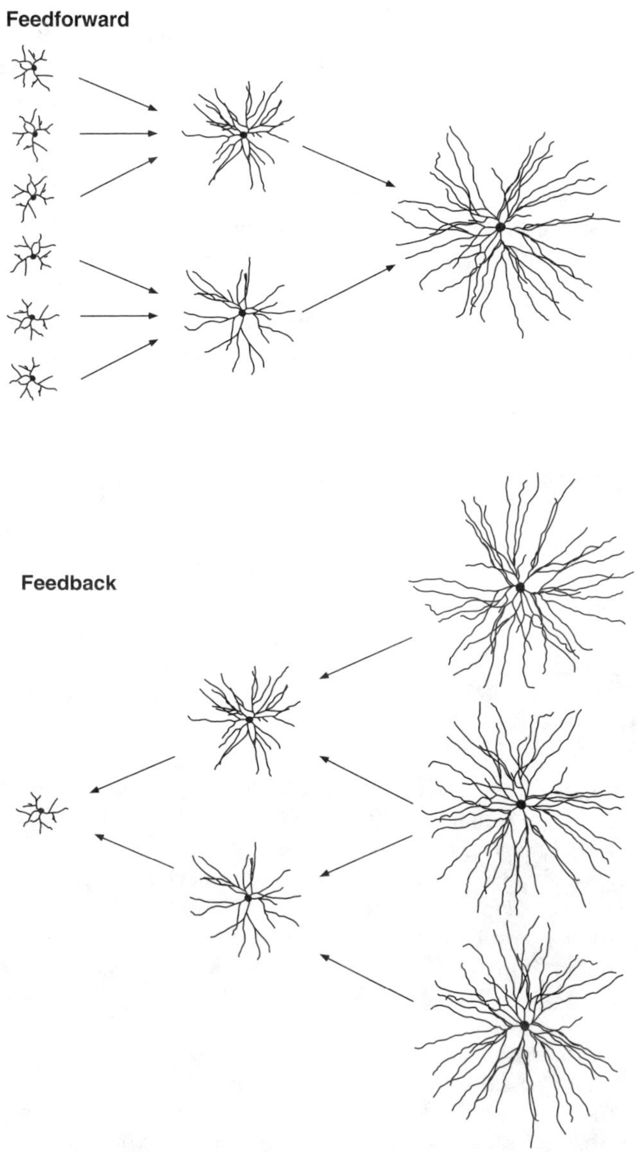

FIGURE 15.9 Stylistic representation of pyramidal neurons in the feedforward and feedback pathways of visual cortex. Pyramidal neurons become increasingly more spinous with progression through the feedforward pathways, with a doubling in the putative number of excitatory inputs integrated by cells in each area. On the contrary, pyramidal cells become less spinous with progression through cortical areas in the feedback direction. Thus, there is a potential for a greater degree of convergence in the feedforward pathway as compared with the feedback pathway.

aspects of visual processing. A look at regional variation in the pyramidal cell phenotype in monkeys reveals how specializations in neuronal structure may have facilitated the expansion of visual cortex in primate evolution. For example, expansion of cortex, and the possible addition of new areas, poses interesting connectional problems in the brain. Enlargement of cortex by the addition of more of the same cortical module results in decreased connectivity between cortical areas[124] and would inevitably result in fragmentation of processing. The addition of new circuitry composed of neurons that have the potential to integrate progressively larger numbers of inputs may be one strategy to overcome this problem. Thus, large spinous pyramidal cells in areas such as TEO, TE, and STP may provide a substrate for cross talk between cortical areas and regulation of coherent visual processing. Such connectivity between visual areas may be important in binding different aspects of visual processing to maintain a single conscious visual percept (see References 125 and 126 for reviews).

ACKNOWLEDGMENTS

The author thanks Rowan Tweedale for proofreading and Alejandra Elston for help with preparing the figures. The work for this chapter was supported by the Australian National Health and Medical Research Council.

REFERENCES

1. Allman, J., Jeo, R., and Sereno, M., The functional organization of the visual cortex in owl monkeys, in *Aotus: The Owl Monkey*, Baer, J.F., Weller, R.E., and Kakoma, I., Eds., Academic Press, San Diego, 1994, 287.
2. Felleman, D.J. and Van Essen, D.C., Distributed hierarchical processing in primate cerebral cortex, *Cereb. Cortex*, 1, 1, 1991.
3. Kaas, J.H., The organization of sensory and motor cortex in owl monkeys, in *Aotus: The Owl Monkey*, Baer, J.F., Weller, R.E., and Kakoma, I., Eds., Academic Press, Orlando, 1994, 331.
4. Kaas, J.H., Theories of visual cortex organization in primates, in *Cerebral Cortex,* Vol. 12: *Extrastriate Cortex in Primates*, Rockland, K., Kaas, J.H., and Peters, A., Eds., Plenum Press, New York, 1997, 91.
5. Rosa, M.G.P., Visuotopic organization of primate extrastriate cortex, in *Cerebral Cortex,* Vol. 12: *Extrastriate Cortex in Primates*, Rockland, K., Kaas, J.H., and Peters, A., Eds., Plenum Press, New York, 1997, 127.
6. Northcutt, R.G. and Kaas, J.H., The emergence and evolution of the mammalian neocortex, *Trends Neurosci.*, 18, 373, 1995.
7. Kaas, J.H., The organization of neocortex in mammals: implications for theories of brain function, *Annu. Rev. Psychol.*, 38, 129, 1987.
8. Kaas, J.H., The evolution of complex sensory systems in mammals, *J. Exp. Biol.*, 146, 165, 1989.
9. Kaas, J.H., Why does the brain have so many visual areas? *J. Cogn. Neurosci.*, 1, 121, 1989.

10. Krubitzer, L.A., The organization of neocortex in mammals: are species differences really so different? *Trends Neurosci.*, 18, 408, 1995.
11. Krubitzer, L.A., How does evolution build a complex brain, in *Evolutionary Developmental Biology of the Cerebral Cortex,* Bock, G. R. and Cardew, G., Wiley, Chichester, U.K., 2000, 206.
12. Mountcastle, V.B., An organizing principle for general cortical function: the unit module and the distributed system, in *The Mindful Brain*, Schmitt, F.O., Ed., MIT Press, Cambridge, MA, 1978, 7.
13. Churchland, P.S. and Sejnowski, T.J., *The Computational Brain*, MIT Press, Cambridge, MA, 1992.
14. Hendry, S.H.C., The anatomy of the cerebral cortex: aspects of neuronal morphology and organization, in *Excitatory Amino Acids and the Cerebral Cortex*, Conti, F. and Hicks, T.P., Eds., MIT Press, Cambridge, MA, 1996, 3.
15. Creutzfeldt, O.D., Generality of the functional structure of the neocortex, *Naturwissenschaften*, 64, 507, 1977.
16. Eccles, J.C., The cerebral neocortex: a theory of its operation, in *Cerebral Cortex,* Vol. 2: *Functional Properties of Cortical Cells*, Jones, E.G. and Peters, A., Eds., Plenum Press, New York, 1984, 1.
17. Gabbott, P.L. and Bacon, S.J., Local circuit neurons in the medial prefrontal cortex (areas 24, 25 and 32) in the monkey: II. Quantitative areal and laminar distributions, *J. Comp. Neurol.*, 364, 609, 1996.
18. Rockel, A.J., Hiorns, R.W., and Powell, T.P.S., The basic uniformity in structure of the neocortex, *Brain*, 103, 221, 1980.
19. Szentagothai, J., The neuron network of the cerebral cortex: a functional interpretation, *Proc. R. Soc. Lond. B*, 201, 219, 1978.
20. Douglas, R.J., Martin, K.A.C., and Whitteridge, D., A canonical microcircuit for neocortex, *Neural Comp.*, 1, 480, 1989.
21. Jones, E.G., Laminar distribution of cortical efferent cells, in *Cerebral Cortex,* Vol. 1: *Cellular Components of the Cerebral Cortex*, Peters, A. and Jones, E.G., Eds., Plenum Press, New York, 1984, 521.
22. DeFelipe, J. and Fariñas, I., The pyramidal neuron of the cerebral cortex: morphological and chemical characteristics of the synaptic inputs, *Prog. Neurobiol.*, 39, 563, 1992.
23. Feldman, D., Morphology of the neocortical pyramidal neuron, in *Cerebral Cortex,* Vol. 1: *Cellular Components of the Cerebral Cortex*, Peters, A. and Jones, E.G., Eds., Plenum Press, New York, 1984, 123.
24. Jones, E.G., GABAergic neurons and their role in cortical plasticity in primates, *Cereb. Cortex*, 3, 361, 1993.
25. Powel, T., Certain aspects of intrinsic organisation of the cerebral cortex, in *Brain Mechanisms and Perceptual Awareness*, Pompeiano, O. and Marsan, C.A., Eds., Raven Press, New York, 1981, 1.
26. Houser, C.R., Vaughn, J.E., Hendry, S.H.C., Jones, E.G., and Peters, A., GABA neurons in cerebral cortex, in *Cerebral Cortex,* Vol. 2: *Functional Properties of Cortical Cells*, Jones, E.G. and Peters, A., Eds., Plenum Press, New York, 1984, 63.
27. Hendry, S.H.C., Schwark, H.D., Jones, E.G., and Yan, J., Numbers and proportions of GABA-immunoreactive neurons in different areas of monkey cerebral cortex, *J. Neurosci.*, 7, 1503, 1987.
28. Elston, G.N. and González-Albo, M.C., Density and distribution of parvalbumin-, calbindin-, and calretinin-immunoreactive neurons in the prefrontal cortex of the owl monkey (*Aotus trivirgatus*): a quantitative comparison with sensory and motor cortical areas, *Brain Behav. Evol.*, in press.

29. DeFelipe, J., Types of neurons, synaptic connections and chemical characteristics of cells immunoreactive for calbindin-D28K, parvalbumin and calretinin in the neocortex, *J. Chem. Neuroanat.*, 14, 1, 1997.
30. Somogyi, P., Tamas, G., Lujan, R., and Buhl, E.H., Salient features of synaptic organisation in the cerebral cortex, *Brain Res. Rev.*, 26, 113, 1998.
31. Elston, G.N., Rosa, M.G.P., and Calford, M.B., Comparison of dendritic fields of layer III pyramidal neurones in striate and extrastriate visual areas of the marmoset: a Lucifer Yellow intracellular injection study, *Cereb. Cortex*, 6, 807, 1996.
32. Elston, G.N. and Rosa, M.G.P., The occipitoparietal pathway of the macaque monkey: comparison of pyramidal cell morphology in layer III of functionally related cortical visual areas, *Cereb. Cortex*, 7, 432, 1997.
33. Elston, G.N. and Rosa, M.G.P., Morphological variation of layer III pyramidal neurones in the occipitotemporal pathway of the macaque monkey visual cortex, *Cereb. Cortex*, 8, 278, 1998.
34. Elston, G.N. and Rosa, M.G.P., Pyramidal cells, patches, and cortical columns: a comparative study of infragranular neurons in TEO, TE, and the superior temporal polysensory area of the macaque monkey, *J. Neurosci.*, 20, RC117, 2000.
35. Elston, G.N., Interlaminar differences in the pyramidal cell phenotype in cortical areas 7m and STP of the macaque monkey, *Exp. Brain Res.*, 138, 141, 2001.
36. Elston, G.N., Benavides-Piccione, R., and DeFelipe, J., The pyramidal cell in cognition: a comparative study in human and monkey, *J. Neurosci.*, 21, RC163, 2001.
37. Elston, G.N., Pyramidal cells of the frontal lobe: all the more spinous to think with, *J. Neurosci.*, 20, RC95, 2000.
38. Elston, G.N., Tweedale, R., and Rosa, M.G.P., Cortical integration in the visual system of the macaque monkey: large scale morphological differences of pyramidal neurones in the occipital, parietal and temporal lobes, *Proc. R. Soc. Lond. Ser. B*, 266, 1367, 1999.
39. Elston, G.N., Tweedale, R., and Rosa, M.G.P., Cellular heterogeneity in cerebral cortex: a study of the morphology of pyramidal neurones in visual areas of the marmoset monkey, *J. Comp. Neurol.*, 415, *33*, 1999.
40. Lund, J.S., Yoshioka, T., and Levitt, J.B., Comparison of intrinsic connectivity in different areas of macaque monkey cerebral cortex, *Cereb. Cortex*, 3, 148, 1993.
41. Jacobs, B., Schall, M., Prather, M., Kapler, L., Driscoll, L., Baca, S., Jacobs, J., Ford, K., Wainwright, M., and Treml, M., Regional dendritic and spine variation in human cerebral cortex: a quantitative study, *Cereb. Cortex*, 11, 558, 2001.
42. Funahashi, S., Bruce, C.J., and Goldman-Rakic, P.S., Neuronal activity related to saccadic eye movements in the monkey's dorsolateral prefrontal cortex, *J. Neurophysiol.*, 65, 1464, 1991.
43. Bruce, C.J. and Goldberg, M.E., Primate frontal eye fields. I. Single neurons discharging before eye saccades, *J. Neurophysiol.*, 53, 603, 1985.
44. Huerta, M.F., Krubitzer, L.A., and Kaas, J.H., Frontal eye field as defined by intracortical microstimulation in squirrel monkeys, owl monkeys, and macaque monkeys: I. Subcortical connections, *J. Comp. Neurol.*, 253, 415, 1986.
45. Huerta, M.F., Krubitzer, L.A., and Kaas, J.H., Frontal eye field as defined by intracortical microstimulation in squirrel monkeys, owl monkeys, and macaque monkeys: II. Cortical connections, *J. Comp. Neurol.*, 265, 332, 1987.
46. Elston, G.N. and Rosa, M.G.P., Complex dendritic fields of pyramidal cells in the frontal eye field of the macaque monkey: comparison with parietal areas 7a and LIP, *NeuroReport*, 9, 127, 1998.

47. Cusick, C.G., The superior temporal polysensory region in monkeys, in *Cerebral Cortex,* Vol. 12: *Extrastriate Cortex in Primates*, Rockland, K., Kaas, J.H., and Peters, A., Eds., Plenum Press, New York, 1997, 435.
48. Gross, C.G., Rodman, H.R., Gochin, P.M., and Colombo, M.W., Inferior temporal cortex as a pattern recognition device, in *Computational Learning and Recognition: Proceedings of the 3rd NEC Research Symposium*, Baum, E., Ed., Society for Industrial and Applied Mathematics, Philadelphia, 1993, 44.
49. Perrett, D.I., Harries, M.H., Mistlin, A.J., Hietanen, J.K., Benson, P.J., Bevan, R., Thomas, S., Oram, M.W., Ortega, J., and Brierley, K., Social signals analyzed at the single cell level: someone is looking at me, something touched me, something moved! *Int. J. Comp. Psychol.*, 4, 25, 1990.
50. Tanaka, K., Inferotemporal cortex and object vision, *Annu. Rev. Neurosci.*, 19, 109, 1996.
51. Yukie, M., Organization of visual afferent connections to inferior temporal cortex, area TE, in the macaque monkey, in *The Association Cortex, Structure and Function*, Sakata, H., Mikami, A., and Fuster, J., Eds., Harwood Academic Publishers, Amsterdam, 1997, 247.
52. Elston, G.N. and Jelinek, H.F., Dendritic branching patterns of pyramidal cells in the visual cortex of the New World marmoset monkey, with comparative notes on the Old World macaque monkey, *Fractals*, 9, 297, 2001.
53. Elston, G.N., Pyramidal cell heterogeneity in the visual cortex of the nocturnal New World Owl monkey (*Aotus trivirgatus*), *Neuroscience*, 117, 213, 2003.
54. Kritzer, M., Cowey, A., and Somogyi, P., Patterns of inter- and intralaminar GABAergic connections distinguish striate (V1) and extrastriate (V2, V4) visual cortices and their functionally specialized subdivisions in the rhesus monkey, *J. Neurosci.*, 12, 4545, 1992.
55. DeFelipe, J., del Río, M.R., González-Albo, M.C., and Elston, G.N., Distribution and patterns of connectivity of interneurons containing calbindin, calretinin and parvalbumin in visual areas of the occipital and temporal lobes of the macaque monkey, *J. Comp. Neurol.*, 412, 515, 1999.
56. Goodchild, A.K. and Martin, P.R., The distribution of calcium-binding proteins in the lateral geniculate nucleus and visual cortex of a New World monkey, the marmoset, *Callithrix jacchus*, *Vis. Neurosci.*, 15, 625, 1998.
57. Kondo, H.T., Hashikawa, T., Tanaka, K., and Jones, E.G., Neurochemical gradient along the monkey occipito-temporal cortical pathway, *NeuroReport*, 5, 613, 1994.
58. Stuart, G.J. and Sackman, B., Active propagation of somatic action potentials into neocortical pyramidal cell dendrites, *Nature*, 367, 69, 1994.
59. Koch, C., Poggio, T., and Torre, V., Retinal ganglion cells: a functional interpretation of dendritic morphology, *Philos. Trans. R. Soc. Lond. Ser. B*, 298, 227, 1982.
60. Elston, G.N., Cortical heterogeneity: implications for visual processing and polysensory integration, *J. Neurocytol.*, 31, 317, 2003.
61. Elston, G.N., Cortex, cognition and the cell: new insights into the pyramidal cell and prefrontal function, *Cereb. Cortex*, in press.
62. Jacobs, B. and Scheibel, A.B., Regional dendritic variation in primate cortical pyramidal cells, in *Cortical Areas: Unity and Diversity,* Schüz, A. and Miller, R., Eds., Taylor & Francis, London, 2002, 111.
63. Colonnier, M.L., The tangential organization of the cortex, *J. Anat.*, 98, 327, 1964.
64. Rockland, K.S., Lund, J.S., and Humphrey, A.L., Anatomical banding of intrinsic connections in striate cortex of tree shrews (*Tupaia glis*), *J. Comp. Neurol.*, 209, 41, 1982.

65. Rockland, K.S. and Lund, J.S., Widespread periodic intrinsic connections in the tree shrew visual cortex, *Science*, 215, 1532, 1982.
66. Rockland, K.S. and Lund, J.S., Intrinsic laminar lattice connections in primate visual cortex, *J. Comp. Neurol.*, 216, 303, 1983.
67. McGuire, B.A., Gilbert, C.D., Rivlin, P.K., and Wiesel, T.N., Targets of horizontal connections in macaque primary visual cortex, *J. Comp. Neurol.*, 305, 370, 1991.
68. Kisvárday, Z.F., Martin, K.A.C., Freund, T.F., Magloczky, Z., Whitteridge, D., and Somogyi, P., Synaptic targets of HRP-filled layer III pyramidal cells in the cat striate cortex, *Exp. Brain Res.*, 64, 541, 1986.
69. Gilbert, C.D. and Wiesel, T.N., Morphology and intracortical projections of functionally characterised neurones in the cat visual cortex, *Nature*, 280, 120, 1979.
70. Fujita, I. and Fujita, T., Intrinsic connections in the macaque inferior temporal cortex, *J. Comp. Neurol.*, 368, 467, 1996.
71. Malach, R., Cortical columns as devices for maximizing neuronal diversity, *Trends Neurosci.*, 17, 101, 1994.
72. Peters, A. and Kaiserman-Abramof, I.A., The small pyramidal neuron of the rat cerebral cortex. The synapses upon dendritic spines, *Z. Zellforsch.*, 100, 487, 1969.
73. Jones, E.G., An electron microscopic study of the terminations of afferent fiber systems onto the somatic sensory cortex of the cat, *J. Anat.*, 103, 595, 1968.
74. Colonnier, M., Synaptic patterns on different cell types in the different laminae of the cat visual cortex, *Brain Res.*, 9, 268, 1968.
75. DeFelipe, J., Conti, F., Van Eyck, S.L., and Manzoni, T., Demonstration of glutamate-positive axon terminals forming asymmetric synapses in cat neocortex, *Brain Res.*, 455, 162, 1988.
76. Kharazia, V.N. and Weinberg, R.J., Glutamate in terminals of the thalamocortical fibers in rat somatic sensory cortex, *Neurosci. Lett.*, 157, 162, 1993.
77. Beaulieu, C. and Colonnier, M., A laminar analysis of the number of round-asymmetrical and flat-symmetrical synapses on spines, dendritic trunks, and cell bodies in area 17 of the cat, *J. Comp. Neurol.*, 231, 180, 1985.
78. Beaulieu, C. and Somogyi, P., Targets and quantitative distribution of GABAergic synapses in the visual cortex of the cat, *Eur. J. Neurosci.*, 2, 296, 1990.
79. Beaulieu, C., Kisvárday, Z., Somogyi, P., Cynader, M., and Cowey, A., Quantitative distribution of GABA-immunopositive and -immunonegative neurons and synapses in the monkey striate cortex (area 17), *Cereb. Cortex*, 2, 295, 1992.
80. Peters, A. and Harriman, K.M., Different kinds of axon terminals forming symmetric synapses with the cell bodies and initial axon segments of layer II/III pyramidal cells. III. Origins and frequency of occurrence of the terminals, *J. Neurocytol.*, 21, 679, 1992.
81. Micheva, K.D. and Beaulieu, C., An anatomical substrate for experience-dependent plasticity of the rat barrel field cortex, *Proc. Natl. Acad. Sci. U.S.A.*, 92, 11834, 1995.
82. Elston, G.N., DeFelipe, J., Arellano, J.I., González-Albo, M.C., and Rosa, M.G.P., Variation in the spatial relationship between parvalbumin immunoreactive interneurones and pyramidal neurones in rat somatosensory cortex, *NeuroReport*, 10, 2975, 1999.
83. Rall, W., Burke, R.E., Holmes, W.R., Jack, J.J.B., Redman, S.R., and Segev, I., Matching dendritic neuron models to experimental data, *Physiol. Rev.*, 72, 159, 1992.
84. Segev, I. and Rall, W., Excitable dendrites and spines: earlier theoretical insights elucidate recent direct observations, *Trends Neurosci.*, 21, 453, 1998.
85. Mainen, Z.F. and Sejnowski, T.J., Influence of dendritic structure on firing pattern in model neocortical neurons, *Nature*, 382, 363, 1996.

86. Mel, B., Why have dendrites? A computation perspective, in *Dendrites*, Stuart, G., Spruston, N., and Häusser, M., Eds., Oxford University Press, New York, 1999, 271.
87. Koch, C., *Biophysics of Computation. Information Processing in Single Neurons,* Oxford University Press, New York, 1999.
88. Helmchen, F., Dendrites as biochemical compartments, in *Dendrites*, Stuart, G., Spruston, N., and Häusser, M., Eds., Oxford University Press, New York, 1999, 161.
89. Poirazi, P. and Mel, B., Impact of active dendrites and structural plasticity on the storage capacity of neural tissue, *Neuron*, 29, 779, 2000.
90. Taylor, W.R., He, S., Levick, W.R., and Vaney, D.I., Dendritic computation of direction selectivity by retinal ganglion cells, *Science*, 289, 2347, 2000.
91. Ashford, J.W. and Fuster, J.M., Occipital and inferotemporal responses to visual signals in the monkey, *Exp. Neurol.*, 90, 444, 1985.
92. Miller, E.K., Li, L., and Desimone, R., Activity of neurons in anterior inferior temporal cortex during a short-term memory task, *J. Neurosci.*, 13, 1460, 1993.
93. Nakamura, K. and Kubota, K., Mnemonic firing of neurons in the monkey temporal pole during a visual recognition memory task, *J. Neurophysiol.*, 74, 162, 1995.
94. Sobotka, S. and Ringo, J.L., Investigation of long term recognition and association memory in unit responses from inferotemporal cortex, *Exp. Brain Res.*, 96, 28, 1993.
95. Miller, E.K. and Cohen, J.D., An integrative theory of prefrontal cortex function, *Annu. Rev. Neurosci.*, 24, 167, 2001.
96. Tanaka, K., Koyama, T., and Mikami, A., Neurons in the temporal cortex change their preferred direction of motion dependent on shape, *NeuroReport*, 10, 393, 1999.
97. Perrett, D.I., Smith, P.A.J., Potter, D.D., Mistlin, A.J., Head, A.S., Milner, A.D., and Jeeves, M.A., Neurons responsive to faces in the temporal cortex: studies of functional organization, sensitivity to identity and relation to perception, *Hum. Neurobiol.*, 3, 197, 1984.
98. Hikosaka, K., Iwai, E., Saito, H., and Tanaka, K., Polysensory properties of neurones in the anterior bank of the caudal superior temporal sulcus of the macaque monkey, *J. Neurophysiol.*, 60, 1615, 1988.
99. Murayama, Y., Fujita, I., and Kato, M., Contrasting forms of synaptic plasticity in monkey inferotemporal and primary visual cortices, *NeuroReport*, 8, 1503, 1997.
100. Young, M.P., Objective analysis of the topological organization of the primate cortical visual system, *Nature*, 358, 152, 1992.
101. Young, M.P., The organization of neural systems in the primate cerebral cortex, *Proc. R. Soc. Lond. Ser. B*, 252, 113, 1993.
102. Ungerleider, L.G. and Mishkin, M., Two cortical systems, in *Analysis of Visual Behavior*, Ingle, D.J., Goodale, M.A., and Mansfield, R.J.W., Eds., MIT Press, Cambridge, MA, 1982, 549.
103. Weller, R.E., Two cortical visual systems in Old World and New World primates, in *Progress in Brain Research*, Hicks, T.P. and Benedek, G., Eds., Elsevier, Oxford, U.K., 1988, 293.
104. Casagrande, V.A. and Kaas, J.H., The afferent, intrinsic and efferent connections of primary visual cortex in primates, in *Cerebral Cortex,* Vol. 10: *Primary Visual Cortex in Primates*, Peters, A. and Rockland, K.S., Eds., Plenum Press, New York, 1994, 201.
105. Rockland, K.S., Bistratified distribution of terminal arbors of individual axons projecting from area V1 to middle temporal area (MT) in the macaque monkey, *Vis. Neurosci.*, 3, 155, 1989.
106. Rockland, K.S. and Virga, A., Organization of individual cortical axons projecting from area V1 (area 17) to V2 (area 18) in the macaque monkey, *Vis. Neurosci.*, 4, 11, 1990.

107. Livingstone, M.S. and Hubel, D.H., Anatomy and physiology of a color system in the primate visual cortex, *J. Neurosci.*, 4, 309, 1984.
108. Montero, V.M., Patterns of connections from the striate cortex to cortical visual areas in superior temporal sulcus of macaque and middle temporal gyrus of owl monkey, *J. Comp. Neurol.*, 189, 45, 1980.
109. Saleem, K.S., Tanaka, K., and Rockland, K.S., Specific and columnar projection from area TEO to area TE in the macaque inferotemporal cortex, *Cereb. Cortex*, 3, 454, 1993.
110. Ungerleider, L.G. and Desimone, R., Projections to the superior temporal sulcus from the central and peripheral field representations of V1 and V2, *J. Comp. Neurol.*, 248, 147, 1986.
111. Van Essen, D.C., Newsome, W.T., Maunsell, J.H.R., and Bixby, J.L., The projections from striate cortex (V1) to areas V2 and V3 in the macaque monkey: asymmetries, areal boundaries, and patchy connections, *J. Comp. Neurol.*, 244, 451, 1986.
112. Rockland, K.S. and Van Hoesen, G.W., Direct temporal-occipital feedback connections to striate cortex (V1) in the macaque monkey, *Cereb. Cortex*, 4, 300, 1994.
113. Rockland, K.S. and Drash, G.W., Collateralized divergent feedback connections that target multiple cortical areas, *J. Comp. Neurol.*, 373, 529, 1996.
114. Wong-Riley, M., Columnar cortico-cortical interconnections within the visual system of the squirrel and macaque monkeys, *Brain Res.*, 162, 201, 1979.
115. Shipp, S. and Zeki, S., The organization of connections between areas V5 and V1 in macaque monkey visual cortex, *Eur. J. Neurosci.*, 1, 309, 1989.
116. Shipp, S. and Zeki, S., The organization of connections between area V5 and V2 in macaque monkey visual cortex, *Eur. J. Neurosci.*, 1, 333, 1989.
117. Lennie, P., Single units and visual cortical organization, *Perception*, 27, 889, 1998.
118. Kaas, J.H., The structural basis for information processing in the primate visual system, in *Visual Neuroscience*, Pettigrew, J.P., Sanderson, K.J., and Levick, W.R., Eds., Cambridge University Press, Cambridge, U.K., 1986, 315.
119. Schiller, P.H., On the specificity of neurons in visual areas, *Behav. Brain Res.*, 76, 21, 1996.
120. Cowey, A. and Stoerig, P., The neurobiology of blindsight, *Trends Neurosci.*, 14, 14, 1991.
121. Rockland, K.S., Elements of cortical architecture: hierarchy revisited, in *Cerebral Cortex,* Vol. 12: *Extrastriate Cortex in Primates*, Rockland, K., Kaas, J.H., and Peters, A., Eds., Plenum Press, New York, 1997, 243.
122. Bullier, J., Girard, P., and Salin, P.-A., The role of area 17 in the transfer of information to extrastriate visual cortex, in *Cerebral Cortex,* Vol. 10: *Primary Visual Cortex in Primates*, Peters, A. and Rockland, K.S., Eds., Plenum Press, New York, 1994, 301.
123. Mountcastle, V.B., Posterior parietal association cortex of the monkey: command functions for operations within extrapersonal space, *Cereb. Cortex*, 5, 377, 1995.
124. Ringo, J.L., Neuronal interconnection as a function of brain size, *Brain Behav. Evol.*, 38, 1, 1991.
125. Crick, F.H.C. and Koch, C., Some reflections on visual awareness, *Cold Spring Harbor Symp. Quant. Biol.*, 55, 953, 1990.
126. Singer, W. and Gray, C.M., Visual feature integration and the temporal correlation hypothesis, *Annu. Rev. Neurosci.*, 18, 555, 1995.
127. Nelson, R.B., Friedman, D.P., O'Neill, J.B., Mishkin, M., and Rottenberg, A., Gradients of protein kinase C substrate phosphorylation in primate visual system in visual memory storage areas, *Brain Res.*, 416, 387, 1987.

128. Jelinek, H.F. and Elston, G.N., Pyramidal neurones in macaque visual cortex: significant phenotypic variation in arbour structure, *Fractals*, 9, 287, 2001.
129. Preuss, T.M. and Goldman-Rakic, P.S., Myelo- and cytoarchitecture of the granular frontal cortex and surrounding regions in the Strepshrine primate *Galago* and the anthropoid primate *Macaca*, *J. Comp. Neurol.*, 310, 429, 1991.
130. Lyon, D.C. and Kaas, J.H., Connectional and architectonic evidence for dorsal and ventral V3, and dorsomedial area in marmoset monkeys, *J. Neurosci.*, 21, 249, 2001.
131. Lyon, D.C. and Kaas, J.H., Evidence from V1 connections for both dorsal and ventral subdivisions of V3 in three species of monkeys, *J. Comp. Neurol.*, 449, 281, 2002.
132. Hof, P.R. and Morrison, J.H., Neurofilament protein defines regional patterns of cortical organization in the macaque monkey visual system: a quantitative immunohistochemical analysis, *J. Comp. Neurol.*, 352, 161, 1995.
133. Kondo, H.T., Tanaka, K., Hashikawa, T., and Jones, E.G., Neurochemical gradients along monkey sensory cortical pathways: calbindin-immunoreactive pyramidal neurons in layers II and III, *Eur. J. Neurosci.*, 11, 4197, 1999.

16 Feedback Connections: Splitting the Arrow

Kathleen S. Rockland

CONTENTS

16.1 INTRODUCTION

Feedback connections belong to the broad category of long distance corticocortical connections; that is, connections that extend beyond a single cortical area. The general organization of corticocortical connections was mapped by extensive tract tracing studies especially during the last third of the 20th century. This, however, is very much a first step; and important questions, such as how different connectional systems interact and how these interactions relate to particular outcomes or behaviors, remain to be determined. An adequate categorization of types and subtypes is still lacking (see Reference 1); and it is highly likely that connections differ in many important ways yet to be characterized, including conduction velocity, postsynaptic localization, and specific receptors. It is interesting that similar questions, of how to move beyond "functional cartography," are being actively considered by researchers working in neuroimaging.[2] In this context, distinctions have been made between structural connectivity (the direct anatomical connections between brain regions), and functional or effective connectivity.[3,4]

At this time, connectional studies still rely heavily on traditional criteria, such as layers of origin and destination, for classification. A by now familiar distinction

0-8493-1243-4/04/$0.00+$1.50

identifies "feedforward" connections as mainly originating from neurons in layer 3 and terminating in layer 4. Neurons of origin of "feedback" connections are believed to have a complementary distribution, in layers 3A and 6, and their terminations are in layer 1.[5,6] In fact, finer analysis reveals many variations and exceptions to this general description; for example, feedforward projecting neurons occur in layer 5.[7,8] In part for this reason, early descriptions of laminar organization only incidentally commented on the complementarity of feedforward and feedback connections, and emphasized instead the different types of connections.[9]

The emphasis on two connectional classes led to the unfortunate extrapolation that each has distinct functional roles. "Feedback" connections, as the term itself implies, were associated with a modulatory role.[10] The terminations in layer 1 were assumed to be targeting distal apical dendrites; and, before the discovery of active conductance mechanisms in the 1990s, this was thought to be less efficacious than inputs nearer the cell soma and axon hillock.[11]

The idea of a modulatory role seemed consistent with another feature of feedback connections; namely, that within the cortical sensory pathways, the feedback connections typically are directed to areas "closer" to the periphery, as if they were correcting or otherwise modifying incoming information. Although attractive in its simplicity, this view remains hypothetical; and, as more data become available, the idea of a strong feedforward-feedback, bidirectional dichotomy seems increasingly limited.

Ideally a new chapter on feedback connections would propose a detailed wiring diagram of the basic circuit or circuits, and some elucidation of the particular function or functions at the level of receptive fields, neural assemblages, and behavior, including state-dependent effects. One hopes a more comprehensive review along these lines will be possible in another few years. In this chapter, however, where the goals are necessarily more modest, the purpose is to review several factors and issues that the author thinks will be important in the reassessment of feedback connections. The overall viewpoint is mainly from an anatomical perspective, as there are already several excellent reviews on physiological aspects.[12] Focus is on work since the mid-1990s (subsequent to References 10 and 13) and, except for one section (Section 16.8), is restricted primarily to the early visual pathway.

16.2 SUBTYPES OF FEEDBACK CONNECTIONS

There is a strong indication, even from anatomical criteria alone, that feedback connections should be further subdivided. One indication, already known from the earliest retrograde tracer experiments, is that the cells of origin occur at two levels,[7,9,10] superficial in layers 2 and 3A, and deep in layer 6 (Figure16.1 and Figure 16.2). This bistratification represents a major difference in terms of both cellular and network properties. Neurons in the superficial layers are "standard" pyramidal neurons, which may or may not comprise a single class; but those in layer 6 include many modified pyramids (Figure 16.3), and constitute an anatomically and presumably functionally mixed population.[14]

The simple criterion of cell size provides further corroboration of heterogeneity. Thus, the more superficial neurons, as a population, are on average larger than those in layer 6 (Figure 16.3 and Figure 16.4; see Reference 15). In area V2, the average

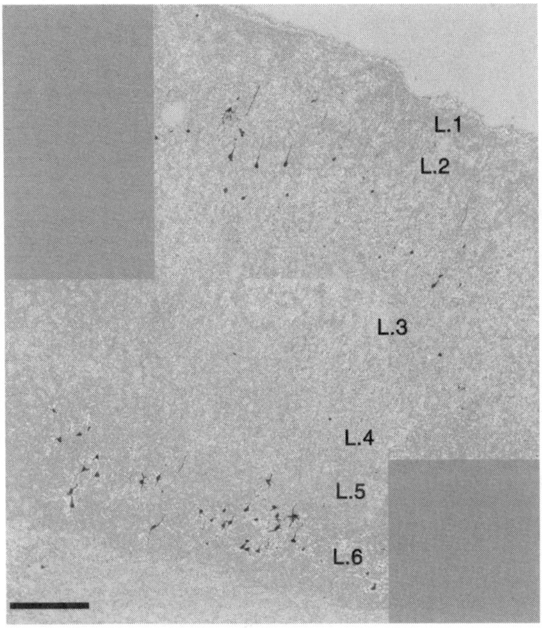

FIGURE 16.1 Photomicrograph of neurons in layers 3A and 6 of area V2, retrogradely labeled by an injection of biotinylated dextran amine (BDA) in area V1. Note typical bistratified lamination of feedback projecting neurons. Anterogradely labeled terminations are evident in layer 4, and labeled axons in layer 6. (BDA is primarily an anterograde tracer, but "sporadic" retrograde labeling can occur, in a projection-dependent and animal-dependent manner). Scale bar = 200 μm.

soma size for supragranular neurons is 104.88 μm^2 (range: 51.72 to 213.36 μm^2) compared with 94.12 μm^2 (range: 32.10 to 187.66 μm^2) for neurons in layer 6. The distribution of cell sizes is asymmetric, in that there are a greater number of large neurons in the superficial layers than in layer 6. In area MT, layer 6 cells are larger than the cells in both layer 6 and layer 3 of area V2 (average: 170.68 μm^2; range 80.10 to 346.93 μm^2). This area-specific difference is unlikely to be due to simple distance effects (i.e., longer axon trajectory between MT and V1), given the size differences observed within V2. (In our material, the number of labeled neurons in layer 3 of MT was small, and not included in these measurements.)

A second indication of subtypes is the variation in laminar terminations of single axons, as reported in several recent studies. Feedback axons from MT to V1 can terminate in layer 1 ($n = 2$), layer 4B ($n = 3$), layers 1 and 4B ($n = 1$), or layers 1, 4B, and 6 ($n = 3$).[16] Similarly, axons projecting from area TE to TEO and posterior inferotemporal cortex[17] had terminations in layers 1 to 3 ($n = 4$), in layer 1 ($n = 1$), in layers 5 and 6 ($n = 2$), or in layers 1 to 3 and 5 and 6 ($n = 2$). For further interpretation, it would be helpful to have a larger sample size and to know the actual parent neuron, or at least the layer of origin, for particular axons. This could be achieved by intracellular, juxtacellular, or (as a compromise) laminar specific injection of tracers.

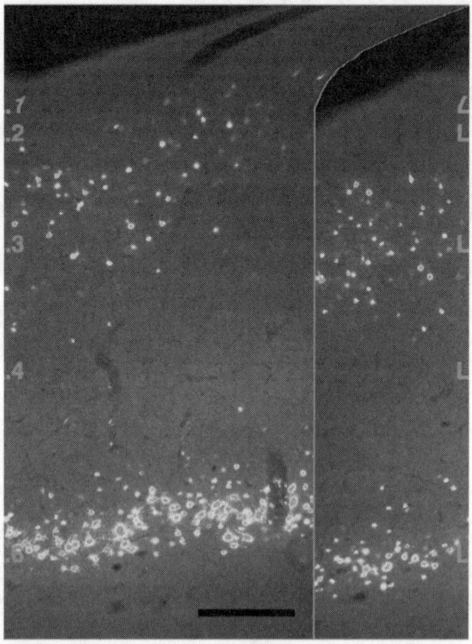

FIGURE 16.2 Photomicrograph of neurons in the upper part of layer 3 and in layer 6 of area V2, retrogradely labeled by an injection of Fast Blue in area V1. Scale bar = 200 μm.

Third, again on morphological grounds, feedback axons exhibit differences in caliber and possibly in terminal specializations. According to a recent report[18] on feedback connections from V2 to V1, one type of axon is small-caliber (0.3 μm), unmyelinated, and with many small terminal specializations distributed over a divergent terminal field (2.0 mm). A second type of profile is thicker (>1.0 μm), myelinated, and has separate clusters of larger boutons. Feedback connections from area V2 to V1 are reported to have overall rapid conduction velocities (median = 3.5 m/s),[19] but with a range of velocities,[20,21] as would be consistent with a range in caliber. Again, additional substantiating data, including the cells of origin, would be important.

Physiological recordings in area V1, after manipulation of feedback from V2 or MT, consistently show multiple effects,[22–24] of both receptive field enhancement and inhibition (see below). There may be several explanations for this range of effects, but among these would be the existence of different types of connections.

Finally, a subpopulation of feedback projecting neurons projects to multiple targets, and may constitute a distinct category. Branching of axon collaterals is well established for feedback axons,[25] by both retrograde and anterograde techniques (from V4 to V2 and V1;[26] from MT to V2 and V1;[7,16] from TE to TEO, V4, and V2[27]). An interesting possibility is that some neurons in layer 3A project to both feedback and feedforward targets. Some terminations from V2 target layer 1 of V4 (figure 15 in Reference 28) and might be collaterals of axons projecting to V1. Double retrograde injections in V1 and V4 would be one way to answer this question.

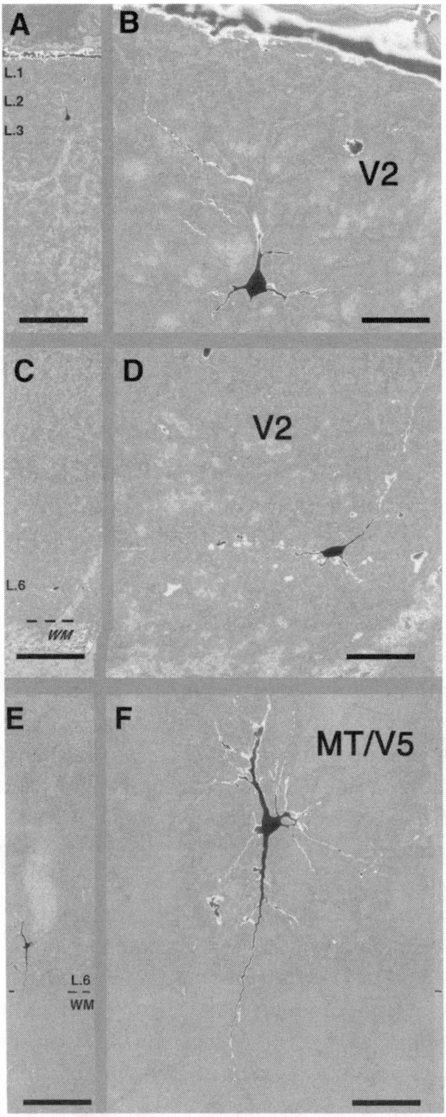

FIGURE 16.3 Photomicrograph of feedback projecting neurons retrogradely labeled by an injection of BDA in area V1. A and (higher magnification) B, Pyramidal neuron in area V2 at the border of layers 2 and 3. C and (higher magnification) D, Pyramidal neuron in layer 6 of V2. E and (higher magnification) F, Inverted pyramidal neuron in layer 6 of MT/V5. Scale bar = 200 μm in A, C, E; 50 μm in B, D, F.

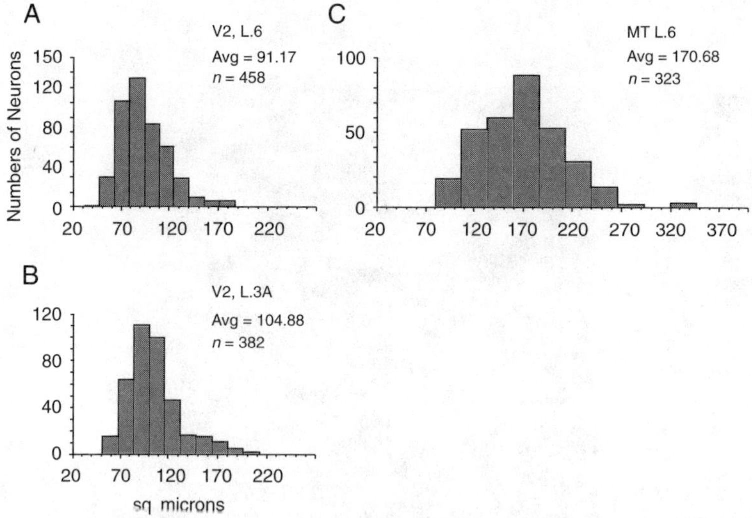

FIGURE 16.4 Histograms of area in square microns (*x*-axis) of cells in layer 6 of area V2 (A), layer 3 of area V2 (B), and in layer 6 of MT (C) in two monkeys. Distributions were determined to be significantly different from each other (Mann–Whitney U, $p < 0.0001$).

16.3 MICROCIRCUITRY OF FEEDBACK CONNECTIONS

There is very little information concerning the microcircuitry of feedback connections, especially in primates. At the light microscopic level, the terminations along a feedback axon are a mix of stalked (*boutons terminaux*) and beaded (*en passant*) profiles. In this regard, feedback axons appear indistinguishable from most feedforward, callosal, or intrinsic corticocortical connections (Figure 16.5 and Figure 16.6; but see section 3 of Reference 6), although continued work may yet identify connection-specific features. A recent electron microscopic (EM) study proposes that there are at least two types of beaded terminations, namely, small beaded terminations that make single, small asymmetric synapses, and larger beaded terminations that make up to four synapses per bouton.[18]

EM observations of connections from V2 to V1 in primates reveal a mixed postsynaptic distribution onto dendritic spines and shafts (respectively, 82 and 18% in Reference 5; or according to Reference 18, 84 and 16% for thin feedback axons, and 77 and 23% for thicker axons). The most detailed EM study so far is of connections from the lateromedial extrastriate area (LM) to area V1 in rats.[29,30] In this system, connections terminating in layer 1 are reported to make excitatory contacts preferentially (98%) with apical dendrites of pyramidal cells, many of which can be identified as feedforward projection neurons. In contrast, feedforward connections from area 17 to LM have a somewhat lower proportion (90%) of contacts onto pyramidal neurons. The apparent discrepancy between results in monkey and in rat may be due to species differences and/or to pathway specializations between LM and V2.

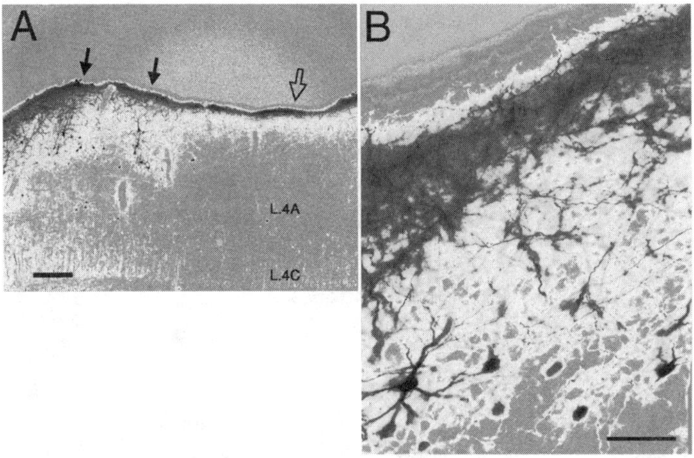

FIGURE 16.5 (Color figure follows p. 142.) Photomicrographs of projections in area V1 labeled by a BDA injection in V2 (injection diameter ≤ 300 μm). The injection anterogradely labels feedback terminations, mainly in layer 1; and a few neurons in V1 have been retrogradely labeled by the same injection. Note the widespread extent of the feedback axons, which extend beyond (hollow arrow) and between patches (solid arrows) of retrogradely filled neurons. Scale bars: 200 μm in A; 50 μm in B.

A provisional conclusion is that feedback connections, at least in primates, contact a mixed population of excitatory and inhibitory neurons. Within layers 1 and 2, this is likely to include a high proportion of apical dendritic targets, but these could belong to either superficial or deeper pyramidal neurons. Different subpopulations are likely to exhibit specific differences in electrophysiological properties; for example, layer 5 intrinsic bursting cells are readily fired by inputs to the apical tuft (in rat visual cortex[31]). Deeper pyramidal neurons in layer 6 are likely to receive feedback terminations at proximal dendritic loci, in layers 6 or 5. This is an important point, as discussions of feedback axons have frequently emphasized contacts onto distal apical dendrites, to the exclusion of other circuitry.

16.4 DIVERGENCE OF FEEDBACK CONNECTIONS

Feedback terminations in layers 1 and 5/6 are typically highly divergent (Figure 16.5 and Figure 16.6). The technique of single axon reconstruction is particularly useful in demonstrating the extent and complexity of individual terminal fields. These are >1.0 mm for connections from V2 to V1, and 6.0 to 8.0 mm for connections from V4, TEO, or MT to V1.[5,6] Feedback connections are more widespread than feedforward, and do not seem tightly bound to functional topography.[5,6] As previously noted, the anatomical divergence may be only "apparent," if it were a consequence of intrinsic inter- or intralaminar within the feedback-projecting area (figure 15 in Reference 28). However, delimited injections in V4 or TEO, to cite a particularly clear example, result in connections which are widespread in V1, to the extent of involving both hemifields (e.g., figures 3, 9, and 12 in Reference 26 and personal observations).

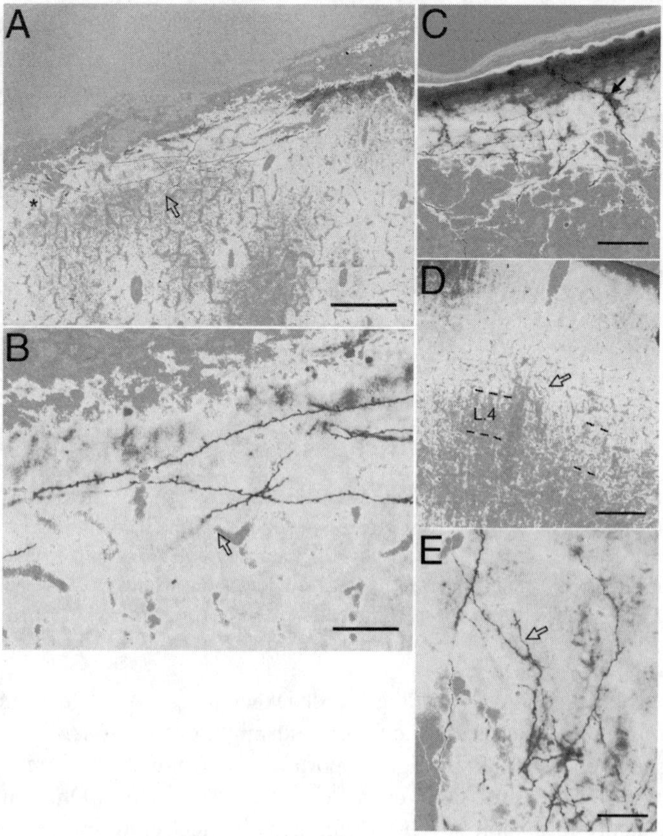

FIGURE 16.6 (Color figure follows p. 142.) Photomicrographs of axons anterogradely labeled by fluor-ruby (FR) injections. A and (higher magnification) B, Single feedback axon from area V4 terminating in layer 1 of V1. Two adjacent sections have been merged in Adobe Photoshop (at the asterisk). C, Feedback axons from V2 terminating in layer 1 of V1. D and (higher magnification) E, Feedforward axons from V2 terminating in layers 3 and 4 of area V4 (from the same brain as the image in C). Note the generally similar mix of stalked and beaded terminations. One segment, indicated by the solid arrow in C, is distinguishable by its higher density of stalked boutons. Hollow arrows mark corresponding points in A and B, D and E. Scale bars: 200 μm in A, D; 50 μm in B, C, E.

As one might expect from the large spread of feedback axons, they have a larger number of terminations than do the feedforward axons. Table 16.1 summarizes, by way of comparison, the terminal field dimensions and number of boutons for 15 "feedforward" axons terminating in MT/V5 and 12 "feedback" axons from MT.

There is some discrepancy between these results and recent reports that feedback projections connect corresponding and equal-sized regions of the visual field in striate and extrastriate cortices.[35,36] These are based on projections from V3 to layer 4B of V1 (figure 5 of Reference 36), and use cholera toxin subunit B (CTB). CTB is considered to be a "bidirectional" tracer, producing both anterograde and retrograde label; but there may also be some collateral transport, from the retrogradely

TABLE 16.1
Axons Projecting to MT/V5 from V1 and from V2

Axon	Source Area	Field Size	No. of Boutons
5-3	V1	1.2 DV × 1.3 AP	220
5-4	V1	1.2 DV × 1.0 AP	516
5-7	V1	2.4 DV × 1.0 AP	452
5-6	V1	0.9 DV	
7-7	V1	0.8 ML	
11-16	V1	1.5 DV × 0.3 AP	327
6-1	V1	1.8 DV × 1.6 ML	
27-1	V1	3.0 DV × 2.5 AP	459
27-2	V1	2.0 DV × 2.25 AP [layer 6]	384
27-3	V1	3.0 DV × 1.25 AP	341
27-4	V1	2.5 DV × 2.75 AP	224
27-5	V1	4.0 DV	624
5-1	V2	1.1 DV × 0.55 ML	343
4-3	V2	0.9 DV × 1.5 ML	139
2-4	V2	0.25 DV × 0.4 ML	470

Axons Projecting from MT

Axon	Target Area	Field Size	No. of Boutons
8-1	V1	>4.5	221
8-2	V1	8.0 DV × 4.0 AP	1671
9-1	V1	4.0 DV × 1.5 AP	1015 (35 in L.4B) [+V2]
9-3	V1	>1.5 DV × 3.6 AP	2139
10-1	V1	>2.0 ML × 4.0 AP	>534
9-5	V1[L.4B]	8.0 ML × 0.5 AP	453
9-6	V1[L.4B]	3.0 ML × 2.0 AP	395
8-4	V2	6.0 ML × 3.10 AP	1582 (1100 L.1, 2; 482, L.5)
8-5	V2	6.0 ML × 2.75 AP	1596 (1358 L.1; 238 L.6)
10-3	V2	~3.2 DV × 3.0 ML	671 (213, L.1; 458, L.6)
10-4	V2	1.5 AP × 1.15 ML	455 (88 asc; 367, L.6)
10-5	V2	>1.7 AP × 1.50 ML	680 (222 sg; 458, L.5, 6)

Abbreviations: DV = dorsoventral; AP = anteroposterior; ML = mediolateral; asc = ascending; sg = supragranular.

Sources: References 16, 33, and 34; data from axons 27-1 to 27-5 are unpublished results.

filled neurons. In this case, the labeled field may include a mix of both feedback and horizontal connections. Another recent report,[37] using an adenoviral-mediated method for anterogradely labeling feedback projections from V2 to V1, confirms the widespread distribution and apparent lack of specificity in relation to functional domains. However, an injection situated within a thick cytochrome oxidase stripe in V2 was noted to produce a more clustered termination pattern than three others that were not so localized.

The widespread dimensions of feedback axons are most compatible with a nontopographic organization, where single axons cross over ocular dominance or orientation domains in area V1. Double-label experiments, where feedback axons from V2 are co-registered with the optical imaging map of orientation domains, support this inference.[37] Interestingly, in the cerebellar cortex, the length of individual parallel fibers (2 to 3 mm) exceeds the width of a single microzone (0.3 to 1.0 mm, depending on species).[38]

16.5 RECIPROCITY OF FEEDFORWARD AND FEEDBACK CONNECTIONS

One of the popular defining characteristics of feedback connections is that they "reciprocate" feedforward connections. This is, however, a complicated issue for a number of reasons. First, as noted above, the feedback connections are more divergent (Figure 16.6).[39–41] The divergence applies also at the level of area-to-area connectivity. In the early visual pathway, a cascade of areas projects back to V1, including some (TEO, TE, and parts of the parahippocampal gyrus) that do not receive direct connections from V1.[27,42]

Second, reciprocity may not occur, or may not always occur at the single neuron level. Although there is evidence that feedback neurons in LM can directly contact feedforward neurons in rat area V1 (see above), there would seem to be ample opportunity for a mix of multiple mono- and polysynaptic loops. To give just one example, feedforward projections from V1 to V2, which are concentrated in layer 4, would not overlap, or only very partially overlap with feedback-projecting neurons in the superficial layers. For these neurons, which contribute to feedback projections to V1, the basal dendrites do not extend into layer 4 (see figures 14 and 33 in Reference 5).

A curious point, relevant to the issue of reciprocity, is that the terminal fields of feedback axons are not uniform. They frequently have morphologically distinct subregions, which are relatively free of terminations, and interspersed small clusters of higher terminal density (Figure 16.7). The terminal-free zones may be related to large-scale cortical modularity. The interspersed clusters, typically <100 μm wide, are more difficult to explain; but may correspond to small "hot spots" of higher synaptic density (efficacy?), where more neurons are contacted and/or more contacts are made. This contrasts to the more usual, "linear" configurations, where terminations occur at 1 to 20 μm intervals along the length of the axon.

16.6 NETWORK ORGANIZATION OF FEEDBACK CONNECTIONS

The idea of feedforward–feedback bidirectional reciprocity, as discussed in the previous section, does not seem to account for what must in actuality be a dense network of interacting connections. The complexity of interactions is hinted at by investigations of temporal aspects. Data are now available from several groups, which have examined parameters such as conduction velocity, response latency, and

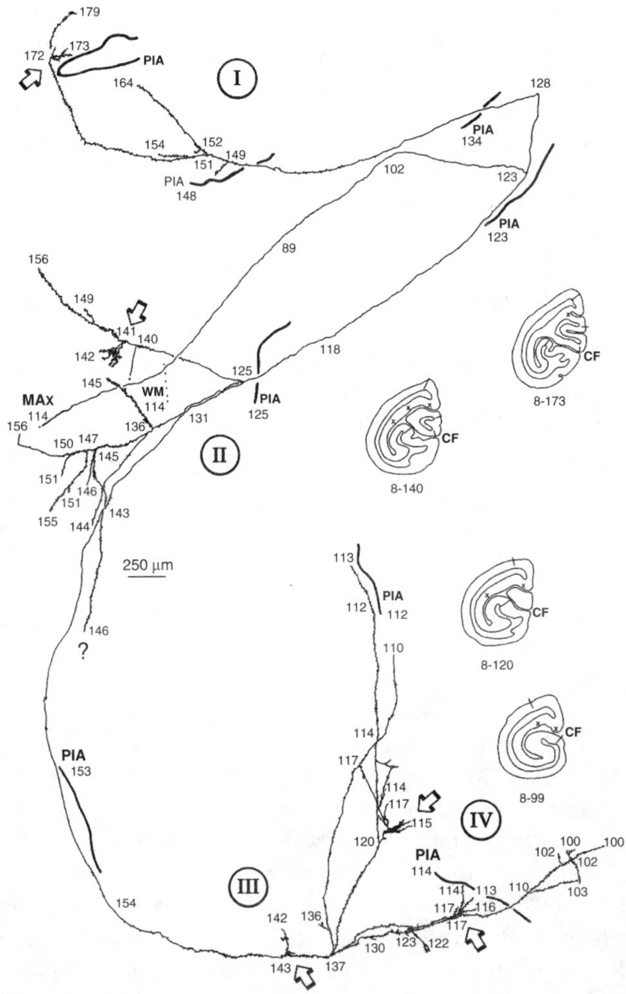

FIGURE 16.7 Camera lucida reconstruction (through 73 sections) of an axon (from case 8) anterogradely labeled by an injection of BDA in area MT (Saimiri). The terminal field is widespread, covering over 8.0 mm dorsoventral and 4.0 mm anteroposterior (AP). Numbers correspond to individual tissue sections (cut at 50 μm), with larger numbers more anterior. The main trunk ("MAx") ascends to layer 1 (at section 123), and bifurcates. Four clusters (I to IV) can be distinguished, separated (especially I and II) by regions of lower bouton number. The five hollow arrows point to small terminal clusters, which are intermingled with more typical longer terminal segments. One branch ("?") is incomplete. The coronal section outlines (at right) indicate the approximate location of portions of the axon (X) at different AP levels. (Modified from Reference 16.)

integration time of connections in different areas. The consensus emerging from these studies is that there is considerable scatter in the response latencies across cortical visual areas, such that response latencies at so-called later stages do not necessarily differ significantly from those in areas V1 and V2.[20,21,43] Moreover, responses evolve over time, so that neurons distributed over several areas can be coactive. A reasonable conclusion is that visual processing takes place in a tightly interconnected assembly of areas.[44]

One recent publication has proposed an organization by triplets rather than pairs of reciprocally connected areas.[45] Simulations based on reciprocal interactions between pairs of areas produced time lags that did not accord with experimental evidence for synchronous dynamics between areas. However, models based on three-way combinations were able to successfully produce zero-lag synchronization, more consistent with experimental findings.

Even without simulations, the connectional architecture is at least as strongly suggestive of a triplet organization as a pair-based one. Within the visual pathway, area V1 projects to V2 and MT, and MT projects back to both areas. Even more compelling, a portion of the feedback connections from MT bifurcate to both V1 and V2.[7,16] Similarly, V4 projects to TEO and posterior TE; and again, TE projects back to both areas, in part via branched collaterals of a single axon.[27] This fact alone is a cautionary reminder not to take too seriously the "pairing" of feedforward–feedback connections, as is often depicted in the shorthand of arrow anatomy.[46]

The importance of integrated network dynamics, although difficult to investigate and difficult to articulate, has become a leitmotif in the recent literature. One article on the role of visual feedback connections concludes, "[The] logic of corticocortical connections … must be conceptualized as a network of interacting areas responding with near-simultaneity, rather than as a pipeline-type architecture."[22]

Another recent review,[47] on the relationship of corticothalamic (CT) and thalamocortical (TC) connections, states, "It is obviously legitimate to consider CT pathways as feedback paths in experimental situations in which sensory stimuli are delivered to a head-restrained or anesthetized animal, but when animals hunt for stimuli, the inverse viewpoint likely prevails. … Amazingly, … [it can be] the incoming inputs that would act as a corrective feedback of what is anticipated." Deschênes et al.[47] take note of various discrepancies generated by too close adherence to the principle of reciprocity in the context of CT and TC connections, and propose an alternative "rule of parity." This continues to endorse the idea of "matching," but suggests that the topographic distribution of CT projections is determined by the branching patterns of different classes of prethalamic afferents, rather than by a direct match with TC projections. Adopting the same approach to sensory feedback connections, it might be fruitful to compare feedback connections not with feedforward, but with some other set (or sets) of connections, such as the pulvinocortical or the intrinsic connections.

16.7 PHYSIOLOGICAL STUDIES OF FEEDBACK CONNECTIONS: EARLY VISUAL AREAS

One of the standard approaches to investigating the functional role of feedback connections is by receptive field alterations in a target area. The most extensive investigations have been of the influence of MT feedback connections. In one study,[22] MT was inactivated by cooling in two anesthetized monkeys, and extracellular recording was conducted in areas V1, V2, and V3 (n = 169 neurons). When moving stimuli were presented to the monkey, a response decrease was shown for 51 neurons, and an increase for 6. When a flashed stimulus was used, the authors report a response decrease for 15 neurons and an increase for 14 (from a pool of 95 recorded neurons). The authors further established that these were early response effects, significant within the first 10 ms of the response.

Another study[24] used a slightly different paradigm, involving drug manipulation rather than inactivation. Simultaneous recordings were carried out of paired neurons in MT and V1, while drugs were applied focally in MT to create local changes in the neuronal responsivity to MT-customized stimuli. The advantage of this procedure is that spontaneous activity remains intact, and there is accordingly less drastic modification of the network circuitry. For 79 cells showing effects in area V1 during drug iontophoresis in MT, 60 (76%) showed changes in response magnitude to a texture stimulus. Magnitude was significantly increased for 29 cells, and decreased for 31. Shifts in response magnitude were not laminar specific, but were cell-type specific. For non-orientation-tuned cells, 75% exhibited opposite polarity changes to the paired MT cell responses, consistent with a suppressive effect. For orientation-tuned cells (simple and complex), 64% showed the same polarity changes, although the proportion was higher for simple cells (82%). The result can be interpreted as a facilitatory effect on most simple cells (Reference 24 and H. Jones, personal communication).

The relationship of receptive field changes to anatomical connections is difficult to establish. The consistent reports of mixed effects — of both increases and decreases in responsivity — might involve modifications of the inhibitory and/or excitatory circuitry; but one would like to know more about the role and distribution of interneurons, about whether the activation throughout the termination field of a feedback axon is uniform or not, and about whether changes in V1 neurons occur in nonretinotopically matched regions. The axonal configurations suggest that the activation field is not uniform (see above, and Figure 16.7), and that there may be nonretinotopic effects. The lack of laminar specificity may seem surprising, given the well-defined anatomical patterns of termination, but presumably results from intrinsic interlaminar processing.

Similar experiments have been carried out in cats (for example, References 48 and 49), and have also resulted in mixed changes, increases or decreases, for a subpopulation of recorded neurons. Because, unlike in primates, both areas 17 and 18 in cats receive strong geniculate input, the degree to which these changes reflect similar mechanisms in the two species is not clear (see also discussion in References 48 and 49).

16.8 LAYER 1 CONNECTIONS FROM HIGHER-ORDER AREAS

Since the early identification of feedback connections, this system has been closely associated with terminations in layer 1; and it has become common to consider as "feedback" or "top-down" any cortical system that terminates exclusively or mainly in layer 1. Examples are terminations from areas in prefrontal cortex to inferotemporal areas, or from perirhinal and parahippocampal cortices to several neocortical areas.[50] As remarked elsewhere,[5,51] however, terminations in layer 1 should not necessarily be grouped as similar on this sole criterion. Olfactory input is logically "feedforward," but terminates in layer 1 of piriform cortex. Similarly, cortical terminations to entorhinal cortex concentrate in layers 1 and 2.

Like feedback connections in the early visual pathway, the function of "top-down" connections is only partially understood. The tendency has been to associate these with higher cortical functions, ranging from figure/ground distinctions, to attention, blindsight, and even consciousness.[52–59]

For top-down connections, physiological assays have by and large used more behavioral paradigms, rather than receptive field analysis. In the case of projections from perirhinal cortex to area TE, several studies have investigated specific functional roles. Disrupting the neural signals from entorhinal and perirhinal cortex by lesioning destroys the responsivity of inferotemporal neurons to a previously learned (prelesion) association between picture pairs, although the ability of neurons to respond to a particular visual stimulus is left intact.[60] As proposed by Miyashita et al., it may be that recognition is carried out by inferotemporal neurons largely by reliance on "feedforward" inputs through extrastriate cortices, whereas their ability to represent the long-term mnemonic code between picture pairs critically depends on what could be called "backward" signals from the limbic areas. It is important to keep in mind, however, that the lesion not only disrupts limbic to inferotemporal connections, but in addition undoubtedly has widespread network effects.

Along similar lines, the time course of perceptual signals and memory-retrieval signals was compared for neurons in inferotemporal and perirhinal cortices while monkeys performed a visual pair-association task.[61] In this paradigm, perceptual signals were reported in inferotemporal cortex (median 77 ms) before perirhinal cortex (median 89 ms); but the reverse holds for mnemonic signals. One interpretation, as suggested by the authors, is that the mnemonic signal is somehow focused in perirhinal cortex and "spreads backward" to area TE. As the authors also state, there are other possible explanations; for example, intra-areal processing within TE or frontotemporal "top-down" signals for voluntary recall.

16.9 AREA V1 AS A HIGHER-ORDER AREA

An interesting perspective on feedback connections is provided by recent reports of nonvisual projections to area V1, particularly in the more peripheral parts of the visual field representation. These have been demonstrated by anterograde tracer injections in auditory and parietal association cortices, which show terminations in layers 1 and 6 of area V1 in the calcarine fissure, and layer 1 and subjacent layers of adjacent area V2.[62,63] Falchier et al.,[64] by making retrograde tracer injections in the calcarine fissure, independently reported labeled neurons in auditory areas as well as in the polysensory area (STP) of the superior temporal sulcus. As neurons were clustered in layer 6, the laminar patterns could be suggestive of "feedback-type" projections.

Recent psychophysical studies provide evidence for sound-induced illusory flash effects,[65] which might relate to these extravisual inputs. Namely, a single flash accompanied by two auditory beeps is perceived by human subjects as two flashes. This multimodal effect, which is manifested more strongly in the peripheral visual field, indicates that activity in early visual cortical areas is modulated by sound, but whether this should be considered "feedback" is not clear.

16.10 CONCLUSIONS

What are feedback connections? At this juncture, the term is widely used to refer to corticothalamic connections, connections from higher-order areas, and those directed to early sensory areas. Arguably, this is too broad, and some revision is long overdue. One option would be to restrict the term *feedback* to connections in the early sensory pathways. Corticothalamic connections and connections from higher-order areas do not, strictly speaking, need any further designators.

Returning to the introductory remarks, further understanding of feedback, in both the narrow and broad senses, requires progress on two fronts. There is certainly an urgent need for more data. We need better information concerning characteristics of feedback neurons, specifics of microcircuitry, and convergence with other connections.

Part of the problem in further understanding feedback connections is the general difficulty of investigating long-distance connections. *In vitro* techniques are only partially useful as they are usually not able to preserve intact an identified parent cell and its full terminations. For the purpose of analyzing connectivity interactions, noninvasive techniques, to allow repeated experimentation with high-resolution tags under different conditions, would be highly desirable. These may not be too far in the future. Axonal pathways can already be visualized; for example, by MRI imaging of manganese chloride or other axonally transported substances in living animals[66] and by genetic labeling in mice.[67]

Second, and not unrelated, is the need for a different conceptual framework, where feedback is considered as part of a dynamic network of interacting components. The implied duality, of feedback–feedforward, is oversimplified. The dichotomies between the "two" systems are not really sharp except in a few special cases and need to be further investigated. Feedback projections originate from neurons in layers 3 as well as 6; they terminate on elements other than distal apical dendrites. Feedforward projections extend beyond layer 4, including layers 1 and 2. Unraveling the basic functional operations will require a richer picture, and a notation, other than arrows, that can represent omnidirectional and diffuse neural functions.[46]

ACKNOWLEDGMENTS

The author thanks Michiko Fujisawa for her assistance with manuscript preparation, and Adrian Knight for assistance with the figures.

REFERENCES

1. Crick, F. and Koch, C., Constraints on cortical and thalamic projections: the no-strong-loops hypothesis, *Nature*, 391, 245, 1998.
2. Friston, K., Beyond phrenology: what can neuroimaging tell us about distributed circuitry? *Annu. Rev. Neurosci.*, 25, 221, 2002.
3. Stone, J.V. and Kötter, R., Making connections about brain connectivity, *Trends Cogn. Sci.*, 6, 327, 2002.
4. Büchel, C. and Friston, K., Assessing interactions among neuronal systems using functional neuroimaging, *Neural Netw.*, 13, 871, 2000.
5. Rockland, K.S., Elements of cortical architecture, in *Cerebral Cortex,* Vol. 12: *Extrastriate Cortex*, Rockland, K.S., Kaas, J.H., and Peters A., Eds., Plenum Press, New York, 1997, 243.
6. Rockland, K.S., Visual cortical organization at the single axon level: a beginning. *Neurosci. Res.*, 42, 155, 2002.
7. Kennedy, H. and Bullier, J., A double-labeling investigation of the afferent connectivity to cortical area V1 and V2 of the macaque monkey, *J. Neurosci.*, 5, 2815, 1985.
8. Sincich, L.C. and Horton, J.C., Divided by cytochrome oxidase: a map of the projections from V1 to V2 in macaques, *Science,* 295, 1734, 2002.
9. Rockland, K.S. and Pandya, D.N., Laminar origins and terminations of cortical connections of the occipital lobe in the rhesus monkey, *Brain. Res.*, 179, 3, 1979.
10. Rockland, K.S., The organization of feedback connections from area V2 (18) to V1 (17), in *Cerebral Cortex,* Vol. 10: *Primary Visual Cortex in Primates,* Peters, A. and Rockland, K.S., Eds., Plenum Press, New York, 1994, 261.
11. Cauller, L.J. and Connors, B.W., Functions of visual distal dendrites: experimental and computational studies of layer I synapses on neocortical pyramidal cells, in *Single Neuron Computation*, McKenna, T., Davis, J., and Zorznetzer, S.F., Eds., Academic Press, San Diego, 1992, 199.
12. Bullier, J., Hupé, J.M., James, A.C., and Girard, P., The role of feedback connections in shaping the responses of visual cortical neurons, *Prog. Brain Res.*, 134, 193, 2001.
13. Salin, P.A. and Bullier, J., Corticocortical connections in the visual system: structure and function, *Physiol. Rev.*, 75, 107, 1995.
14. Tombol, T., Layer VI cells, in *Cerebral Cortex,* Vol. 4: *Association and Auditory Cortices*, Peters, A. and Jones, E.G., Eds., Plenum Press, New York, 1984, 479.
15. Kaskan, P. and Rockland, K.S., Size as a measure of heterogeneity of feedback projecting neurons in areas V2 and MT of macaque cortex, *Soc. Neurosci. Abstr.*, 27, 1632, 2001.
16. Rockland, K.S. and Knutson, T., Feedback connections from area MT of the squirrel monkey to areas V1 and V2, *J. Comp Neurol.*, 425, 345, 2000.
17. Suzuki, W., Saleem, K.S., and Tanaka, K., Divergent backward projections from the anterior part of the inferotemporal cortex (area TE) in the macaque, *J. Comp. Neurol.*, 422, 206, 2000.
18. Anderson, J. and Martin, K.A.C., Parallel pathways from V2 to layer 1 of V1 in the macaque monkey, *FENS For. Abstr.*, 144, 2002.
19. Girard, P., Hupé, J.M., and Bullier, J., Feedforward and feedback connections between areas V1 and V2 of the monkey have similar rapid conduction velocities, *J. Neurophysiol.*, 85, 1328, 2001.

20. Nowak, L.G. and Bullier, J., The timing of information transfer in the visual system, in *Cerebral Cortex,* Vol. 12: *Extrastriate Cortex,* Rockland, K.S., Kaas, J., and Peters, A., Eds., Plenum Press, New York, 1997, 205.
21. Swadlow, H.A., Information flow along neocortical axons, in *Time and the Brain,* Miller, R., Ed., Harwood Academic Publishers, Amsterdam, 2000, 131.
22. Hupé, J.M., James, A.C., Girard, P., Lomber, S.G., Payne, B.R., and Bullier, J., Feedback connections act on the early part of the responses in monkey visual cortex, *J. Neurophysiol.*, 85, 134, 2001.
23. Hupé, J.M., James, A.C., Girard, P., and Bullier, J., Response modulations by static texture surround in area V1 of the macaque monkey do not depend on feedback connections from V2, *J. Neurophysiol.,* 85, 146, 2001.
24. Jones, H.E., Wang, W., Andolina, I.M., Grieve, K.L., Salt, T.E., and Sillito, A.M., Spatial characteristics and laminar organization of feedback effects from MT to V1, *Soc. Neurosci. Abstr.*, 27, 1633, 2001.
25. Bullier, J., Axonal bifurcation in the afferents to cortical areas of the visual system, in *Visual Neuroscience,* Pettigrew, J.D., Sanderson, K.J., and Levick, W.R., Eds., Cambridge University Press, Cambridge, U.K., 1986, 239.
26. Rockland, K.S., Saleem, K.S., and Tanaka, K., Divergent feedback connections from areas V4 and TEO in the macaque, *Vis. Neurosci.,* 11, 579, 1994.
27. Rockland, K.S. and Drash, G.W., Collateralized divergent feedback connections that target multiple cortical areas, *J. Comp. Neurol.,* 373, 529, 1996.
28. Rockland, K.S., Configuration, in serial reconstruction, of individual axons projecting from area V2 to V4 in the macaque monkey, *Cereb. Cortex,* 2 353, 1992.
29. Johnson, R.R. and Burkhalter, A., Microcircuitry of forward and feedback connections within rat visual cortex, *J. Comp. Neurol.*, 368, 383, 1996.
30. Johnson, R.R. and Burkhalter, A., A polysynaptic feedback circuit in rat visual cortex, *J. Neurosci.*, 17, 7129, 1997.
31. Rhodes, P.A. and Llinas, R.R., Apical tuft input efficacy in layer 5 pyramidal cells from rat visual cortex, *J. Physiol.*, 536, 167, 2001.
32. Rockland, K.S. and Virga, A., Terminal arbors of individual "feedback" axons projecting from area V2 to V1 in the macaque monkey: a study using immunohistochemistry of anterogradely transported *Phaseolus vulgaris-leucoagglutinin, J. Comp. Neurol.,* 285, 54, 1989.
33. Rockland, K.S., Bistratified distribution of terminal arbors of individual axons projecting from area V1 to middle temporal area (MT) in the macaque monkey, *Vis. Neurosci.,* 3, 155, 1989.
34. Rockland, K.S., Morphology of individual axons projecting from area V2 to MT in the macaque, *J. Comp. Neurol.,* 355, 15, 1995.
35. Angelucci, A., Levitt, J.B., and Lund, J.S., Anatomical origins of the classical receptive field and modulatory surround field of single neurons in macaque visual cortical area V1, in *Progress in Brain Research,* Vol. 136, Azmitia, E.C. et al., Eds., Elsevier Science, Amsterdam, 2002, 373.
36. Angelucci, A., Levitt, J.B., Walton, E.J., Hupé, J.M., Bullier, J., and Lund, J.S., Circuits for local and global signal integration in primary visual cortex, *J. Neurosci.*, 22, 8633, 2002.
37. Stettler, D.D., Das, A., Bennett, J., and Gilbert, C.D., Lateral connectivity and contextual interactions in macaque primary visual cortex, *Neuron,* 36, 739, 2002.
38. Ito, M., Cerebellar long-term depression: characterization, signal transduction, and functional roles, *Physiol. Rev.,* 81, 1143, 2001.

39. Shipp, S. and Zeki, S., The organization of connections between areas V5 and V1 in macaque monkey, *Eur. J. Neurosci.*, 1, 333, 1989.
40. Ungerleider, L.G. and Desimone, R., Cortical connections of visual area MT in the macaque, *J. Comp. Neurol.*, 248, 190, 1986.
41. Weller, R.E. and Kaas, J.H., Retinotopic patterns of connections of area 17 with visual areas V-II and MT in macaque monkeys, *J. Comp. Neurol.*, 220, 253, 1983.
42. Rockland, K.S. and Van Hoesen, G.W., Direct temporal-occipital feedback connections to striate cortex (V1) in the macaque monkey, *Cereb. Cortex,* 4, 300, 1994.
43. Raiguel, S., Lagae, L., Gulyas, B., and Orban, G.A., Response latencies of visual cells in macaque areas V1, V2 and V5, *Brain Res.,* 493, 155, 1989.
44. Paradiso, M.A., Perceptual and neuronal correspondence in primary visual cortex, *Curr. Opin. Neurobiol.,* 12, 155, 2002.
45. Chawla, D., Friston, K.J., and Lumer, E.D., Zero-lag synchronous dynamics in triplets of interconnected cortical areas, *Neural Netw.*, 14, 727, 2001.
46. Schott, G.D., Illustrating cerebral function: the iconography of arrow, *Philos. Trans. R. Soc. Lond. B,* 355, 1789, 2000.
47. Deschênes, M., Veinante, P., and Zhang, Z.-W., The organization of corticothalamic projections: reciprocity versus parity, *Brain Res. Rev.*, 28, 286, 1998.
48. Martinez-Conde, S., Cudeiro, J., Grieve, K.L., Rodriguez, R., Rivadulla, C., and Acuna, C., Effects of feedback projections from area 18 layers 2/3 to area 17 layers 2/3 in the cat visual cortex, *J. Neurophysiol.*, 82, 2667, 1999.
49. Wang, C., Waleszczyk, W.J., Burke, W., and Dreher, B., Modulatory influence of feedback projections from area 21a on neuronal activities in striate cortex of the cat, *Cereb. Cortex*, 10, 1217, 2000.
50. Lavenex, P., Suzuki, W., and Amaral, D., Perirhinal and parahippocampal cortices of the macaque monkey: projections to the neocortex, *J. Comp. Neurol.*, 447, 394, 2002.
51. Wellman, B.J. and Rockland, K.S., Divergent cortical connections to entorhinal cortex from area TF in the macaque, *J. Comp. Neurol.,* 389, 361, 1997.
52. Grossberg, S. and Raizada, R.D.S., Contrast-sensitive perceptual grouping and object-based attention in the laminar circuits of primary visual cortex, *Vision Res.*, 40, 1413, 2000.
53. Hupé, J.M., James, A.C., Payne, B.R., Lomber, S.G., Girard, P., and Bullier, J., Cortical feedback improves discrimination between figure and background by V1, V2 and V3 neurons, *Nature*, 394, 784, 1998.
54. Lamme, V.A.F., Blindsight: the role of feedforward and feedback corticocortical connections, *Acta Psychol.*, 107, 209, 2001.
55. Lamme, V.A.F., Super, H., and Spekreijse, H., Feedforward, horizontal, and feedback processing in the visual cortex, *Curr. Opin. Neurobiol.*, 8, 529, 1998.
56. Pascual-Leone, A. and Walsh, V., Fast back projections from the motion to the primary visual area necessary for visual awareness, *Science,* 292, 510, 2001.
57. Pollen, D.A., On the neural correlates of visual perception, *Cereb. Cortex,* 9, 4, 1999.
58. Tomita, H., Ohbayashi, M., Nakahara, K., Hasegawa, I., and Miyashita, Y., Top-down signal from prefrontal cortex in executive control of memory retrieval, *Nature,* 401, 699, 1999.
59. Vidyasagar, T.R., Gating of neuronal responses in macaque primary visual cortex by an attentional spotlight, *NeuroReport*, 9, 1947, 1998.
60. Miyashita, Y., Okuno, H., Tokuyama, W., Ihara, T., and Nakajima, K., Feedback signal from medial temporal lobe mediates visual associative mnemonic codes of inferotemporal neurons, *Cogn. Brain Res.*, 5, 81, 1996.

61. Naya, Y., Yoshida, M., and Miyashita, Y. Backward spreading of memory-retrieval signal in the primate temporal cortex, *Science,* 291, 661, 2001.
62. Rockland, K.S. and Ojima, H., Calcarine area V1 as a multimodal convergence area, *Soc. Neurosci. Abstr.*, 27, 1342, 2001.
63. Rockland, K.S. and Ojima, H., Multisensory convergence in calcarine visual areas in macaque monkey, *Int. J. Psychophysiol.,* in press.
64. Falchier, A., Clavagnier, S., Barone, P., and Kennedy, H., Anatomical evidence of multimodal integration in primate striate cortex, *J. Neurosci.,* 22, 5749, 2002.
65. Shams, L., Kamitani, Y., Thompson, S., and Shimojo, S., Sound alters visual evoked potentials in humans, *NeuroReport*, 12, 3849, 2001.
66. Saleem, K.S., Pauls, J.M., Augath, M., Trinath, T., Prause, B.A., Hashikawa, T., and Logothetis, N.K., Magnetic resonance imaging of neuronal connections in the macaque monkey, *Neuron,* 34, 685, 2002.
67. Zubair, M., Watanabe, E., Fukada, M., and Noda, M., Genetic labeling of specific axonal pathways in the mouse central nervous system, *Eur. J. Neurosci.*, 15, 807, 2002.

Index

A

B

C

D

E

F

J

K

L

M

N

O

P

R

S

T

U

V

W

X

Y

Z